Design of Intelligent Applications Using Machine Learning and Deep Learning Techniques

Design of Intelligent Applications Using Machine Learning and Deep Learning Techniques

Edited by
Ramchandra S. Mangrulkar
Antonis Michalas
Narendra M. Shekokar
Meera Narvekar
Pallavi V. Chavan

CRC Press is an imprint of the
Taylor & Francis Group, an **informa** business
A CHAPMAN & HALL BOOK

First edition published 2021
by CRC Press
6000 Broken Sound Parkway NW, Suite 300, Boca Raton, FL 33487-2742

and by CRC Press
2 Park Square, Milton Park, Abingdon, Oxon, OX14 4RN

CRC Press is an imprint of Taylor & Francis Group, LLC

ISBN: 978-0-367-67979-8 (hbk)
ISBN: 978-0-367-67989-7 (pbk)
ISBN: 978-1-003-13368-1 (ebk)

Typeset in Times
by codeMantra

Contents

Preface vii
Editors ix
Contributors xi

1. **Data Acquisition and Preparation for Artificial Intelligence and Machine Learning Applications** 1
Kallol Bosu Roy Choudhuri and Ramchandra S. Mangrulkar

2. **Fundamental Models in Machine Learning and Deep Learning** 13
Tatwadarshi P. Nagarhalli, Ashwini M. Save, and Narendra M. Shekokar

3. **Research Aspects of Machine Learning: Issues, Challenges, and Future Scope** 37
Reena Thakur, Mayur Tembhurney, and Dheeraj Rane

4. **Comprehensive Analysis of Dimensionality Reduction Techniques for Machine Learning Applications** 61
Archana Vasant Mire, Vinayak Elangovan, and Bharti Dhote

5. **Application of Deep Learning in Counting WBCs, RBCs, and Blood Platelets Using Faster Region-Based Convolutional Neural Network** 77
Nirav Jain, Shail Shah, Ramchandra S. Mangrulkar, and Pankaj Sonawane

6. **Application of Neural Network and Machine Learning in Mental Health Diagnosis** 99
Aniruddha Das, Enakshie Prasad, and Sindhu Nair

7. **Application of Machine Learning in Cardiac Arrhythmia** 115
Gresha S. Bhatia, Shefali Athavale, Yogita Bhatia, Tanya Mohanani, and Akanksha Mittal

8. **Advances in Machine Learning and Deep Learning Approaches for Mammographic Breast Density Measurement for Breast Cancer Risk Prediction: An Overview** 125
Shivaji D. Pawar, Kamal Kr. Sharma, and Suhas G. Sapate

9. **Applications of Machine Learning in Psychology and the Lifestyle Disease Diabetes Mellitus** 145
Ruhina Karani, Dharmik Patel, Akshay Chudasama, Dharmil Chhadva, and Gaurang Oza

10. **Application of Machine Learning and Deep Learning in Thyroid Disease Prediction** 155
Aditi Vora, Ramchandra S. Mangrulkar, Narendra M. Shekokar, and Meera Narvekar

11. **Application of Machine Learning in Fake News Detection** 165
Smita Bhoir, Jyoti Kundale, and Smita Bharne

12. **Authentication of Broadcast News on Social Media Using Machine Learning** 185
Smita Sanjay Ambarkar, Narendra M. Shekokar, Monika Mangla, and Rakhi Akhare

13. Application of Deep Learning in Facial Recognition 195
Jimit Gandhi, Aditya Jeswani, Fenil Doshi, Parth Doshi, and Ramchandra S. Mangrulkar

14. Application of Deep Learning in Deforestation Control and Prediction of Forest Fire Calamities 209
Muskan Goenka and Ramchandra S. Mangrulkar

15. Application of Convolutional Neural Network in Feather Classifications 223
Milind Shah, Keval Nagda, Anirudh Mukherjee, and Pratik Kanani

16. Application of Deep Learning Coupled with Thermal Imaging in Detecting Water Stress in Plants 233
Saiqa Khan, Meera Narvekar, Anam Khan, Aqdus Charolia, and Mushrifah Hasan

17. Machine Learning Techniques to Classify Breast Cancer 245
Drashti Shah and Ramchandra S. Mangrulkar

18. Application of Deep Learning in Cartography Using UNet and Generative Adversarial Network 257
Deep Gandhi, Govind Thakur, Pranit Bari, and Khushali Deulkar

19. Evaluation of Intrusion Detection System with Rule-Based Technique to Detect Malicious Web Spiders Using Machine Learning 273
Nilambari G. Narkar and Narendra M. Shekokar

20. Application of Machine Learning to Improve Tourism Industry 289
Krutibash Nayak and Saroj Kumar Panigrahy

21. Training Agents to Play 2D Games Using Reinforcement Learning 309
Harshil Jhaveri, Nishay Madhani, and Narendra M. Shekokar

22. Analysis of the Effectiveness of the Non-Vaccine Countermeasures Taken by the Indian Government against COVID-19 and Forecasting Using Machine Learning and Deep Learning 321
Akash Shah, Romil Shah, Manan Gandhi, Rashmil Panchani, Govind Thakur, and Kriti Srivastava

23. Application of Deep Learning in Video Question Answering System 353
Mansi Pandya, Arnav Parekhji, Aniket Shahane, Palak V. Chavan, and Ramchandra S. Mangrulkar

24. Implementation and Analysis of Machine Learning and Deep Learning Algorithms 373
Samip Kalyani, Neel Vasani, and Ramchandra S. Mangrulkar

25. Comprehensive Study of Failed Machine Learning Applications Using a Novel 3C Approach 403
Neel Patel, Prem Bhajaj, Pratik Panchal, Tanmai Prabhune, Pankaj Sonawane, and Ramchandra S. Mangrulkar

Index 421

Preface

Recently, machine learning (ML) and deep learning (DL) have drawn huge interest from scientists and technology designers throughout the world and are considered as the future of computing. A subfield of ML and DL techniques involves the use of systems, called neural networks, to recognize and understand patterns in data, helping a machine carry out tasks in a manner similar to humans. In essence, it helps the machine arrive at conclusions, make decisions, and predict events while using reasoning like a human being. This concept starts from the learning mechanism of human beings and is artificially implemented through the models of different categories like supervised, unsupervised, and reinforcement. This way, the intelligent models are effectively designed and are fully investigated bringing in practical applications in many fields such as health care, agriculture, and security. In practice, specific applications of ML and DL have proven their feasibility, convenience, and promising benefits. However, ML and DL can only be successfully applied in the context of data computing and analysis. Today, ML and DL have brought potential development in detection and prediction. Intelligent applications are widely used nowadays for the prediction of the diseases in the healthcare domain. On the other side, prediction of the quality and productivity of the crops in the agriculture domain is possible through predictive models. This technological development has brought a revolution in these domains. Apart from these domains, ML and DL are found useful in analyzing the social behavior of humans. These models stand at their best against different network attacks. With the advancements in the amount and type of data available for use, it became necessary to build a means to process the data and that is where deep neural networks prove their importance. These networks are capable of handling a large amount of data having a background in finance, documents as well as images to name a few. More importantly, the results produced by these techniques have proved to be closer to human results than any other method available at present. This book attempts to provide a wide range of research and development work under the umbrella of Intelligent Computing.

The volume is comprised of 25 chapters, providing different applications of ML and DL specifically in healthcare, agriculture, and social network domains. An overview of data acquisition and preparation for artificial intelligence, ML, and DL is given in Chapter 1. Chapter 2 describes fundamental models of ML and DL. Chapter 3 deals with the research aspects of ML, followed by challenges, issues, and future research directions. Chapter 4 describes the comprehensive analysis of dimensionality reduction techniques for ML applications and also the probabilistic models of ML. Chapters 5–10 give the applications of ML in various healthcare issues and how the models have been applied to resolve those issues under consideration. Chapter 5 gives the applications of DL in identifying RBCs, WBCs, and platelets using a convolutional neural network. Chapter 6 describes the applications of ML in mental health diagnosis. Chapter 7 gives the applications of ML in cardiac arrhythmia and also insights on ML applications in thyroid disease prediction. Chapter 8 describes ML and DL techniques to predict breast cancer risk prediction techniques. Chapter 9 describes the applications of ML in psychology, lifestyle diseases, and space-occupying lesions. Chapter 10 describes the applications of ML in thyroid disease prediction. This book has also considered applications other than healthcare applications. Chapter 11 gives the applications of ML in Fake news detection. Nowadays, the usage of social media is drastically increasing. The authenticity of the information is important over the network. This has been presented in Chapter 12, focusing on the authentication of broadcast news on social media. A wide range of DL applications in different domains is covered in this book. Chapter 13 gives facial recognition applications of DL. Chapter 14 gives the applications of DL in deforestation control and prediction of forest fire calamities. Chapter 15 gives the feather classification through DL approach. Chapter 16 defines thermal imaging in detecting water stress in plants, which is another application of DL. Chapter 17 gives ML technique to detect breast cancer. Chapter 18 describes DL application in cartography using UNET and GAN. Chapter 19 describes intrusion detection, a threat to security. Apart from this, Chapters 20 and 21, respectively, describe the other applications using ML and DL such as tourism and 2D games. Chapter 22 provides a detailed analysis of the effectiveness of the non-vaccine countermeasures taken by the Indian Government. Chapter 23 describes the applications of the video question answering system using DL. Chapter 24 gives the implementation and analysis of various ML and DL algorithms under consideration. Finally, Chapter 25 gives a comprehensive study of failed and semi-failed ML applications.

We would like to express our deep gratitude to the authors for their contributions. It would not have been possible to reach this proposal submission without their contribution. As the editors, we hope that this book will stimulate further research in ML and DL. Special thanks go to our publisher, CRC Press/Taylor & Francis Group.

We hope that this book will present promising ideas and outstanding research contributions supporting further development of ML and DL approaches by applying intelligence in various applications.

Editors

Dr. Ramchandra S. Mangrulkar received his PhD in Computer Science and Engineering from SGB Amravati University, Amravati, in 2016, and currently he is working as Associate Professor in the Department of Computer Engineering at Dwarkadas J. Sanghvi College of Engineering, Mumbai, Maharashtra, India. Prior to this, he was working as Associate Professor and Head in the Department of Computer Engineering, Bapurao Deshmukh College of Engineering, Sevagram, Maharashtra, India. He has published a significant number of papers and book chapters in field-related journals and conferences and has also participated as a session chair in various conferences and conducted various workshops on Network Simulator and LaTeX. He also received certification of appreciation from DIG Special Crime Branch Pune and Superintendent of Police and broadcasting media gives wide publicity for the project work guided by him on the topic "Face Recognition System". He also received 3.5 lakhs grant under Research Promotion Scheme of AICTE for the project "Secured Energy Efficient Routing Protocol for Delay Tolerant Hybrid Network". He is an active member of the Board of Studies in various universities and autonomous institutes in India.

Dr. Antonis Michalas received his PhD in Network Security from Aalborg University, Denmark, and currently, he is working as Assistant Professor in the Department of Computing Science at Tampere University of Technology, Faculty of Computing and Electrical Engineering. Prior to this, he was working as Assistant Professor in Cyber Security at the University of Westminster, London. Earlier, he was working as a postdoctoral researcher at the Security Lab at the Swedish Institute of Computer Science in Stockholm, Sweden. As a postdoctoral researcher at the SCE Labs, he was actively involved in National and European research projects. He has published a significant number of papers in field-related journals and conferences and has also participated as a speaker in various conferences and workshops. His research interests include private and secure e-voting systems, reputation systems, privacy in decentralized environments, cloud computing, trusted computing, and privacy-preserving protocols in participatory sensing applications.

Dr. Narendra M. Shekokar received his PhD in Engineering (Network Security) from NMIMS University, Mumbai, and he is working as Professor and Head in the Department of Computer Engineering at SVKM's Dwarkadas J. Sanghvi College of Engineering, Mumbai (autonomous college affiliated to the University of Mumbai). He was a member of the Board of Studies at the University of Mumbai for more than 5 years, and he has also been a member of various committees at the University of Mumbai. His total teaching experience is 23 years. He is a PhD guide for 8 research fellows and more than 25 students at the postgraduate level. He has presented more than 65 papers at international and national conferences and has also published more than 25 research papers in renowned journals. He received the Minor Research Grant twice from the University of Mumbai for his research projects. He has delivered an expert talk and chaired a session at numerous events and conferences.

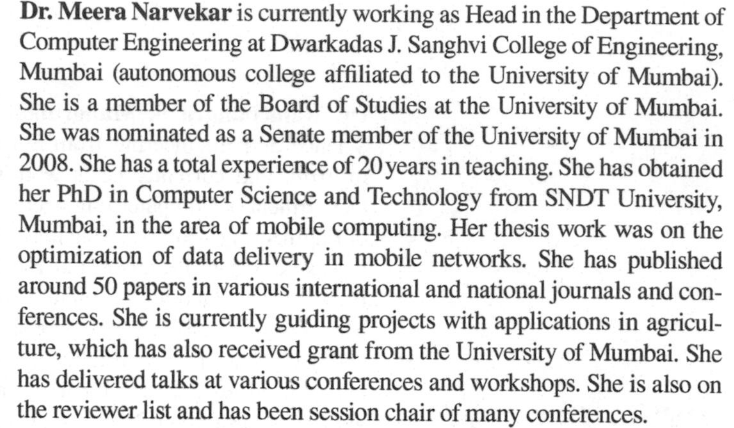

Dr. Meera Narvekar is currently working as Head in the Department of Computer Engineering at Dwarkadas J. Sanghvi College of Engineering, Mumbai (autonomous college affiliated to the University of Mumbai). She is a member of the Board of Studies at the University of Mumbai. She was nominated as a Senate member of the University of Mumbai in 2008. She has a total experience of 20 years in teaching. She has obtained her PhD in Computer Science and Technology from SNDT University, Mumbai, in the area of mobile computing. Her thesis work was on the optimization of data delivery in mobile networks. She has published around 50 papers in various international and national journals and conferences. She is currently guiding projects with applications in agriculture, which has also received grant from the University of Mumbai. She has delivered talks at various conferences and workshops. She is also on the reviewer list and has been session chair of many conferences.

Dr. Pallavi V. Chavan received her PhD in Computer Science and Engineering from RTM Nagpur University, and she is working as Associate Professor in the Department of Information Technology, RAIT Nerul, Navi Mumbai, India. Prior to this, she was working as Assistant Professor and Head in the Department of Computer Engineering, Bapurao Deshmukh College of Engineering, Sevagram, Maharashtra, India. Her areas of research are visual cryptography and secret sharing. She also interestingly works with image processing and soft computing. She is the recipient of the UGC Workshop Grant twice for the conduction of national-level workshops. She is also a recipient of CSIR seminar grant for the conduction of national-level seminars. Her subjects of interest are theory of computation, database management system and artificial neural network and fuzzy logic. She is the follower of the spiritual approach of Brahma Kumaris for Rajyoga Meditation.

Contributors

Rakhi Akhare
Department of Computer Engineering
Lokmanya Tilak College of Engineering
Navi Mumbai, India

Smita Sanjay Ambarkar
Department of Computer Engineering
Lokmanya Tilak College of Engineering
Navi Mumbai, India

Shefali Athavale
Department of Computer Engineering
Vivekanand Education Society's Institute of Technology
Mumbai, India

Pranit Bari
Department of Computer Engineering
Dwarkadas J. Sanghvi College of Engineering
Mumbai, India

Prem Bhajaj
Department of Computer Engineering
Dwarkadas J. Sanghvi College of Engineering
Mumbai, India

Smita Bharne
Department of Computer Engineering
Ramrao Adik Institute of Technology
Mumbai, India

Gresha S. Bhatia
Department of Computer Engineering
Vivekanand Education Society's Institute of Technology
Mumbai, India

Yogita Bhatia
Department of Computer Engineering
Vivekanand Education Society's Institute of Technology
Mumbai, India

Smita Bhoir
Department of Computer Engineering
Ramrao Adik Institute of Technology
Mumbai, India

Aqdus Charolia
Department of Computer Engineering
M.H. Saboo Siddik College of Engineering
Mumbai, India

Palak Chavan
Department of Computer Engineering
Dwarkadas J. Sanghvi College of Engineering
Mumbai, India

Dharmil Chhadva
Department of Computer Engineering
Dwarkadas J. Sanghvi College of Engineering
Mumbai, India

Kallol Bosu Roy Choudhuri
Artificial Intelligence and Analytics (AIA)
Cognizant Technology Solutions
Kolkata, India

Akshay Chudasama
Department of Computer Engineering
Dwarkadas J. Sanghvi College of Engineering
Mumbai, India

Aniruddha Das
Department of Computer Engineering
Dwarkadas J. Sanghvi College of Engineering
Mumbai, India

Khushali Deulkar
Department of Computer Engineering
Dwarkadas J. Sanghvi College of Engineering
Mumbai, India

Bharti Dhote
Department of Computer Engineering
Sinhgad Institute of Technology
Lonavala, India

Fenil Doshi
Department of Computer Engineering
Dwarkadas J. Sanghvi College of Engineering
Mumbai, India

Parth Doshi
Department of Computer Engineering
Dwarkadas J. Sanghvi College of Engineering
Mumbai, India

Vinayak Elangovan
Penn State University
Abington, Pennsylvania

Deep Gandhi
Department of Computer Engineering
Dwarkadas J. Sanghvi College of Engineering
Mumbai, India

Jimit Gandhi
Department of Computer Engineering
Dwarkadas J. Sanghvi College of Engineering
Mumbai, India

Manan Gandhi
Department of Computer Engineering
Dwarkadas J. Sanghvi College of Engineering
Mumbai, India

Muskan Goenka
Department of Computer Engineering
Dwarkadas J. Sanghvi College of Engineering
Mumbai, India

Mushrifah Hasan
Department of Computer Engineering
M.H. Saboo Siddik College of Engineering
Mumbai, India

Nirav Jain
Department of Computer Engineering
Dwarkadas J. Sanghvi College of Engineering,
Mumbai, India

Aditya Jeswani
Department of Computer Engineering
Dwarkadas J. Sanghvi College of Engineering
Mumbai, India

Harshil Jhaveri
Department of Computer Engineering
Dwarkadas J. Sanghvi College of Engineering
Mumbai, India

Samip Kalyani
Department of Computer Engineering
Dwarkadas J. Sanghvi College of Engineering
Mumbai, India

Pratik Kanani
Department of Computer Engineering
Dwarkadas J. Sanghvi College of Engineering
Mumbai, India

Ruhina Karani
Department of Computer Engineering
Dwarkadas J. Sanghvi College of Engineering
Mumbai, India

Anam Khan
Department of Computer Engineering
M.H. Saboo Siddik College of Engineering
Mumbai, India

Saiqa Khan
Department of Computer Engineering
Dwarkadas J. Sanghvi College
of Engineering
Mumbai, India

Jyoti Kundale
Department of Computer Engineering
Ramrao Adik Institute of Technology
Mumbai, India

Nishay Madhani
Department of Computer Engineering
Dwarkadas J. Sanghvi College
of Engineering
Mumbai, India

Monika Mangla
Department of Computer Engineering
Lokmanya Tilak College of Engineering
Navi Mumbai, India

Archana Vasant Mire
Department of Computer Engineering
Terna Engineering College
Navi Mumbai, India

Akanksha Mittal
Department of Computer Engineering
Vivekanand Education Society's Institute of
Technology
Mumbai, India

Tanya Mohanani
Department of Computer Engineering
Vivekanand Education Society's Institute of Technology
Mumbai, India

Anirudh Mukherjee
Department of Computer Engineering
Dwarkadas J. Sanghvi College of Engineering
Mumbai, India

Tatwadarshi P. Nagarhalli
Department of Computer Engineering
Vidyavardhini's College of Engineering and Technology Virar, India

Keval Nagda
Department of Computer Engineering
Dwarkadas J. Sanghvi College of Engineering
Mumbai, India

Sindhu Nair
Department of Computer Engineering
Dwarkadas J. Sanghvi College of Engineering
Mumbai, India

Nilambari G. Narkar
Department of Computer Engineering
Dwarkadas J. Sanghvi College of Engineering
Mumbai, India

Krutibash Nayak
School of Computer Science and Engineering
VIT-AP University
Amaravati, India

Gaurang Oza
Department of Computer Engineering
Dwarkadas J. Sanghvi College of Engineering
Mumbai, India

Pratik Panchal
Department of Computer Engineering
Dwarkadas J. Sanghvi College of Engineering
Mumbai, India

Rashmil Panchani
Department of Computer Engineering
Dwarkadas J. Sanghvi College of Engineering
Mumbai, India

Mansi Pandya
Department of Computer Engineering
Dwarkadas J. Sanghvi College of Engineering
Mumbai, India

Saroj Kumar Panigrahy
School of Computer Science and Engineering
VIT-AP University
Amaravati, India

Arnav Parekhji
Department of Computer Engineering
Dwarkadas J. Sanghvi College of Engineering
Mumbai, India

Dharmik Patel
Department of Computer Engineering
Dwarkadas J. Sanghvi College of Engineering
Mumbai, India

Neel Patel
Department of Computer Engineering
Dwarkadas J. Sanghvi College of Engineering
Mumbai, India

Shivaji D. Pawar
Department of Computer Science and Engineering
Lovely Professional University
Jalandhar, India

Tanmai Prabhune
Department of Computer Engineering
Dwarkadas J. Sanghvi College of Engineering
Mumbai, India

Enakshie Prasad
Department of Psychology, Gargi College
University of Delhi
New Delhi, India

Dheeraj Rane
Department of Computer Science and Engineering
Medicaps University
Indore, India

Suhas G. Sapate
Department of Computer Science and Engineering
Annasaheb Dange College of Engineering and Technology
Ashta, India

Ashwini M. Save
Department of Computer Engineering
Dwarkadas J. Sanghvi College of Engineering
Mumbai, India

Akash Shah
JPMorgan Chase & Co.
Mumbai, India

Drashti Shah
Department of Computer Engineering
Dwarkadas J. Sanghvi College of Engineering
Mumbai, India

Milind Shah
Department of Computer Engineering
Dwarkadas J. Sanghvi College of Engineering
Mumbai, India

Romil Shah
Department of Computer Engineering
Dwarkadas J. Sanghvi College of Engineering
Mumbai, India

Shail Shah
Department of Computer Engineering
Dwarkadas J. Sanghvi College of Engineering
Mumbai, India

Aniket Shahane
Department of Computer Engineering
Dwarkadas J. Sanghvi College of Engineering
Mumbai, India

Kamal Kr. Sharma
School of Electronics and Electrical Engineering
Lovely Professional University
Jalandhar, India

Pankaj Sonawane
Department of Computer Engineering
Dwarkadas J. Sanghvi College of Engineering
Mumbai, India

Kriti Srivastava
Department of Computer Engineering
Dwarkadas J. Sanghvi College of Engineering
Mumbai, India

Mayur Tembhurney
Department of Computer Science and Engineering
S. B. Jain Institute of Technology, Management & Research
Nagpur, India

Govind Thakur
Department of Computer Engineering
Dwarkadas J. Sanghvi College of Engineering
Mumbai, India

Reena Thakur
Department of Computer Science and Engineering
Jhulelal Institute of Technology
Nagpur, India

Neel Vasani
Department of Computer Engineering
Dwarkadas J. Sanghvi College of Engineering
Mumbai, India

Aditi Vora
Department of Computer Engineering
Dwarkadas J. Sanghvi College of Engineering
Mumbai, India

1

Data Acquisition and Preparation for Artificial Intelligence and Machine Learning Applications

Kallol Bosu Roy Choudhuri
Cognizant Technology Solutions

Ramchandra S. Mangrulkar
University of Mumbai

CONTENTS

1.1 Introduction 2
1.2 Reference Architecture 2
1.2.1 Data Sources 3
1.2.2 Data Storage 3
1.2.3 Batch Processing 3
1.2.4 Real-Time Message Ingestion 3
1.2.5 Stream Processing 3
1.2.6 Machine Learning 4
1.2.7 Analytical Data Store 4
1.2.8 Analytics and Reports 4
1.2.9 Orchestration 4
1.3 Data Acquisition Layer 4
1.3.1 File Systems 5
1.3.2 Databases 5
1.3.3 Applications 5
1.3.4 Devices 6
1.3.5 Enterprise Data Gateway 6
1.3.6 Field Gateway 6
1.3.7 Data Integration Services 6
1.3.8 Data Ingestion Services 6
1.4 Data Ingestion Layer 6
1.4.1 Data Storage Layer 7
1.4.2 Landing Layer 8
1.4.3 Cleansed Layer 8
1.4.4 Processed Layer 8
1.4.5 Data Processing Layer 8
1.4.6 Data Processing Engine 8
1.4.7 Data Processing Programs 9
1.4.8 Scheduling Engine 9
1.4.9 Scheduling Scripts 9
1.5 Data Quality and Cleansing Layer 9
1.5.1 Master Data Management (MDM) System 10
1.5.2 Master Data Management (MDM) Referencing Programs 10
1.5.3 Data Quality Check Programs 10
1.5.4 Rejected/Quarantined Layer 11
Bibliography 11

1.1 Introduction

This chapter introduces the essential concepts of data acquisition, ingestion, data quality, cleansing and preparation. Data forms the basis of all decision-making processes. AI and ML being heavily dependent on accurate and reliable data, these stages are important prerequisites to build any AI and ML application. Before data can be effectively fed into a ML model or an AI algorithm, the data goes through the following stages such as:

- Data acquisition
- Data ingestion
- Data quality and cleansing

In the following sections of this chapter, we will understand what each of the above stages in data processing means. We will also study the various components and concepts involved in successfully executing each phase of the data processing.

1.2 Reference Architecture

In the next few sections, we will look at each of the above stages in more detail. Before proceeding further, let us look at the below diagram that depicts the general reference architecture for data collection, ingestion, cleansing and preparation for AI and ML applications (Figure 1.1).

Figure 1.1 gives the details about the different stages. The first stage is to establish data ingestion components from the source, followed by cleansing the raw data and then preparing the data for training a ML model or feeding it as an input for AI applications. The various components of the reference architecture are listed below:

- Data sources
- Data storage
- Batch processing
- Real-time message ingestion
- Stream processing
- ML
- Analytical data store

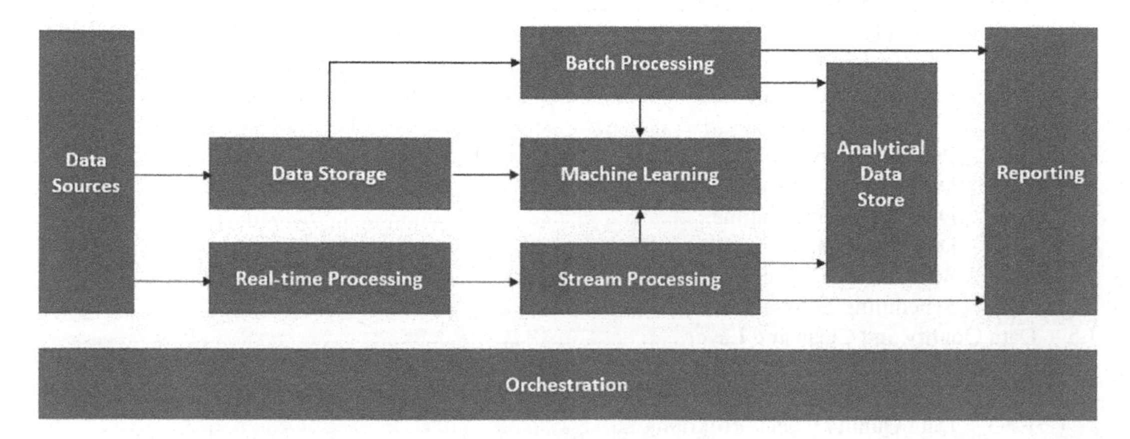

FIGURE 1.1 AI/ML data processing reference architecture.

- Analytics and reports
- Orchestration

In the below sections, we will look at each of these data processing components.

1.2.1 Data Sources

For AI/ML applications, data can be sourced from various types of sources. The data is first collected (acquired), followed by storage and further processing. Usually, data can come from the following types of source systems:

- Application data stores such as relational databases (SQL Server, Oracle, MySQL and so on), or it can also be NoSQL databases (Cassandra, MongoDB, PostgreSQL and so on)
- Log files and flat files generated by various types of business applications, monitoring and logging software (Splunk, ELK and so on)
- IoT devices that produce data

1.2.2 Data Storage

Data acquired from various sources comes in various types of file formats such as XML, JSON, CSV, ORC, AVRO and Parquet to name a few popular file types. In AI/ML applications, the data is so huge that usually the data is stored in the form of files on affordable commodity storage media governed by powerful file system management software. This type of data is known as big data, and the underlying hardware storage media coupled with the file system management software is known as a data lake. Hadoop is an example of a big data lake system.

1.2.3 Batch Processing

The acquired data keeps accumulating in the data lake. At regular intervals, the accumulated data is processed by scheduled workflows and processes. These scheduled workflows and processes are known as batch processes, while the process of scheduled processing of the accumulated data at regular intervals is known as batch processing. Owing to the massive amount of data to be processed, often, parallel computing is employed to achieve efficiency and speed. Apache Spark is an example of a big data parallel computing engine.

1.2.4 Real-Time Message Ingestion

At times, usually in the case of IoT device data, there is an immediate need to process the data. This type of data is processed immediately in order to extract vital time-sensitive information, such as remote monitoring of manufacturing equipment. In such cases, it will not be enough to have scheduled workflows waiting to process the incoming data at regular intervals. Instead, processes are triggered as soon as new data arrives. This process of immediate (near real time) data processing is known as real-time processing.

Sometimes, a trade-off between batch processing and real-time processing may be required for certain systems. In such cases, the concept of micro-batching is employed. Micro-batching is a technique where the scheduled workflows are executed at shorter intervals of, say, every 5 minutes (instead of hourly or daily).

1.2.5 Stream Processing

Once the real-time messages have been ingested from IoT device sources or applications, there may be requirements to operate directly on the real-time data. The processing of data in real time enables the discovery of critical time-sensitive information, such as predicting a health emergency from a heart patient's Fitbit data in real time. Apache Storm and Spark Streaming are the examples of stream data processing components.

1.2.6 Machine Learning

The processed data is fed into ML components for the purpose of model training and tuning. Often, it may not be possible to feed the raw data into a ML model. There may be many inaccuracies, outliers, string values, special characters and so on that if addressed before the data can be used by a ML process. The ML components use the cleansed and processed data for various learning purposes.

1.2.7 Analytical Data Store

After the data has been cleansed, the data is stored in a central data store for consumption by various downstream processes. Analytical data stores are often referred to as data warehouses where processed data is properly organized and generally stored in the form of fact dimensional models. Data from the tables in the analytical data stores is used in various downstream reports. Examples of some modern popular analytical data stores are SQL Data Warehouse, Amazon Redshift and Google BigQuery.

1.2.8 Analytics and Reports

The output of ML components may be stored in data lakes, as well as in analytical data stores. Various types of analysis can be performed on the data, and reports may then be built. Databricks is a popular cloud-based data analytics platform built on the Spark engine that can be leveraged to explore data by writing small but powerful data wrangling scripts in languages such as Python, Scala and R. Another popular data analytics tool that allows users to visually manipulate the data is Alteryx, whereas reporting tools such as Tableau and Power BI may be used to build quick reports and dashboards.

1.2.9 Orchestration

The orchestration components are used to schedule various workflows and trigger the execution of processes in a predetermined sequence. For example, first, the data ingestion workflow should be run, followed by data cleansing, and then, the data processing workflows should be run. Proper order of execution is very important to achieve the desired results. Orchestration components help to achieve this proper sequence of executing the workflow one after another. In a nutshell, the orchestration module ties together all the data processing and data movement executing components in the AI/ML applications.

Now that we have understood the various components in the reference architecture, let us look at the details of each layer in the reference architecture.

1.3 Data Acquisition Layer

In this section, we will learn about the acquisition of data from various sources. Data acquisition is the process of collecting data from various source systems. Figure 1.2 depicts the general set-up for the data acquisition layer.

The above diagram depicts various components of the data acquisition layer. These components help in establishing connectivity to various source systems and transferring data from various sources to the data ingestion layer. In other words, the main purpose of the data acquisition layer is to gather data from multiple source systems and transfer data generated in the sources to the data ingestion layer, where the acquired data will be ingested for further processing. The various components of the data acquisition layer are listed below:

- File systems
- Databases

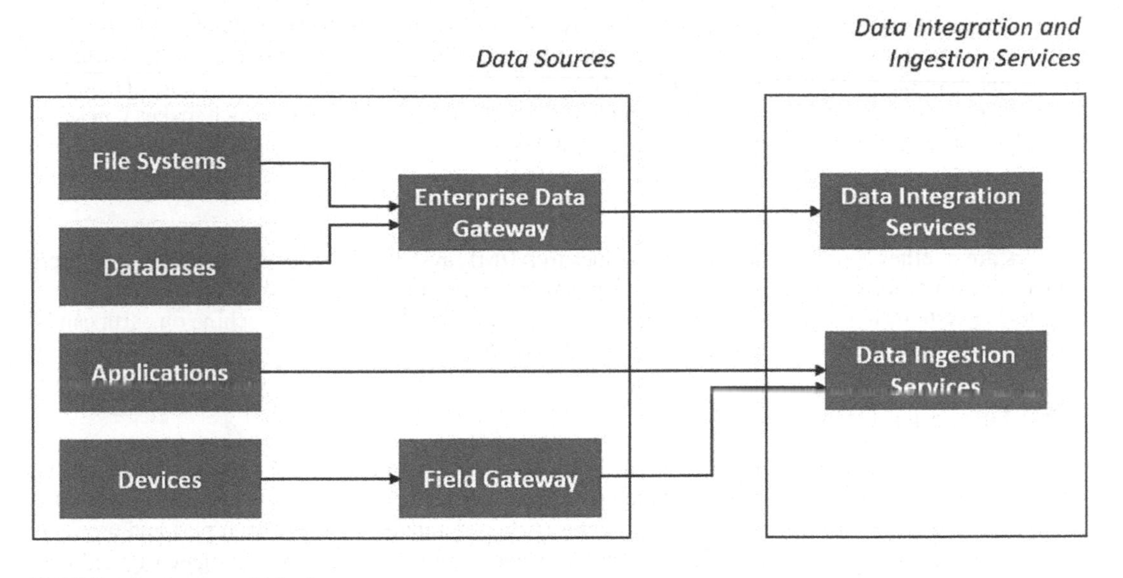

FIGURE 1.2 Data acquisition layer.

- Applications
- Devices
- Enterprise data gateway
- Field gateway
- Data integration services
- Data ingestion services

In the below sections, we will look at each of the above data acquisition components.

1.3.1 File Systems

File systems are a staple when it comes to gathering data for intelligent AI/ML systems. In the modern-day workplace and homes, a wealth of information gets generated by multiple applications. The data generated by various applications is stored in local computer disks, cloud storages (such as OneDrive and Google Drive) and enterprise data lake file systems (such as HDFS – Hadoop Distributed File System). For example, various log files generated by our Windows or Mac computers may get uploaded (with user consent) into Microsoft or Apple data lakes for analysing various aspects of operating systems performance, crashes and so on.

1.3.2 Databases

In most schools, colleges and workplaces, databases are used to store user data and information. Then, there are applications that store data in local computer databases or enterprise databases. These datasets may be acquired for a plethora of information. For example, in bank, there are several computer systems running in its premises or in its data centre, and every moment, millions of transactions take place. AI/ML applications can tap into these transactions to derive real-time information and take action such as detecting fraudulent credit card transactions.

1.3.3 Applications

Taking the example of large retailers, their Point of Sales (POS) application systems generate a humungous volume of data. Data can be sourced from these applications in order to be fed into AI/ML applications. For example, an intelligent customer relationship management system may predict the likes and dislikes of a customer based on their purchase history and recommend products to the customer. You might have noticed the "Movies you may like" feature in online movie streaming apps. Data from applications is gathered to be processed and fed into intelligent AI/ML systems, which makes it possible to implement such recommender systems in applications.

1.3.4 Devices

Devices are another source of data when it comes to AI/ML systems. Nowadays, data collection from devices occurs everywhere around us. Data from smart electric meters, Dish TV set-top boxes, patient monitoring systems in hospital ICUs, moving trains and buses and literally any machine on earth can be connected to the Internet, and data can be acquired from them.

1.3.5 Enterprise Data Gateway

An enterprise data gateway is a component that exposes on-premise data to the outside world. In other words, it is a piece of software that is installed on a computer inside a company's network and then it is configured to expose data from on-premise systems such as databases over a chosen port and protocol. Systems sitting outside the company's network can connect with the enterprise data gateway and read data from a database that resides in the company's network.

1.3.6 Field Gateway

A field gateway is a device (along with embedded software) that usually sits between a group of devices and the Internet. The field gateway helps the devices behind it to connect to the Internet so that important data can be acquired from these devices. A typical setting would be a remote power plant situated in a remote mountain valley. Critical data from devices attached to electrical machines in the power plant can be acquired through a field gateway.

1.3.7 Data Integration Services

Data integration service components are a class of software that helps to build data pipelines, commonly known as workflows, for moving and transforming data from one system to another. These systems often offer sophisticated visual integrated development environments where one can drag and drop data manipulation activities on a drawing canvas, and within a few hours, a powerful workflow can be built. Some examples of popular data integration software are Informatica, Talend, DataStage and SQL Server Integration Services. With the advent of cloud, data integration software is also available on popular cloud-based systems such as AWS Glue, Azure Data Factory and Google Data Fusion.

1.3.8 Data Ingestion Services

Data ingestion services are ubiquitous Application Programming Interfaces (API), reliable messaging systems and Web-based endpoints that can be invoked from source systems to post or upload data. Consider a trucking company, with hundreds of trucks travelling around a country. An embedded device on the truck continuously sends data regarding the truck's location to a central tracking dashboard, so that at any given point in time, the truck location is visible on the dashboard. Open-source messaging platforms such as Apache Kafka or proprietary platforms such as Azure Service Bus and IBM MQ are some examples of platforms that support data ingestion in real time.

1.4 Data Ingestion Layer

In this section, we will learn about the ingestion of acquired data from various sources (Figure 1.3).

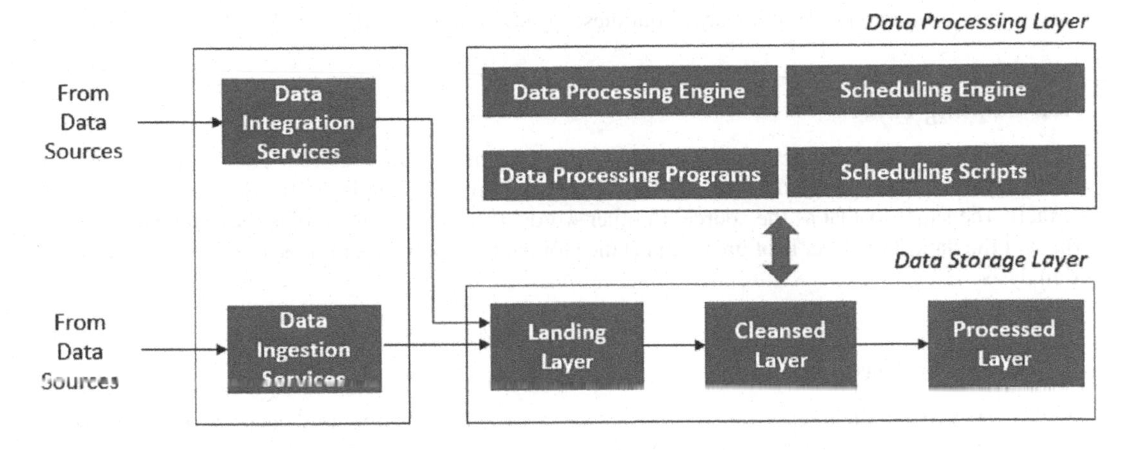

FIGURE 1.3 Data ingestion layer.

The data ingestion layer is responsible for receiving the data from various sources via the data integration and ingestion services. This layer typically comprises a data lake, usually with distributed file systems such as HDFS, AWS S3, Google Cloud Storage or Microsoft Azure Data Lake Storage. The data is first read or received by the data integration and ingestion services and then written to the data storage layer. Thereafter, the data is processed or ingested by various processing engines. The overall components together constitute the data ingestion layer. The various components of the data ingestion layer are listed below:

- Data storage layer
  - Landing layer
  - Cleansed layer
  - Processed layer
- Data processing layer
  - Data processing engine
  - Data processing programs
  - Scheduling engine
  - Scheduling scripts

In the below sections, we will look at each of the above data ingestion components.

1.4.1 Data Storage Layer

The first integral part of the data ingestion layer is the data storage layer. The storage layer provides the storage management software and physical storage medium in which ingested data is stored. Usually, a robust file management system software (like HDFS) sits on top of cheap commodity hardware disks.

The data storage layer may be logically subdivided into three types of areas depending upon the type of data that is stored:

- Landing layer
- Cleansed layer
- Processed layer

In the next three sections, we will learn about these layers in more detail.

1.4.2 Landing Layer

The landing layer is the first storage location where incoming data is stored. Hence, it is known as landing layer for the ingested data. It is to be noted that, in the landing layer, the data is usually stored in exactly the same format as the source. In other words, the landing layer holds the unaltered original replica of the data. This layer is of great importance for data science explorations as it offers the raw data for analysis.

1.4.3 Cleansed Layer

The cleansed layer is the next storage area, after the landing layer. Often, raw data may not be suitable for consumption directly by advanced AI/ML processes. Hence, various Extract Transform Load (ETL) and Extract Load Transform (ELT) processes are employed to pick up data from the landing layer and they perform some basic data cleansing activities on the raw data. Once the raw data has been properly cleansed, i.e. made suitable for further consumption, the data is stored in the cleansed layer. The cleansed layer may be thought of as an intermediate storage area. In the industry, the cleansed layer may also be referred to as a staging layer.

1.4.4 Processed Layer

After the data has been cleansed, it is further processed in order to refine the data for use within AI/ML applications. Once further refinement of the already cleansed data is done, the data is then stored in the processed layer. The processed layer stores data that can be readily used by AI/ML applications.

1.4.5 Data Processing Layer

The second integral part of the data ingestion layer is the data processing layer. The data processing layer is constituted of the data processing engines and underlying physical compute resources using which ingested data is processed. Usually, a powerful data processing engine (like Spark/Hadoop MapReduce) runs on a cluster of cheap commodity machines or virtual machines to provide a robust data processing environment.

The data processing layer may be logically subdivided into four types of components depending upon the type of computation capability offered:

- Data processing engine
- Data processing programs
- Scheduling engine
- Scheduling scripts

In the next four sections, we will learn about these components in more detail.

1.4.6 Data Processing Engine

The data processing engine provides computing power and capabilities to process large amounts of data. Examples of popular big data processing engines are Hadoop MapReduce and Spark. Data processing programs run on these data processing engines. In turn, the data processing engines manage the execution of the programs that operate on the data. These engines take care of all aspects of executing and managing the programs that process the data.

1.4.7 Data Processing Programs

The data processing programs are pieces of code written to operate on the data. Popular examples of coding languages used for writing big data programs are Pig Latin, Scala and Java. Once the code has been written, the programs are compiled and then deployed on machine clusters managed by data processing engines. To give a common industry example, one would often talk about deploying Scala or Java programs on Spark clusters. When the time comes, the deployed programs are run on the Spark cluster in order to process data at a massively parallel scale.

1.4.8 Scheduling Engine

Scheduling engines are software components that can be configured to schedule the execution of data processing programs. In other words, if we want that a particular program should start running at midnight every day, then the schedule can be configured in a scheduling engine. Popular examples of scheduling engines are Apache Airflow and Oozie.

1.4.9 Scheduling Scripts

Scheduling scripts are pieces of code, often written in a tool-specific scripting language. For example, an Apache Airflow script may be written to specify the running schedule for a Scala or Java program deployed on a Spark cluster. Such pieces of code that are used to define the running schedule for data processing programs or jobs are known as scheduling scripts.

1.5 Data Quality and Cleansing Layer

In this section, we will learn about the data cleansing- and data quality-related aspects (Figure 1.4).

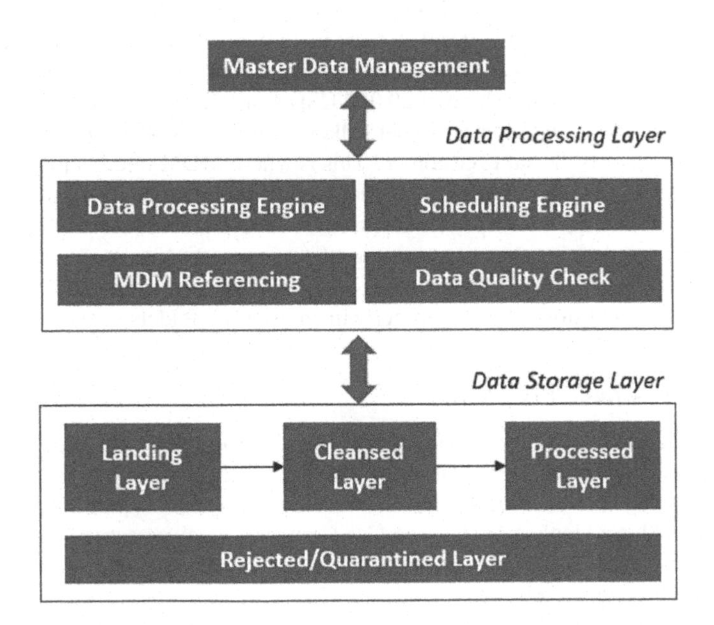

FIGURE 1.4 Data quality and cleansing layer.

The data quality and cleansing layer is responsible for checking the quality of the data received from various sources. Even after performing all types of data refining, there may be scenarios that have to be handled for ensuring the quality of the ingested data meets certain standards. These standards are subjective and may depend from one organization to another. For example, one organization may define that all their percentage data be corrected to two decimal places, whereas another organization may want their percentage data to be rounded up to the nearest whole number. What is right for one company may not work for another. This is where the data quality and cleansing layer comes into the picture.

This layer typically comprises of the following components:

- Master data management (MDM) system
- MDM referencing programs
- Data quality check programs
- Rejected/quarantined layer

In the below sections, we will look at each of the above components in more detail.

1.5.1 Master Data Management (MDM) System

Every organization uses a central database system for maintaining master data. These systems contain master lists of various types of data such as employee records, product lists, departments, vendors, suppliers and list of equipment. Having a central list of unique information helps organizations to refer the central lists across their organization.

All ingested data is referenced against the data in a company's MDM systems. This is done to make sure that stray data does not flow into the system and jeopardize the integrity of the data in the system. Hence, it is extremely important to subject all data to be checked against central MDM records.

In the next section, we will see how an MDM referencing program plays a vital role in the data quality process.

1.5.2 Master Data Management (MDM) Referencing Programs

In order to understand the significance of MDM referencing programs, let us study a scenario where a Jewellery Company regularly collects sales data from its POS machines. Every time, a customer makes a payment, his or her details such as name, address and phone number are captured.

Let us say, a customer makes a purchase of 20,000 USD at a Jewellery store located in one city in the month of September, and then, the same customer makes a purchase of 45,000 USD at a store in another city. How can we say that it is the same customer? This is where MDM referencing programs come into the picture. By using various personally identifiable information data such as a mobile number, the sales data can be uniquely associated with a particular customer. These techniques are simple yet powerful. It helps companies segment their customers and engage them in targeted marketing campaigns.

Thus, one of the main tasks of MDM referencing programs is to properly associate the incoming data with existing MDM data and store the data properly in appropriate folders or tables in the storage area.

1.5.3 Data Quality Check Programs

In order to understand the significance of data quality check programs, let us study a scenario where a Gas Supply Company regularly collects data from its authorized dealers. Let us say, there are exactly 1542 dealers in a country and each dealer has a unique dealer code assigned to them serially starting from number 1 to 1542. Suddenly one day, the Gas Company receives some data for a dealer with a dealer code 1543. Note that the company does not have any dealer with a code 1543 in their MDM system. The largest dealer code in their MDM system is 1542.

The data processing job or data quality check program should be able to intercept this situation and mark the record with dealer code 1543 as "invalid data".

1.5.4 Rejected/Quarantined Layer

All the data that does not conform to an organization's standards or does not match with a company's MDM data may be marked as rejected data. These rejected or non-conformant data are stored in a separate storage area referred to as the reject/quarantine layer.

Periodically, efforts are made to address the piled-up data in the rejected/quarantined layer. Often, data is either sent back to source system owners for correction. Sometimes, even after honest efforts, there may be unresolved data lying in the rejected/quarantined zone – in such an eventuality, the data may be purged after completing the requisite retention period as per organizational policies.

BIBLIOGRAPHY

1. Migliorini, M., Castellotti, R., Canali, L., & Zanetti, M. (2020). Machine learning pipelines with modern big data tools for high energy physics. *Computing and Software for Big Science*, 4(1). doi: 10.1007/s41781-020-00040-0.
2. Boehm, M., Kumar, A., & Yang, J. (2019). Data management in machine learning systems. *Synthesis Lectures on Data Management*, 14(1), 1–173. doi: 10.2200/s00895ed1v01y201901dtm057.
3. Narvekar, M., & Fargose, P. (2015). Daily weather forecasting using artificial neural network. *International Journal of Computer Applications*, 121(22), 9–13. doi: 10.5120/21830-5088.
4. Bhowmick, K., & Narvekar, M. (2022). A semi-supervised clustering based classification model for classifying imbalanced data streams in the presence of scarcely labelled data. *International Journal of Business Intelligence and Data Mining*, 1(1), 1. doi: 10.1504/ijbidm.2022.10034300.
5. Dighe, P., Chavan, M., Gaikwad, A., Chavan, P., & Anandan, A. (2020). Revamping real estate document storage with blockchain. *International Journal of Blockchains and Cryptocurrencies*, 1(1), 1. doi: 10.1504/ijbc.2020.10034707.
6. Kushwaha, A., Chavan, P., & Kumar Singh, V. (2020). COVID-19 data analysis and innovative approach in prediction of cases. *Big Data Analytics and Artificial Intelligence against COVID-19: Innovation Vision and Approach*, 78, 91–115. doi: 10.1007/978-3-030-55258-9_6.
7. Srivastava, K., & Shekokar, N. (2020). Design of machine learning and rule based access control system with respect to adaptability and genuineness of the requester. *EAI Endorsed Transactions on Pervasive Health and Technology*, 6(24), 1–12. doi: 10.4108/eai.24-9-2020.166359.
8. Jhaveri, M., Jhaveri, D., & Shekokar, N. (2015). Big data authentication and authorization using SRP protocol. *International Journal of Computer Applications*, 130(1), 26–29. doi: 10.5120/ijca2015906862.
9. Shinde, G. R., Kalamkar, A. B., Mahalle, P. N., & Dey, N. (2020). *Data Analytics for Pandemics*. CRC Press: Boca Raton, FL. doi: 10.1201/9781003095415.
10. Mahalle, P. N., Sable, N. P., Mahalle, N. P., & Shinde, G. R. (2020). Data analytics: COVID-19 prediction using multimodal data. In Joshi, A., Dey, N., & Santosh, K. C. (Eds), *Intelligent Systems and Methods to Combat COVID-19* (pp. 1–10). Springer: Singapore. doi: 10.1007/978-981-15-6572-4_1.
11. Chaki, J., & Dey, N. (2020). Pattern analysis of genetics and genomics: A survey of the state-of-art. *Multimedia Tools and Applications*, 79(15–16), 11163–11194. doi: 10.1007/s11042-019-7181-8.
12. Save, A., & Shekokar, N. (2018). Analysis of cross domain sentiment techniques. In *International Conference on Electrical, Electronics, Communication Computer Technologies and Optimization Techniques, ICEECCOT 2017*, Vol. 2018-January, pp. 935–943. Institute of Electrical and Electronics Engineers Inc. doi: 10.1109/ICEECCOT.2017.8284637.

2

Fundamental Models in Machine Learning and Deep Learning

Tatwadarshi P. Nagarhalli
Vidyavardhini's College of Engineering and Technology

Ashwini M. Save and Narendra M. Shekokar
D. J. Sanghvi College of Engineering

CONTENTS

2.1 Introduction 13
2.2 Classification of Machine Learning Models 15
  2.2.1 Supervised Learning 15
  2.2.2 Unsupervised Learning 15
  2.2.3 Semi-Supervised Learning 16
  2.2.4 Reinforcement Learning 16
2.3 Fundamental Supervised Learning Models 17
  2.3.1 Regression 17
  2.3.2 Classification 19
    2.3.2.1 Logistic Regression 19
    2.3.2.2 Support Vector Machines 20
  2.3.3 Classification–Regression 20
    2.3.3.1 Decision Tree 21
    2.3.3.2 Random Forest 22
    2.3.3.3 Artificial Neural Network 22
  2.3.4 Implementation Code Snippet for Classification and Classification–Regression Techniques 24
2.4 Fundamental Unsupervised Learning Models 25
  2.4.1 k-means Clustering 25
  2.4.2 Apriori Algorithm 26
2.5 Fundamental Deep Learning Models 27
  2.5.1 Autoencoder 28
  2.5.2 Recurrent Neural Network 29
  2.5.3 Convolutional Neural Network 31
References 33

2.1 Introduction

The resurgence of AI has revolutionised the whole of the computing industry, which in turn has revolutionised almost all the possible sectors of industry. AI is the ability of the machines to think and learn in order to solve a problem by making smart decisions [1]. AI has given birth and rise to many new subfields and area of studies like ML, robotics, computer vision, natural language understanding and expert systems [2].

Generally, all computer programs are written to solve a specific problem. To solve the designated problem, the computer follows the instructions given in these programs. These instructions are followed as it is; that is, a program written to perform the addition of numbers will certainly not be able to perform any other arithmetic operations. In other words, the process of solving the problem is explicitly coded in the program. But, in the real-world scenario, explicitly coding for all the different contingencies is very difficult if not impossible. For example, if we want to differentiate cats from dogs, it is impossible to explicitly code every possible combinations and measurements of cats and dogs for the purpose of differentiation.

Hence, there is a requirement of a way through which the computer can work and solve problems beyond what has been explicitly told to it. This is where ML comes in. ML is an important subfield of AI. ML is a field of study where the computer or the machine is empowered to learn, with or without any supervision, and reproduce these learnings in solving problems, which are difficult for the traditional methods. The learning and the reproduction of learning here include three very important components: remembering, adapting and generalising [3]. While learning the machine makes certain general rules or general assumptions, this is generalising. Generalising is the most important characteristic of a ML method. When the problem is posted to the machine, it should be able to remember the general rules which it had learned and adapt its learning to fit with the current situation in order to solve the given problem. Different ML methods have different ways of generalising their learnings, or remembering or adapting.

This task of remembering and generalisation becomes challenging when the concepts to be remembered and generalised become complex. This is especially true when we are dealing with the understanding of large texts or while processing images. This is where DL comes in. DL is a specialisation of ML employed when the concepts to be learned are complex or deeper insights are expected to be extracted [4].

Over a period of time, many methods and algorithms for ML and DL have been proposed, which make use of the learning feature to make the systems more responsive, accurate and smart. This learning by the different ML and DL techniques is through past experiences. For developing these experiences, the algorithms are provided with large amount of data. Different algorithms have different ways of gaining the knowledge or experiences from the given data. As DL methods are required to learn complex concepts, the data requirement for these methods is very high.

Some of the fundamental and basic ML algorithms include linear regression, logistic regression, support vector machine (SVM), random forest (RF), artificial neural networks (ANNs), decision trees (DTs), k-means clustering and Apriori algorithm [5,6]. Some basic and important DL algorithms include recurrent neural network (RNN) and convolutional neural network (CNN) [4,7].

The high-performance computing at our disposal has propelled and fuelled great interest in experimentation and implementation of ML and DL methods in different domains and different applications. Another reason for this interest is the ease through which these different learning methods can be implemented using different libraries like scikit-learn, TensorFlow and Keras [8]. Different ML and DL methods have been adapted in many different domains and applications including natural language processing, medical diagnosis, document classification, fault detection, bioinformatics, recommendation systems, speech recognition, image analysis and video analysis [9].

Even though ML and DL techniques provide a way to make systems smart, there are many inherent challenges and issues that need to be considered [10]. All the learning techniques are dependent on the data that has been provided to it for learning. It can be safely said that to a very large extent, the learning is as good as the data provided and accuracy of the systems are directly related to the learning. So, the availability of proper data is a very important factor in ML and DL [11]. But, the quality of the data cannot be guaranteed, and because of this, systems implementing the learning methods may be undermined [12]. Even if the availability of the data is ensured, the data itself might be unbalanced, or some values might be missing. Similarly, there are many other issues and challenges like overfitting and underfitting of data and feature selection in ML, which have dogged the field of learning [13]. These issues and challenges need further exploration for a better understanding of the challenges faced while applying the learning methods for real-world applications.

So, the chapter proposes to provide a detailed view of the fundamental models in ML and DL, while concurrently exploring the different issues and challenges. Apart from giving the understanding of these basic

ML and DL models, the chapter proposes to give a snippet of implementation on real-world applications, so that the readers can appreciate both the theoretical and the practical aspects of ML and DL techniques.

After the introduction of ML and DL concepts, with their significance, issues and challenges in this section, different ML and DL models are introduced in the next section. Further, detailed theoretical concepts of the different important ML models will be explored along with the different issues and challenges. Along with explaining the theoretical concepts, an implementation snippet will be given for a real-world application with possible solutions for the issues and challenges. Similarly, the theoretical concepts with issues and challenges along with implementation snippet will be given for fundamental DL methods.

2.2 Classification of Machine Learning Models

ML models or algorithms can be broadly categorised into four types. These categorisations have been done based on the type of data given for training and the presence of reward, if any. Based on these two parameters, ML models can be classified as supervised learning, unsupervised learning, semi-supervised learning and reinforcement learning [14,15]. Figure 2.1 shows the classification of ML models based on the said parameters.

2.2.1 Supervised Learning

In supervised learning approach, the training data provided to the algorithms contains both the feature vectors and their expected output. In technical terms, we call this labelled training data. The use of giving the expected output for training is that when the models train itself using the training data, it can check how well it has learnt by comparing its predicted output with the actual output. Based on this information, the models can optimise their learning to get better results.

The supervised learning methods can be further classified into two types based on the type of output produced by the methods. Regression-based methods are expected to produce a finite continuous numerical value, and, on the other hand, the classification algorithms classify the given inputs into distinct groups called classes. There are certain methods like DT and neural network (NN), which can work for predicting a continuous value or a distinct class [16].

2.2.2 Unsupervised Learning

Unsupervised learning methods, on the other hand, are provided with unlabelled training data. That is, the expected output is not provided in the training data. These types of methods make use of the internal characteristics of the training data features in order to create groups of similar data or to extract insights [17].

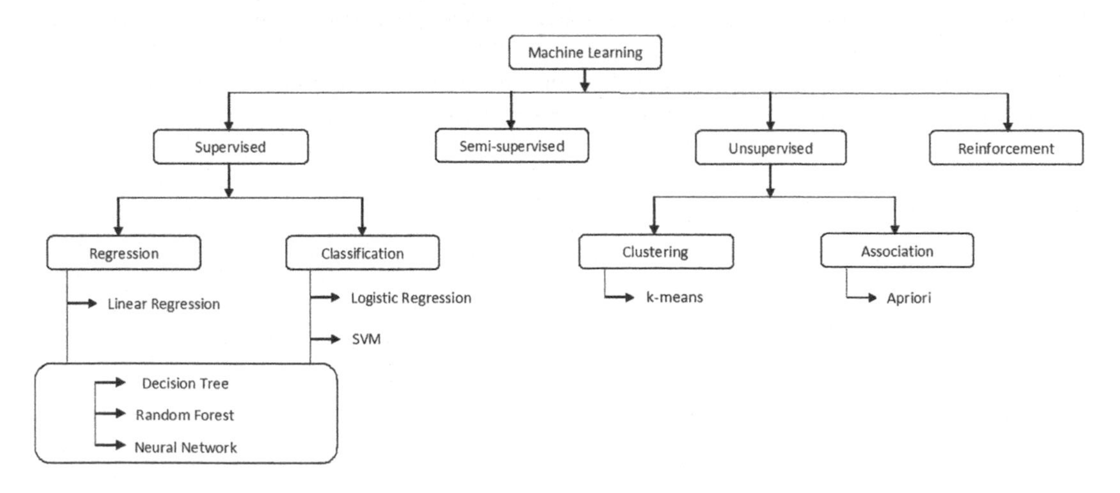

FIGURE 2.1 Classification of machine learning models.

Even unsupervised methods can be categorised into two types based on the type of information acquired. The methods that try to group together data points or data objects based on their similarity are called clustering algorithms, whereas the methods that try to find some associations among the data points are called association methods.

2.2.3 Semi-Supervised Learning

It needs to be understood that, generally, the learnings of supervised learning algorithms are more sophisticated and the predictions are much more accurate than unsupervised learning methods. But, this does not mean that unsupervised learning methods are useless. There are many applications where unsupervised learning methods are required and perform much better. Also, the supervised learning methods require labelled or annotated training data, development of which is an expensive proposition [18].

In order to work around this problem of annotation of training data and to take benefits of both supervised and unsupervised techniques, semi-supervised methods were introduced. The semi-supervised learning methods produce results or predictions similar to the supervised learning methods. But the advantage of semi-supervised learning techniques over supervised learning techniques is that the semi-supervised methods can work with unlabelled data in the presence of very less labelled training data.

In semi-supervised learning approach, the unlabelled training data is studied first. This unlabelled data is provided labels by using statistical methods or heuristic-based approaches. Once the labels have been identified for the unlabelled data, any supervised learning models can be used for regression or classification.

2.2.4 Reinforcement Learning

The analogy that best describes reinforcement learning is the carrot and stick approach. If the prediction of a learning method is satisfactory, then it is rewarded, and if it is not as expected, then it is not rewarded or might be penalised. Any supervised or unsupervised learning approach can have the reinforcement learning element by having a feedback mechanism, which works on the principle of rewards [19].

Generally, it can be seen that supervised learning approaches produce more accurate results in the short term compared to other learning methods. But, over a period of time, any technique equipped with reinforcement learning outmatches all the other learning approaches that do not contain reinforcement learning feature.

Figure 2.2 shows the general working of a reinforcement learning method, where once the learning agent acts on the environment it gets a feedback in terms of reward.

It is important to turn the focus on the fundamental models in supervised learning and unsupervised learning approaches. The reason for this is that there are no specific algorithms for semi-supervised learning and reinforcement learning. Both of these types of learning are dependent upon the supervised or unsupervised learning models.

Generally, in ML, the important task or the learning is to identify the relationship between the dependent variables (output) and the independent variables (input features) and generate a generalised rule

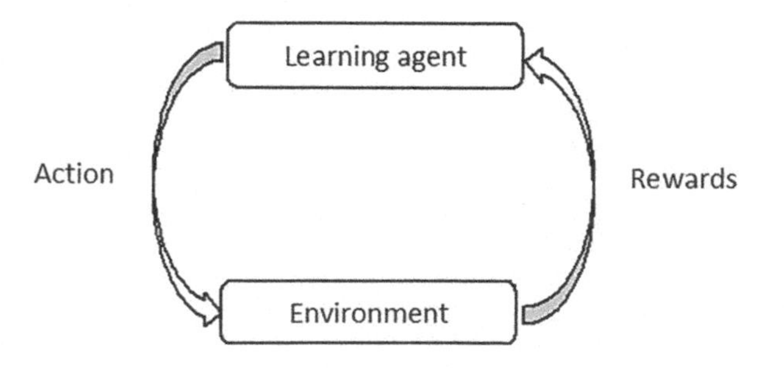

FIGURE 2.2 Reinforcement learning.

or the hypothesis function. Once the hypothesis function has identified output, any new objects or data points can be obtained using this hypothesis function.

2.3 Fundamental Supervised Learning Models

As already mentioned, supervised learning methods require labelled or annotated training data. Based on the type of prediction expected of the model, the models can be further categorised into regression and classification techniques.

2.3.1 Regression

The distinctive property of the regression analysis is that the expected output after regression analysis is a finite continuous numerical value. The most basic regression technique is simple linear regression [20].

In simple linear regression, the training data points can be plotted on a two-dimensional plane as there is only one input feature. If this is the case, then its hypothesis function can be given by using a straight line function. It is given as

$$y = ax + b$$

where

x is the input feature value

a and b are constants that can be calculated as follows:

$$a = \frac{\left(\sum y\right)\left(\sum x^2\right) - \left(\sum x\right)\left(\sum xy\right)}{n\left(\sum x^2\right) - \left(\sum x\right)^2}$$

$$b = \frac{n\left(\sum xy\right) - \left(\sum x\right)\left(\sum y\right)}{n\left(\sum x^2\right) - \left(\sum x\right)^2}$$

The input feature values and its given output for the training data can be visualised as shown in Figure 2.3 [20], where the grey dots represent the input data point features and the straight line is the hypothesis function of a simple linear regression.

The metric to test the accuracy of a linear regression model is the least mean squared error. There is a variant of simple linear regression called multivariate or multiple linear regression. In multivariate linear regression, there are two or more than two input features as the independent variables for one dependent variable. The hypothesis function for multivariate linear regression can be given as

$$y = b + a_1x_1 + a_2x_2 + a_3x_3 + \cdots + a_nx_n$$

One of the important issues with regard to regression analysis is the correlation among the independent variables [21]. Because of this correlation, it becomes difficult to find the optimal hypothesis function. The simplest solution for this problem is to combine the correlated input features to form a new input feature. This forms the fundamental of feature engineering.

Implementation:

For the implementation of linear regression, the Boston housing dataset has been used, which is available at Kaggle [22]. This dataset contains 13 input features including crime rate and average number of

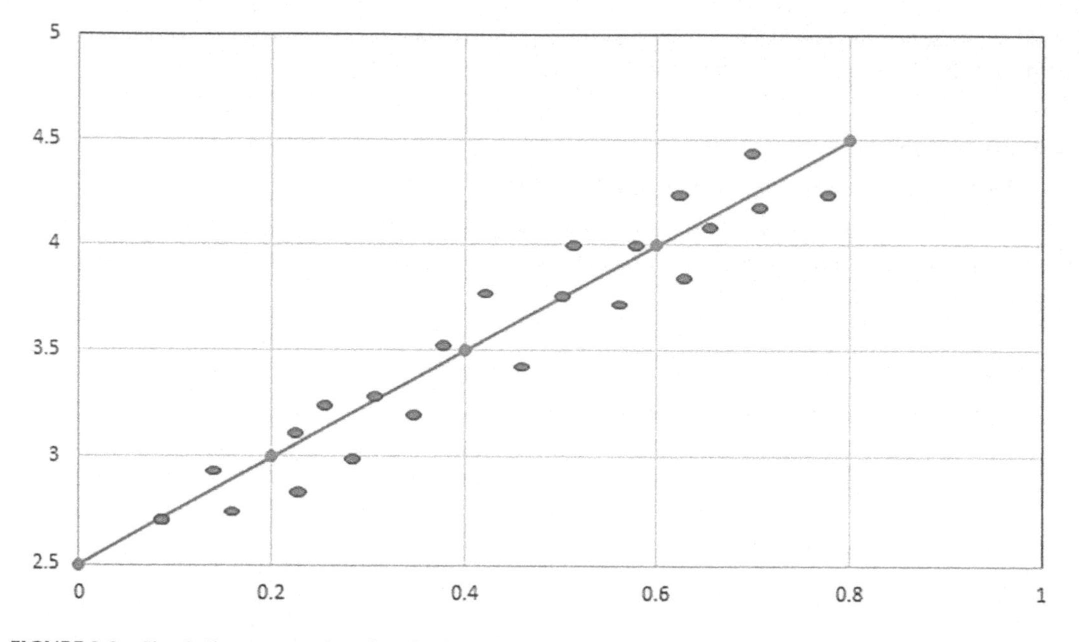

FIGURE 2.3 Simple linear regression visualisation.

rooms. The object is to find the housing price for the given input features. The linear regression has been implemented using the sklearn ML package. Almost all the ML techniques have been implemented in this package. In this case, 'LinearRegression' package under the 'linear_model' of 'sklearn' package has been used.

Apart from processing the data to make it fit for the ML algorithm, the process of implementation of all the fundamental ML models remains the same. First, the input and the output data with their respective labels is prepared and stored in a variable. Then, the data is split into two parts: one for training and the other for testing. The ratio is decided by using 'test_size'. The keyword 'random_state' ensures that split remains constant on all computing machines. After data splitting, the ML model is provided with the training data for training. Then, the model is asked to predict unknown data points and finally accuracy has been found out.

In the present case, the linear regression model is producing a least mean squared error of 4.861, which is on a higher side. The data can be worked upon to decrease the error value and increase the accuracy of the model.

Simple Linear Regression Implementation Snippet:

```
#Linear Regression on Boston Housing Dataset
#Import relevant packages
from sklearn.linear_model import LinearRegression
import pandas as pd
from sklearn.model_selection import train_test_split
from sklearn.metrics import mean_squared_error
import math
lr=LinearRegression() #Instance of Linear Regression Function

#Read Dataset
df_train=pd.read_csv("D:/Projects/Reg_train.csv", sep= ',')
df_test=pd.read_csv("D:/Projects/Reg_test.csv", sep= ',')

#Preparing input feature X and output Y
X=df_train.drop('ID', axis=1)
```

```
X=X.drop('medv', axis=1)
Y=df_train['medv']

#Splitting of training and testing data
X_train,X_test,Y_train,Y_test=train_test_split(X,Y,random_state=0,test_
size=0.2)
lr.fit(X_train,Y_train) #Running Linear Regression
y_pred=lr.predict(X_test)
score=mean_squared_error(Y_test,y_pred)
print(math.sqrt(score))

Output:
4.861035032958114
```

2.3.2 Classification

In simple words, classification is learning a class or distinctive group through examples. For example, if we want to identify whether a particular image belongs to a cat or dog or human, this is a classification problem. Here, the cat, god and human form the distinctive groups or classes. The classification models have studied the input features and identified the class to which the given data point might belong to. This process is called classification [14]. There are two fundamental classification models: logistic regression and SVMs.

To showcase the effectiveness of the different techniques, the heart attack prediction dataset available at Kaggle [23] has been used for both the classification models. This dataset contains input features like age, sex, chest pain location, chest pain type and resting blood pressure. And, the prediction task is to identify whether heart attack is possible or not for the given input features. Also, simple accuracy matrix has been showcased in order to understand the accuracy of the models. The implementation code snippet has been given in Section 2.3.4.

2.3.2.1 Logistic Regression

Logistic regression is one of the most basic classification techniques. It analyses the relation between the input features and categorical dependent variable. The prediction probability is estimated by fitting the data points on the logistic curve [24]. The logistic curve is an 'S'-shaped curve, which is called the sigmoid curve. It means that the analysis output might be that if it is placed on the plane, the values below a certain threshold are converted to '0' and the values above the threshold are taken as '1'. Figure 2.4 [25] shows the sigmoid curve.

As logistic regression converts the output to either 0 or 1, it is also called binary classifier. And it is generally used only for binary classification. The hypothesis function for logistic regression is given as [26]

$$y = \frac{e^{ax+b}}{1+e^{ax+b}}$$

where

x is the input feature
a and b are constants

Like linear regression, even logistic regression is non-tolerant with respect to the correlation among input features. Apart from this, another important issue with logistic regression is that it generally requires higher number of data points for training, compared to other ML algorithms in order to give better results.

Implementation:
For the implementation of logistic regression, the 'LogisticRegression' package under the 'linear_model' of 'sklearn' package is to be used. The accuracy of logistic regression on the given data is 85.25%.

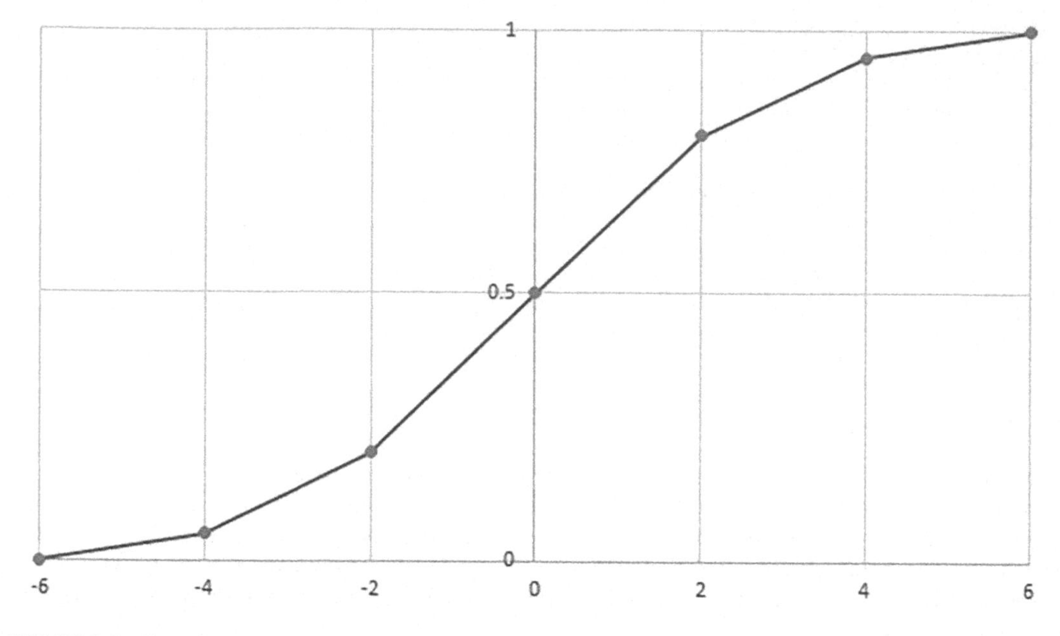

FIGURE 2.4 Sigmoid curve.

2.3.2.2 Support Vector Machines

Sometimes, SVMs are considered to be the best stock classifiers. The reason for this is that it can be easily understood visually. If all the training data points containing two classes are plotted on a plane with respect to the feature vectors, it is believed that more often than not they take their positions closer to their own type, thus forming two separate groups. SVMs draw a line between these two groups called hyperplane. When an unknown data is to be classified, it is checked on which side of the hyperplane it falls to in order to identify the class [20]. Figure 2.5 [27] shows the visual representation of how the SVMs divide the two classes.

The position of the hyperplane that divides the two classes is identified by calculating the distance between the data points of two different classes which are closest to each other. These data points participating in determining the position of the hyperplane are called support vectors. SVMs are also called maximum margin separators as they try to maximise the distance between the opposing support vectors.

One of the major issues of SVMs is that they were developed for binary classification. It can work with multiclass classification with certain modifications [28]. Another issue with it is that it works well when the data is linearly separable, that is if data can be separated by a single straight line. This problem can be solved by using certain kernel function like Gaussian kernels [29]. Kernel function increases the dimension of the data so that the data becomes linearly separable.

Implementation:
For the implementation of SVMs, the 'svm' package of 'sklearn' is to be imported. The accuracy of SVMs on the given data is 59.02%.

2.3.3 Classification–Regression

There are some important models in supervised learning, which can predict both finite continuous numerical values and distinct classes. That is, they can be used for both regression and classification purposes. Some of the important models under this category are DT, ANN and RF.

The implementation snippet has been given for the classification problem. For comparative analysis, the heart attack prediction dataset used for the previous section has been used again for implementation. The code snippet has been given in Section 2.5.4.

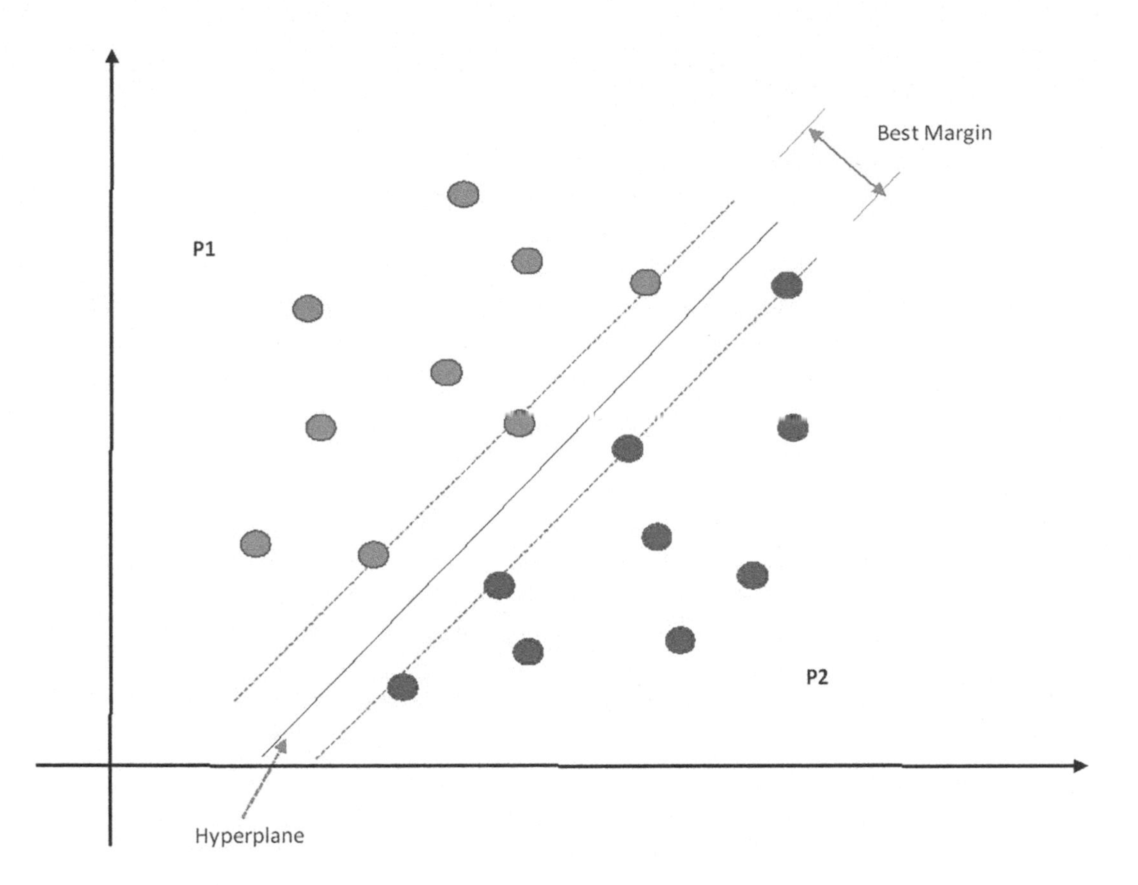

FIGURE 2.5 Support vector machines.

2.3.3.1 Decision Tree

DT is one of the most commonly used supervised learning techniques. DT classifiers work by separating one feature at a time creating a tree structure, which is used for making decision regarding classification or regression [20].

As shown in Figure 2.6 [30] for a given dataset, the DT is constructed by asking questions about the features given. Step by step, the whole DT is constructed.

Here, the important thing to be identified is which feature becomes the root node and which comes after that [31]. The hierarchy of features nodes is decided by different DT algorithms like Iterative Dichotomiser 3 (ID3), classification and regression tree (CART) and C 4.5. For example, the ID3 algorithm utilises the information gain and dataset impurity in order to identify which feature gets precedence over others, whereas CART makes use of the Gini index to identify the best-suited feature over others.

The formula for finding the best feature node in ID3 is given as

$$\text{Gain}\ (T,x) = \text{Entropy}\ (T) - \text{Entropy}(T,x)$$

where

x is individual feature and T is a set of features in the dataset

Here,

$$\text{Entropy} = -\sum p_j \log_2 (p_j)$$

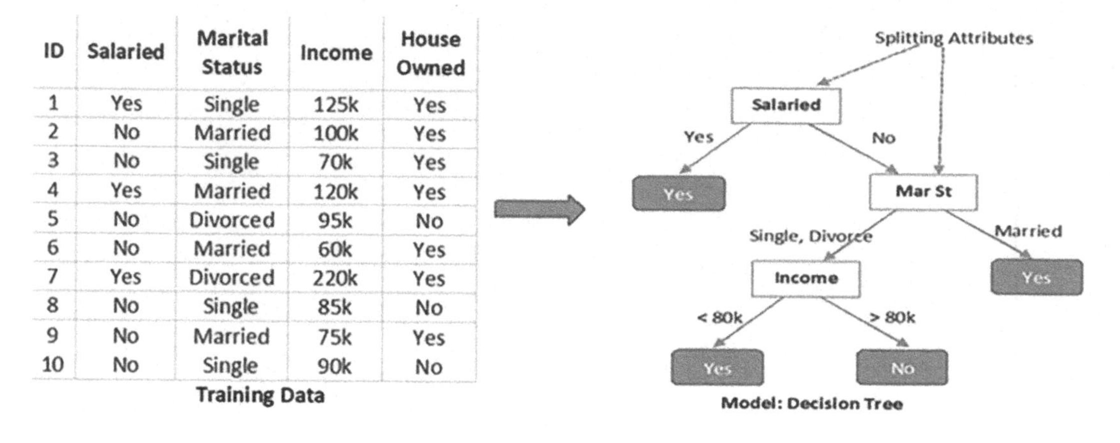

ID	Salaried	Marital Status	Income	House Owned
1	Yes	Single	125k	Yes
2	No	Married	100k	Yes
3	No	Single	70k	Yes
4	Yes	Married	120k	Yes
5	No	Divorced	95k	No
6	No	Married	60k	Yes
7	Yes	Divorced	220k	Yes
8	No	Single	85k	No
9	No	Married	75k	Yes
10	No	Single	90k	No

Training Data

FIGURE 2.6 Decision tree classifier.

However, for CART, it is given as

$$\text{Gini Gain} = 1 - \sum p_j^2$$

There are a number of issues with regard to DT classifiers, which disrupt the proper working of the DT classifier. These issues include the presence of continuous-valued attributes and mission values, which need further investigation [32].

Implementation:

For the implementation of the ID3 algorithm, the 'DecisionTreeClassifier' package under the 'sklearn.tree' package is to be imported. The accuracy of the ID3 DT classifier on the given data is 78.69%.

2.3.3.2 Random Forest

RF is an ensemble-based learning method. Instead of applying an algorithm only once on a dataset, ensemble algorithms apply the algorithm multiple times on a subset of a dataset. Here, either the same algorithms can be applied multiple times or different algorithms may also be applied. These algorithms are applied not on the complete dataset, rather on a subset of the whole dataset. These subsets can be created by using different combinations of features and the number of data points [33].

RF algorithms traditionally use many DT classifiers on the subset of the given dataset. Here, all the different DT classifiers will produce and predict a class label for the unknown data point. In RF technique, the class label that has been predicted maximum number of times is accepted as the final class label for the unknown data point [34,35].

Figure 2.7 [36] shows the working of RF ensemble model.

Apart from basic ML issues like overfitting and underfitting, other important issues that affect RF algorithm include inconsistency in subsamples and inconsistencies in average estimators [37].

Implementation:

For the implementation of RF algorithm, the 'RandomForestClassifier' package under the 'sklearn.ensemble' package is to be imported. The accuracy of RF classifier on the given data is 88.52%.

2.3.3.3 Artificial Neural Network

ANN design tries to mimic and is loosely based on the collaborative and non-linear learning aspect of the human neurological system. Like human neurons, even neurons in ANN carry specific information and collaboratively help each other in order to achieve a target [38]. Figure 2.8 shows the basic structure of an ANN.

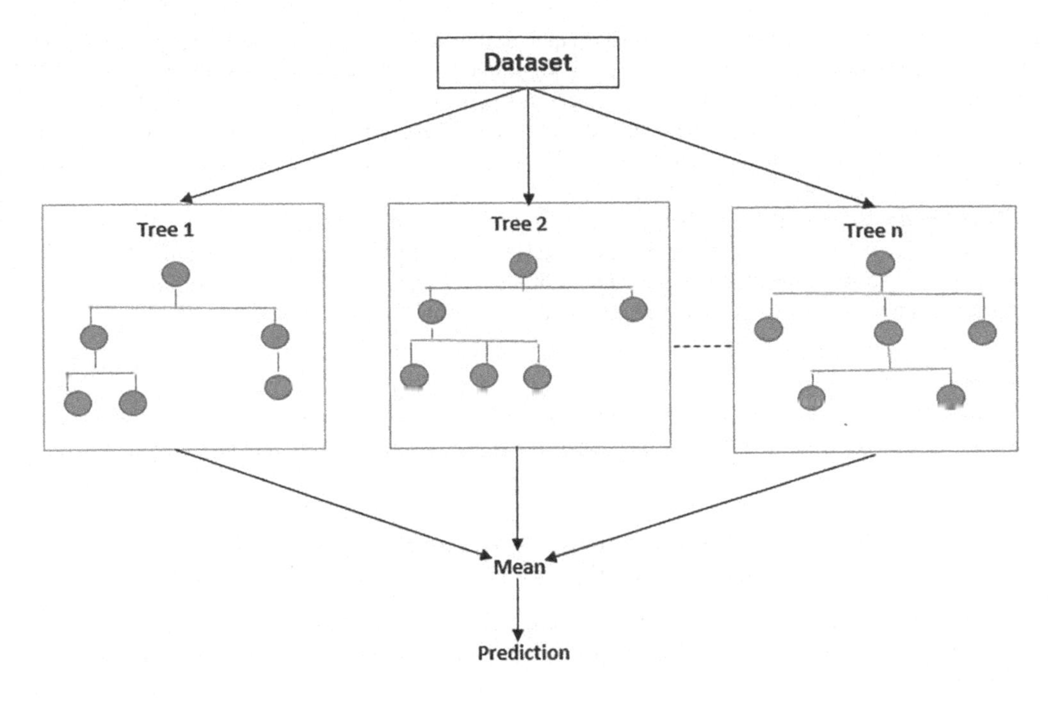

FIGURE 2.7 Random forest model.

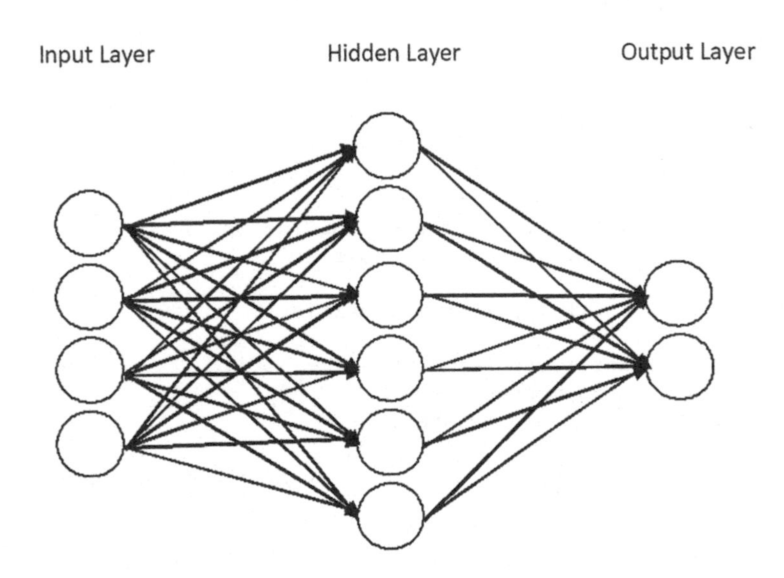

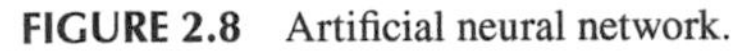
FIGURE 2.8 Artificial neural network.

A typical ANN contains two layers, hidden layer and output layer. If a NN contains more than one hidden layer, then it is called deep neural network. A constant called weight is associated with each edge in the network. And, each neuron of the two layers has a constant called a bias associated with it. The ANN carries information and performs learning by manipulating these two constants.

Initially, these constants are assigned at random. When the input feature values are given at each input neuron, this value is multiplied with the respective weights of the edges. This multiplied value is then added with the bias value of the respective neuron. In this way, the information is carried forward. This type of network is called feedforward neural network [39].

In NNs, the number of output layer neurons is equal to the number of classes. The neuron that produces the highest probability is the class label for the given input. In training, if the probability score of class labels does not match that of the actual, then an error is recorded. This error is propagated backwards so that all the weights and biases can be adjusted and recalculated. This propagating of error backwards, towards the input, is called the backpropagation algorithm or technique. The backpropagation algorithm is a very fundamental unit of a NN [40].

ANN is highly prone to perform badly. This is because of issues like noisy dataset. Also, for better learning, the NNs require large amount of data for training compared to other ML techniques. Also, one should be very careful while choosing the number of hidden layers and number of neurons in the hidden layer as it plays a very important part in the optimal working of NN [41].

Implementation:
For the implementation of the ANN algorithm, the 'MLPClassifier' package under the 'sklearn.neural_network' package is to be imported. The accuracy of ANN on the given data is 55.74%.

2.3.4 Implementation Code Snippet for Classification and Classification–Regression Techniques

A sample implementation code snippet for the classification and classification–regression techniques discussed in the previous two sections has been given in order to showcase the ease through which these ML techniques can be implemented.

Implementation Code Snippet

```
#Heart Attack Prediction - Kaggle
import pandas as pd
from sklearn.ensemble import RandomForestClassifier
from sklearn.linear_model import LogisticRegression
from sklearn.model_selection import train_test_split
from sklearn.metrics import accuracy_score
from sklearn.ensemble import GradientBoostingClassifier
from sklearn.tree import DecisionTreeClassifier
from sklearn import svm
from sklearn.neural_network import MLPClassifier

rf=RandomForestClassifier(random_state=0)
lr=LogisticRegression(random_state=0)
dt=DecisionTreeClassifier(random_state=0)
sv=svm.SVC()
nn=MLPClassifier(solver='lbfgs', alpha=1e-5, hidden_layer_sizes=(5, 2),
random_state=0)

df=pd.read_csv("D:/Projects/heart.csv", low_memory=False)

x=df.drop("target",axis=1)
y=df["target"]

X_train,X_test,Y_train,Y_test=train_test_split(x,y,random_state=0,test_
size=0.2)

rf.fit(X_train,Y_train)
lr.fit(X_train,Y_train)
dt.fit(X_train,Y_train)
sv.fit(X_train,Y_train)
nn.fit(X_train,Y_train)
```

```
y_pred1=rf.predict(X_test)
y_pred2=lr.predict(X_test)
y_pred3=dt.predict(X_test)
y_pred4=sv.predict(X_test)
y_pred5=nn.predict(X_test)
print('Accuracy Scores ---- ')
print('Logistic Regression: ', accuracy_score(Y_test,y_pred2))
print('Support Vector Machine: ', accuracy_score(Y_test,y_pred4))
print('Decision Tree: ', accuracy_score(Y_test,y_pred3))
print('Random Forest: ', accuracy_score(Y_test,y_pred1))
print('Neural Network: ', accuracy_score(Y_test,y_pred5))
```

Output:

```
    Accuracy Scores ----
Logistic Regression: 0.8524590163934426
Support Vector Machine: 0.5901639344262295
Decision Tree: 0.7868852459016393
Random Forest: 0.8852459016393442
Neural Network: 0.5573770491803278
```

It can be seen that for the given heart attack prediction dataset that contains both continuous-valued features and features with categorical values, the RF model produces the highest accuracy with 88.52% among the fundamental ML models discussed. The accuracy of other models can also be increased by working on the data to make it more suitable for a particular ML model.

2.4 Fundamental Unsupervised Learning Models

As mentioned, the unsupervised learning methods train on dataset, which does not contain target values or classes. The unsupervised learning techniques make use of the relationship among the features and their values in order to learn new insights. In other words, the internal characteristics of the dataset are leveraged in order to learn new information. There are two fundamental unsupervised learning models: k-means clustering and Apriori algorithm.

2.4.1 k-means Clustering

Clustering algorithms like k-means are partitioning-based methods, where the training data is portioned into separate groups called clusters. k-means specifically is centroid-based technique. The 'k' in k-means is the number of clusters to be created and 'mean' because at every iteration the mean of all values is found out [42].

In k-means clustering, initially, the data points are distributed randomly among number of clusters. Suppose if there are ten data points and the number of clusters to be made is 3, then two clusters will have three data points each and the third cluster will have four data points. Then, the mean of all the data points in respective clusters is identified, and this mean is the provisional centroid of the cluster. Now, the distance between data points and centroid is found out, and the data point with the list distance with a particular centroid joins that particular cluster. This operation is repeated till all the data points have settled in a particular partition and there is no change in the centroid value and the data point position [43].

Though the working principle of k-means clustering is quite simple, there are certain issues that magnify the problem while implementing the technique. For one, there is a possibility that some clusters might be empty; that is, no data points are allocated to a particular cluster. The k-means technique is unable to handle this type of problem. Also, k-means cannot deal with outliers making the algorithm work weirdly in certain cases [44].

Implementation:

In order to showcase the working of k-means cluster using scikit-learn ML package, the dataset has been created in the program itself. Here, three clusters with a total of 500 data points with random values have been created. Then, the k-means clustering is performed by using the sklearn library. Finally, the output has been displayed. For the implementation of k-means algorithm, the 'KMeans' package under the 'sklearn.cluster' package is to be imported.

k-means Clustering Implementation Snippet [45]:

```
#K-Means
from sklearn.cluster import KMeans
import matplotlib.pyplot as plt
from sklearn.datasets import make_blobs

# create dataset
X, y = make_blobs(
       n_samples=500, n_features=2,
       centers=3, cluster_std=0.5,
       shuffle=True, random_state=0
)
km = KMeans(
       n_clusters=3, init='random',
       n_init=10, max_iter=300,
       tol=1e-04, random_state=0
)
y_km = km.fit_predict(X)
# plot the 3 clusters
plt.scatter(
       X[y_km == 0, 0], X[y_km == 0, 1],
       s=50, c='lightgreen',
       marker='s', edgecolor='black',
       label='cluster 1'
)
plt.scatter(
       X[y_km == 1, 0], X[y_km == 1, 1],
       s=50, c='orange',
       marker='o', edgecolor='black',
       label='cluster 2'
)
plt.scatter(
       X[y_km == 2, 0], X[y_km == 2, 1],
       s=50, c='lightblue',
       marker='v', edgecolor='black',
       label='cluster 3'
)
# plot the centroids
plt.scatter(
       km.cluster_centers_[:, 0], km.cluster_centers_[:, 1],
       s=250, marker='*',
       c='red', edgecolor='black',
       label='centroids'
)
plt.legend(scatterpoints=1)
plt.grid()
plt.show()
```

Figure 2.9 shows the output of the implemented code. It can be seen that after iterations, the different data points have positioned themselves closest to one of the three centroids marked by '*'.

2.4.2 Apriori Algorithm

Apriori algorithm was put forward by Rakesh Agrawal and Ramakrishnan Srikant [46] in 1994. Apriori algorithm is another fundamental algorithm used under unsupervised learning. Apriori algorithm is a

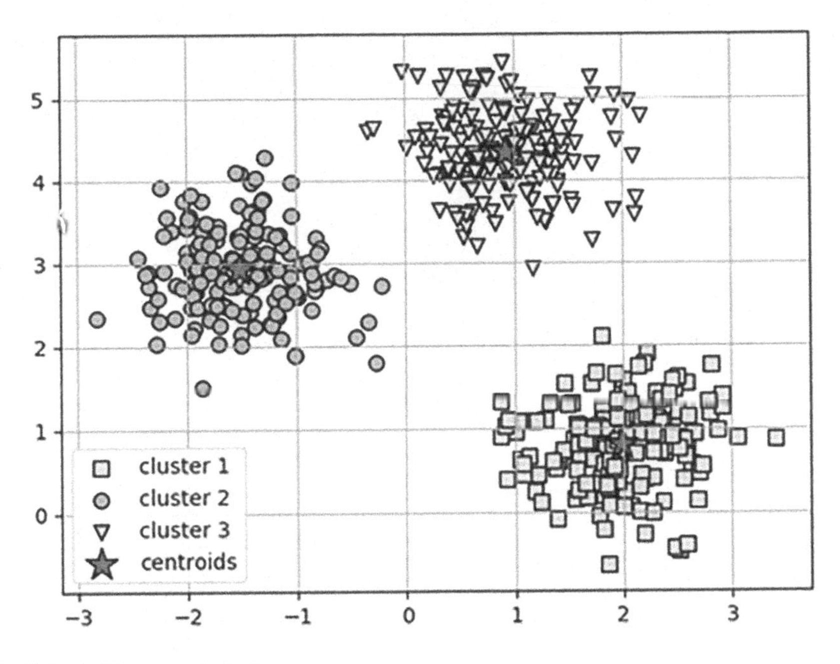

FIGURE 2.9 Output of k-means clustering.

frequent item set and association rule mining method used for extracting information from data. Generally, the Apriori algorithm identifies the association rules highlighting the trend of the given dataset.

The Apriori algorithm uses bottom-up approach in order to find the frequent item set in the given dataset. For this, it also makes use of breadth-first search and hash tree [47]. For example, consider the dataset with the following item set with a support threshold of 3:

$$\{\{1,2,3,4\},\{1,2,3\},\{1\},\{2,3\},\{2\},\{3,4\},\{1,2,4\},\{1,4\}\}$$

For the given set, it can be seen that there are four unique items {1, 2, 3 and 4}. First, it is found out how many times these items have appeared in the given set.

$$\{1\}=5,\ \{2\}=5,\ \{3\}=4,\ \{4\}=4.$$

As all the items exceed the support threshold value, we can say that all are frequent sets. Next, the frequency of all possible pairs is found out.

$$\{1,2\}=3,\ \ \{1,3\}=2,\ \ \{1,4\}=2,\ \ \{2,3\}=3,\ \ \{2,4\}=2,\ \ \{3,4\}=2.$$

As only two pairs {1, 2} and {2, 3} equal or exceed the support threshold, it can be said that these two pairs are frequent item sets. Next, if a set is created with these three elements, it will create a value of 2, which is below the support threshold and hence unacceptable. Finally, it can be said that item sets {1, 2} and {2, 3} are the most frequent item sets.

2.5 Fundamental Deep Learning Models

As ML is an important subfield of AI, in a similar way DL is a very important subfield of ML. Figure 2.10 shows the relationship among the smart field of AI, ML and DL.

Specifically, DL is an important specialisation of ANN which is a ML model. Generally, it is accepted that a basic ANN contains only one hidden layer; to improve the learning and to learn deep insights, the

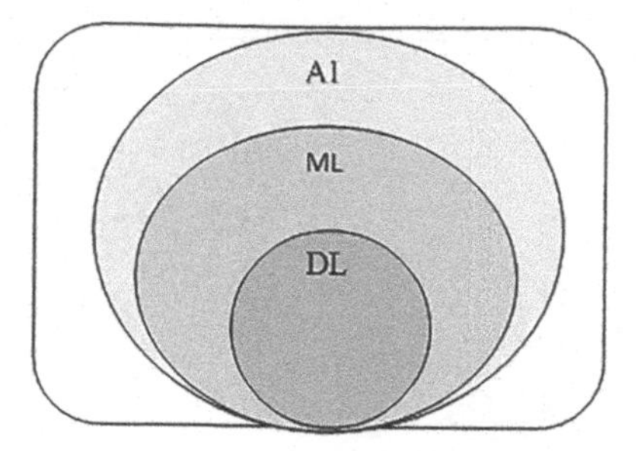

FIGURE 2.10 Artificial intelligence–machine learning–deep learning.

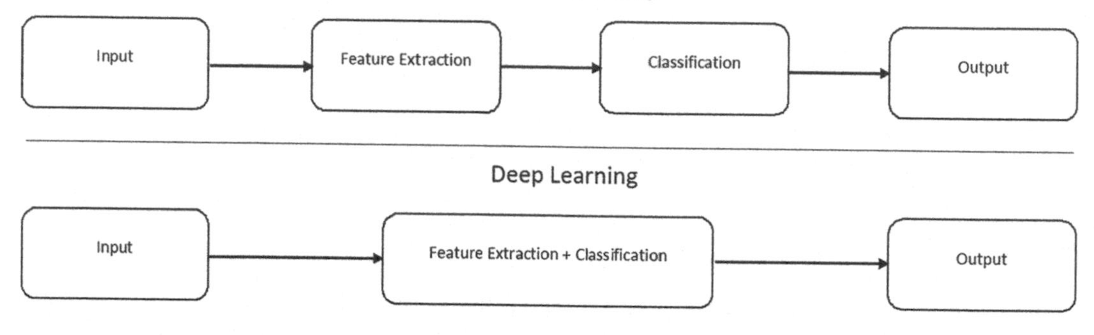

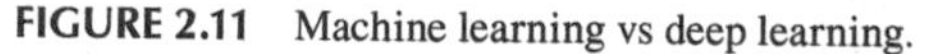

FIGURE 2.11 Machine learning vs deep learning.

number of hidden layers can be increased. A DL network is a special case of NN where the number of hidden layers is large in number; that is, they use deep layered neural network [48].

The main distinguishing factor between ML and DL is with regard to the important step of extracting features. Feature extraction is a very important step because ML algorithms cannot be applied directly on a raw data like a text, or image or video. This raw data has to be processed and relevant information with which the algorithms can work is extracted, and this process is called feature extraction. Before applying any ML algorithm, the data has to undergo this important step of feature extraction, whereas DL techniques do not require this additional step, as the techniques extract features by themselves [49].

In DL, both the feature extraction task and classification task are handled by the DL algorithm themselves. This is one of the biggest advantages of DL over ML. The downside to this is that the data requirement for learning in DL is higher. In other words, it can be said that the DL algorithms work well if the data given to them is sufficiently large, at least in thousands if not more. But, it is also true that over a period of time, DL systems prove to be more accurate and efficient than ML techniques. This is the major reason why most of the research in the field of learning is being done in DL.

Figure 2.11 [49] shows the fundamental difference between ML and DL.

The fundamental DL models include autoencoder, RNN and CNN. Over the years, in-depth research has been carried out on these basic techniques and many improvements have been proposed over these fundamental DL techniques.

2.5.1 Autoencoder

Autoencoder is a special type of ANN. Autoencoder generally works in two phases: encoding and decoding. In the encoding process, the input data's dimensions are compulsorily reduced, thus generating a coded representation of the given data. While decoding, using this encoded data the original data, as close to original as possible, representations are developed or created [50]. For this purpose, in the encoding phase, the number of neurons in a particular layer is always less than its preceding layer. And, as in

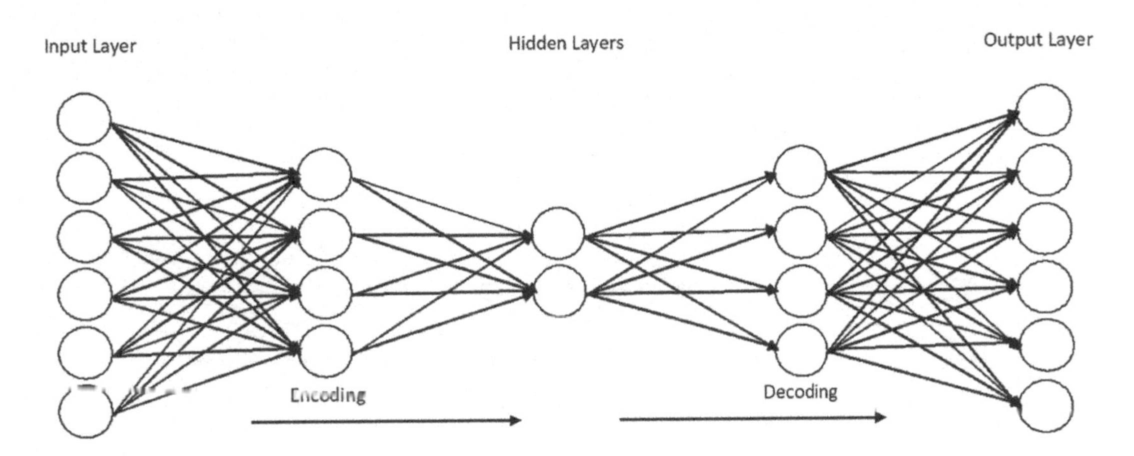

FIGURE 2.12 Structure of a typical autoencoder.

the decoding phase, the original form of the data is to be generated, the number of neurons in a particular layer is always more than its preceding layer.

Figure 2.12 [51] shows general representation of autoencoder structure. Here, the 'a_2' layer is the code generated from the input layer and it is used for generating the output layer. In the figure, there are two hidden layers apart from the code layer, which is also a hidden layer. Here, only one hidden layer has been applied for encoding and decoding each. These numbers can be increased according to the research and the input–output data.

Autoencoders are generally unsupervised learning methods. But, in recent times using pipeline methods, they have been extensively used for classification task as well [52]. Autoencoders have been extensively researched and have been adapted in many different domains including security and sentiment analysis. This research has also led to many modifications and improvements over stock autoencoders. This has led to denoising autoencoder, sparse autoencoder and contrastive autoencoder [53].

2.5.2 Recurrent Neural Network

RNN is another special type of deep feedforward neural network with a speciality that it has a feedback mechanism. This feedback mechanism shares the output of the neurons with others [4]. Figure 2.13 [54] shows the general structure of a RNN.

A value at a particular neuron in ANN is input multiplied by weight added with the bias at the neuron, which is given as

$$V = i_1 * w_1 + b_1$$

For RNN, the output of the previous neuron is also added. Thus, we get:

$$V = i_1 * w_1 + b_1 + o_{i-1}$$

Because of this feedback mechanism, RNN can store state information. And, because it can store state information and as this feedback mechanism works at the neuron level as shown in Figure 2.14 [55], RNN has been popular while working with temporal data like working with texts or videos [56]. RNN has also been used for other applications as well like intrusion detection [57] and image analysis [58].

There are two major issues with regard to RNN. First, because of the output feedback mechanism, these are end-to-end networks; thus, the processing of information is sequential. Because of this, the training time requirement for RNNs is exponentially high. One possible solution was given by Google in their white paper on machine translation [59], where with the help of attention mechanism and the way network was set up, the RNN was able to process in parallel.

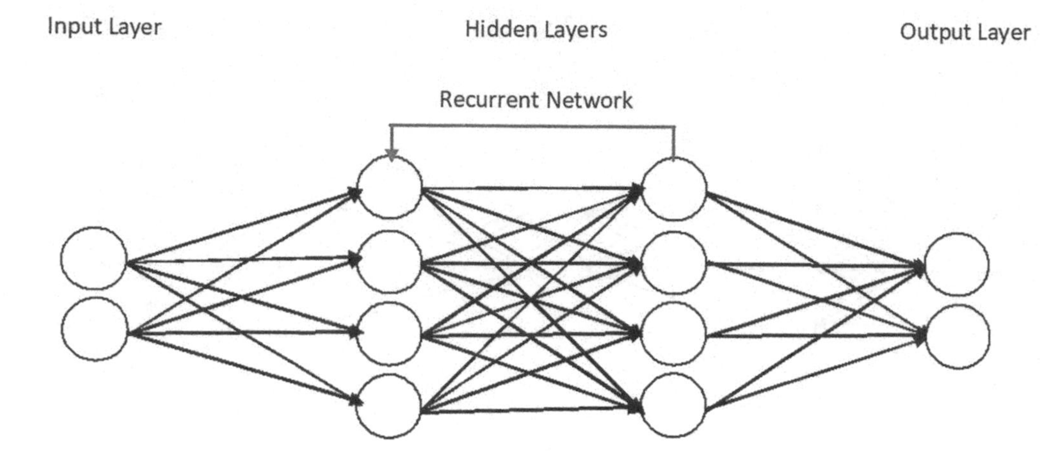

FIGURE 2.13 Recurrent neural network.

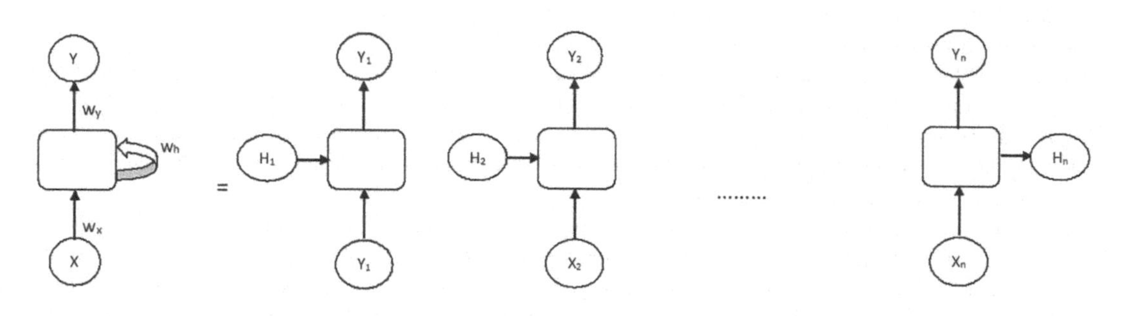

FIGURE 2.14 Feedback mechanism in recurrent neural network.

The second major issue with RNN is that it is supposed to hold temporal information but as the network moves ahead, the temporal information gets blurred. In other words, it can be said that while processing and learning a long sentence, the words that are at the start of the sentence are not remembered by the network. This is detrimental when processing large sentences and paragraph for any applications like sentiment analysis. In other words, the RNN is unable to hold long-term dependencies.

To solve this problem of the ability to hold long-term dependencies, LSTM was introduced [60]. LSTM is a RNN where each neuron has a gated structure in order to hold information for a bit longer [61]. The LSTM unit contains a cell, an input gate, an output gate and a forget gate.

The basic idea behind LSTM is to decide which information to store and which information to throw away; this way, critical information can be retained over a longer duration. The first thing the LSTM has to do is to decide which information is to be discarded. This is done using sigmoid function, which is also the forget gate. The next step is to decide which information is to be retained and stored. This is done using the sigmoid and tanh functions, and this is called the input gate. And finally, the LSTM has to decide what kind of output is to be given out. This is done using the sigmoid function, which is called the output layer [62].

But, the presence of these many gates in LSTM pushes the training time requirement even more exponentially higher than simple RNN, which itself required high training time. To solve this problem, gated recurrent unit (GRU) was introduced. GRU has only two gates: update gate and reset gate [63]. This makes GRU time complexity wise less expensive than LSTM. Figure 2.15 [63] gives the illustration of the internal structure of simple RNN, LSTM and GRU.

Implementation:

The implementation snippet provides a basic way through which RNN can be implemented using the Keras package. Keras is high-level package, which works on TensorFlow. In the Keras package, keras.

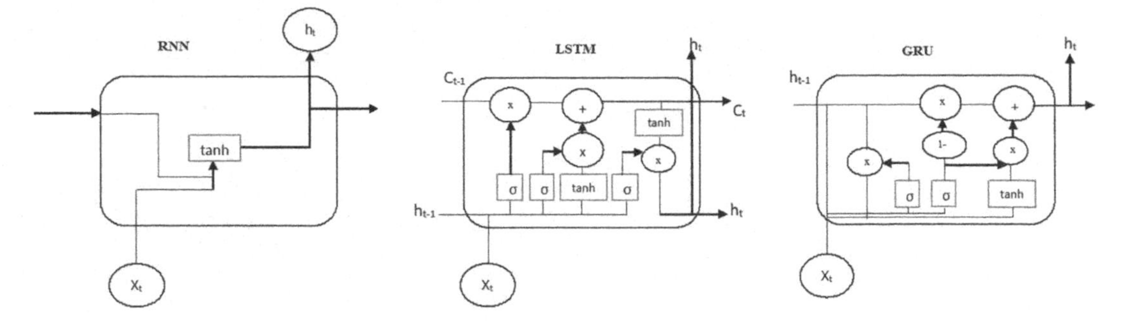

FIGURE 2.15 RNN, LSTM and GRU internal structure.

layers has the classes for SimpleRNN, LSTM and GRU, which need to be imported. Here, first training and test data has to be prepared in X_train, y_train, X_test and y_test. After the data has been prepared, the RNN can be implemented as given in the snippet.

The whole implementation of LSTM and GRU also remains the same with only three differences, which are as follows:

1. from keras.layers import LSTM or from keras.layers import GRU, instead of keras.layers import SimpleRNN.
2. model.add(LSTM or model.add(gru_model, instead of model.add(SimpleRNN.
3. model_rnn=KerasClassifier(build_fn=lstm or model_rnn=KerasClassifier(build_fn=gru_model, instead of model_rnn=KerasClassifier(build_fn=vanilla.

Recurrent Neural Network Implementation Snippet:

```
from keras.models import Sequential
from keras.layers import Dense, SimpleRNN, Activation
from keras import optimizers
from keras.wrappers.scikit_learn import KerasClassifier

def vanilla_rnn():
        model = Sequential()
        model.add(SimpleRNN(50, input_shape = (49, 1), return_sequences = False))
        model.add(Dense(46))
        model.add(Activation('softmax'))

        adam = optimizers.Adam(lr=0.001)
        model.compile(loss = 'categorical_crossentropy', optimizer = adam,
        metrics = ['accuracy'])
        return model

model_rnn = KerasClassifier(build_fn = vanilla_rnn, epochs = 200, batch_size =
50, verbose = 1)
model_rnn.fit(X_train,y_train)
y_pred_rnn = model_rnn.predict(X_test)
y_test_ = np.argmax(y_test, axis = 1)
print(accuracy_score(y_pred_rnn,y_test_))
```

2.5.3 Convolutional Neural Network

CNN is one of the most extensively researched and used DL models. The CNN generally works on only images. In other words, the input data to a CNN should be similar to an image. Similar to an image

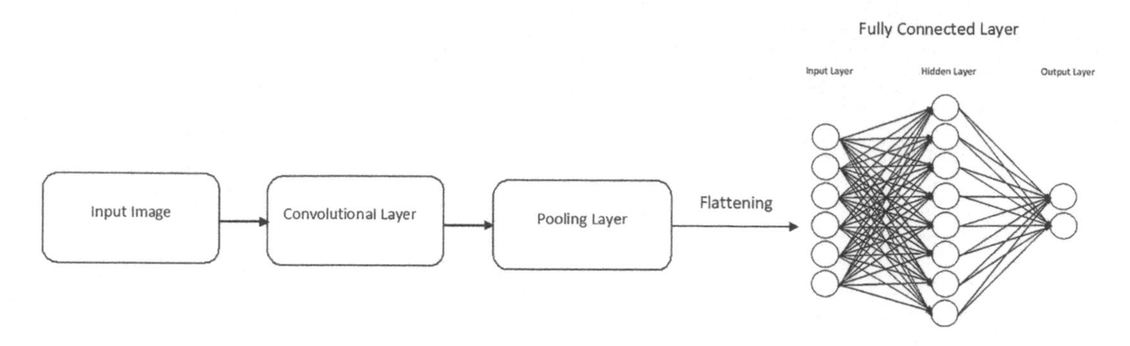

FIGURE 2.16 Convolutional neural network layers.

means the data should be given in a matrix form and any changes or any substitution in the values of the matrix should render the data useless. This is the reason why formal databases with different features cannot be given to CNN for learning any insights as if any two rows are interchanged there is no distortion in final insights, whereas, in an image, even a change in one pixel value distorts the whole image. This is also true for texts; even a single change of alphabet distorts the whole message. So, CNN works on only images or texts. Their capabilities have been extended to include video analysis as well [64–66].

There are two important components in CNN, namely the convolutional layer and the pooling layer. Figure 2.16 [67] shows the basic structure and components of a CNN.

The convolutional layer is responsible for the feature extraction task. This is done by multiplying the weights of the convolutional layer window with the pixel value of the image and then taking an average. Sliding this window over the whole image gives a new convoluted image or the filtered image. Generally, in the convolutional layer, the rectified linear unit (ReLU) activation function is used, which converts all the negative values into zero. Thus, only the prominent features are extracted during convolution [68].

On the other hand, the purpose of the pooling layer is to reduce the computing complexity. There are a number of convoluted images produced after the feature extraction task by one convolutional layer, and a number of convolutional layers can be applied. So the number of filtered images increases exponentially. So, in order to reduce the computing complexity pooling, especially max-pooling layer is used. Max-pooling reduces the size of the filtered images while allowing only prominent features to go for further processing [69].

For the classification purpose, a fully connected layer is used, which is a simple multilayer perceptron neural network. In order to give input to this fully connected layer, the image has to be converted into a vector, and this process is called flattening. The output of the fully connected layer is the class for the given input.

Implementation:

The implementation snippet showcases a basic way of implementing a CNN using Keras. The keras.layers contains the package Conv2D for performing convolution on images. For performing convolution on texts, Conv1D should be used. In the given model, two convolutional layers with 32 and 64 filters have been applied with the activation function being ReLU. Also, one max-pooling and two fully connected layers have been applied.

Convolutional Neural Network Implementation Snippet:

```
import keras
from keras.models import Sequential
from keras.layers import Dense, Dropout, Flatten
from keras.layers import Conv2D, MaxPooling2D

model = Sequential()
```

```
model.add(Conv2D(32, kernel_size=(3, 3),
       activation='relu',
       input_shape=(28,28,1)))
model.add(Conv2D(64, (3, 3), activation='relu'))
model.add(MaxPooling2D(pool_size=(2, 2)))
model.add(Dropout(0.25))
model.add(Flatten())
model.add(Dense(128, activation='relu'))
model.add(Dropout(0.5))
model.add(Dense(num_classes, activation='softmax'))
model.compile(loss=keras.losses.categorical_crossentropy,
       optimizer=keras.optimizers.Adadelta(),
       metrics=['accuracy'])
model.fit(x_train, y_train,
       batch_size=128,
       epochs=12,
       verbose=1,
       validation_data=(x_test, y_test))
score = model.evaluate(x_test, y_test, verbose=0)
print('Test loss:', score[0])
print('Test accuracy:', score[1])
```

In CNN, the research part is the finalisation of the architecture. That is, the number and position of convolutional and max-pooling layers, the number of filters, the activation functions and the number of fully connected layers are all a matter of research and generally depend on the input.

Extensive research has been and is being carried out with respect to CNN, which has led to the development of improved versions of CNN like region-based convolutional neural network (RCNN) [70], Fast RCNN [71] and Faster RCNN [72]. But, there is one major issue with CNNs: they cannot work with temporal data. This has led to the development of a hybrid system called recurrent convolutional neural network. Here, the convolutional layer has been replaced by recurrent layer, which enables the technique to understand temporal data [73].

In conclusion, it can be said that as in the new millennium data is the oil, the effective usage of this data is a very important factor. The subfields of AI, ML and DL provide a smart way of the understanding of data and extraction of useful information [74,75]. In today's time, there is hardly anyone who has not been touched, in one way or the other, by systems that have their core in ML or DL. Hence, an attempt has been made to introduce the fundamental models in ML and DL.

REFERENCES

1. Russell S. J. and P. Norvig. 2003. *Artificial Intelligence: A Modern Approach*, 2nd Edition. Pearson Education Inc: London.
2. Bullinaria J. A. 2005. IAI: The Roots, Goals and Sub-Fields of AI. https://www.cs.bham.ac.uk/~jxb/IAI/w2.pdf. (Accessed 27th March, 2020).
3. Marsland S. 2009. *Machine Learning: A Algorithmic Perspective*. CPC Press: Boca Raton, FL.
4. Goodfellow I., Y. Bengio, and A. Courville. 2015. *Deep Learning*. MIT Press: Cambridge, MA.
5. Smola A. and S. V. N. Vishwanathan. 2008. *Introduction to Machine Learning*. Cambridge University Press: Cambridge.
6. Mohammed M. M. Z. E., M. B. Khan and E. B. M. Bashier. 2017. *Machine Learning: Algorithms and Applications*. CPC Press: Boca Raton, FL.
7. LISA lab. 2015. *Deep Learning Tutorial*, Release 0.1. University of Montreal: Montreal.
8. Nguyen G., S. Dlugolinsky, M. Bobák, et al. 2019. Machine learning and deep learning frameworks and libraries for large-scale data mining: A survey. *Artificial Intelligence Review*, vol. 52, no. 1, pp. 77–124.
9. Shinde P. P. and S. Shah. 2018. A review of machine learning and deep learning applications. *IEEE Fourth International Conference on Computing Communication Control and Automation (ICCUBEA)*, Pune, India.
10. Mitchel T. M. 1997. *Machine Learning*. McGraw-Hill Science/Engineering/Math: New York.

11. L'Heureux A., K. Grolinger, H. F. ElYamany, and M. A. M. Capretz. 2017. Machine learning with big data: Challenges and approaches, *IEEE Access*, vol. 5, pp. 1–9.
12. Stewart M. 2019. https://towardsdatascience.com/the-limitations-of-machine-learning-a00e0c3040c6. (Accessed 25th April, 2021).
13. Schelter S., F. Biessmann, T. Januschowski, D. Salinas, S. Seufert, and G. Szarvas. 2018. On challenges in machine learning model management. *IEEE Computer Society Technical Committee on Data Engineering*, vol. 41, pp. 5–15.
14. Alpaydin E. 2010. *Introduction to Machine Learning,* 2nd Edition. The MIT Press: Cambridge, MA.
15. Ayodele T. O. Types of Machine Learning Algorithms. https://www.researchgate.net/publication/221907660_Types_of_Machine_Learning_Algorithms. (Accessed 27th March, 2020).
16. Saravanan R. and P. Sujatha. 2018. A state of art techniques on machine learning algorithms: A perspective of supervised learning approaches in data classification. *IEEE Second International Conference on Intelligent Computing and Control Systems (ICICCS)*, Madurai, India, pp. 945–949.
17. Jothi R. 2018. A comparative study of unsupervised learning algorithms for software fault prediction. *IEEE Second International Conference on Intelligent Computing and Control Systems (ICICCS)*, Madurai, India, pp. 741–745.
18. Zhu X. and A. Goldberg. 2009. *Introduction to Semi-Supervised Learning.* IEEE Morgan & Claypool Publishers: San Rafael, CA, pp. 1–130.
19. Jeerige A., D. Bein and A. Verma. 2019. Comparison of deep reinforcement learning approaches for intelligent game playing. *IEEE 9th Annual Computing and Communication Workshop and Conference (CCWC)*, Las Vegas, NV, pp. 366–371.
20. Harrington P. 2012. *Machine Learning in Action.* Manning Publications: Shelter Island, NY.
21. Frost J., Multicollinearity in Regression Analysis: Problems, Detection, and Solutions. https://statisticsbyjim.com/regression/multicollinearity-in-regression-analysis/. (Accessed 27th March, 2020).
22. Kaggle. Boston Housing Dataset for Linear Regression. Available at: https://www.kaggle.com/henriqueyamahata/boston-housing-with-linear-regression. (Accessed 27th March, 2020).
23. Kaggle Heart Attack Prediction Dataset. Available at: https://www.kaggle.com/imnikhilanand/heart-attack-prediction. (Accessed 27th March, 2020).
24. Park H. 2013. An introduction to logistic regression: From basic concepts to interpretation with particular attention to nursing domain. *Journal of Korean Academy of Nursing,* vol. 43, no. 2, pp. 154–164.
25. Logistic regression. https://en.wikipedia.org/wiki/Logistic_regression. (Accessed 27th March, 2020).
26. Swamy H. 2019. How to Deploy a Logistic Regression Model in GCP. https://medium.com/analytics-vidhya/insiders-view-on-logistic-regression-and-how-do-we-deploy-regression-model-in-gcp-as-batch-c62a64563210. (Accessed 27th March, 2020).
27. Carrasco O. C. Support Vector Machines for Classification. https://towardsdatascience.com/support-vector-machines-for-classification-fc7c1565e3. (Accessed 27th March, 2020).
28. Liu B., Z. Hao and X. Yang. 2005. Nesting support vector machinte for muti-classification machinte read machine. *IEEE International Conference on Machine Learning and Cybernetics*, vol. 7, pp. 4220–4225.
29. Guo, H., W. Wang and C. Men. 2009. A novel learning model-Kernel granular support vector machine. *IEEE International Conference on Machine Learning and Cybernetics*, Baoding, China, pp. 930–935.
30. http://mines.humanoriented.com/classes/2010/fall/csci568/portfolio_exports/lguo/decisionTree.html. (Accessed 27th March, 2020).
31. Chopda J., H. Raveshiya, S. Nakum and V. Nakrani. 2018. Cotton crop disease detection using decision tree classifier. *IEEE International Conference on Smart City and Emerging Technology (ICSCET)*, Mumbai, India, pp. 1–5.
32. http://pandamatak.com/people/anand/771/html/node29.html. (Accessed 27th March, 2020).
33. Breima L. 2002. *Random Forests.* Springer Machine Learning, vol. 45, pp. 5–32.
34. Yi H., Q. Xiong, Q. Zou, R. Xu, K. Wang and M. Gao. 2019. A novel random forest and its application on classification of air quality. *IEEE 8th International Congress on Advanced Applied Informatics (IIAI-AAI),* Toyama, Japan, pp. 35–38.
35. Suresh Kumar M., V. Soundarya, S. Kavitha, E. S. Keerthika and E. Aswini. 2019. Credit card fraud detection using random forest algorithm. *IEEE 3rd International Conference on Computing and Communications Technologies (ICCCT)*, Chennai, India, pp. 149–153.
36. https://deepai.org/machine-learning-glossary-and-terms/random-forest. (Accessed 27th March, 2020).

37. Tang C., D. Garreau, and U. Von Luxburg. 2018. When do random forests fail? *32nd Conference on Neural Information Processing Systems (NeurIPS 2018)*, Montréal, Canada, pp. 1–11.
38. Dacombe J. An Introduction to Artificial Neural Networks (with Example). https://medium.com/@jamesdacombe/an-introduction-to-artificial-neural-networks-with-example-ad459bb6941b. (Accessed 27th March, 2020).
39. Zupan J. 1994. Introduction to Artificial Neural Network (ANN) methods: What they are and how to use them. *Acta Chimica Slovenica*, vol. 41, no. 3, pp. 327–352.
40. Ozyilmaz L. and T. Yildirim. 2003. Artificial neural networks for diagnosis of hepatitis disease. *IEEE Proceedings of the International Joint Conference on Neural Networks*, vol. 1, Portland, OR, pp. 586–589.
41. Golovko V., M. Komar and A. Sachenko. 2010. Principles of neural network artificial immune system design to detect attacks on computers. *IEEE International Conference on Modern Problems of Radio Engineering, Telecommunications and Computer Science (TCSET)*, Lviv, Ukraine, pp. 237–237.
42. Tan P., M. Steinbach, A. Karpatne, and V. Kumar. 2019. *Introduction to Data Mining*, 2nd Edition. Person Publication: London.
43. Han J., M. Kamber and J. Pei. 2012. *Data Mining: Concepts and Techniques*, 3rd Edition. Elsevier Morgan Kaufmann Publishers: Amsterdam.
44. *Data Mining: The Hypertextbook*. http://www.hypertextbookshop.com/dataminingbook/public_version/contents/chapters/chapter004/section002/blue/page003.html. (Accessed 28th March, 2020).
45. Li L. K-Means Clustering with Scikit-Learn. https://towardsdatascience.com/k-means-clustering-with-scikit-learn-6b47a369a83c. (Accessed 28th March, 2020).
46. Agrawal R. and R. Srikant. 1994. Fast algorithms for mining association rules. *Proceedings of the 20th VLDB Conference*, Santiago, Chile, pp. 487–499.
47. Hegland M. 2005. The Apriori Algorithm–a Tutorial. WSPC/Lecture Notes Series, pp. 1–51.
48. Goularas D. and S. Kamis. 2019. Evaluation of deep learning techniques in sentiment analysis from twitter data. *IEEE International Conference on Deep Learning and Machine Learning in Emerging Applications (Deep-ML)*, Istanbul, Turkey, pp. 12–17.
49. Oppermann A. Artificial intelligence vs. machine learning vs. deep learning. https://www.deeplearning-academy.com/p/ai-wiki-machine-learning-vs-deep-learning. (Accessed 29th March, 2020).
50. Chen Z., C. K. Yeo, B. S. Lee and C. T. Lau. 2018. Autoencoder-based network anomaly detection. *IEEE Wireless Telecommunications Symposium (WTS)*, Phoenix, AZ, pp. 1–5.
51. Jordan J. Introduction to Autoencoders. https://www.jeremyjordan.me/autoencoders/. (Accessed 29th March, 2020).
52. Siddiqua A. and G. Fan. 2019. Asymmetric supervised deep autoencoder for depth image based 3D model retrieval. *IEEE Visual Communications and Image Processing (VCIP)*, Sydney, NSW, pp. 1–4.
53. Janod K., M. Morchid, R. Dufour, G. Linarès and R. De Mori. 2017. Denoised bottleneck features from deep autoencoders for telephone conversation analysis. *IEEE/ACM Transactions on Audio, Speech, and Language Processing*, vol. 25, no. 9, pp. 1809–1820.
54. Valkov V. Making a Predictive Keyboard Using Recurrent Neural Networks: TensorFlow for Hackers (Part V). https://medium.com/@curiousily/making-a-predictive-keyboard-using-recurrent-neural-networks-tensorflow-for-hackers-part-v-3f238d824218. (Accessed 29th March, 2020).
55. Khuong B. The Basics of Recurrent Neural Networks (RNNs). https://medium.com/towards-artificial-intelligence/whirlwind-tour-of-rnns-a11effb7808f. (Accessed 29th March, 2020).
56. Bhatia Y., A. Bajpayee, D. Raghuvanshi and H. Mittal. 2019. Image captioning using google's inception-resnet-v2 and recurrent neural network. *IEEE Twelfth International Conference on Contemporary Computing (IC3)*, Noida, India, pp. 1–6.
57. Chowdhury N. and M. A. Kashem. 2008. A comparative analysis of feed-forward neural network and recurrent neural network to detect intrusion. *IEEE International Conference on Electrical and Computer Engineering*, Dhaka, Bangladesh, pp. 488–492.
58. Li P., L. Peng, J. Cai, X. Ding and S. Ge. 2017. Attention based RNN model for document image quality assessment. *IEEE 14th IAPR International Conference on Document Analysis and Recognition (ICDAR)*, Kyoto, Japan, pp. 819–825.
59. Wu Y., M. Schuster, Z. Chen, et al. Google's Neural Machine Translation System: Bridging the Gap between Human and Machine Translation. https://research.google/pubs/pub45610/. (Accessed 29th March, 2020).

60. Hochreiter S. and J. Schmidhuber. Long short term memory. *Neural Computation* vol. 9, no. 8. http://www.bioinf.jku.at/publications/older/2604.pdf. (Accessed 25th April, 2020).
61. Nguyen M. Illustrated Guide to LSTM's and GRU's: A Step by Step Explanation. https://towardsdatascience.com/illustrated-guide-to-lstms-and-gru-s-a-step-by-step-explanation-44e9eb85bf21. (Accessed 29th March, 2020).
62. Olah C. Understanding LSTM Networks. https://colah.github.io/posts/2015-08-Understanding-LSTMs/. (Accessed 25th April, 2020).
63. Kostadinov S. Understanding GRU Networks. https://towardsdatascience.com/understanding-gru-networks-2ef37df6c9be. (Accessed 29th March, 2020).
64. Li L., L. Xiao, N. Wang, G. Yang and J. Zhang. 2017. Text classification method based on convolution neural network. *3rd IEEE International Conference on Computer and Communications (ICCC)*, Chengdu, China, pp. 1985–1989.
65. Yue G. and L. Lu. 2018. Face recognition based on histogram equalization and convolution neural network. *IEEE 10th International Conference on Intelligent Human-Machine Systems and Cybernetics (IHMSC)*, Hangzhou, China, pp. 336–339.
66. Benkrid K. and S. Belkacemi. 2002. Design and implementation of a 2D convolution core for video applications on FPGAs. *IEEE Third International Workshop on Digital and Computational Video, DCV*, Clearwater Beach, FL, pp. 85–92.
67. SuperDataScience Team. Convolutional Neural Networks (CNN): Summary. https://www.superdatascience.com/blogs/convolutional-neural-networks-cnn-summary/. (Accessed 29th March, 2020).
68. O'Shea K. T. and R. Nash. An Introduction to Convolutional Neural Networks. https://www.researchgate.net/publication/285164623_An_Introduction_to_Convolutional_Neural_Networks. (Accessed 29th March, 2020).
69. Wang B., Y. Liu, W. Xiao, Z. Xiong and M. Zhang. 2013. Positive and negative max pooling for image classification. *IEEE International Conference on Consumer Electronics (ICCE)*, Las Vegas, NV, pp. 278–279.
70. Cai D., X. Sun, N. Zhou, X. Han and J. Yao. 2019. Efficient mitosis detection in breast cancer histology images by RCNN. *IEEE 16th International Symposium on Biomedical Imaging (ISBI 2019)*, Venice, Italy, pp. 919–922.
71. Ullah A., H. Xie, M. O. Farooq and Z. Sun. 2018. Pedestrian detection in infrared images using fast RCNN. *IEEE Eighth International Conference on Image Processing Theory, Tools and Applications (IPTA)*, Xi'an, China, pp. 1–6.
72. Zou J. and R. Song. 2018. Microarray camera image segmentation with faster-RCNN. *IEEE International Conference on Applied System Invention (ICASI)*, Chiba, Japan, pp. 86–89.
73. Lai S., L. Xu, K. Liu, and J. Zhao. 2015. Recurrent convolutional neural networks for text classification. *Proceedings of the Twenty-Ninth Association for the Advancement of Artificial Intelligence Conference on Artificial Intelligence*, Austin, Texas, pp. 2267–2273.
74. Kotia J., A. Kotwal, R. Bharti and R. Mangrulkar. 2021. "Few shot learning for medical imaging". In Das S., Das S., Dey N., Hassanien AE. (eds) *Machine Learning Algorithms for Industrial Applications*. Studies in Computational Intelligence, vol. 907. Springer: Cham. doi: 10.1007/978-3-030-50641-4_7.
75. Sharma B. and R. Mangrulkar. 2019. Deep learning applications in cyber security: A comprehensive review, challenges and future prospects. *International Journal of Engineering Applied Science and Technology*, vol. 4, no. 8: pp. 148–159.

3

Research Aspects of Machine Learning: Issues, Challenges, and Future Scope

Reena Thakur
Jhulelal Institute of Technology

Mayur Tembhurney
S. B Jain Institute of Technology, Management & Research

Dheeraj Rane
Medicaps University

CONTENTS

3.1 Introduction 38
 3.1.1 The ML Approach 38
3.2 Issues 40
3.3 Challenges 42
 3.3.1 ML Challenges Originating from BD 43
  3.3.1.1 Volume 43
  3.3.1.2 Variety 46
  3.3.1.3 Velocity 47
  3.3.1.4 Veracity 48
 3.3.2 ML Challenges Originating from Wireless Sensor Networking 49
  3.3.2.1 ML-Based 49
  3.3.2.2 Infrastructure Update 49
  3.3.2.3 ML-Based Network Slicing 50
  3.3.2.4 Standard Datasets and Environments for Research 50
  3.3.2.5 Theoretical Guidance for Algorithm Implementation 50
  3.3.2.6 Transfer Learning 50
 3.3.3 ML Challenges Originating from Blockchain 50
  3.3.3.1 Suitability 50
  3.3.3.2 Infrastructure 51
  3.3.3.3 Privacy 51
  3.3.3.4 Memory 51
  3.3.3.5 Implementation 51
  3.3.3.6 Security 51
  3.3.3.7 Quantum Resilience 51
 3.3.4 ML Challenges Originating from IoT 51
 3.3.5 ML Challenges Originating from Bioinformatics 52
  3.3.5.1 ML Algorithms for Bioinformatics 52
 3.3.6 Accuracy 53
 3.3.7 Complexity 53
  3.3.7.1 Algorithm Computational Efficiency 53

3.4 Future Scope 53
3.4.1 Agriculture 53
3.4.2 Integrative Framework for Anticancer Drug Prediction 54
3.4.3 Medical Disease Diagnosis 55
3.4.4 Stock Market Analysis 55
3.4.5 IoT 55
3.4.6 Health Data Linkage 57
3.4.7 Deep Learning 57
3.4.8 Educational Technology 58
References 58

3.1 Introduction

The purpose of this chapter "Research Aspects of Machine Learning: Issues, Challenges, and Future Scope" is to provide a mature approach to the basics of ML, which is one of the hot topics recently for the researchers. This chapter is originally meant to present the various issues and challenges using ML. It is based on certain empirical statements, which are apparently simple and naive. Moreover, it focuses on various research avenues using ML. Today, ML is an important part of many business applications and research ventures, ranging from medical diagnosis and care to social networking with friends.

ML is a tool for converting information into knowledge. In the past 50 years, there has been an explosion of data. This volume of data is ineffective till the author reviews it and locates the patterns hidden inside. By considering the complex data, ML techniques are used to recognize the valuable hidden pattern which one would have other difficulty identifying. It is possible to use hidden trends and knowledge about a topic to predict future events and carry out all kinds of detailed decision-making.

ML is an AI application that includes the ability to learn and develop naturally from history without being computer vision. The aim of ML is to create computer programs that can access the data and use this to train on its own.

ML is nothing but the empirical study of models as well as algorithms which modern computers practice to accomplish the research objectives without using specific instructions, focusing on that rather than patterns and inferences. This is considered to be a branch of AI. ML techniques construct a statistical equation based on available data, known as "training data," so that predictions or assumptions can be made without the function being specifically programmed (Bishop 2006).

ML is an AI application that allows the machine to understand as well as to improve automatically from experience without explicit programming. Generally, the aim of ML is to be acquainted with data structure and implant it into the ML models through which individuals can be aware and use it for the specific task. The main task of ML is that it programs the computer to minimize the operation criterion using past experience or training data.

3.1.1 The ML Approach

ML is indeed an attractive alternative, so the totally automated evolution of the process will enable the function to keep pace with the growing application increase in supply. Therefore, with the exception of the rules of the industry, ML determines the best decisions straight from the data to be able to explicitly code the laws of hardcoding decisions. This transition from rules-based strategic decisions to ML-based decision-making implies that the decisions will become more correct and will strengthen over time. With limited handholding, you can be confident that the ML program can deliver optimal decisions.

The data forms the framework in ML for discovering knowledge into the issue at hand. ML includes empirical training data to determine the right plan of action for each new process, in order to decide whether to approve each new application. For example, to really get involved with ML for acceptance of loans, you begin by compiling the training data also for 1000 loans that were issued. Such training data contains the inputs for each new loan, as well as the known outcome over whether each loan has been credited on time. In addition, the input data contains a series of categorical and numerical metrics components capturing the important aspects of each submission, such as credit score, age, and profession of the applicant.

The proposed chapter comprises three sections, followed by subsections, including different issues, challenges, and research avenues using ML. This chapter provides readers with basic theoretical knowledge about the topic.

The first part followed by the subsections of the chapter is devoted to thorough issues. The issues of complete face authentication system.

Another problem, credit card fraud identification where a credit card customer and a month's estimate of transactions for a customer determine what purchases the customer has made and those that were not.

Another issue is the need to get individual datasets that have the quality and size of samples required to train province-of-the-art ML methods. Another issue is face detection, where a multiple hundreds of digital photographs digital photo album identifies those photographs which include a given individual person.

Brand recommendation is a problem in which a consumer's buying history and a broad inventory of goods classify certain items that would attract the consumer and are likely to buy.

Diagnosis of disease where symptoms of particular patient matched with other patient symptoms which was already stored in database from which prediction of disease of the patient has been diagnosed, another issue is stock market issue where the price movement occurs continuously, on behalf of past and current price to decide whether individual have to make decision to bought, hold or sold the stock.

User segmentation is again an issue to identify the user's activity history.

Most challenges and issues include the use of ML and Internet of Things (IoT) for physically handicapped (PH). It inherits the existing fundamental issues of IoT and ML. The detailed analysis has been introduced throughout the sections and subsections.

Secondly, it focuses on the challenges explicitly or vastly applicable to ML in the BD perspective, relates with all dimensions (*V*) such as volume, velocity, variety, and veracity, and gives an outline of how developing approaches are reacting to them and in the case of data mining, networking, and other areas. Nevertheless, in BD, large datasets pose a number of challenges as conventional algorithms have not been developed to fulfill these requirements. In recent years, the networks, tools, and resources in networking systems are becoming more diverse and heterogeneous with the fast growth for new Internet and communication technologies in mobile. Because of traditional networks' inherently distributed features, however, ML methods for monitoring and operating networks are difficult to incorporate and deploy. This is an IoT challenge to ensure as much as possible that the dataset that will be used for ML is free of human differences. Moreover, the resource-related challenges are also explained such as data storage, space constraint, computing power, and training time. Evaluation metrics challenges are accuracy, complexity, and algorithm computational efficiency. Challenges in the areas of organization and searching have led to the modern field of "data mining"; "bioinformatics" has created statistical and computational problems in biology and medicine.

Finally, to increase the utility of the chapter, research avenues using ML have been added. Throughout the text, various tables have been used. The chapter has a specific perspective, and the author recognized research challenges and opportunities in the BD perspective, networking, and data mining. Moreover, health care, deep learning (DL), education, banking, social media, and weather forecasting are the areas with ML. ML is used in the field of procedural content generation, computational intelligence, text mining, Internet traffic classification, and classification of traffic in greynet or darknet networks. More areas are prediction of rainfall using linear regression, Kaggle Breast Cancer Wisconsin Diagnosis using logistic regression, detection of handwritten digits using logistic regression, support vector machine (SVM) to identify facial features in C++, and implementation of Film Recommenders Program. For building mathematical models, ML uses the theory of statistics, and the essential job is making implication from the illustration. In ML, computer science (CS) worked in two ways: Firstly for the training, this needs a different ML algorithm to resolve and optimize the problem, in addition to process and sort the huge quantity of data, which generally individuals have from different sources. Secondly, when the learning process of a model has been completed, the learning of a model from a huge quantity of data, its algorithmic solution and representation should be efficient and able to give the predicted result, the effectiveness of learning the model and proficient use of algorithms are important. ML not only solves the database problem, but it is used in AI. Intelligence means a type of system that will have the learning capability in a changing environment.

This section is an attempt to bring together many of the important research avenues using ML. As a result, the authors hope that this chapter will appeal not just to academicians but also to researchers and practitioners in a wide variety.

This chapter starts with an introduction to ML and why it has been so critical in the latest days. The rest of the chapter is organized with an exhaustive explanation of each problem as various issues using ML. This chapter is intended to serve as knowledge for all the researchers, students, and teachers. The final section of the proposed chapter shall focus on the research opportunities with ML that can be carried out, with developing techniques and growing data.

3.2 Issues

The following are the issues that can be solved by using ML algorithms:

1. ML has to learn and know instantaneously how or when to convert pairs from English to Hindi sentences.
2. ML affects computer vision areas, speech recognition, the answering of natural language questions, robotic control, and computer dialog systems. There is an issue of identifying input and output as well.
3. ML has also attained a noticeable part in other areas of information retrieval, bioinformatics, spam detection, and credit card fraud detection and database consistency.
4. Nevertheless, many ML researchers believe that the current neuroscience interpretation is also very weak as part of interest in learning process engineering.
5. The pipeline uses the backbone of the neural network (NN) to detect and define the faces in input images.
6. Problems regarding the design and consistency of the data.
7. Supervised algorithms, softmax algorithm, k-nearest neighbor (KNN), naive Bayes algorithm, SVM, and random forest (RF) have been useful for the stock cost forecast.
8. Several data may be gathered from the fault machine and then converted to useful one.
9. This is yet another issue about the application of condition monitoring ML, which is solved by various ML methods.
10. A further practical problem is to determine the dependence in the input image among features when the sample data is input.

There is any description of ML that is arguable. ML from a scientific viewpoint is the analysis of effective learning to make use of past experience for future behavior. ML from an engineering point of view is the study of various algorithms to construct computer programs automatically and efficiently from the training data, for instance, a metadata of converting databases in which English strings paired to Hindi sentences. ML therefore has to learn and know instantaneously how or when to convert pairs from English to Hindi sentences.

Currently, this way is obtained using the latest technology machine translation systems. ML has been an important solution to the classical problems of most AIs. ML affects computer vision areas, speech recognition, the answering of natural language questions, robotic control, and computer dialog systems. ML has also attained a noticeable part in other areas of information retrieval, bioinformatics, spam detection, and credit card fraud detection and database consistency.

Some would make the argument that ML is redefining the author's perception of the procedure of making software.

ML is an AI type that lets a computer learn from data rather than an explicit program. However, ML is not just a tool. When the algorithms absorb training data, they can then create more detailed models depending on available evidence. The performance produced when one trains ML method through the facts is a ML model. When you provide an input to a model after training, you will be given an output. A specific algorithm, for example, can create forecasting models. When you then provide data to the forecasting model, a prediction will be provided depending on the statistics that learned the model.

ML has become a modern basis for much of the informatics research. Although commonly seen as a view of computer vision, ML is often strongly associated with classical statistics. Of course, they share a common math language. Yet in general, they research very different applications commonly used in ML studies, which are the objective of CS or electrical engineering.

Neuroscience also has a significant relationship with ML. Moreover, the human brain is the strongest learning system that somebody knows about. Nevertheless, many ML researchers believe that the current neuroscience interpretation is also very weak as part of interest in learning process engineering. In addition, the understanding process of learning advances further when humans specifically engineer structures rather than attempting to change the brain. Also, very simple neural computational properties including how short-term memory is processed are still quite widely known. This seems clear that transistors can do everything neurons can do too. So humans can better understand the transistors than neurons. Neuroscience's principal contribution to ML practice is the artificial neural network (ANN) as a model of computation. ANN is an important method in the development of learning systems, whether or not they are specifically connected to real neurons. The authors now consider having tried to describe ML a variety of fundamental issues.

It presents a full face authentication method in Kurnianggoro et al. (2018). The pipeline uses the backbone of the NN to detect and define the faces in input images. In the experiment result, it is shown that the device can accurately detect the face and recognize faces with impressive outcomes. As explained and illustrated in the experiment, any fresh faces in a given picture can be inserted to the database to make the machine recognize those faces. The author would like to consolidate both the identification and definition network to the overhead computation for future work (Kurnianggoro et al. 2018).

Today, there are many obstacles that impede rapid incorporation of ML into health care. Some of the biggest challenges seem to be the capacity to get datasets of patients that contain the quality and sample size required to train a province-of-the art ML algorithm. Owing to strict privacy and protection laws protecting patient data, do not have any quick access to data, distribute and transfer. In addition, due to problems regarding the design and consistency of the data, typically much effort is required to clean up in the preparation for ML analyses.

ML has seen immense use and has developed into NNs and DL, but with all of them, the main concept is almost the same. The author gives a smooth insight into the way ML is applied. There are different ways, approaches, and strategies available, in different circumstances imaginable, to manage and overcome different problems, reduced to supervised ML only, and attempts only to clarify the basics of this complex process (Pahwa et al. 2019) (Figure 3.1).

The author in Kumar et al. (2018) proposed supervised algorithms, softmax algorithm, KNN, naive Bayes algorithm, SVM, and RF, which have been useful for the stock cost forecast. The findings reveal that for large datasets, the RF algorithm outdoes all other efficiency algorithms, but when the dataset size is smaller to almost half of the source, the naive Bayes algorithm represents the best accuracy results. Reducing the number of technical indicators often decreases the accuracy of growing stock market trends algorithm (Kumar et al. 2018).

Currently, recommendation systems are a really popular method used for commercial reasons. The study only explains some of the methods used to formulate recommendations. The emphasis is on neighborhood-based algorithms and material, since they are the most commonly used command creation methods. An effort was made to build algorithms that measure the prediction rating by accessing public use databases (Nawrocka et al. 2018).

ML is reflected in the following data. Several data may be gathered from the fault machine. First-ever question is how to allocate the data between the two system statuses. The ML algorithms can be used as classifications (Negandhi et. al. 2019). It can, however, distinguish between the two types of data. The classifier for ML distinguishes the data by setting the boundary between two distinct forms. ML can set very versatile limits.

ML proposed by the author as classical ML of classifier implies that the data is iid (Theodoridis and Koutroumbas 2008). For example, when non-iid data will be applied as content, the universal VC

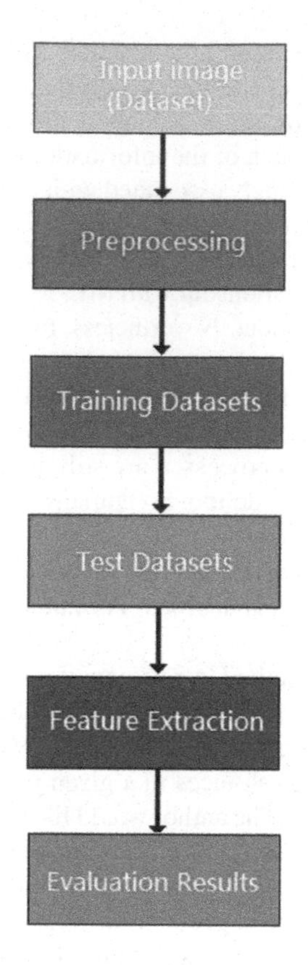

FIGURE 3.1 Flowchart of methodology (Pahwa et al. 2019).

boundary, consistency property, and risk management minimization principle require theoretical modification and validation when using a SVM.

Learning by machine only can address the vector data. Typically, ground-collected data is not vector. For condition monitoring, an early processing is required. This is yet another issue about the application of condition monitoring ML. The process of translating the facts into a type of ML, which can manage in condition monitoring, is called feature extraction. ML success depends so much on the extraction of the feature, rather than the ML techniques themselves.

A further practical problem is to determine the dependence in the input image among features when the sample data is input. For example, structured data, features that are associated or connected in the function vector do not require modification of the ML principle.

3.3 Challenges

There are several challenges that a ML practitioner can face when designing an application to carry them to production from scratch.

1. Collecting Data

   This is the step to organize the data before storing it in the database. This involves understanding of BD, where it plays an important role.
2. Less Training Knowledge amount

   You need to validate whether the quantity is appropriate for use once the data is obtained.
3. Non-representative evidence from training

   To generalize well, the training data should be indicative of the new cases.

4. Weak Data Quality
   In fact, don't begin training the model directly; the most significant step is to analyze data. But the information that is obtained may not be ready for training; certain samples, for example, are abnormal from others with outliers or missing values.
5. Irrelevant/Unwanted Features
   If a large number of irrelevant features and appropriate relevant characteristics are included in the training data, the ML system will not give the results as anticipated. One of the significant elements is needed.
6. The Training Data Overfit
   Overfitting happens at the point where the model is overly unpredictable as opposed to the noisiness of the training dataset.
7. Underfitting the data from training
   Underfitting, as opposed to overfitting, typically happens when the model is too simplistic to understand the data's basic structure.
8. Off-line learning and model deployment
   There are other challenges of ML that are explained in detail as follows.

3.3.1 ML Challenges Originating from BD

Massive datasets, however, present a variety of challenges since conventional algorithms have not been developed to meet needs. For example, some ML methods were built for lesser datasets, assuming a whole dataset could fit into memory. Another presumption is also that, at the time of preparation, the whole dataset is available for review. BD violates all assumptions, making conventional algorithms inoperable or hampering their act considerably. A variety of methods for adapting ML algorithms to effort with huge datasets have been developed: Studies are new, distributed processing contexts like Hadoop (Shvachko et al. 2010) as well as creative concepts like MapReduce (Dean et al. 2008). In an attempt with BD to address the complexities of ML, ML sections such as online learning and DL were also familiarized with this.

BD is also defined in terms of its dimensions, called its Vs. Early BD concepts concentrated on three Vs (Li et al. 2019); nevertheless, now a more widely accepted definition relies on four Vs (Pahwa et al. 2019). This is more significant to notice that there are other Vs in the research, too. For example, worth is frequently added as a fifth *V* (Kumar et al. 2018). Value (*V*) is, on the other hand, interpreted as the efficient BD processing objectives (Ren et al. 2019), moreover not as the properties significant BD on its own. The author therefore only considers the four dimensions that describe BD. It gives a chance to specifically link encounters to the defining features of BD, clearly considering the source and purpose of each one. Figure 3.2 describes the challenges of ML as well as integrates every challenge with a particular BD aspect.

3.3.1.1 Volume

The BD's primary characteristic is volume, which is the quantity, data scale, and size. In the sense of ML, the scale may be calculated either directly as the number of entries or test results in a dataset, or vertically by the number of features or features found within. Therefore, the quantity is proportional to the data form: A smaller group of very complicated data sources can be treated similarly to a larger percentage of simple data (Kurnianggoro et al. 2018). It is maybe the simplest aspect to describe in BD, though this is the source of several challenges or issues. The following are the subsections to address the volume-caused challenges in ML.

3.3.1.1.1 Processing Efficiency

One of the key challenges caused by BD computing is the basic idea that volume or scale adds complexity to computation. As the scale gets high, even insignificant functions might become expensive. For example, the typical SVM algorithm has O(m3) complexity of training time and O(m2) space complexity

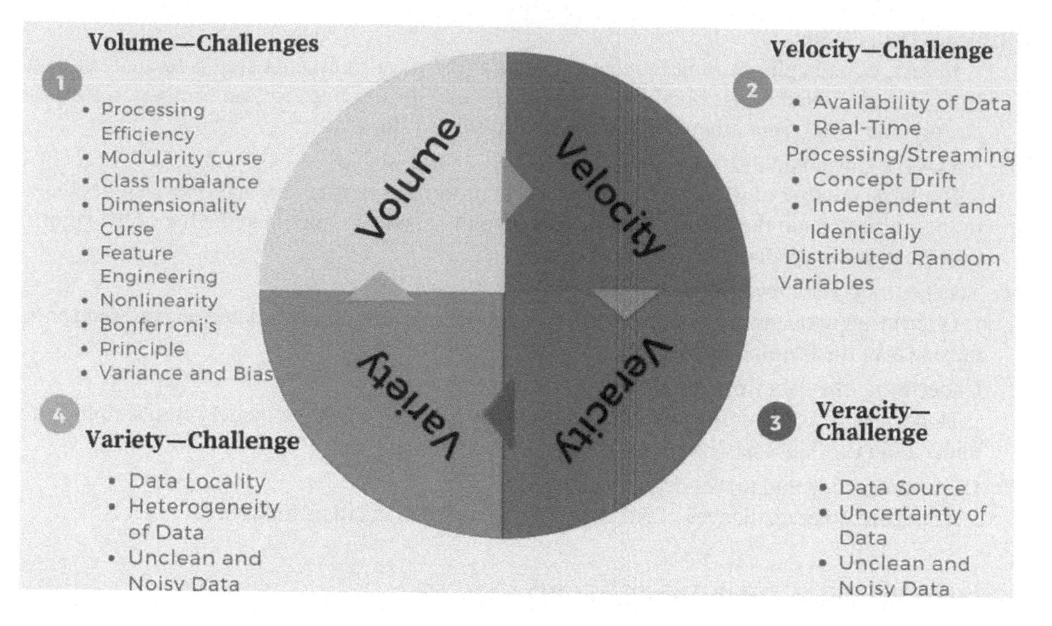

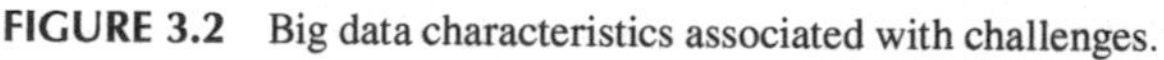
FIGURE 3.2 Big data characteristics associated with challenges.

TABLE 3.1

Algorithm Computational Efficiency

Algorithm	Regression/Classification	Training	Prediction
Decision tree (DT)	R+C	O(n2p)O(n2p)	O(p) O(p)
Random forest (RF)	R+C	O(n2pn trees) O(n2pn trees)	O(pn trees) O(pn trees)
Random forest (RF)	R (regression) Breiman implementation	O(n2pn trees) O(n2pn trees)	O(pn trees) O(pn trees)
Random forest (RF)	C (classification) Breiman implementation	O(n2√pntrees) O(n2pn trees)	O(pntrees) O(pn trees)
Extremely randomized trees (ERT)	R+C	O(npn trees) O(npn trees)	O(npn trees) O(npn trees)
Linear regression (LR)	R	O(p2n+p3) O(p2n+p3)	O(p) O(p)
SVM (Kernel)	R+C	O(n2p+n3) O(n2p+n3)	O(nsvp) O(nsvp)
k-nearest neighbors (KNN)	R+C	--	O(np) O(np)
Nearest centroid (NC)	C	O(np) O(np)	O(p) O(p)
Neural network (NN)	R+C	--	O(pnl1+nl1nl2+...) O(pnl1+nl1nl2+...)
Naive Bayes (NB)	C	O(np) O(np)	O(p) O(p)

(Alhajri et al. 2018) in which m represents the training sets number. A rise in size m would therefore have a dramatic impact on the time as well as memory required to train the SVM algorithm, and it was algorithmically impracticable on huge datasets. Also, several other ML algorithms display time complexity, which is high, as shown in Table 3.1. The time required to perform the calculations for all the algorithms will therefore increase exponentially with increase in data size, and it might consider the algorithms to be inoperable for very huge data. In addition, as data volume increases, algorithm output becomes even more reliant on the infrastructure, which is used to save and retrieve facts. With the growth in data size (Cambria et al. 2014), data partitioning, concurrent data structures and positioning, and data recycling are becoming more significant resilient distributed datasets (RDDs) (Cambria et al. 2014), which are indeed a sample of a modern paradigm for large cluster in memory computing; RDDs are introduced within the Spark cluster computing framework (Mikolov et al. 2013). Hence, data size not only affects the overall outcome, but also contributes to the necessity to reconsider the standard architecture used to build and enforce algorithms.

3.3.1.1.2 Modularity Curse

Various learning techniques are based on the premise that the data being processed can actually be preserved on a disk in the memory, and in a single file (Oquab et al. 2014). Many algorithm categories are based on strategies and key components that rely on statement validity. But when data size tends to cause this principle to crumble, it affects entire algorithm families. Such challenge is termed the modularity curse (Yuan 2016). MapReduce, a distributed programming framework for computer vision large amounts of data by running on a thousand nodes in parallel, is one of the methods proposed as a mechanism for the modularity curse. In Kau et al. (2015), the author discussed challenges in BD for MapReduce. The three algorithms' key types that experience a modularity expletive when trying to use the MapReduce standard comprise gradient descent iterative graphs and algorithms for optimizing expectations. Consequently, whereas few algorithms like k-means algorithm may be modified to solve the modularity curse by distributed and parallel computing, some are either bound with other concepts, or perhaps even unusable to them.

3.3.1.1.3 Class Imbalance

When datasets grow wider, the idea that only the data is distributed equally among all groups is also fragmented (Milletari et al. 2016). It relates to a challenge called class imbalance—if datasets contain data from groups with varying incidence probabilities, the throughput of a ML technique may be negatively impacted. Such an issue is particularly prevalent when a larger sample is expressed in some groups and a few in others. Class inequality is different to BD. This is the focus of study for over a decade (Yosinski et al. 2015). Studies by the author in Yosinski et al. (2015) will illustrate that the strictness of the issue of the inequality depends on the assignment complexity, the class difference strength, and the total training collection size. In Yosinski et al. (2015), the same authors have shown that algorithms of NNs, decision trees (DTs), and SVM are all actually vulnerable to imbalances of class. Consequently, the unobstructed BD activity without resolving the class inequality will yield insufficient results. Furthermore, the author considered gamification to be an extreme class inequality and illustrated its negative impacts on Watson ML. Subsequently, the likelihood of a class difference arising in the BD sense is high because of data size. In addition, the potential impacts of class imbalances on ML are severe due to the extreme composite problems embedded in such data.

3.3.1.1.4 Dimensionality Curse

Another problem linked to the BD volume is the dimensionality curse (Boustani et al. 2005), which leads to the challenges associated while working in huge dimensional feature space. This dimensionality precisely defines the variety of functions or attributes available throughout the dataset. The Hughes effect (Escudero et al. 2013) specifies that the predictive potential and efficacy of an algorithm decrease with increasing dimensionality for a static scale training set. However, the accuracy and efficiency of ML algorithms reduce as the number of features increases. It can be asserted by destroying the similarity-based logic, which depends on many ML algorithms (Boustani et al. 2005). Appropriately, because there are more prospective elements, the larger the quantity of data available to explore the occurrence, the better will be the potential for huge dimensionality. Additionally, dimensionality influences computational performance: ML algorithms' space and time complexity is closely associated with data dimensionality.

3.3.1.1.5 Feature Engineering

High dimensionality is strongly correlated with a particular capacity challenge. Feature engineering is really a way of developing structures to make ML work better, usually through domain knowledge. As such, among the most time-intensive ML preprocessing tasks is the collection of the most suitable features. Accordingly, as the scale of the dataset grows, the challenges associated with feature engineering do so in a way close to dimensionality. Feature engineering is relevant to the selection of features, while feature engineering develops innovative features in an effort to enhance learning objectives; selection of features is intended to choose the features that are most appropriate. While selection of features decreases dimensionality and consequently has the potential to. ML time is challenging in large datasets due to random errors and incidental endogeneity. Generally, both the collection of selection of features and the technologies are quite very important in the intelligence of BD, and similarly, they become more complicated.

3.3.1.1.6 Nonlinearity

The size of data presents challenge when implementing rising methodologies for evaluating dataset characteristics and performance of the algorithms. Nonetheless, the validity of several metrics and approaches is founded on assumptions collection, with the very famous linearity assumption (Dahl et al. 2014). An experiment performed by (Tan et al. 2015) discussed that nonlinearity very negatively affects the efficiency of NNs and logistic regression. BD challenge also arises through the challenges related with the linearity assessment which is also measured by technology like scatterplots; such as the example of BD, however, the huge amounts of point generally makes a massive cloud, creating it hard to define relationships as well as to assess linearity (Dahl et al. 2014). Thus, the complexity of determining linearity and the presence of nonlinearity present the issues for implementing ML algorithms in association with BD.

3.3.1.1.7 Bonferroni's Principle

This principle of Codella et al. (2017) is such that when individual searches for a particular kind of occurrence in a given volume of data, the probability of such occurrence being discovered is strong. Moreover, such events are not spurious, implying that they do not really have any origin and are therefore irrelevant in a dataset. Also, sometimes, the spurious correlation is a statistical challenge (Kurnianggoro et al. 2018). Bonferroni's correction principle offers a way, in statistics, to prevent certain fictitious positive checks within a dataset. It implies that any individual phenomenon should be evaluated at a meaning level of α/m Milletaris et al. (2016) when evaluating m hypotheses with a desired α value. The author in Chalapathy et al. (2016) recently addresses the effect and occurrence of erroneous BD correlations. The authors show that the majority of correlations appear to be null despite a big enough number. Therefore, it is important to consider having a means of preventing these false-positive rates in the sense of ML with BD.

3.3.1.1.8 Variance and Bias

ML is built on the concept of generalization; representations can be generalized to allow for estimation and forecasting through interpretations and manipulations of data. Error may also be divided into two components: variance with bias (Ribeiro et al. 2016). Variance describes an ability of learner to identify random events, while bias defines a capability of learner to understand the incorrect fact (Boustani et al. 2005). Ideally, to achieve an effective performance, both the variance and the bias error should be significantly reduced. When data volume increases, however, the trainer should become nearly biased toward the training set, which may not be able to simplify sufficiently with newly occurred data. Furthermore, care should be taken when in context with BD, as partiality can be imposed, thus undermining the capability to generalize.

3.3.1.2 Variety

The Big Data (BD) variety defines not just the structural variability of a dataset as well as the data types it includes, but rather the diversity of what they represent, their semantic interpretation (Nawrocka et al. 2018), and their references. While not as regular as with other dimensions of the *V*, the problems identified with this dimension have an important impact.

3.3.1.2.1 Data Locality

Data locality is the variety-related challenge, and the author described it in Codella et al. (2017). Yet again, algorithms for ML believe that the whole dataset is stored in the memory/maybe on only the device file (Yuan 2016). Nonetheless, in the BD context, it cannot be feasible because of the size; the data does not really blend into memory and is usually spread over large quantities of files that exist in various objective locations. Classic ML will involve transfer of data to the computing location first. Transmission would result in latency processing of large datasets, which could trigger huge network traffic. Subsequently, an approach has arisen to bring data into computation rather than to bring data into computation on the basis of the assumption that carrying computation is not costly, in addition to bandwidth and time, than transmitting information. With BD, the approach is particularly popular. BD is also used by the MapReduce paradigm—for through map tasks performing its local data, map tasks are performed at the nodes in which data resides. Similarly, this model suits a huge number of NoSQL data repositories; they preserve data for a large number of nodes as distributed storage structures and then use the MapReduce paradigm to bring data computation (Dean et al. 2008). However, as already described, when operating with strongly iterative algorithms, MapReduce-based strategies experience problems. Physical position is a non-issue with small samples; moreover with BD, a major problem is the locality of data, which needs to be solved in any productive BD system.

3.3.1.2.2 Heterogeneity of Data

Big analytics of data also requires the aggregation of various data from multiple sources. Such data may be varied in nature of type of data, structure, model of data, and grammar. Two major types of heterogeneity can be recognized: heterogeneity of the syntactic and the semantic. Syntactic heterogeneity applies to data type variation, file formats, encryption, data model, and the like. Such syntactic differences must be resolved to conduct analytics with integrated datasets (Nawrocka et al. 2018). ML also involves preprocessing as the first step and cleaning as the second step in order to create data that fits within a particular model. These data are likely to be structured differently, even with data coming from all different channels. However, the data that is to be analyzed is of different types, for example, objects that need to be interpreted together with numerical and categorical data. It creates problems to ML algorithms, which are unstructured to identify different types of illustrations through individual time and make effective, coherent generalizations. Semantic variability refers to varying definitions and meanings (Shvachko et al. 2010). In particular, statistical heterogeneity and semantic heterogeneity have always been interesting research fields, but they have gained renewed interest with the appearance of BD (Shvachko et al. 2010). Typically, the business benefit of data processing includes correlating various datasets, and convergence is essential to the implementation of ML over these datasets.

3.3.1.2.3 Unclean and Noisy

Data as per the author in Dahl et al. (2014) has its individual number of different characteristics, which is used for identification:

- State decides the quality of data for evaluation.
- Position of the data relative to a tangible location.
- Population defines the identities and its collection of shared attributes, which together form the dataset.

BD is usually defined as unconditional based on the resources and time needed to acquire them readily available for analysis. These often arise from a variety of places, often from unknown cultures. Combining those properties also results in BD being represented as dirty. The author called such data as data with noise in (Stanford Vision Lab, ImageNet); they involve different types of mistakes in calculation, missing value, outliers, and inconsistent data. Remember that the author in (Stanford Vision Lab, ImageNet) found that the big challenge of BD analysis is noisy data.

3.3.1.3 Velocity

BD is identified by not only with the speed with which the data is generated, but with the level at it that can be analyzed. Also, with the pervasiveness of mobile and real-time devices, BD velocity has become a critical aspect in emerging innovations such as smart cities.

3.3.1.3.1 Availability of Data

Traditionally, numerous strategies to ML have relied on the availability of data, meaning that the whole dataset was presumed to be available until learning starts. However, a necessity of this nature cannot be met in the context of data streams, by which new data arrives continuously. In addition, even data that arrives at non-real-time intervals will present an issue. A model usually trains from the training set in ML and executes the knowledgeable function on new data, for example, classification of forecasting. In this case, the model does not immediately classify new data but provides an already trained feature on updated information in the opposite direction. Those models need to be retrained to incorporate the knowledge embedded in new data. They can become obsolete without retraining and tend to represent the system behavior. Therefore, algorithms may support incremental learning in line with changing information (Salah et al. 2019), also defined as the sequential learning that is characterized as the ability of algorithm to adjust learning depending on an implementation of newly occurred data to be able to reorient the entire dataset. Sequential learning method means that not only the whole set of training must be presented before the start of the learning process, but also the additional data will be processed upon arrival. While incremental learning is a fairly old term, owing to the complexity of adapting certain algorithms to the continual arrival of data (Velankar et al. 2018), it is still an active research subject.

3.3.1.3.2 Streaming or Real-Time Processing

When the availability of data previously addressed is compared to the challenge, conventional ML methods are not built to manage persistent data streams (Kurnianggoro et al. 2018), resulting in yet another velocity aspect problem, a need for real-time processing. Data availability relates to that to modify the ML model as new research occurs, in real-time processing applies a need to process data in close-real-time, quickly arriving. The market advantage of real-time processing technologies is one's ability to generate instant response; these solutions were particularly interested in algorithmic managers selling, video surveillance, and fraud identification (Kau et al. 2015). Therefore, the significance of real-time processing manifested itself in the introduction of a variety of streaming systems in today's age of mobile platforms, IoT with sensor, such as Yahoo's S4 (Krizhevsky et al. 2012a) and Twitter's Storm (Chawathe 2018).

3.3.1.3.3 Concept Drift

Big Data (BD) is a time varying application which involve huge amount of data generated within fractions of seconds. All data need to be captured for processing called data set. Additionally, this is not possible to acquire the entire dataset before processing, so it is very difficult to be verified if an existing fact follows the similar distribution as upcoming data. It brings with it another fascinating problem in big data machine learning that is concept drift (Yuan 2016), which can even be generally associated with modifications in the targeted output's restricted distribution provided the input, whereas input distribution itself may appear unchanged (Lan et al. 2018). In particular, this causes a problem emerging when ML models are developed by past data that no further precisely represents current data distribution (Min et al. 2017). Windowing strategy implies more applicable recent data, which is not always true (Lan et al. 2018). Different forms of concept drift occur: progressive, incremental, recurrent, and rapid (Yan et al. 2017), and each with its individual set of problems. Moreover, the issues usually exist in rapid technological changes when design drift occurs and in efficiently managing the model transformation during these changes. However, BD's emergence and existence have increased the speed of its incidence and made certain past methodologies inaccessible. The author in Xiang et al. (2017) performed research on the impact of BD of high dimensions on present concept drift mitigation techniques.

3.3.1.3.4 Independent and Identically

Distributed Random Variables: The random variables are identically distributed and independent in ML (Sun et al. 2019). This simplifies the methods behind and enhances convergence. The Markov process implies that the distribution of probability of the future state entirely influenced on the present state (Sun et al. 2019).

3.3.1.4 Veracity

Big Data (BD's) veracity corresponds to not only the reliability of data from the dataset but also to the implicit inaccuracy of required data sources as IBM has described (Kurnianggoro et al. 2018).

BD'sauthenticity and effectiveness together define the component of veracity (Li et al. 2019), but moreover face a variety of challenges that are explained in the following subsections.

3.3.1.4.1 Data Source

Data provenance is a method of tracking and monitoring all locations and their movements to the source of the data. Documented data, or the original source data, is often used to identify a source of processing error since it explains all of the steps, transactions, and operations that invalid data has gone through, providing useful knowledge for machine learning. So capturing and maintaining this metadata is important (Nawrocka et al. 2018). Nonetheless, as Wang et al. (Li et al. 2014) pointed out, the origin dataset itself is too broad in the sense of BD, and while these data have an excellent sense for ML, the scale of such metadata poses a collection of challenges for itself. While some methods had developed to capture authentication used for particular data processing, concepts, like the Map and Reduce Provenance, are established for MapReduce by means of Hadoop enhancement. Subsequently, because provenance data includes an approach to assess BD's veracity, it means to align the computational overhead and expense through the veracity benefit.

3.3.1.4.2 Uncertainty of Data

Nowadays, data is collected in several ways regarding various phases in life; furthermore, methods and techniques used to collect data will generate confusion, thus affecting the veracity of the dataset. For example, emotional data is gathered over social media; nevertheless, such data is extremely significant since it provides useful understandings through subjective knowledge, and the data itself is indefinite. The data reliability and precision are not conclusive; for this reason, it depends solely on human verdict (Rajesh and Naveen Kumar 2017). The absence of perspective within the data makes learning from all of this impossible for a ML algorithm. The data provided by crowdsourcing, especially data collected through data shows, contains an even greater uncertainty degree than data on sentiment (Nawrocka et al. 2018). Another time, ML algorithms are not intended to handle such a type of incorrect data, which results in several challenges.

3.3.1.4.3. Unclean and Noisy

Moreover, data can also be noisy, in addition to being imprecise. The labels or textual information relevant to the results, for example, may be unreliable, or readings may be erroneous. It is different from incomplete data from the computational point of view; taking an ambiguous image is diverse to taking the false image while it may produce analogous outcomes.

3.3.2 ML Challenges Originating from Wireless Sensor Networking

3.3.2.1 ML-Based

Heterogeneous Backhaul/Fronthaul Management: Numerous backhaul/fronthaul technologies will concur in future wireless networks (Yan et al. 2017), together with wired backhaul/fronthaul such as cable and fiber in addition to wireless backhaul/fronthaul such as sub-6GHz band. Every result has specific bandwidth and different energy consumption, so backhaul/fronthaul management is critical for the efficiency of the entire system.

3.3.2.2 Infrastructure Update

Recent wireless network services and infrastructure should be developed to plan for the introduction of ML-based communications networks. For instance, graphics processing unit-equipped servers can be installed at the edge of the network to implement DL-based channel estimation, localization, and resource allocation. To accomplish in-time data analysis, in addition to proper network, sufficient storage capacity devices are required. Network function virtualization will be used in mobile networks; further, it integrates the roles of the device and network, and later, it can be introduced as software.

3.3.2.3 ML-Based Network Slicing

Network slicing can be promoted by individual academia as well as industry by means of cost-effective methods to serve diverse use cases (Xiang et al. 2017; Sun et al. 2019). The core of network slicing to assign suitable resources like on-demand storage, routing, radio resources, and backhaul/fronthaul to make sure the quality necessities of numerous slices with respect to slice separation limitations.

Initially, ML is used to understand mapping through delivering services to resource sharing plans; thus, a new type slice of network will be quickly developed. Secondly, knowledge concerning resource management strategies for specific utilization cases within one setting serves as valuable information in a different setting, which is to improve learning procedure by using transfer learning. In recent times, the authors of (Chen et al. 2019) have applied deep reinforcement learning to network slicing and then benefitted using simulations.

3.3.2.4 Standard Datasets and Environments for Research

Awareness of some basic challenges in wireless networks in the direction of investigator pay more focus to learning algorithm models with unbiased comparison of different ML-based approaches. These challenges should be combined with associated unlabeled or labeled data for unsupervised learning-based or supervised learning-based methods, related to open MNIST (Modified National Institute of Standards and Technology) dataset, which is normally used in computer vision.

3.3.2.5 Theoretical Guidance for Algorithm Implementation

The efficiency of ML algorithms is believed and influenced by the selection of hyperparameters such as learning rate and loss functions. It is a time-consuming task to try different hyperparameters directly, particularly when the learning rate used for modeling under a static set of hyperparameters is extensive. In addition, new issues are the theoretical study of the size of the dataset required for training, performance limitations of DL techniques, as well as the ability to generalize different learning models. Given that stability is now one of the key features of communication systems, it is important to carry out comprehensive theoretical studies to ensure that ML-based techniques continue to work well throughout the functional system.

3.3.2.6 Transfer Learning

Transfer learning aims to transfer acquired information from one specific task to another. The learning process can be accelerated in new environments by extracting training learning models from scratch, and even with a small amount of data, the ML algorithm can perform well. For this reason, transfer learning is crucial to the accurate operation of learning models that take into account the expense of training without prior experience. Network operators can use transfer learning to solve new, but related, problems with cost-effective manners. However, as noted in Li et al. (2014), undesirable effects of previous experience on output of the system may be discussed, and further research is required.

3.3.3 ML Challenges Originating from Blockchain

Researchers have been looking at these developments around the world, and still, the combination of BT and ML avoids numerous obstacles (Singh et al. 2020). The integration is now in its early stages. There are so many unanswered challenges and concerns to be resolved. The author addresses challenges such as appropriateness, infrastructure, safety, protection, memory, execution, and resilience to quantity.

3.3.3.1 Suitability

Blockchain technology is a feasible option where there is no confidence in the source of the data and many entities are large in the decentralized environment. If output is needed, then a better alternative is a simple database. Therefore, the blockchain architecture must always be recognized in every application before it is used (Salah et al. 2019).

3.3.3.2 Infrastructure

Blockchain particular network infrastructure and hardware enhance the throughput of several applications based on blockchains, which may involve network management, hardware mining, communication protocols, and distributed storage (Velankar et al. 2018). However, solutions designed to be used in blockchain have been under research.

3.3.3.3 Privacy

The data generated by blockchain devices is available to all the blockchain nodes (Chawathe 2018). This leads to possible privacy issues about data that must be kept confidential or private. The use of private blockchains, managed access, and encryption could solve these types of issues. However, the implementation of the ML models placed barriers to predictions and analytics on such restricted data.

3.3.3.4 Memory

The size of blockchain is continuing to expand as further adding blocks to it. Additionally, all nodes must store the entire system, creating a major memory limit on those computers. A rise in the size of chain also affects efficiency. In addition, processing irrelevant features often spoils the resources of computing. Blockchains are undefined; therefore, for implementation purposes, there is a major issue occurring in storage.

3.3.3.5 Implementation

An extremely vast amount of transactions would involve a blockchain network. The development of this blockchain size raises possible issues, for example, requirement of high Internet speed, which cannot be minimized, and due to network issue transactions challenges. Therefore, to fulfill the unavoidable requirement, the inclusion of blocks as well as transactions requires to be reduced.

3.3.3.6 Security

Blockchains are decentralized and vulnerable to safety issues. The biggest common element is that, because of attacks, the consensus protocol may be abused, such that the processing capacity of a few farms decides the blocks to be added to the network. In public blockchains, this presents a particular concern; from these attacks, private versions are not affected because they have identified each node, with a suitable consensus protocol in place.

3.3.3.7 Quantum Resilience

For blockchains, hashing algorithms are used and would finally be broken down for quantum computers. Since it incorporates techniques with certain types with cryptography, a blockchain is also at danger from this. That will impact all of the aspects that make blockchains a viable storage system. Quantum cryptography, as the basic security of quantum communication, will enhance the safety of the blockchain network (Figure 3.3).

3.3.4 ML Challenges Originating from IoT

The lack of freshwater supplies all over the place of the world has developed a necessity to make full usage of them. IoT technologies, focused on data collection and intelligence technology of application-specific

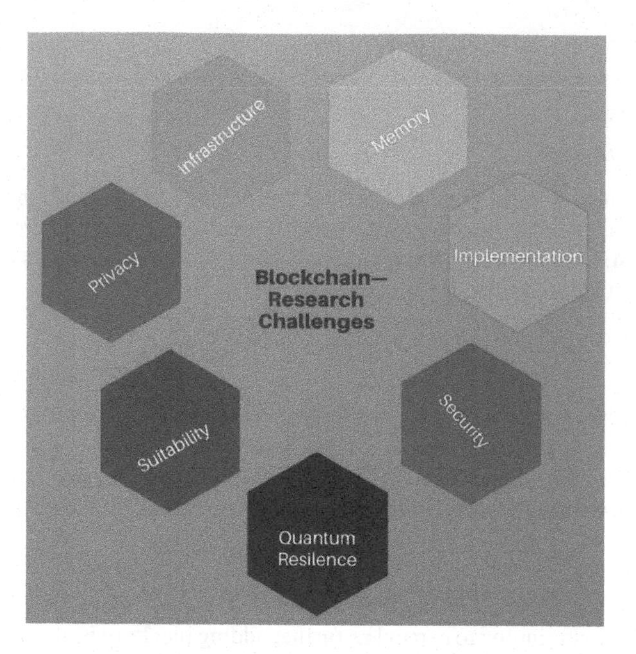

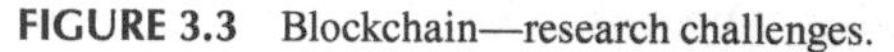

FIGURE 3.3 Blockchain—research challenges.

devices, cross the differences among the world of cyber security. IoT-based smart agricultural management tool helps to achieve optimal use of water supplies in the smart agriculture environment. The author presents an open-source smart system highly innovative to forecast a field's irrigation requirements using ground parameter sensing such as temperature, soil moisture, and environmental changes along with meteorological conditions and data of weather forecast from the Internet. The sensor devices that are interested in environmental sensing and soil consider temperature and soil moisture, ultraviolet (UV) light radiation, air temperature, and crop field relative moisture. As sensor node data is collected wirelessly through the cloud using the online network and browser-based knowledge visualization and decision support system, real-time information insights are created based on sensor data analysis and weather forecast data. The complete network was developed and deployed on a large scale. The system provides for a closed-loop supply of water control to realize the fully autonomous irrigation system. This method is completely automated, and the findings of the analysis are highly encouraging (Goap et al. 2018).

The IoT and ML were applied widely in personalized healthcare (PH) (Qi et al. 2017). In particular, these systems are a set of interconnected IoT devices that in some cases perform advanced analytics, remote monitoring, preventive analysis diagnosis, and surgery (Pawar et al. 2008; Kau et al. 2015; Ermes et al. 2008; Riano et al. 2012). Several technology companies have invested in building a PH system focused on ML.

3.3.5 ML Challenges Originating from Bioinformatics

ML is efficient; for instance, it has state-of-the-art applications and is seen in the examples of articles. ML is a model that is driven by data. It does not require a complete understanding of the relevant system. Using ML can improve overall prediction with accuracy, rather than observing its mechanisms behind the various patterns, based on the two examples: standard ML application to condition observing. This advantage is useful to the situation where it is difficult to examine the physical history of the data.

3.3.5.1 ML Algorithms for Bioinformatics

The following section describes the ML algorithms for bioinformatics.

Clustering: A primary objective of classification is the classification of the data into various categories so that cluster participants are nearer to one another than participants of several other groups.

In bioinformatics, standard clustering algorithms are grid-based, hierarchical, model-based, relocation partition, and density-based clustering (Krizhevsky et al. 2012a).

Classification: A primary objective of classification is the classification of the data into groups, which are already predefined. Standard bioinformatics classification algorithms are SVMs, DT, KNN, and Bayesian network (Krizhevsky et al. 2012b).

Deep Neural Networks (DNNs): A multilayer nonlinear ANN is the DL model. Deep belief network (DBN), stacked autoencoder (SAE), and multilayer perceptron (MLP) are known as DNNs (Min et al. 2017). Restricted Boltzmann machines, autoencoders, and perceptrons are the generally used building blocks for SAE, DBN, and MLP, respectively.

Convolutional Neural Networks (CNNs): CNN is similar to a feedforward ANN consisting of convolution layers, nonlinear layers, and pooling layers. Specifically, in image recognition, CNN has good results. Now, AlexNet is a common CNN proposed in Lan et al. (2018), with eight layers.

Recurrent Neural Networks (RNNs): RNN is a cyclic linked ANN, with the aim of internal states (memory) that can process the sequential information. Often used in RNN are building blocks, for example, perceptron and long short-term memory (LSTM).

Emerging Architectures: Several emerging DL models have been introduced recently, for example, multidimensional recurrent neural networks, RNN, convolutional autoencoder, and deep spatiotemporal neural networks. Some ML problems are illustrated in Webb (2018). The data volume and quality must over fit a model, and data cleaning and storage plays a major part in bioinformatics. Training data could be distorted, and the resulting intelligence would still not be unbiased. Whether the ML algorithms can clarify predictions persists in a black box.

3.3.6 Accuracy

This can be hard not just for a computer but also for a person to grasp. The constant repetition of the terms used in the humorous sentences makes it difficult to train models of sentiment analysis effectively. Two people must exchange common subjects, preferences, and historical details to make sarcasm available.

Word uncertainty is another pitfall you will face while working on a question of sentiment analysis. The question of word uncertainty is the impossibility of predefining polarity, since the polarity of certain words depends heavily on the meaning of the sentence.

3.3.7 Complexity

A model trained on a higher-dimensional dataset is computationally less efficient.

3.3.7.1 Algorithm Computational Efficiency

Table 3.1 shows the computational efficiency of various algorithms.

3.4 Future Scope

3.4.1 Agriculture

A crucial element for designing a smart farming system is soil moisture. A variety of environmental factors influence the soil moisture, for example, soil temperature, air humidity, air temperature, and UV. The weather forecast accuracy has significantly improved with advances in technology, and the climate forecast facts can be used to predict modifications in topsoil humidity.

The author in Goldstein et al. (2018) presented guidelines for irrigation based on the ML algorithm, backed by the encysted expertise of the agronomist. This was found that gradient boosted regression tree with 93% accuracy in irrigation plan or recommendation prediction was the best regression model. The model developed is useful for the management of irrigation by the agronomist. The author in Roopaei et al. (2017) suggested an intelligent, thermal imaging-based irrigation monitoring device. The proposed method uses drone-mounted thermal imaging camera.

Several earlier irrigation schemes did not take weather forecast details (e.g., precipitation) into account when making irrigation decisions. This results in a waste of freshwater, electricity, and harvest growth loss (caused by extra water) once a rain is immediately followed by crop watering. To manage these situations, IoT-based systems will be responsible for improved agricultural predictive analytics by using weather forecasting information on the Internet. Still, thanks to the advent of satellite imaging technology, for the increasing weather forecasting accuracy, the author in Goap et al. (2018) identified an IoT-based intelligent irrigation system including a hybrid ML approach to soil moisture prediction. The methodology that is proposed utilized the data from nearly past sensors and forecasted weather data to predict moisture of soil for upcoming days. Soil moisture's expected benefit is greater in terms of their being an individual device prototype. The prototype system is economical and building upon open-source technologies. The default system makes it a smart device, and it can be further adapted to different application areas. We expect to carry out a water-saving experiment in the future on the basis of a proposed algorithm with several nodes together with decreasing cost of the scheme.

3.4.2 Integrative Framework for Anticancer Drug Prediction

The authors have shown that clinical data mining is a new research area aimed at using the resources of ML and data mining to uncover the biological pattern. In addition, the work area on cogenomics seeks to classify and interpret cancer-related genes and thus assists in genotype diagnosis.

While numerous methods for classification have been suggested in the literature, the selection of genes remains a major curse. Cancer is a heterogeneous disorder consisting of multiple subtypes. Therefore, there is an immediate need to build programs or approaches that can assist in early diagnosis and traditional cancer prognosis. Numerous modern methods related to cancer research have developed over the past decade. Scientists used numerous biological and analytical methods for the early detection of the type of cancer. The large collection of cancer data has hiked work in this area. Several ML techniques were used to determine whether or not the tumor is cancerous. So the proposed approach is an effort to reconcile classification problems for cancerous genomic profiles in order to solve the future problems. This approach is based on the principle of using ML algorithms, SVM and NN. Result offers comparative output analysis of the model when the sample size is varied. However, the sample size improves the efficiency of the model too, which indicates a positive trend toward the model's robustness and adaptability. This approach can be expanded to incorporate integrative structure for the prediction of anticancer drugs (Yuan 2016) in the future.

Real-time face recognition is introduced, which helps to identify the face of a human which can be used to authenticate and identify the individual. Face emotion detection is carried out with the help of vector machine classifiers, which are able to precisely distinguish various emotion types. This will improve the accuracy of both emotion detection and face recognition by increasing the number of images throughout training. The time taken for detection is slightly lower, and therefore, the machine yields a smaller amount of runtime even with high accuracy. Forthcoming research involves implementing the system in Android that will increase the availability of the system for more users (Rajesh and Naveen Kumar 2017) (Figure 3.4).

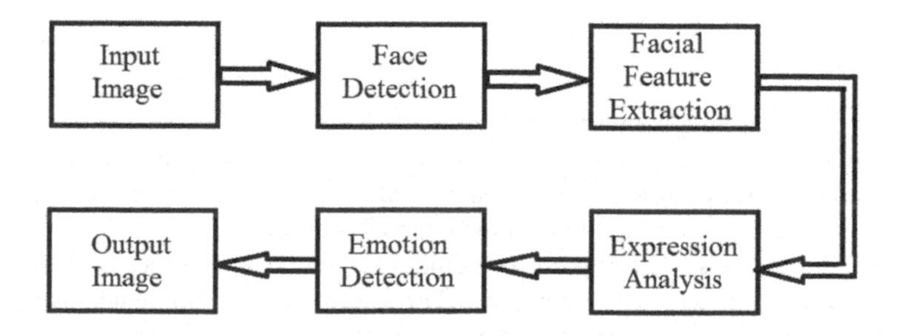

FIGURE 3.4 Block diagram of emotion detection system.

In Zhou et al. (2018), the author performed a survey of something—rhythms for face detection—and evaluated their representatives: hair-like histogram of oriented gradient-support vector machine (HoG-SVM) and AdaBoost cascade as standard approaches, and S3FD and quicker RCNN as DL methods for low-class images. The author conducts the deterioration of the given models when altering the amount of brightness glare or blur. The results from the trial showed that together with hand-crafted features that are intensely trained prone to inputs of low-quality outcome, the scale-variant architecture of multilayer NN extraction features may support the recognition of fuzzy and small faces in comparison with the scale-invariant structure. The author findings will promote additional quality-invariant face detectors for real-world applications (Zhou et al. 2018).

3.4.3 Medical Disease Diagnosis

This study supports the use of ML algorithms in prediction of disease and its early detection. The model developed as per the proposed method shows greater precision than the existing models to the author's best understanding (Riano et al. 2012; Theodoridis and Koutroumbas 2008; Kurnianggoro et al. 2018; Rajesh and Naveen Kumar 2017). The precision rate of this proposed approach hits 87.10% for heart-related disease detection with the help of logistic regression, 85.71% for prediction of diabetes using SVM (linear kernel), and 98.57% for the detection of breast cancer using AdaBoost classifier. The project's potential reach and enhancement include automating the measures, for example, function selection, data munging, and model fitting for better predictive accuracy. Using pipeline layout for preprocessing data will further support in achieving better outcomes (Kohli and Arora 2018).

The authors presented a computer-assisted framework to evaluate the severity of chronic HBV liver fibrosis. Compared to 19 current models and other three ML algorithms, RFC with the 9 indicators has the potential to enhance the accuracy of liver fibrosis severity evaluation, especially for S2 and S3 stages. It will be kept in mind that for the development of classifiers, the consistency of the training samples is essential. In addition, forthcoming studies, based on broad datasets like physical layer image information and serum maker, are needed to improve diagnostic accuracy and facilitate its clinical application (Li et al. 2019).

3.4.4 Stock Market Analysis

Investor confidence plays a large part in the stock market. User-generated online textual content offers a valuable source for representing investors predicting stock values and psychology as a supplement to market data. The author incorporates the analysis of sentiment into a method of ML focused on SVM. In addition, the author has taken into account the day of week impact by creating more accurate and practical sentiment catalogs. Factual results show, after adding sentiment variables, the precision of forecasting the way of movement of the SSE 50 Index. And the pattern, meanwhile, lets investors make smarter choices. These findings also show that sentiment is likely to have useful knowledge around the specific benefit prices and can be considered the best leading index direction on the stock market (Ren et al. 2019). The author was considering extending the time span for potential research that means more textual documents are scrapped from the Internet. And, it is imperative to make the voluminous data more effective in real time.

3.4.5 IoT

The author developed a ML method for the classification of enclosed environments, which is based on the RF signal measurements in real time. Several classification methods for ML were studied using different RF features including help vector machines, DTs, and KNN. The results confirm that a ML method using weighted KNN method, channel transfer frequency (FCF), and channel transfer function coherence (CTF) outclasses the other approaches in defining the inside type environment with a classification of 99.3% accuracy. The prediction period was <10 μs, which confirms that the algorithm adopted is a good candidate for real-time implementation situations. The special effects of this approach are to promote well-organized deployment in diverse channels of IoT applications. The aim was to demonstrate the methods and focus on the advantages of using ML as an innovative idea and a flexible method for the classification of indoor environments. Consequently, outcomes were found on the basis of measurements performed in a static situation with a neat and clean line-of-sight module. The forthcoming work of the author will develop for the findings of this research and expand the results, which will be obtained by investigating other possibilities comprising time-varying channels and non-line-of-sight, based on stationary channel measurements. The ability of DL to remove the essential features of this additional complex inside radio channel scenarios (AlHajri et al. 2018) will be explored.

Whereas, the fields of endeavor of BD analytics are still maturing, the actual condition of ML in the treatment of dementia leftovers after the present state-of-the-art approaches. However, numerous researches have suggested applications capable of providing assuring biomarkers of dementia or of proposing diagnostic actions in combination with ML approaches capable of outperforming current procedures. In addition to growths in fields of ML opening earlier inaccessible pathways, many areas of study still remain largely unexplored. The integration of modern ML applications in the fields of NLP, time-series-based analysis, and computer vision has allowed the capacity for novel and complex modeling of different problem spaces.

Natural Language Processing: In the analysis or selection of sustainable development (SD), recent literature concerning the application of natural language processing (NLP) algorithms has shown promise by analyzing semantic fluidity over continuous system development life cycle (SS). NLP systems are passive and non-intrusive while preserving recorded performance metrics equivalent to existing defined prestage selection practices present processes within primary or social attention formations. Nevertheless, with the current iteration state-of-the-art defined as covering other growing and innovative technologies behind the pace, the NLP is still at its beginning (Cambria et al. 2014). The author supports such opinion by suggesting the three paradigms of the present and future course of study through their envisaged evolution of NLP science. In the first field in syntactic analysis, the utilization of phrase, co-occurrence frequency in language analysis and punctuation, remains state-of-the-art defined by NLP. Although the next big move in NLP development is expected to be integrating semantics and sentics of terms and phrases into NLP applications, the development of NLP technology has extensive implications, despite the continued ubiquity of NLP applications, for example, voice assistants inside mobile phones and household appliances. Current applications for the diagnosis of dementia using NLP rely on word bags or word2vec (Mikolov et al. 2013) approaches for encoding of word, succeeding the principal NLP research standard as described above. The implementation of semantic encoding of terms and phrases applies specifically to uses examining patient syntactic ambiguity. Reduced semantic patients fluency with SD may be specifically related to the difficulty of semantic encoding relative to those who do not have SD. Other avenues of study consist of time-series analysis methods of consecutive words contained by sentences. Nowadays, the term order of appearance is largely overlooked. Through his utilization of sliding window relational temperature analysis (Goap et al. 2018), Lopez somewhat touches on this idea, but flattening the features of mentioned time series removes the scope for analysis of word order. These time-series-based analytical methods require the integration during the interview process of the changing conditions exhibited by dementia subjects. Some papers note a significant energy drain and endurance in dementia topics, as interviews usually tend to result in early interview termination.

Computer Vision: The main problem nowadays facing diagnosis based on neuroimaging comprises standardizing the picture before classification. The necessity for standardized functionality across different MRI equipment and brain configurations, current feature encoding is strongly based on regional GM densities, image flattening into non-spatial vectors, or ROI measurements. Application, for example, CNNs referring to spatial relationships among neighboring pixels allow the possible ROIs identification to the result of classification training simultaneously. Popular computer vision applications include the newly developed methodology for the transfer of pretrained large-scale classification of image models for usage in different alternative ML tasks with restricted monitoring datasets (Oquab et al. 2014). Research areas in neuroimaging include superb and innovative environments where these procedures may be influenced for better ROI detection and smooth the diagnosis of Alzheimer's disease. Nonetheless, the black box existence of NN methodologies theoretically limits these applications. The understandability of these methodologies remains an ongoing opportunity of study across different fields (Yosinski et al. 2015), one such possible area being health informatics. ROI encoding features of mesh representation produce unusual domains, for example, in electroencephalogram (EEG). Because the utilization of normal domain ML techniques that value regional features for example CNNs produces inadequate, limited representations of that domain. Irregular ML domain applications like graph-based CNNs (Rajesh et al. 2017) will give a better match by accommodating irregular node-based domain interaction. The segmentation feature-enhanced applications are also open for further research, with such a deep dependency on the edge measurements of different ROIs.

3.4.6 Health Data Linkage

Through the continuous integration of clinical data offering the potential of accurate, standardized, and comprehensive medical care records, the use of comprehensive tests such as EEG, MRI, and cognitive evaluations can be combined through patient information for long term from electronic health records (EHRs) approving toward the development of full term and full concept of diagnostic help systems. Although some research work is tried with the help of statistical methods (Boustani et al. 2005) and ML methods (Escudero et al. 2013), there has been little study going on within this research path.

Nowadays, heart diseases are a very critical issue for human beings. There is a common need for an automated system to predict heart disease in its early stage to reduce the chances of heart damages. Automated systems will be useful for physicians to diagnose the diseases of patients effectively, and they will also be helpful for monitoring the health of the patient. For an automated system, there are two essential things: one is feature selection and another is prediction. With the help of proper feature selection, a person can predict heart disease precisely. For selecting the features, the author can use a random search algorithm, hybrid grid search algorithm, etc. So, in the forthcoming days, firstly, researchers have to use search algorithm, and then, for the prediction, the researcher can apply ML techniques for better results in the prediction of heart disease (Katarya and Srinivas 2020).

3.4.7 Deep Learning

ImageNet Large Scale Visual Recognition Challenge (ILSVRC) in 2012 through SuperVision has been transformed image recognition in machine vision through deep CNNs. Therefore, numerous implementations of deep CNNs have dominated subsequent challenges. In medical fields, for example, drug discovery (Dahl et al. 2014), categorization of patients (Tan et al. 2015), imaging (Codella et al. 2017), biomedical text mining (Liu et al. 2015), and EHRs (Chalapathy et al. 2016), DL was used. With DL demonstrating the ability to represent complex structures in further fields, integration of these DL architectures with dementia healthcare implementations has potential for new attitudes for nowadays research. However, there is criticism of several aspects of DL methodologies, which restrict widespread utilization in fields like medicine. Criticisms arise from the common NN paradigm from which architectures, for example, RNNs, CNNs, and deep belief networks derive. Transparency problems or the design of models "black box" restricts the authentication of trained models to solely experimental proofs derived from unseen test results. Consequently, to signify applicable aspects of the dataset through the dependence on a model is not always assured as shown by Ribeiro et al. (2016). Nevertheless, the disadvantage of these possible datasets bias problems, as suggested by Ribeiro, remains an important concern in health informatics, with a strong focus on reducing the residual bias in population demographics built on most of the literature reviewed.

Future research may include social media data, ratings, and blogs that affect the stock market from a very long period of time and consideration of a large number of news data instances.

For this function, the algorithm based on user similitude and object similarity was used. Better results were obtained for the analyzed datasets using the user similarity test. The errors considered—RMSE (root mean square error) and IEA (international energy agency)—were smaller for this sort of algorithm than for methods focused on similar artifacts. What's more, this algorithm in a shorter time allowed for more recommendations. Nevertheless, based on these findings, it cannot be said that the similar user approach is superior in all systems. In the assessed scenario, a certain number of minimum ratings given by users are guaranteed for the datasets. This situation is generally very unusual, and often, you need to build a recommendation for users who have received only a few ratings. Further work on the recommendation system problem can involve an attempt to introduce hybrid methods to minimize errors in the estimation of prediction values.

3.4.8 Educational Technology

A lack of research competitiveness from different countries has been described as a consequence of this study, although there is geographical diversity. Although an obstacle may have been the English medium, there is still a wide variety of variations that can inspire potential researchers to take part in this line of analysis. ML is continuously evolving, and more outcomes in the foregoing themes from various contexts will certainly benefit the mature sector.

REFERENCES

Alhajri, Mohamed I., Nazar T. Ali, and Raed M. Shubair. 2018. "Classification of indoor environments for IoT applications: A machine learning approach." *IEEE Antennas and Wireless Propagation Letters*.

Bishop, Christopher M. 2006. *"Bishop: Pattern Recognition and Machine Learning"* Springer: Berlin/Heidelberg.

Boustani, Malaz, Christopher M. Callahan, Frederick W. Unverzagt, Mary G. Austrom, Anthony J. Perkins, Bridget A. Fultz, Siu L. Hui, and Hugh C. Hendrie. 2005. "Implementing a screening and diagnosis program for dementia in primary care." *Journal of General Internal Medicine*. doi: 10.1111/j.1525-1497.2005.0126.x.

Cambria, Erik, and Bebo White. 2014. "Jumping NLP curves: A review of natural language processing research." *IEEE Computational Intelligence Magazine*. doi: 10.1109/MCI.2014.2307227.

Chawathe, Sudarshan. 2018. "Monitoring blockchains with self-organizing maps." In *Proceedings -17th IEEE International Conference on Trust, Security and Privacy in Computing and Communications and 12th IEEE International Conference on Big Data Science and Engineering, Trustcom/BigDataSE 2018*. doi: 10.1109/TrustCom/BigDataSE.2018.00283.

Chen, Xianfu, Zhifeng Zhao, Celimuge Wu, Mehdi Bennis, Hang Liu, Yusheng Ji, and Honggang Zhang. 2019. "Multi-tenant cross-slice resource orchestration: A deep reinforcement learning approach." *IEEE Journal on Selected Areas in Communications*. doi: 10.1109/JSAC.2019.2933893.

Codella, N. C.F., Q. B. Nguyen, S. Pankanti, D. A. Gutman, B. Helba, A. C. Halpern, and J. R. Smith. 2017. "Deep learning ensembles for melanoma recognition in dermoscopy images." *IBM Journal of Research and Development*. doi: 10.1147/JRD.2017.2708299.

Dahl, Gordon B., Katrine V. Løken, and Magne Mogstad. 2014. "Peer effects in program participation." *American Economic Review*. http://dx.doi.org/10.1257/aer.104.7.2049.

Dean, Jeffrey, and Sanjay Ghemawat. 2008. "MapReduce: Simplified Data Processing on Large Clusters." *Communications of the ACM*. doi: 10.1145/1327452.1327492.

Ermes, Miikka, Juha Parkka, Jani Mantyjarvi, and Ilkka Korhonen. 2008. "Detection of daily activities and sports with wearable sensors in controlled and uncontrolled conditions." *IEEE Transactions on Information Technology in Biomedicine*. doi: 10.1109/TITB.2007.899496.

Escudero, Javier, Emmanuel Ifeachor, John P. Zajicek, Colin Green, James Shearer, and Stephen Pearson. 2013. "Machine learning-based method for personalized and cost-effective detection of Alzheimer's disease." *IEEE Transactions on Biomedical Engineering*. doi: 10.1109/TBME.2012.2212278.

Goap Amarendra, Deepak Sharma, A. K. Shukla, and C. Rama Krishna. 2018. "An IoT based smart irrigation management system using machine learning and open source technologies." *Computers and Electronics in Agriculture*. doi: 10.1016/j.compag.2018.09.040.

Goldstein, Anat, Lior Fink, Amit Meitin, Shiran Bohadana, Oscar Lutenberg, and Gilad Ravid. 2018. "Applying machine learning on sensor data for irrigation recommendations: Revealing the agronomist's tacit knowledge." *Precision Agriculture*. doi: 10.1007/s11119-017-9527-4.

Kau, Lih Jen, and Chih Sheng Chen. 2015. "A smart phone-based pocket fall accident detection, positioning, and rescue system." *IEEE Journal of Biomedical and Health Informatics*. doi: 10.1109/JBHI.2014.2328593.

Kohli, Pahulpreet Singh, and Shriya Arora. 2018. "Application of machine learning in disease prediction." In *2018 4th International Conference on Computing Communication and Automation, ICCCA 2018*. doi: 10.1109/CCAA.2018.8777449.

Krizhevsky, Alex, Ilya Sutskever, and Geoffrey E. Hinton. 2012a. "ImageNet classification with deep convolutional neural networks." *Advances in Neural Information Processing Systems*. doi: 10.1061/(ASCE)GT.1943-5606.0001284.

Krizhevsky, Alex, Ilya Sutskever, Geoffrey E. Hinton, Sergey Levine, Chelsea Finn, Trevor Darrell, Pieter Abbeel, et al. 2012b. "ImageNet classification with deep convolutional neural networks alex." *Proceedings of the 31st International Conference on Machine Learning (ICML-14)*, Beijing.

Kumar, Indu, Kiran Dogra, Chetna Utreja, and Premlata Yadav. 2018. "A comparative study of supervised machine learning algorithms for stock market trend prediction." In *Proceedings of the International Conference on Inventive Communication and Computational Technologies, ICICCT 2018*. doi: 10.1109/ICICCT.2018.8473214.

Kurnianggoro, Laksono, and Kang Hyun Jo. 2018. "Towards an integrated method of detection and description for face authentication system." In *Proceedings-2018 11th International Conference on Human System Interaction, HSI 2018*. doi: 10.1109/HSI.2018.8430774.

Lan, Kun, Dan Tong Wang, Simon Fong, Lian Sheng Liu, Kelvin K. L. Wong, and Nilanjan Dey. 2018. "A survey of data mining and deep learning in bioinformatics." *Journal of Medical Systems*. doi: 10.1007/s10916-018-1003-9.

Li, Rongpeng, Zhifeng Zhao, Xianfu Chen, Jacques Palicot, and Honggang Zhang. 2014. "TACT: A transfer actor-critic learning framework for energy saving in cellular radio access networks." *IEEE Transactions on Wireless Communications*. doi: 10.1109/TWC.2014.022014.130840.

Li, Naiping, Jinghan Zhang, Sujuan Wang, Yongfang Jiang, Jing Ma, Ju Ma, Longjun Dong, and Guozhong Gong. 2019. "Machine learning assessment for severity of liver fibrosis for chronic Hbv based on physical layer with serum markers." *IEEE Access*. doi: 10.1109/ACCESS.2019.2923688.

Liu, Shengyu, Buzhou Tang, Qingcai Chen, and Xiaolong Wang. 2015. "Effects of semantic features on machine learning-based drug name recognition systems: Word embeddings vs. manually constructed dictionaries." *Information (Switzerland)*. doi: 10.3390/info6040848.

Mikolov, Tomas, Kai Chen, Greg Corrado, and Jeffrey Dean. 2013. "Efficient estimation of word representations in vector space." In *1st International Conference on Learning Representations, ICLR 2013-Workshop Track Proceedings*, Scottsdale, Arizona.

Milletari, Fausto, Nassir Navab, and Seyed Ahmad Ahmadi. 2016. "V-net: Fully convolutional neural networks for volumetric medical image segmentation." In *Proceedings-2016 4th International Conference on 3D Vision, 3DV 2016*. doi: 10.1109/3DV.2016.79.

Min, Seonwoo, Byunghan Lee, and Sungroh Yoon. 2017. "Deep learning in bioinformatics." *Briefings in Bioinformatics*. doi: 10.1093/bib/bbw068.

Nawrocka, Agata, Andrzej Kot, and Marcin Nawrocki. 2018. "Application of machine learning in recommendation systems." In *Proceedings of the 2018 19th International Carpathian Control Conference, ICCC 2018*. doi: 10.1109/CarpathianCC.2018.8399650.

Negandhi P., Y. Trivedi, and R. Mangrulkar. 2019. "Intrusion detection system using random forest on the NSLKDD dataset". In: Shetty N., Patnaik L., Nagaraj H., Hamsavath P., and Nalini N. (eds), *Emerging Research in Computing, Information, Communication and Applications*. Advances in Intelligent Systems and Computing, vol. 906. Springer: Singapore. doi:10.1007/978-981-13-6001-5_43.

Nguyen, Anh, Jason Yosinski, and Jeff Clune. 2019. "Understanding neural networks via feature visualization: A survey." *Lecture Notes in Computer Science (Including Subseries Lecture Notes in Artificial Intelligence and Lecture Notes in Bioinformatics)*. doi: 10.1007/978-3-030-28954-6_4.

Oquab, Maxime, Leon Bottou, Ivan Laptev, and Josef Sivic. 2014. "Learning and transferring mid-level image representations using convolutional neural networks." *In Proceedings of the IEEE Computer Society Conference on Computer Vision and Pattern Recognition*. doi: 10.1109/CVPR.2014.222.

Pahwa, Kunal, and Neha Agarwal. 2019. "Stock market analysis using supervised machine learning." In *Proceedings of the International Conference on Machine Learning, Big Data, Cloud and Parallel Computing: Trends, Prespectives and Prospects, COMITCon 2019*. doi: 10.1109/COMITCon.2019.8862225.

Pawar, Tanmay, N. S. Anantakrishnan, Subhasis Chaudhuri, and Siddhartha P. Duttagupta. 2008. "Impact of ambulation in wearable-ECG." *Annals of Biomedical Engineering*. doi: 10.1007/s10439-008-9526-8.

Qi, Jun, Po Yang, Geyong Min, Oliver Amft, Feng Dong, and Lida Xu. 2017. "Advanced internet of things for personalised healthcare systems: A survey." *Pervasive and Mobile Computing*. doi: 10.1016/j.pmcj.2017.06.018.

Rahul Katarya and Polipireddy Srinivas. 2020. "Predicting heart disease at early stages using machine learning: A survey" *Proceedings of the International Conference on Electronics and Sustainable Communication Systems (ICESC 2020)*, ISBN: 978-1-7281-4108-4.

Rajesh, K. M., and M. Naveen Kumar. 2017. "A robust method for face recognition and face emotion detection system using support vector machines." In *2016 International Conference on Electrical, Electronics, Communication, Computer and Optimization Techniques, ICEECCOT 2016*. doi: 10.1109/ICEECCOT.2016.7955175.

Ren, Rui, Desheng Dash Wu, and Desheng Dash Wu. 2019. "Forecasting stock market movement direction using sentiment analysis and support vector machine." *IEEE Systems Journal*. doi: 10.1109/JSYST.2018.2794462.

Riaño, David, Francis Real, Joan Albert López-Vallverdú, Fabio Campana, Sara Ercolani, Patrizia Mecocci, Roberta Annicchiarico, and Carlo Caltagirone. 2012. "An ontology-based personalization of health-care knowledge to support clinical decisions for chronically Ill patients." *Journal of Biomedical Informatics*. doi: 10.1016/j.jbi.2011.12.008.

Ribeiro, Marco Tulio, Sameer Singh, and Carlos Guestrin. 2016. "'Why should i trust you?' Explaining the predictions of any classifier." *In Proceedings of the ACM SIGKDD International Conference on Knowledge Discovery and Data Mining*. doi: 10.1145/2939672.2939778.

Roopaei, Mehdi, Paul Rad, and Kim Kwang Raymond Choo. 2017. "Cloud of things in smart agriculture: Intelligent irrigation monitoring by thermal imaging." *IEEE Cloud Computing*. doi: 10.1109/MCC.2017.5.

Salah, Khaled, M. Habib Ur Rehman, Nishara Nizamuddin, and Ala Al-Fuqaha. 2019. "Blockchain for AI: Review and open research challenges." *IEEE Access*. doi: 10.1109/ACCESS.2018.2890507.

Shvachko, Konstantin, Hairong Kuang, Sanjay Radia, and Robert Chansler. 2010. "The hadoop distributed file system." In *2010 IEEE 26th Symposium on Mass Storage Systems and Technologies, MSST2010*. doi: 10.1109/MSST.2010.5496972.

Singh, Sushil Kumar, Shailendra Rathore, and Jong Hyuk Park. 2020. "BlockIoTIntelligence: A Blockchain-enabled intelligent IoT architecture with artificial intelligence." *Future Generation Computer Systems*. doi: 10.1016/j.future.2019.09.002.

Sun, Yaohua, Mugen Peng, Shiwen Mao, and Shi Yan. 2019. "Hierarchical radio resource allocation for network slicing in fog radio access networks." *IEEE Transactions on Vehicular Technology*. doi: 10.1109/TVT.2019.2896586.

Tan, Jie, Matthew Ung, Chao Cheng, and Casey S. Greene. 2015. "Unsupervised feature construction and knowledge extraction from genome-wide assays of breast cancer with denoising autoencoders." *In Pacific Symposium on Biocomputing*, Singapore.

Theodoridis, S., and K. Koutroumbas. 2008. *"Pattern Recognition"*, 4th Edn. Academic Press: London. Semen. Revue de Sémio-Linguistique Des Textes et Discours.

Velankar, Siddhi, Sakshi Valecha, and Shreya Maji. 2018. "Bitcoin price prediction using machine learning." *In International Conference on Advanced Communication Technology, ICACT*. doi: 10.23919/ICACT.2018.8323676.

Webb, Sarah. 2018. "Deep learning for biology." *Nature*. doi: 10.1038/d41586-018-02174-z.

Xiang, Hongyu, Wenan Zhou, Mahmoud Daneshmand, and Mugen Peng. 2017. "Network slicing in fog radio access networks: Issues and challenges." *IEEE Communications Magazine*. doi: 10.1109/MCOM.2017.1700523.

Yan, Zhipeng, Mugen Peng, and Chonggang Wang. 2017. "Economical energy efficiency: An advanced performance metric for 5G systems." *IEEE Wireless Communications*. doi: 10.1109/MWC.2017.1600121WC.

Yosinski, Jason, Jeff Clune, Anh Nguyen, Thomas Fuchs, and Hod Lipson. 2015. "Understanding Neural Networks through Deep Visualization." *Deep Learning Workshop*, 31st International Conference on Machine Learning, Lille, France, 2015.

Yuan, F. Q. 2016. "Critical issues of applying machine learning to condition monitoring for failure diagnosis." In *IEEE International Conference on Industrial Engineering and Engineering Management*. doi: 10.1109/IEEM.2016.7798209.

Zhou, Yuqian, Ding Liu, and Thomas Huang. 2018. "Survey of face detection on low-quality images." In *Proceedings - 13th IEEE International Conference on Automatic Face and Gesture Recognition, FG 2018*. doi: 10.1109/FG.2018.00121.

4

Comprehensive Analysis of Dimensionality Reduction Techniques for Machine Learning Applications

Archana Vasant Mire
Terna Engineering College

Vinayak Elangovan
Penn State University

Bharti Dhote
Sinhgad Institute of Technology

CONTENTS

4.1 Introduction ... 61
4.2 Missing Value Ratio ... 63
4.3 Low Variance Filter ... 64
4.4 High Correlation Filter ... 64
4.5 Principal Component Analysis ... 66
4.6 Independent Component Analysis ... 67
4.7 Backward Feature Elimination ... 68
4.8 Forward Feature Construction ... 70
4.9 Singular Value Decomposition ... 71
4.10 Random Forest ... 74
4.11 Conclusion ... 75
References ... 75

4.1 Introduction

ML techniques are applied on real-world data. Our daily digital activities create a huge amount of data daily. In fact, 90% of the data in the world has been generated in the last few years. Our digital logs in the form of likes, shares and posts, friend list, fan following on social networking sites are traced and processed to improve personalized user experiences. Purchase history and browsing history on online shopping sites are processed to suggest new products in the form of advertisements. It may be Facebook, Amazon, YouTube or Flipkart, and every site traces and analyses browsing history. All these data are processed in seconds by these merchandise. Real-world data, such as digital color images, speech signals, and MRI scans, usually has a high dimensionality. This increase in data generation and collection makes data visualization and drawing inferences more challenging. This chapter briefly reviews the glitches associated while handling this huge high-dimensional data. It also discusses various techniques applied to reduce the dimension of data.

The number of samples increases proportionately with the number of features. If the number of features increases, we will need to have all combinations of feature values well represented in our sample. This makes the model more complex. A model that is trained with a large feature set gets more dependent on the training data and leads to overfitting. This resultant trained model shows poor performance

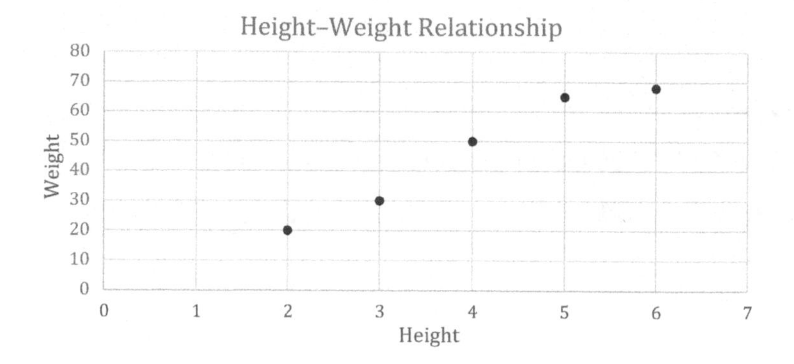

FIGURE 4.1 Scatterplot of weight vs. height.

on real data, beating the purpose. Avoiding overfitting is one of the motivations for applying dimensionality reduction. The reduction in training features results in lesser assumptions made by the ML model and makes it simple. Unwanted effects of high dimensionality are reduced using dimensionality reduction techniques. Hence, it has proven importance in many domains [1].

Data visualization is another challenge that gets more complex with the data dimensions. One of the most common data visualization techniques is scatterplot. Consider a dataset with two variables, height and weight. The simplest way to visualize the relationship between two variables is height and weight. Scatterplot between height and weight is displayed in Figure 4.1. This figure shows high correlation between height and weight. Based on this plot, we can predict a linear regression model between weight and height.

If we have 50 features, we can have $50(50-1)/2=1225$ different scatterplots. These different scatterplots do not make much sense to make any decisions. In such cases, one needs to make decision about subset of features ($p \ll 50$), which may capture sufficient information of original set of variables. For example, data consisting of weight measured in K_g and *Pounds* conveys the same information and can be represented using one-dimensional single line rather than drawing it as two-dimensional scatterplot. There are many advantages of dimensionality reduction techniques. Some of them are addressed below.

1) ML models capture real effects as well as random effects. Hence, a dataset with a large number of features may lead to overfitting. Dimensionality reduction techniques avoid overfitting.
2) Dimensionality reduction techniques remove some of the unimportant features. Hence, ML model accuracy improves with the reduction in misleading data.
3) Reduction in redundant features and noise results in reduction in computational complexity, leading to faster algorithms.
4) Dimensionality reduction techniques transform data into a simple interpretable representation with lesser dimension. This helps to visualize the data properly.

The intrinsic dimensionality of data is the minimum number of parameters needed to account for the observed properties of the data [2]. Preferably, this compact representation should have a dimensionality that corresponds to the intrinsic dimensionality of the data. Frequently, researchers have used dimensionality techniques to improve the performance of their proposed algorithms [1–6]. Dimensionality reduction may help even in passive image forensic analysis [7], where the first digit probability distribution of the first two digits may be enough for analysis purpose. In the subsequent sections, we have discussed these techniques in detail. Every dimensionality reduction technique may have its own benefits while applying to a specific problem domain. In one problem domain, the goal behind dimensionality reduction techniques may be just to visualize the data separation clearly, while in other domains, it may be efficient and faster training. Hence, every algorithm is explained using a different dataset. Algorithms that directly address correlations among features without making any modifications are explained using the same examples. The advantages and disadvantages of these dimensionality reduction techniques are also discussed in this chapter.

4.2 Missing Value Ratio

Often, real-world data is noisy and inhomogeneous. Data may have been missed during collection or extraction. ML models may get trained inaccurately due to these missing values. This may result in wrong and biased predictions. Hence, when the dataset has a large number of missing values, this technique is used. In this technique, features that have a large number of missing values above some threshold with them are dropped.

Consider the pseudo-state geographical information available in Table 4.1. We have tried to create clusters of these states using the k-means clustering algorithm. For simplicity, we have skipped the names of these states. As it is categorical information, it will not affect clustering. Total *NULL* entries for feature *population* are 2. Hence, 0.33% of *population* features are null.

We have applied k-means clustering on this dataset to cluster data in three classes. Figure 4.2 shows the result of clustering. Figure 4.2a shows the result of k-means clustering by considering all three features *longitude, latitude* and *population.* As 0.33% of *population* features are null, we have removed *population* feature and applied k-means clustering on the remaining two features *longitude* and *latitude,*

TABLE 4.1

Dataset for State Geographical Data with Missing Values

State	Longitude	Latitude	Population	Cluster
Maharashtra	−103.77	44.97	100000	0
Madhya Pradesh	−96.80	62.40	100,015	0
Tamilnadu	2.40	46.75	NaN	1
Kerala	−2.53	54.01	100,023	1
Karnataka	10.40	51.15	NaN	1
Haryana	133.11	−25.45	100,050	2

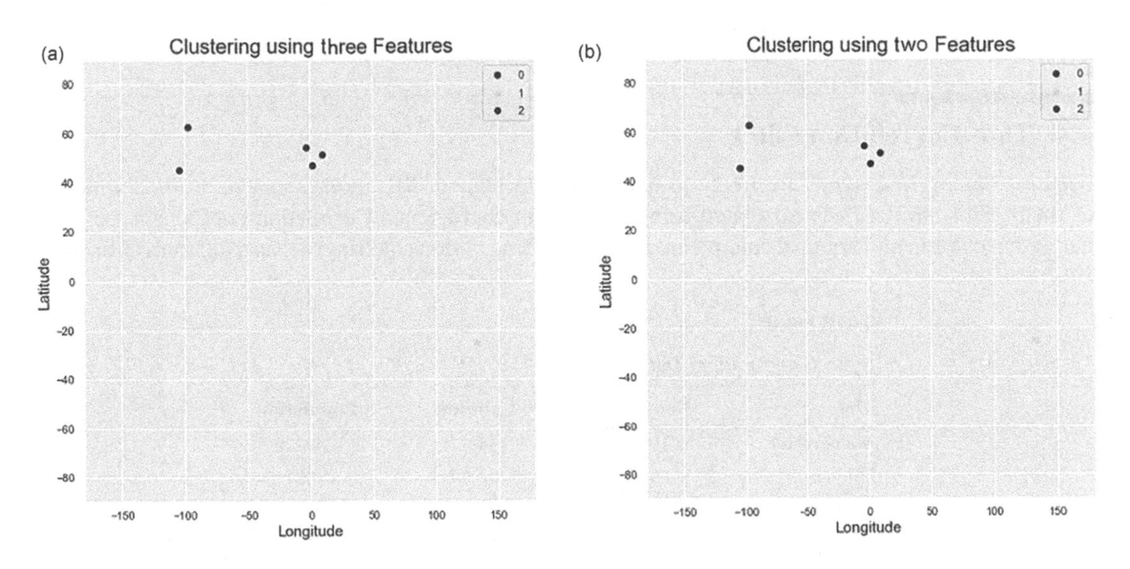

FIGURE 4.2 Clustering on state geographical data. (a) k-means clustering by using three features and (b) k means clustering by using two features.

and the resultant clusters are shown in Figure 4.2(b). One can observe in Figure 4.2(b) that data gets clustered well even after the removal of *population* features.

This technique is very simple to understand and apply. However, it may result in losing some important features. Other approaches such as dropping missing values, filling missing values with statistics and filling missing values with ML approaches are also applied. This approach is more dependent on the type of dataset, and the accuracy of this approach improves with the experience of handling data.

In Ref. [8], the authors have concentrated on the missing value processing techniques in the cluster monitoring applications. They proposed a hybrid multiple imputation framework, which was different from the conventional models. For a random missing pattern, their proposed model was able to impute the missing data with a model based on data-driven and combination architecture. They used deep neural network as the data model and extracted deep features from the data. Deep features were calculated by regression techniques and used to estimate the missing data. In Ref. [9], the authors analysed the impact of missing data on pattern classification problems. In their work, they have analysed many techniques for handling missing values.

4.3 Low Variance Filter

If a particular feature value does not change in a given data sample, its contribution will be least in the decision-making. Ideally, feature value may change slightly but its contribution in the decision-making may still remain least compared with the other features. Hence, in this technique, all features with lower variance (below threshold level) are removed. Usually, data is normalized before applying this technique. If we consider the pseudo-state geographical data in Table 4.2 and calculate its variance using Equation 4.1, we get the following variance for each of the features.

Variance: *Longitude* = 7486.385977, *Latitude* = 1033.996737, *Population*=350.000000

$$\mathrm{Var}_x = \frac{\sum\left(\bar{X} - X_i\right)^2}{n-1} \tag{4.1}$$

By applying k-means clustering with $k=3$, and using all three features *longitude, latitude* and *population,* our dataset gets clustered as shown in Figure 4.2(a). Variance of feature *population* is the least. If we remove *population* features and apply k-means clustering with only *longitude* and *latitude*, then once again we get clusters as shown in Figure 4.2(b). From Figure 4.2, we can observe that removing a low variance feature *population* does not affect the clustering.

4.4 High Correlation Filter

Features having like trends are likely to have like information. ML models can be trained with one of them. This similarity is calculated using correlation coefficients. Correlation coefficients between numerical column and nominal column are computed using various metrics such as Pearson's chi-square

TABLE 4.2

State Geographical Data

State	Longitude	Latitude	Population
Maharashtra	−103.77	44.97	100000
UP	−96.80	62.40	100015
Tamilnadu	2.40	46.75	100,020
Kerala	−2.53	54.01	100,023
Punjab	10.40	51.15	100,030
Haryana	133.11	−25.45	100,050

value and Pearson's product moment coefficient. In this method, if high correlation is observed among two features, one of the features is removed. This technique also needs normalized data.

Covariance between features X and Y can be calculated using Equation 4.2, where $\overline{X}$ and $\overline{Y}$ represent the mean of feature X and the mean of feature Y, respectively.

$$\text{COVar}_{x,y} = \frac{\Sigma\left(\overline{X} - X_i\right)\left(\overline{Y} - Y_i\right)}{n-1} \tag{4.2}$$

Covariance matrix between features X and Y can be defined as shown in Equation 4.3. Covariance between features X, Y is the same as the covariance between features Y, X. Hence, the covariance matrix is diagonally symmetrical. The number of rows and columns of the covariance matrix is equal to the number of features available in the dataset. Covariance is calculated among each pair of features.

$$\text{COVar}M_{x,y} = \begin{bmatrix} \text{Var}_{x,x} & \text{Cov}_{x,v} \\ \text{Cov}_{y,x} & \text{Var}_{y,y} \end{bmatrix} \tag{4.3}$$

Table 4.3 shows some of the sample entries from the ship data available at Ref. [10]. Figure 4.3 shows covariance matrix calculated for features in Refs [10,11], using Equations 4.1 and 4.2.

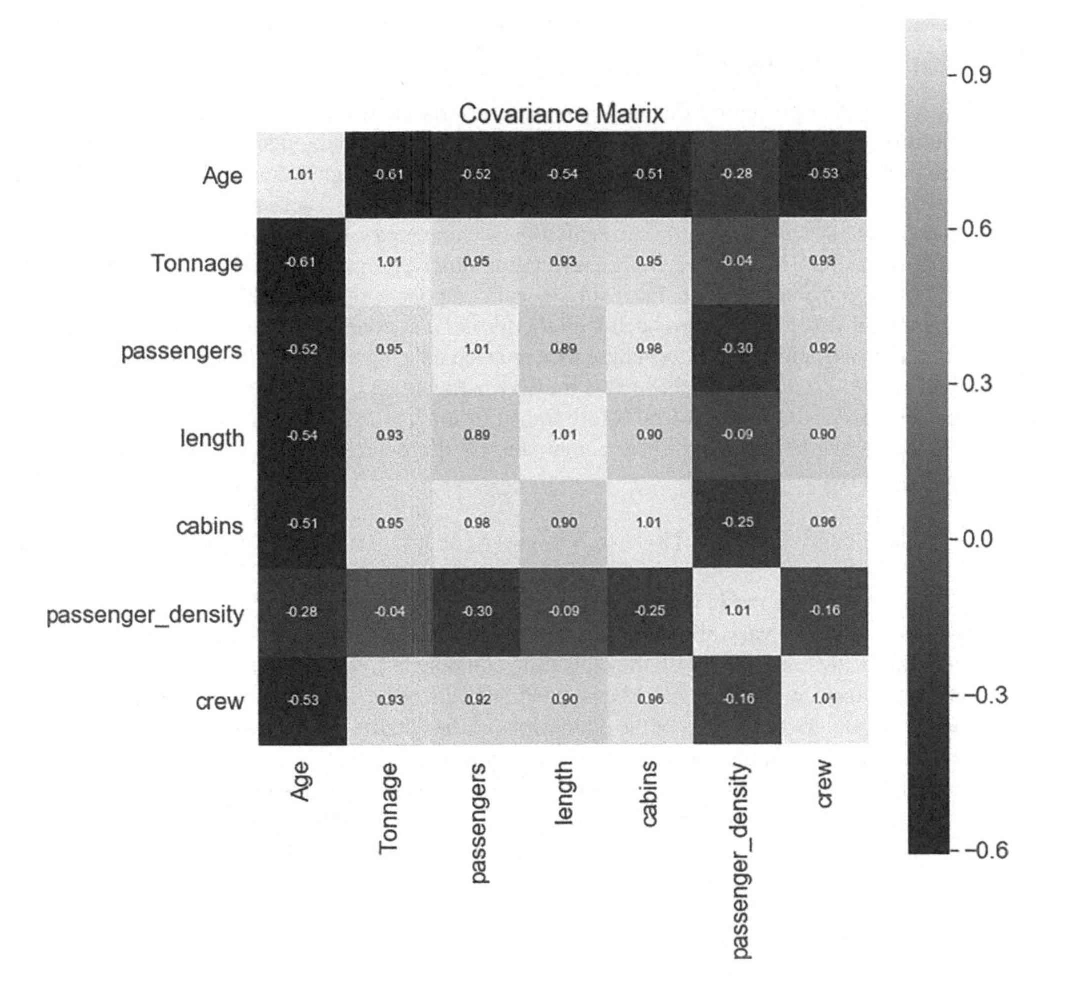

FIGURE 4.3 Covariance matrix.

TABLE 4.3
Ship Data

Ship_Name	Cruise_Line	Age	Tonnage	Passengers	Length	Cabins	Passenger_Density	Crew
Journey	Azamara	6	30.277	6.94	5.94	3.55	42.64	3.55
Quest	Azamara	6	30.277	6.94	5.94	3.55	42.64	3.55
Celebration	Carnival	26	47.262	14.86	7.22	7.43	31.80	6.70
Conquest	Carnival	11	110.000	29.74	9.53	14.88	36.99	19.10
Destiny	Carnival	17	101.353	26.42	8.92	13.21	38.36	10.00

In Figure 4.3, we can observe that the correlation of features *cabins*, *length*, *tonnage*, *passenger* and age is higher with respect to the feature *crew*. Hence, these features can be used to apply multiple regression models for predicting *crew*. Mean squared error (MSE) generated by applying this regression model is 0.955. Figure 4.3 also shows that the correlation between *tonnage* with passenger and *tonnage* with *cabins* is also very high, i.e. 0.95. Hence, we have removed *tonnage* and applied multiple regression model on the remaining features. Multiple regression model used to predict *crew* using *passenger, length* and *cabins* generates MSE as 0.986. Thus, the removal of one feature that has high correlation with other features creates very less impact on MSE value.

4.5 Principal Component Analysis

PCA [12] is a statistical procedure that orthogonally converts the original $m \times n$ dataset into a new set of $m \times n$ coordinates. This new set of coordinates, called principal components (PCs), is formed using eigenvectors of the covariance matrix of the original dataset. The magnitude and direction of variance captured by each axis are measured using eigenvectors and eigenvalues. These axes should be orthogonal to each other in data space. Hence, the correlation between any pair of eigenvectors and eigenvalues is maintained as zero. PCA is arranged in the descending order of eigenvalues. Hence, the first PC contributes the largest possible variance. Each subsequent component shows the highest possible variance under the constraint that it is not correlated with the former components. The first few components are enough to maintain sufficient data information with reduced data dimension. The new coordinates do not represent real system-produced variables. This transformation is subtle to the relative scaling of original data. Hence, data needs to be normalized before transforming it to PCA. As shown in Equation 4.4, PC_i is a variable that is synthesized as a linear combination and it determines the maximum variance and direction of the dataset.

$$PC_i = w_{1,i}(X_1) + w_{2,i}(X_2) + w_{3,i}(X_3), \dots + w_{n,i}(X_n) \tag{4.4}$$

where PC_i is the ith PC, X_i is the ith feature and $w_{1,i}$ is the weight vector.

PC_1 captures the highest variability of the dataset, and succeeding components capture the residual variation without being correlated with the preceding components. As each successive component captures the residual variance, the correlation between PCs should be zero. The total number of PCs that can be constructed is equal to the number of the dimensions of the dataset. Out of these, only the first few are useful. Table 4.4 shows a sample dataset from iris flower database [13].

This dataset consists of four different features such as *petal length, petal width, sepal length* and *sepal width* of three different types of flowers. Figure 4.4 shows the scatterplot of this dataset. Figure 4.4(a) shows the scatterplot using the first two PCs. One can easily visualize that the data is linearly separable using the first two PCs. Figure 4.4(b) shows the clustering results by applying k-means clustering on the first two PCs. This figure shows that data of class *iris_versicolar* is perfectly clustered while there is some misclassification in *iris_setosa* and *iris_verginica.*

PCA is frequently used in filtering noisy datasets, image processing, image compression, image forensics, signal processing, etc. There are some limitations for the application of PCA. PCA needs minimum

TABLE 4.4
Flower Dataset

Sepal Length	Sepal Width	Petal Length	Petal Width	Target
5.1	3.5	1.4	0.2	Iris setosa
4.9	3.0	1.4	0.2	Iris setosa
4.7	3.2	1.3	0.2	Iris setosa
4.6	3.1	1.5	0.2	Iris setosa
5.0	3.6	1.4	0.2	Iris setosa

150 observations for its application [14], all features should be correlated and need to have constant multivariate normal relationship without significant outliers, low variance axes are assumed as noise and may be discarded, etc.

In Ref. [15], the authors combined Bayesian classification, PCA and linear discriminant analysis and proposed PCA-LDA-BC classification algorithm for intrusion detection. In Ref. [16], the authors applied a DL model for the early detection of diabetic retinopathy. First, they normalized the raw dataset using standard scalar methods and extracted the most significant features using PCA for applying DL models. In Ref. [17], the authors showed that classification of high-dimensional, imbalanced data is a very exciting challenge in ML, which may reduce the performance of classification algorithms. They used techniques for balancing data and improved the performance of link prediction models by using PCA.

4.6 Independent Component Analysis

ICA has its origin in information theory [18]. It is one of the most widely used dimensionality reduction techniques. It is based on the assumption that given variables are linear mixtures of some unknown variables. It decomposes independent factors that are linearly mixed with several sources. Hence, this algorithm is called the ICA of the observed data. If variables are independent, it means that they are not dependent on other variables. For example, the number of students in a class is independent of the number of subjects offered to the class. For ICA, an example of blind source separation or cocktail party is frequently discussed in the literature. In this cocktail party problem, a room is assumed to be full of N number of speakers speaking simultaneously. A room also has N microphones kept at different places. Each microphone captures audio signals of different intensities coming from different speakers. The difference in signal strength is due to the distance between speaker and microphone. Decomposition of this mixed signal can be done through ICA.

ICA has many application areas such as image denoising, medical signal processing, facial expression recognition, brain signal interfacing, time series analysis, clustering, classification, voice recognition and speaker recognition. However, it has many limitations such as the number of observed mixtures of signals and of independent components is always the same. ICA can be applied only when independent components have non-Gaussian distribution. Independent components are always assumed as statistically independent of each other, etc. Just like PCA, ICA is computed using eigenvalues and eigenvectors of signals. Still they share a major difference. PCA is arranged in the descending order of variance, and ICA does not focus on variance. Abstract algorithm of ICA can be explained as follows:

1. Normalize mixed signal X by subtracting the mean from it
2. Whiten X in such a way that the covariance matrix of the whitened signal will be equal to the identity matrix.
3. Initialize weight vector W with random data
4. Update W and normalize it.
5. Iterate Step 4 until the solution converges.
6. Compute dot product of W and X to get independent sources

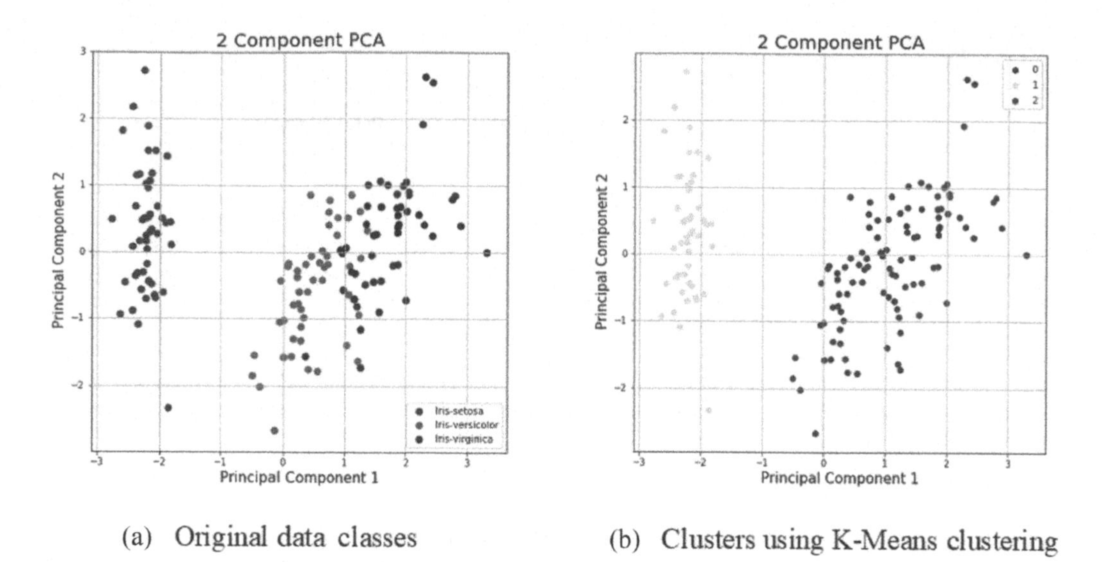

(a) Original data classes

(b) Clusters using K-Means clustering

FIGURE 4.4 Data representation using PCA.

The whitening signal X needs eigenvalue decomposition of its covariance matrix as shown in Equation 4.5.

$$\tilde{x} = ED^{-1/2}E^{T}x \tag{4.5}$$

where D is diagonal matrix of eigenvalues (λ) of the covariance matrix of input signal X and E is an orthogonal matrix of eigenvectors. The solution is said to be converged when WW^T is nearly equal to 1. Figure 4.5 shows real sources, mixtures and predicted sources by applying ICA on mixtures. Python source code can be found at Ref. [19].

In Ref. [20], the authors analysed the motor current signature for broken bar fault detection of induction motors (IMs). They obtained Fourier-domain spectral signals from input and its autocorrelation function. ICA of the resultant signal was used for further analysis. Their main contribution was IM fault diagnosis in a three-dimensional space using ICA-based feature extraction. In Ref. [21], the authors studied vibration signals for multiple fault detection using single-channel signals. They observed that different signal sources cannot be decomposed in the same frequency using variational mode decomposition (VMD). Hence, they used robust ICA as a solution to this problem. They used the fourth-order cumulant of restructured signals from VMD and robust ICA as fault indexes. Further, they used fuzzy c-means clustering for fault detection.

4.7 Backward Feature Elimination

Initially, this technique uses all n input features to apply a selected classification algorithm [22]. During each next iteration, one input feature is removed at a time and the model is trained with the remaining $n-1$ input features. The removal of one feature that results in the least increase in the error rate is removed from the dataset. This leads to the dataset with $n-1$ features. The process is continued with $n-2$ features and repeated so on. Each kth iteration creates the model trained with $n-k$ features. Some threshold is applied on the minimum number of features required to train the model and error rate. To further explain this concept, we have applied multiple regression on the dataset from Ref. [10]. Table 4.5 shows the MSE and R^2 errors generated by applying multiple regression on different sets of features.

The first row of this table shows the MSE and R^2 errors generated by using all six features in the regression model. The remaining rows show the error values generated by applying regression on the

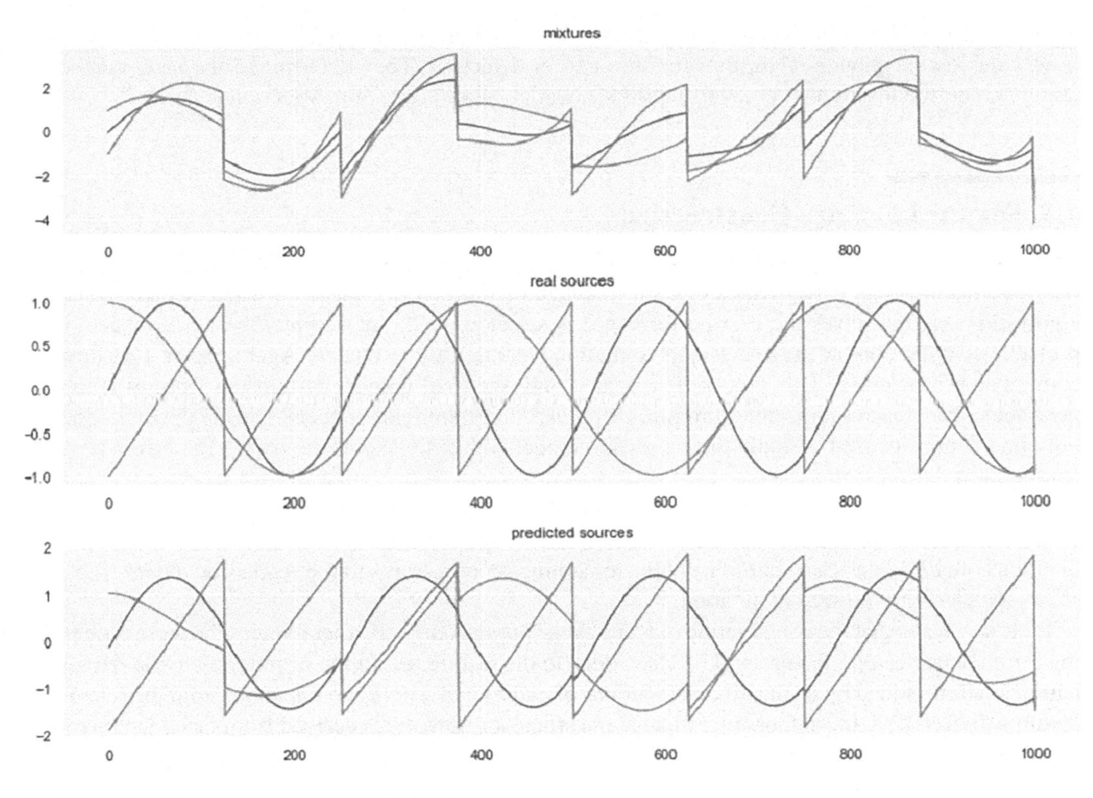

FIGURE 4.5 ICA on manually generated signals.

TABLE 4.5

Error Results after Removing One Feature at a Time

Feature Removed	MSE Train	MSE Test	R² Train	R² Test
-	0.954	0.884	0.920	0.929
Age	0.954	0.884	0.920	0.928
Tonnage	0.965	0.886	0.919	0.929
Passengers	1.009	0.963	0.916	0.922
Length	1.014	0.994	0.915	0.920
Cabins	1.251	1.667	0.896	0.865
Passenger_Density	0.954	0.883	0.920	0.929

remaining features, after removing one at a time, as mentioned in the respective cell. Table 4.5 shows that the removal of features *age* decreases regression error. Hence, this feature is removed. Regression model is reapplied considering the remaining five features, and one feature is removed at a time. In the next iteration, the removal of feature *passenger_density* results in the lowest increase in error rate as MSE train: 0.955, MSE test: 0.889, R^2 train: 0.920 and R^2 test: 0.928. This increase in error is very low compared to the original regression model, applied with all six features. Hence, this feature can also be reduced. Next iterations are continued with the remaining features until some threshold value for the number of features is reached, without compromising regression error.

Medical databases are high-dimensional databases. Classification analysis may generate inaccurate results on training dataset with irrelevant medical features. Hence, in Ref. [23,24], the authors found an optimum feature subset by proposing a feature selection approach. Their results proved that support vector machine (SVM) ranking with a backward search approach improves classification accuracy.

The accuracy of Naive Bayes classifiers improved with their proposed approach. In Ref. [24], the authors used *k*-nearest neighbours algorithm for skin cancer detection. They determined the least number of features required for high accuracy using the sequential backward feature selection technique.

4.8 Forward Feature Construction

This is the reverse of the process to the backward feature elimination [25]. Initially, training starts with only one feature. The performance of each individual feature is evaluated for a selected classification algorithm. One that gives the best performance is selected. In the next cycle, the performance of all probable combinations of the selected and remaining features are evaluated. Again, the best performing combination is selected. This process is repeated until the total number of selected features reaches a threshold value. Once again, data from Ref. [10] is used to explore this concept. During the first iteration, only one feature is used to apply the regression model. Table 4.6 shows the regression errors for each feature used during regression.

We can observe that the feature *cabins* generates the least MSE for regression. Hence, we will select *cabins* as a first feature and will add it to the final feature set. In the next iteration, we will apply all combinations of *cabins* and one feature from the remaining set of features to the regression model. Table 4.7 shows the errors in the second iteration.

Table 4.7 shows that the combination of features *cabins* and *length* further reduces the regression training error. Hence, *length* feature will be also added to the feature set. In the next iteration, the remaining features will be added by trying all combinations of *cabins* and *length* and one more from the remaining features. In Ref. [26], the authors investigated the efficiency of features derived from facial landmarks in facial expression recognition. They used features such as distance vectors extracted from expressive and neutral states of the face. The observed feature vector contains elements that are comparatively unusable in expression recognition. They applied forward sequential feature selection and obtained the most effective feature subset. The selected features were further classified using a multiclass SVM with 89.9% mean class recognition accuracy.

TABLE 4.6

Error Results Using Single Feature Regression (Iteration 1)

Feature Removed	MSE Train	MSE Test	R^2 Train	R^2 Test
Age	9.201	8.352	0.232	0.3260
Tonnage	1.574	1.911	0.869	0.846
Passengers	1.892	2.117	0.842	0.829
Length	2.076	2.941	0.827	0.763
Cabins	1.220	1.099	0.898	0.911
Passenger_Density	11.695	12.298	0.023	0.008

TABLE 4.7

Error Results Using Cabins and One More Feature (Iteration 2)

Feature Removed	MSE Train	MSE Test	R^2 Train	R^2 Test
Cabins, age	1.196	1.054	0.900	0.915
Cabins, tonnage	1.095	1.107	0.909	0.911
Cabins, passengers	1.157	1.098	0.903	0.911
Cabins, length	1.047	1.008	0.913	0.919
Cabins, passenger_density	1.096	1.049	0.908	0.915

TABLE 4.8

Kids Rating for Games

Cricket	Tennis	Football	Chess	Monopoly
1	1	1	0	0
3	3	3	0	0
4	4	4	0	0
5	5	5	0	0
0	2	0	4	4
0	0	0	5	5
0	0	0	5	5

4.9 Singular Value Decomposition

This technique decomposes the original dataset into its three special matrices [27]. Just like PCA and ICA, it also uses the concept of eigenvalue and eigenvectors to calculate three constituent matrices $U\Sigma V^T$. These components are called the left singular vector, singular value matrix and right singular vector, respectively. It makes it easy to explore, analyse and simplify the information content of the dataset. It removes the redundant information from the dataset. There are many application areas for SVD such as video background removal, image compression, spectral clustering and image recovery. This technique is further explored with the data from Table 4.8. Table 4.8 represents ratings given by different kids for various games such as *cricket, tennis, football, chess* and *monopoly.* Rating is based on a scale from 0 to 5 with 5 being highest and 0 being lowest rating assigned by any kid to any game. These games can be broadly classified as outdoor games and board games.

Data from Table 4.8 is divided into SVD $U\Sigma V^T$ as explained below.

$$\begin{bmatrix} 1 & 1 & 1 & 0 & 0 \\ 3 & 3 & 3 & 0 & 0 \\ 4 & 4 & 4 & 0 & 0 \\ 5 & 5 & 5 & 0 & 0 \\ 0 & 2 & 0 & 4 & 4 \\ 0 & 0 & 0 & 5 & 5 \\ 0 & 0 & 0 & 5 & 5 \end{bmatrix} = \begin{bmatrix} 0.13 & 0.02 & -0.01 \\ 0.41 & 0.07 & -0.03 \\ 0.55 & 0.09 & -0.04 \\ 0.68 & 0.11 & -0.05 \\ 0.15 & -0.59 & 0.65 \\ 0.07 & -0.73 & -0.67 \\ 0.07 & -0.29 & 0.32 \end{bmatrix} \times \begin{bmatrix} 12.4 & 0 & 0 \\ 0 & 9.5 & 0 \\ 0 & 0 & 1.3 \end{bmatrix} \times \begin{bmatrix} 0.56 & 0.59 & 0.56 & 0.09 & 0.09 \\ 0.12 & -0.02 & 0.12 & -0.69 & -0.69 \\ 0.40 & -0.80 & 0.40 & 0.09 & 0.09 \end{bmatrix}$$

Matrix U is called a user-to-concept similarity matrix. Hence, the first column of U matrix indicates interest of each kid in outdoor games and the second column indicates interest of students in board games. Matrix Σ indicates the strength of concept. Hence, the strength of outdoor games is more as compared to board games. This means that kids are more interested in outdoor games as compared to board games. The last column V^T indicates feature-to-concept strength. This shows that games *cricket, tennis* and *football* can be more identified as outdoor activities, while games such as *chess* and *monopoly* are less identified as outdoor activities.

Process to find SVD $U\Sigma V^T$ of matrix $A = \begin{bmatrix} 3 & 0 \\ 4 & 5 \end{bmatrix}$ is explained below.

To find V, one needs to find eigenvalues and eigenvectors of $A^T A$

$$A^T A = \begin{bmatrix} 3 & 4 \\ 0 & 5 \end{bmatrix} \begin{bmatrix} 3 & 0 \\ 4 & 5 \end{bmatrix} = \begin{bmatrix} 25 & 20 \\ 20 & 25 \end{bmatrix}$$

$$|A^T A - \lambda I| = 0, \quad \left| \begin{bmatrix} 25-\lambda & 20 \\ 20 & 25-\lambda \end{bmatrix} \right| = 0, \quad (25-\lambda)(25-\lambda) - 400 = 0$$

$$\lambda_1 = \sigma_1^2 = 45, \quad \lambda_2 = \sigma_2^2 = 5, \quad \sigma_1 = \sqrt{45}, \quad \sigma_2 = \sqrt{5}$$

$$\left[A^T A - \lambda I\right][X] = 0$$

For $\lambda_1 = 45$

$$\begin{bmatrix} 25-45 & 20 \\ 20 & 25-45 \end{bmatrix} \begin{bmatrix} x_1 \\ x_2 \end{bmatrix} = \begin{bmatrix} 0 \\ 0 \end{bmatrix}$$

$$\begin{bmatrix} -20 & 20 \\ 20 & -20 \end{bmatrix} \begin{bmatrix} x_1 \\ x_2 \end{bmatrix} = \begin{bmatrix} 0 \\ 0 \end{bmatrix}$$

Using row reduction $R_1 = R_1 - R_2$

$$\begin{bmatrix} 0 & 0 \\ 20 & -20 \end{bmatrix} \begin{bmatrix} x_1 \\ x_2 \end{bmatrix} = \begin{bmatrix} 0 \\ 0 \end{bmatrix}$$

$20x_1 - 20x_2 = 0$

$x_1 = x_2$

$x_1 = 1 \qquad x_2 = 1$

For $\lambda_1 = 5$

$$\begin{bmatrix} 25-5 & 20 \\ 20 & 25-5 \end{bmatrix} \begin{bmatrix} x_1 \\ x_2 \end{bmatrix} = \begin{bmatrix} 0 \\ 0 \end{bmatrix}$$

$$\begin{bmatrix} 20 & 20 \\ 20 & 20 \end{bmatrix} \begin{bmatrix} x_1 \\ x_2 \end{bmatrix} = \begin{bmatrix} 0 \\ 0 \end{bmatrix}$$

Using row reduction $R_1 = R_1 - R_2$

$$\begin{bmatrix} 0 & 0 \\ 20 & 20 \end{bmatrix} \begin{bmatrix} x_1 \\ x_2 \end{bmatrix} = \begin{bmatrix} 0 \\ 0 \end{bmatrix}$$

$20x_1 + 20x_2 = 0$

$x_1 = -x_2$

$x_1 = -1 \qquad x_2 = 1$

Singular vectors are unit vectors. Hence, eigenvectors are converted into unit vectors by dividing them by the square root of summation of squared vector values.

$$v_1 = \frac{1}{\sqrt{2}} \begin{bmatrix} 1 \\ 1 \end{bmatrix} \quad v_1 = \frac{1}{\sqrt{2}} \begin{bmatrix} -1 \\ 1 \end{bmatrix}$$

$$\text{Right singular vector} = V^T = \frac{1}{\sqrt{2}} \begin{bmatrix} 1 & 1 \\ -1 & 1 \end{bmatrix}$$

$$u_1 = \frac{1}{\sigma_1} A v_1 = \frac{1}{\sqrt{45}} \begin{bmatrix} 3 & 0 \\ 4 & 5 \end{bmatrix} \times \frac{1}{\sqrt{2}} \begin{bmatrix} 1 \\ 1 \end{bmatrix} = \frac{1}{\sqrt{90}} \begin{bmatrix} 3 \\ 9 \end{bmatrix} = \frac{1}{\sqrt{10}} \begin{bmatrix} 1 \\ 3 \end{bmatrix}$$

$$u_2 = \frac{1}{\sigma_2} A v_2 = \frac{1}{\sqrt{5}} \begin{bmatrix} 3 & 0 \\ 4 & 5 \end{bmatrix} \times \frac{1}{\sqrt{2}} \begin{bmatrix} -1 \\ 1 \end{bmatrix} = \frac{1}{\sqrt{10}} \begin{bmatrix} -3 \\ 1 \end{bmatrix} = \frac{1}{\sqrt{10}} \begin{bmatrix} -3 \\ 1 \end{bmatrix}$$

$$\text{Left singular vector} = U = \frac{1}{\sqrt{10}} \begin{bmatrix} 1 & -3 \\ 3 & 1 \end{bmatrix}$$

$$U \Sigma V^T = \frac{1}{\sqrt{10}} \begin{bmatrix} 1 & -3 \\ 3 & 1 \end{bmatrix} \begin{bmatrix} \sqrt{45} & 0 \\ 0 & \sqrt{5} \end{bmatrix} \frac{1}{\sqrt{2}} \begin{bmatrix} 1 & 1 \\ -1 & 1 \end{bmatrix}$$

The unique information stored in original matrix A can be measured using the rank of an original matrix. Higher the rank, more is the information content of a matrix. Eigenvectors give directions of maximum spread and variance of data present in matrix A. A high-rank matrix can be converted into a low-rank matrix without losing much image information. Figure 4.6 shows the different versions of image represented using different ranks. One can observe that there is very low visual degradation of an image.

In Ref. [28], the authors estimated the parameters of an analogue circuit model. They used the SVD of voltage samples from the oscilloscope. Singular values and corresponding right eigenvectors were used by them for prediction. In Ref. [29], the authors identified vascular lumen by using SVD. They applied SVD filtering on velocity maps computed using an autocorrelation method.

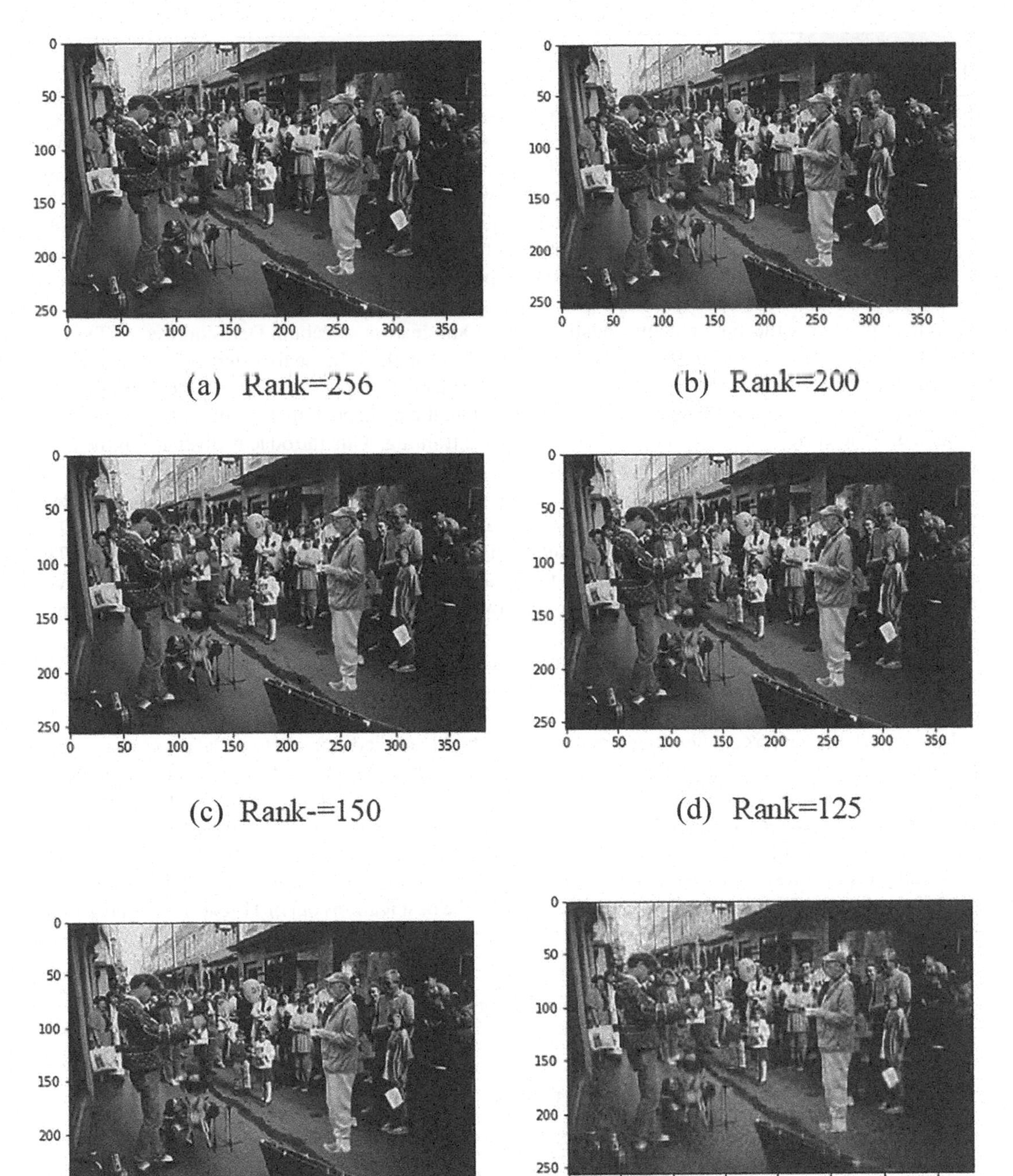

FIGURE 4.6 Low-rank representation of an image. (a) Rank=256, (b) rank=200, (c) rank=150, (d) rank=125, (e) rank=100 and (f) rank=50.

4.10 Random Forest

Random forest (RF) [30], also called an ensemble tree, is used as a supervised classifier. It can also be used for feature selection. Initially, a set of trees is created to predict target classes. In subsequent phases, usage statistics of each feature is used to find the subset of most explanatory features. For example, we can create a huge set of trees with very low height (two levels). Each tree may have been trained with a small portion of total number of features (three features). If a feature is frequently identified as best split, it is more likely to get retained. In the RF algorithm, based on the feature usage statistics, a score is calculated for each feature. This score tells the relative predictive quality of each feature. RFs can be used to classify credit scores, disease prediction, suspicious activities, etc. It can also be used to build applications for image classification, recommendation system and feature selection. Decision tree (DT) is the basic building block of a RF. To build a RF, several steps are applied. In the first step, bootstrap aggregation also called bagging is applied to the given dataset. During bagging, training instances are sampled many times to generate random datasets. During this step, it may happen that certain data instances get repeated while certain instances may not be used while training. This introduces diversity in the database. In the next step, multiple DTs are built using a random subset of features. The individual DTs can be built using the Gini index, gain ratio, information gain or any other impurity measures. Each tree is built using an independent random sample. During classification, a sample is tested by using all the DTs. Votes assigned by each DT are accumulated, and a highly voted decision is assigned to the sample. The contribution of each feature in the DT is used to calculate the relevance score. This relevance score can be used to find the least useful feature. Figure 4.7 shows the importance score of various features generated by applying RF to the data from Ref. [13].

Dataset was divided as 70% for training and 30% for testing. The accuracy of the model is 0.933. Figure 4.7 shows that sepal_width has the least importance in decision-making. Hence, we have removed this feature and reapplied RF. The resultant model accuracy increases to value 0.952.

In Ref. [31], the authors applied RF classifier only on highly ranked features to make predictions. They obtained very promising results on ten benchmark datasets. They presented a highly accurate method for predicting ten different diseases. In Ref. [32], the authors used RFs to define similarity between eleven well-characterized nanomaterials. They defined similarity using only those attributes showing the highest correlation with the activity. For further improvement, they used recursive feature elimination. They achieved their best result with the balanced accuracy of 0.8, using a reduced RF model. Features redox potential, zeta potential and dissolution rate came out as the best features out of 11 features for predicting impact on biological nanomaterial activity.

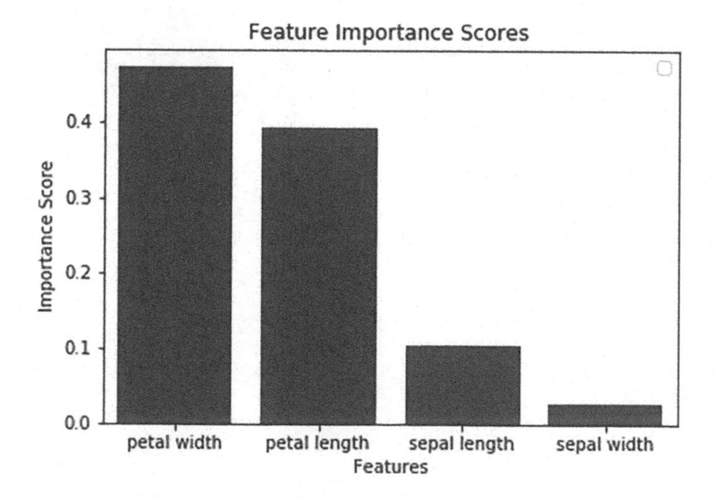

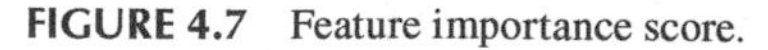
FIGURE 4.7 Feature importance score.

4.11 Conclusion

ML techniques have proven applications to various fields starting from finance to mechanical engineering. Huge data dimension is a curse to the ML techniques, and dimensionality reduction techniques are possible solutions to this curse. Mostly, correlation among features is used to reduce data dimension. Techniques such as low variance filters and high covariance filters use one or two features at a time for analysis. These techniques directly use variance/covariance to remove features. Dimensionality reduction techniques such as forward/backward sequential removal techniques analyse the impact of one feature at a time. However, variance and covariance have an indirect impact on the performance of these techniques. Techniques such as PCA, ICA and SVD transform this covariance in different domains. This domain conversion helps to visualize the data on lower dimensions.

Although techniques such as PCA, ICA and SVD are heavily used in image and audio processing, the authors have neglected its main significance in dimensionality reduction. Often, RF is overlooked for the dimensionality reduction. Combination of multiple dimensionality reduction techniques may lead to better performance.

REFERENCES

1. L. O. Jimenez and D. A. Landgrebe. 1998. Supervised classification in high dimensional space: Geometrical, statistical, and asymptotical properties of multivariate data. *IEEE Transactions on Systems, Man, and Cybernetics, Part C (Applications and Reviews)* 28, no. 1: 39–54.
2. D. L. Donoho and C. Grimes. 2003. Hessian eigenmaps: Locally linear embedding techniques for high-dimensional data. *Proceedings of the National Academy of Sciences* 100, no. 10: 5591–5596.
3. G. E. Hinton and R. R. Salakhutdinov. 2006. Reducing the dimensionality of data with neural networks. *Science* 313, no. 5786: 504–507.
4. S. Lafon and A. B. Lee. 2006. Diffusion maps and coarse-graining: A unified framework for dimensionality reduction, graph partitioning, and data set parameterization. *IEEE Transactions on Pattern Analysis and Machine Intelligence* 28, no. 9: 1393–1403.
5. S. T. Roweis and L. K. Saul. 2000. Nonlinear dimensionality reduction by locally linear embedding. *Science* 290, no. 5500: 2323–2326.
6. O. Arfaoui and M. Sassi. 2011. Fuzzy clustering of large-scale data sets using principal component analysis. *IEEE International Conference on Fuzzy Systems*, Taipei, Taiwan, 683–690.
7. A. V. Mire, S. B. Dhok, N. J. Mistry, and P. D. Porey. 2016. Tampering localization in digital image using first two digit probability features. In: Satapathy S., Mandal J., Udgata S., Bhateja V. (eds) *Information Systems Design and Intelligent Applications.* Advances in Intelligent Systems and Computing 435. Springer, New Delhi. doi.org/10.1007/978-81-322-2757-1_15.
8. J. Lin, N.H. Li, M. A. Alam, and Y. Ma. 2020. Data-driven missing data imputation in cluster monitoring system based on deep neural network. *Applied Intelligence* 50: 860–877. https://doi.org/ 10.1007/s10489-019-01560–y.
9. P. J. Garcia, L. Jose, L. Sancho-Gomez, J. L. Sancho-Gomez, and A. R. Figueiras-Vidal 2010. Pattern classification with missing data: A review. *Neural Computing and Applications* 19, no. 2: 263–282.
10. http://users.stat.ufl.edu/~winner/data/cruise_ship.dat.
11. https://medium.com/towards-artificial-intelligence/machine-learning-model-for-recommending-the-crew-size-for-cruise-ship-buyers-6dd478ad9900.
12. F. R. S. Karl Pearson. 1901. LIII. On lines and planes of closest fit to systems of points in space, *The London, Edinburgh, and Dublin Philosophical Magazine and Journal of Science* 2, no. 11: 559–572. doi: 10.1080/14786440109462720.
13. https://www.kaggle.com/uciml/iris.
14. J. Pallant. 2010. *SPSS Survival Manual: A Step by Step Guide to Data Analysis Using SPSS*, 4th Edition. Open University Press/McGrawHill, Maidenhead.
15. Z. Shen, Y. Zhang, and W. Chen. 2019. A Bayesian classification intrusion detection method based on the fusion of PCA and LDA. *Security and Communication Networks.* https://doi.org/ 10.1155/2019/6346708.

16. T. R. Gadekallu, N. Khare, and S. Bhattacharya. 2020. Early detection of diabetic retinopathy using PCA-firefly based deep learning model. *Special Issue on Deep Neural Networks and Their Applications Electronics* 9, no 2: 274. doi: 10.3390/electronics9020274.
17. N. S. Ankita. 2019. Improved link prediction using PCA. *International Journal of Analysis and Applications* 17, no. 4: 78–585. doi: 10.28924/2291-8639.
18. H. Aapo, K. Juha, and O. Erkki. 2001. *Independent Component Analysis*, 1st Edition. John Wiley & Sons, New York.
19. https://github.com/corymaklin/ida/blob/master/ica.ipynb.
20. J. E. Garcia-Bracamonte, J. M. Ramirez-Cortes, J. de Jesus Rangel-Magdaleno, P. Gomez-Gil, H. Peregrina-Barreto, and V. Alarcon-Aquino. 2019. An approach on MCSA-based fault detection using independent component analysis and neural networks. *IEEE Transactions on Instrumentation and Measurement* 68, no. 5: 1353–1361.
21. X. Bi, S. Cao, and D. Zhang. 2019. A variety of engine faults detection based on optimized variational mode decomposition-robust independent component analysis and Fuzzy C-mean clustering. *IEEE Access* 7: 27756–27768.
22. D. Kostrzewa and R. Brzeski. 2018. The data dimensionality reduction in the classification process through greedy backward feature elimination. *Man-Machine Interactions 5: 5th International Conference on Man-Machine Interactions, ICMMI 2017*, Kraków, Poland. Advances in Intelligent Systems and Computing 659, Springer, Cham.
23. S. Balakrishnan, R. Narayanaswamy, N. Savarimuthu, and R. Samikannu. 2008. SVM ranking with backward search for feature selection in type II diabetes databases. *IEEE International Conference on Systems, Man and Cybernetics*, Singapore, 2628–2633.
24. A. Jeswani, S. More, K. Kapoor, S. Sheikh, and R. Mangrulkar. 2020. Document summarization using graph based methodology. *International Journal of Computer Applications Technology and Research* 9, no. 8: pp. 240–245. doi: 10.7753/ijcatr0908.1005.
25. B. Mahapatra and S. Patnaik. 2015. Data reduction in MANETs using forward feature construction technique. *International Conference on Man and Machine Interfacing (MAMI)*, Bhubaneswar, 1–3.
26. C. Gacav, B. Benligiray, and C. Topal. 2016. Sequential forward feature selection for facial expression recognition. *24th Signal Processing and Communication Application Conference (SIU)*, Zonguldak, 1481–1484.
27. G. W. Stewart. 1993. On the early history of the singular value decomposition. *SIAM Review* 35, no 4: 551–566, doi: 10.1137/1035134.
28. O. Datcu, R. Hobincu, and C. Macovei. 2019. Singular value decomposition to determine the dynamics of a chaotic regime oscillator. *International Semiconductor Conference (CAS)*, Sinaia, Romania: 119–122.
29. R. Nagaoka and H. Hasegawa. 2019. Identification of vascular lumen by singular value decomposition filtering on blood flow velocity distribution. *Journal of Medical Ultrasonics* 46: 187–194. doi: 10.1007/s10396-019-00928-4.
30. T. K. Ho. 1995. Random decision forests. *Proceedings of 3rd International Conference on Document Analysis and Recognition*, 1: 278–282. doi: 10.1109/ICDAR.1995.598994.
31. A. M. Zahangir, M. Saifur Rahman, and S. Rahman. 2019. A random forest based predictor for medical data classification using feature ranking author links open overlay panel. *Informatics in Medicine Unlocked* 15: 100180.
32. A. Bahl, B. Hellack, M. Balas, A. Dinischiotu, J. Brinkmann, and B. Y. Renard. 2019. Recursive feature elimination in random forest classification supports nanomaterial grouping. *NanoImpact* 15: 100179.

5

Application of Deep Learning in Counting WBCs, RBCs, and Blood Platelets Using Faster Region-Based Convolutional Neural Network

Nirav Jain, Shail Shah, Ramchandra S. Mangrulkar, and Pankaj Sonawane
Dwarkadas J. Sanghvi College of Engineering

CONTENTS

5.1 Introduction 77
5.2 Convolutional Neural Network (CNN) 79
 5.2.1 Activation Function 79
 5.2.2 Convolutional Function 79
 5.2.3 Pooling Function 80
 5.2.4 Fully Connected Layer 81
5.3 Region-Based Convolutional Neural Network (RCNN) 81
 5.3.1 RCNN Architecture 82
5.4 Fast RCNN 82
5.5 Faster RCNN 83
 5.5.1 Region Proposal Network (RPN) 84
 5.5.2 Faster RCNN Architecture 84
5.6 Implementation 85
 5.6.1 Steps for Implementation 85
 5.6.2 Some Problems, Solutions, and Suggestions 96
5.7 Viability of the Solution 97
References 98

5.1 Introduction

Since the last few decades, neural networks (NN) have become quite popular due to the advancements in DL. This popularity has led to many new and robust algorithms in the field of computer vision. New doors have been opened, and new paths have been laid down for further advancement in this field. This technology is finding its way into many real-life applications. One such application that this chapter will be discussing is the CBC test using Faster RCNN.

In clinical pathology tests such as CBC, the precise counting of blood cells is crucial. Such tests have been of immense importance in diagnosing a large number of diseases such as infections, inflammations, dengue, malaria, and even deadly diseases like cancer, leukemia (Daqqa et al. 2017, 638–643), and bone marrow failure. Due to its application in many essential areas, automating or, more precisely, speeding up the process of CBC tests seems to be of immense importance (Cruz et al 2017; Habibzadeh and Fevens 2013, 263–274). The WBC count in the blood directly reflects the strength of an immune system. They are considered to be the defense blocks of the body. An enormous change in WBC count can negatively affect the body. A very high WBC count occurs when the body tries to fight a disease, causing problems like fevers and night sweats. In contrast, a lower WBC count indicates a lower number of defense blocks to defend you against the invading disease, making you an easy target for such diseases.

Traditionally, techniques such as "Coulter Counters and Laser Flow Cytometry" were used, which required a very complicated, costly, and time-consuming system for performing the CBC test. Modern techniques such as automated hematology counter are very expensive to be used in rural areas. The automated CBC test uses the latest bioengineering techniques to give an accurate and faster result compared to the manual CBC test. Sysmex XT-2000i is the latest machine used for calculating the CBC. It is the successor of the famous XE-2100 and examines blood samples at the rate of 80 samples/hour. Also, counting cells using image processing (Kaur and Garg 2016, 2574–2577; Meimban et al 2018, 50–53) alone is not feasible as it fails to generalize the model. Hence, an effective replacement of the above methods is required.

The blood count is the first step toward discovering a disease or any unnatural behavior by the body. The blood consists of three major types of cells: RBCs or erythrocytes, WBCs or leukocytes, and blood platelets or thrombocytes. With the CBC, the disease is diagnosed, and then, the doctors take further actions accordingly. The CBC is required during a patient's treatment to monitor the progress and the condition of the patient (George-Gay and Parker 2003, 96–114; Meintker and Krause 2013, 641–650).

A standard CBC contains the following information:

1. RBC Test:
   - A. The number of RBCs in the blood sample
   - B. The amount of hemoglobin present in the blood sample
   - C. The percentage of the volume of RBC in the total volume of the blood sample (hematocrit)
   - D. The measurement of the mean size of the RBC known as mean corpuscular volume
   - E. The calculated average of hemoglobin known as mean corpuscular hemoglobin
   - F. The hemoglobin concentration
   - G. The distribution of the RBCs
   - H. The measurement of young RBCs in the blood sample.
2. WBC Test:
   - A. The number of WBCs in the blood sample
   - B. Depending on the number of WBCs in the sample, further count of individual cells is required
3. Blood Platelet Test:
   - A. The number of blood platelets in the blood sample
   - B. The average size of the blood platelets known as mean platelet volume
   - C. The distribution of the blood platelets in the blood.

There are many ways to calculate CBC but are slow, and others are rather too expensive for the common people. There are two ways of conducting CBC, automated and manual. Both these methods have their share of disadvantages. For instance, the machines used for automated counting are costly. Also, the minute variations in the cells' shape and irregularities are not detected in this method.

On the other hand, manual techniques are time-consuming wherein a pathologist counts the number of cells under a microscope and approximates them for the total count. The hemocytometer is used to perform blood calculations manually in the laboratory. It requires a trained professional to check the blood sample under a microscope and deliver accurate results. It requires more time than the automatic system as a person is involved, and thus, the throughput reduces considerably. A rate of 3–4 samples/hour is achieved by manual tests depending on the individual's speed. Flagging of a laboratory test result demands a labor-intensive manual examination of a blood smear. It is also not possible to create comments on red cell morphology. Platelet lumps are counted as one resulting in low platelet counts. Consequently, the results are error-prone. Furthermore, it is expensive. Hence, an efficient solution to the problem is desired.

The work presented uses the power of a CNN to develop the model to count the blood cells. It uses an optimized version of CNN, known as the Faster RCNN, which processes the images quickly and efficiently. So, this chapter first discusses how DL algorithms can be used to solve the discussed problem. The chapter then explains each of the algorithms, namely RCNN, Fast RCNN, and Faster RCNN, in brief. It then explores the type of algorithm suitable for our problem and the various aspects of data collection, data preprocessing, creating the model, training, testing, and evaluating our models, which are the steps required for successful training of a DL model. The chapter then discusses the solution to the various problems that are faced during or after training the model. The chapter provides a complete methodology to accomplish the above steps. Finally, the chapter discusses the viability of the above solution in the real world.

5.2 Convolutional Neural Network (CNN)

CNN is a part of deep neural network, mainly aimed at developing machine learning (ML) models on visual imagery, which is also referred to as the shift-invariant artificial neural network (SIANN) due to a shared weight architecture. CNN is a concept developed by comparison to a biological process, and the connectivity resembles that of the animal visual cortex. CNN is a fully connected network of nodes, which means every node is connected to every other node in the next layer. CNN has different layers, which is the key to identifying the distinguishing features in the images. The completeness of the networks makes it less susceptible to overfitting of data. CNN uses a special type of linear operation, named convolution instead of traditional matrix multiplication in at least one of their layers (Krizhevsky et al. 2017, 1097–1105) as shown in Figure 5.1.

5.2.1 Activation Function

CNN, unlike other DL models, does not only use activation functions like

Tanh, sigmoid, rectified linear unit (ReLU), or Leaky ReLU

But the convolutional layers or the hidden layers in the CNN also use two different functions, namely

- Convolutional function
- Pooling function

5.2.2 Convolutional Function

Convolutional function is a linear function used as an activation function in CNN. It takes in two tensors and returns a modified third tensor. It can be viewed as the application of a filter to the image in order to generate a new image.

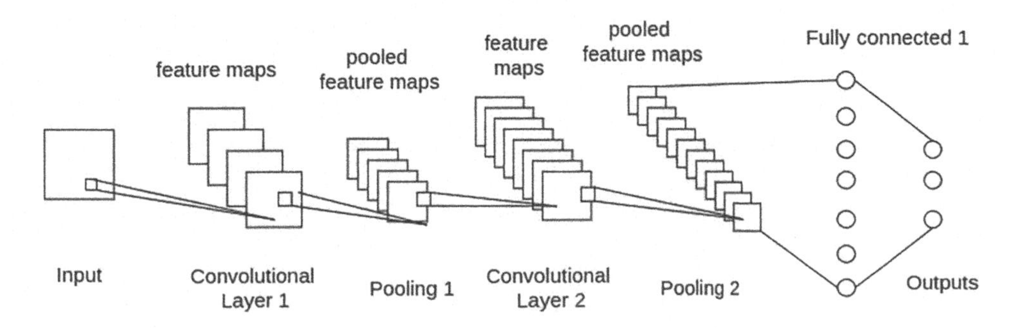

FIGURE 5.1 Convolutional operation.

The second tensor is known as the kernel, which is generally smaller as compared to the input image. The kernel moves across the image and generates a resultant image based on the dot product operation.

The mathematical formula for the convolutional function can be as follows:

$$(f*g)(i) = \sum_{j=1}^{m} g(j) f(i - j + m/2)$$

Figure 5.2 depicts the working of a convolutional function applied on an image.

5.2.3 Pooling Function

Pooling is used to make assumptions regarding the features of the input image, based on sample discretization. It is another type of function used in the CNN. It is majorly used to decrease the input matrix's dimensionality, which means if the input matrix is $n \times n$, the pooling will reduce it to $m \times m$ where $m < n$.

There are two types of pooling:

- **Max Pooling**: The maximum element is taken from the matrix of kernel size and is added to the output image. In Figure 5.3, the kernel size is 4.

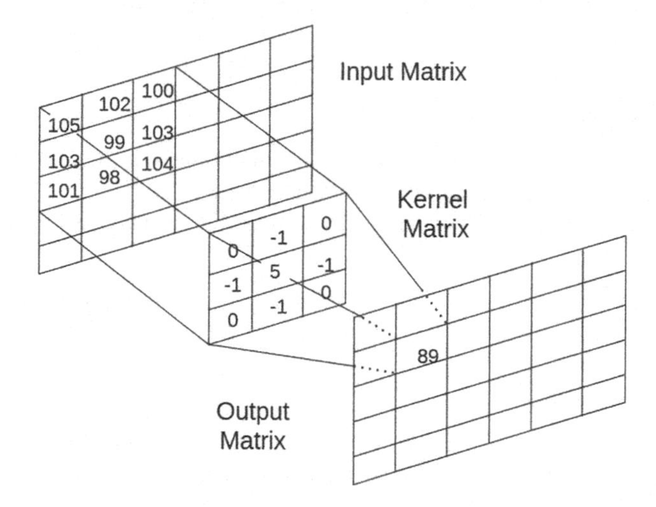

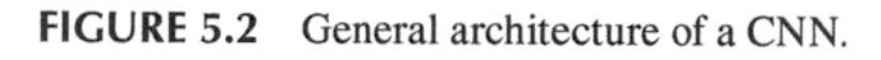
FIGURE 5.2 General architecture of a CNN.

Max-pooling

13	19	44	32
7	11	45	33
56	46	23	39
44	30	34	37

19	45
56	39

FIGURE 5.3 Max-pooling operation.

Min-pooling

13	19	44	32
7	11	45	33
56	46	23	39
44	30	34	37

→

7	32
30	23

FIGURE 5.4 Min pooling operation.

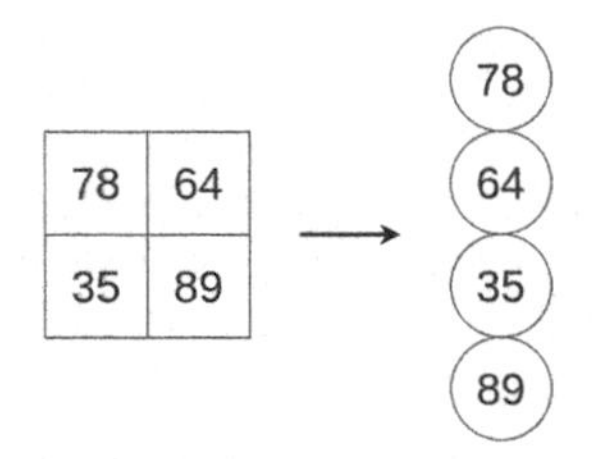

FIGURE 5.5 Flattening operation.

- **Min Pooling**: The minimum element is taken from the matrix of kernel size and is added to the output image. In Figure 5.4, the kernel size is 4.

5.2.4 Fully Connected Layer

The fully connected layer takes the output of the last pooling or convolutional layer as an input. It then flattens the resulting tensor into a single vector to ease the process of training for the further NN. The result is then fed to the fully connected layers.

The word flatten means that the final matrix is reduced from a three-dimensional matrix to a one-dimensional vector, as shown in Figure 5.5.

The final few fully connected networks behave as an artificial neural network (ANN) and have the same functionality as an ANN.

Each layer calculates the following:

$$g(Wx+b)$$

where

x is the input vector with dimension [m, 1]

W is the weight matrix with dimension [m, n] where m is the number of neurons in the previous layer and n is the number of neurons in the next layer

b is the bias vector with dimension [m, 1]

g is the activation function (usually ReLU)

5.3 Region-Based Convolutional Neural Network (RCNN)

RCNNs or regions with CNN features are modified ML algorithms that aim to detect multiple objects in a single image. RCNN divides the whole image into regions based on a selective search algorithm, a fixed algorithm that does not learn or adapt itself (Zhang et al 2020; Weng 2017).

The RCNN algorithm follows the given steps to classify data and detect object:

1. Divide the image into regions with the help of the selective search algorithm.
2. A pre-trained CNN is then applied on the output of the selective search algorithm and CNN converts the region into a suitable form for further processing. Then, it generates an output of classified features of the regions.
3. Results from several regions are combined to train the support vector machine (SVM) for the classification of objects, where a certain SVM decides whether an object belongs in a particular category or not.
4. All the features and bounding-box labels of all the regions are combined to train a linear regression model.

5.3.1 RCNN Architecture

Figure 5.6 depicts the general architecture of RCNN. There are various algorithms used in order to identify the regions optimally and to increase the accuracy of RCNN, namely:

1. **Non-Max Suppression**: Non-max suppression deals with the extra bounding boxes associated with the same object. It follows a *greedy* approach and recursively sorts the bounding boxes based on the confidence scores and discards the one with the lowest score. It ignores the boxes with high Intersection over Union (IoU) (>0.5) with the selected box. This helps in the removal of excess bounding boxes for the same object.
2. **Hard Negative**: The regions with no object, that is, the area possessing the background, noise, and other parts, except objects, are termed as negative regions. Some of these negative regions are easy to classify, like the region containing the background, known as "easy negatives," while other negatives that contain some parts of the object and noise are classified as "hard negatives." The "hard negatives" usually decrease the accuracy of the model. Thus, after classifying them, they can be used to train and evaluate the model to decrease the false positives and thus increase the model's accuracy.

5.4 Fast RCNN

The major drawback of RCNN is that it is a very slow approach as 2000 regions are passed on to the CNN for feature extraction and classifying the image. A faster algorithm known as Fast RCNN

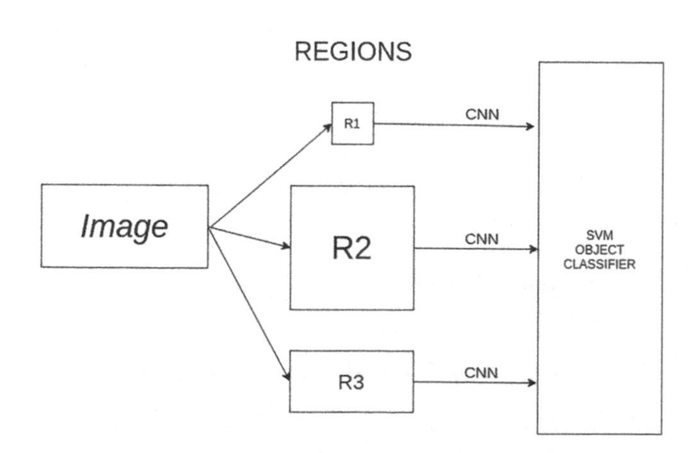

FIGURE 5.6 Architecture of RCNN.

was developed to overcome this drawback. Fast RCNN decreases the computational time to a very great extent. Instead of feeding the regions to the CNN (in the case of RCNN), the input image itself is fed to the CNN. This way, it saves a considerable amount of computational time (Girshick 2015, 1440–1448).

The steps involved in the process of object detection with Fast RCNN are as follows:

1. The input image (size-independent) is fed to a CNN.
2. The CNN extracts the convolutional feature map from the input image and extracts regions of interest or ROIs from the feature map.
3. The regions obtained from the feature map are then passed on to the ROI pooling layer, which resizes them to a constant size.
4. The output of the ROI pooling layer needs to be of the same size to transfer the output to the fully connected layer for further processing.
5. The softmax classifier, along with linear regression, is used to classify the final output obtained.

Note that in Fast RCNN, the input image is fed to the network, unlike RCNN, where the region proposals are fed to the network, as shown in Figure 5.7.

5.5 Faster RCNN

Faster RCNN is an improvised version of the Fast RCNN. Object detection by the Faster RCNN network is composed of a feature extraction network, which is usually a pre-trained CNN. This pre-trained model is followed by two subnetworks, which are trained on the data.

The first subnetwork is a region proposal network (RPN), which is used to predict the object proposals or the regions from the input image. It is an alternative to the selective search algorithm, which is used in the Fast RCNN network. The second subnetwork is a classifier and is used to predict the actual class of the object. Faster RCNN uses ROI pooling and classifier and bounding-box regressor similar to that of the Fast RCNN (Ren and Sun 2015, 1137–1149). Hence, Faster RCNN is a much faster and efficient algorithm than the previous RCNNs. The following sections explain in detail the working of the RPN and the complete architecture of Faster RCNN.

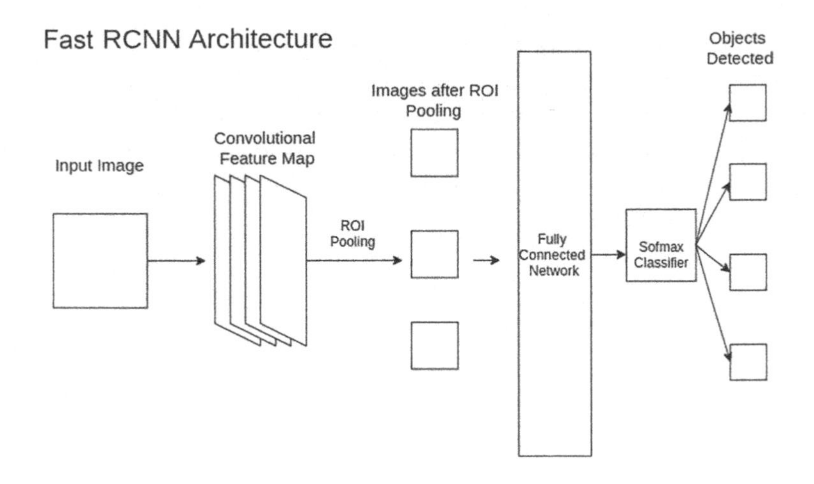

FIGURE 5.7 Architecture of Fast RCNN.

5.5.1 Region Proposal Network (RPN)

It is an intelligent substitute for the more naive selective search algorithm. It uses a CNN, which is used to recognize regions. It not only just decreases the per-image time from 2 s (in case of selective search) to 10 ms in RPN but also shares some features with the detection stage, which improves the efficiency of the overall algorithm.

The steps involved in RPN are as follows:

1. The input image is resized to 600×1000 dimensions. Then, it is forwarded to the backbone CNN (in this chapter, ResNet_101), which gives an output in the form of HxW, which is smaller than the original image and generally depends on the stride used in the backbone CNN (16 in the proposed methodology).
2. RPN uses the concept of anchors, which are boxes of different scales and aspect ratios. They are used to predict the presence and location of an object in the image.
3. Every output pixel of the feature map consists of nine anchor boxes. These anchor boxes each contain a prediction indicating whether an object is present or not (foreground or background) and a prediction of the anchor boxes' coordinates, which are height, width, and center (h, w, x, y).
4. The last part of the RPN is the predictor. There are two types of predictor used in the RPN: the first is the classifier used to predict whether an object is present in the image or not, and the second is the regressor used to predict the coordinates of the anchors used.
5. The RPN's output contains both the classification and the corresponding bounding-box coordinates, which are then forwarded to the ROI pooling layer for further processing, as shown in Figure 5.8.

The RPN's output is the prediction of whether the anchor box, having coordinates given by the regressor (output of RPN), contains an object or not.

5.5.2 Faster RCNN Architecture

The steps involved in object detection by Faster RCNN are as follows:

1. Input image is passed through a CNN, which returns the feature map of the image.
2. The feature map is then passed to the RPN to get the region proposals for further processing.
3. The RPN processes the feature map and returns the coordinates of the proposal and whether it contains the object.
4. The RPN output is then passed to the ROI pooling layer, which converts all the proposals to the same size to make it feasible for the fully connected network.
5. The output of the ROI pooling is passed on to the fully connected layer.
6. The fully connected network output is then passed onto the softmax layer to classify the proposals as objects of different classes. Also, a regressor is used to improve the coordinates of the bounding boxes of the proposals, as shown in Figure 5.9.

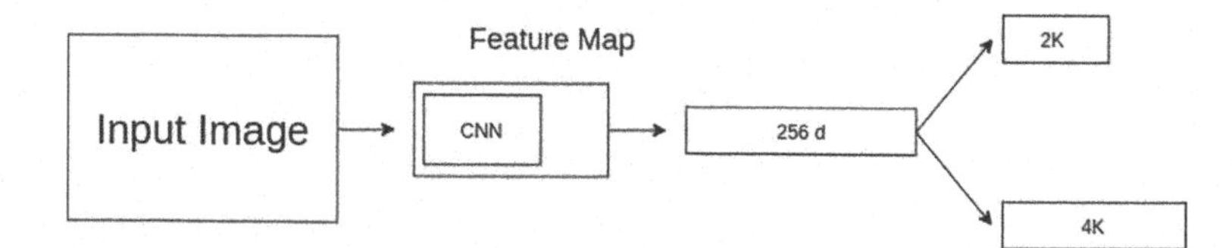

FIGURE 5.8 Architecture of RPN.

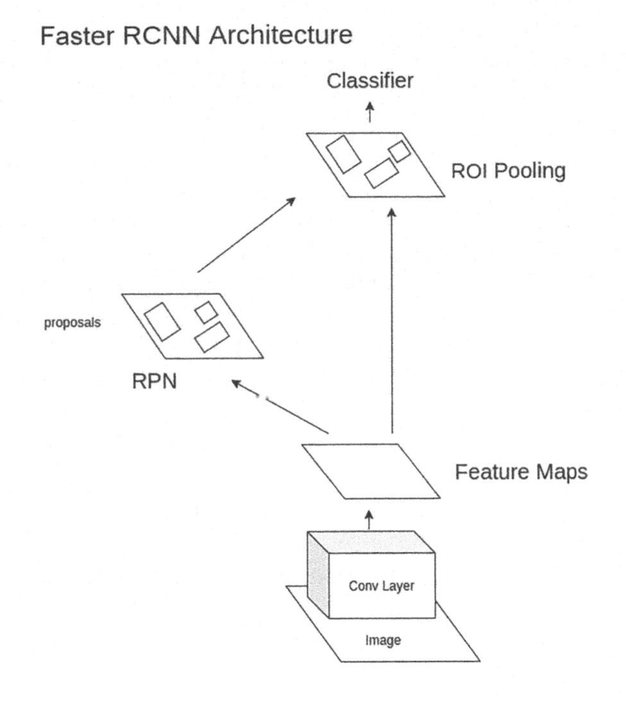

FIGURE 5.9 Architecture of Faster RCNN.

5.6 Implementation

As the chapter has discussed the problem, the different types of CNNs that can be used, and why the proposed methodology uses Faster RCNN, let's dive straight into its implementation. This chapter won't be going into too much detail about how Faster RCNN is actually implemented. Instead, this work will be using already implemented Faster RCNN architecture (in TensorFlow) by industry specialists. This section of the chapter gives detailed step-by-step instructions on using TensorFlow Object Detection API to train the model.

Before starting with the implementation, let's first understand what TensorFlow Object Detection API is. It is an open-source framework built by Google Brain, a DL artificial intelligence research team at Google, for making the tedious task of object detection easier. It is built on top of TensorFlow (Python library for DL) and provides many easy-to-use tools to implement DL models without the need to write huge amounts of code. It makes the task of constructing, training, and deploying new models way easier than implementing it from scratch. Also, it is implemented by industry specialists and is open source. Hence, the chances of getting an error are way too low. Even if there are any errors, the open-source community is always there to help. It supports a variety of CNN architectures like Faster RCNN and single shot detector (SSD). Still, it is highly recommended that the reader implements the Faster RCNN architecture (or at least a basic implementation of the architecture) if he/she understands it completely and has the required knowledge. The following section of the chapter discusses the necessary steps required for implementing the solution.

5.6.1 Steps for Implementation

- **Step 1: Collecting the dataset**
  Download the dataset prepared for the paper "Machine learning approach of automatic identification and counting of blood cells" (Alam and Islam 2019, 103–108). This dataset consists

of images that have been hand-labeled and annotated, and each image is of size 640×480. The training set consists of 300 images, whereas the validation and testing set consists of 60 images. As the dataset is very small, this chapter uses a subset of training data for validation.

- **Step 2: Cleaning the dataset**
  The first step after collecting the dataset is to clean the dataset according to the problem needs. As always, the first step is cleaning the dataset by quickly checking the bounding boxes and the distribution of RBCs, WBCs, and blood platelets. The dataset used in this chapter is already clean to a great extent. The only thing left is to augment the dataset. This is necessary as the dataset used for training is very limited. As the image's orientation does not matter for the problem being discussed, the dataset is augmented by flipping the images in every possible direction. Thus, the size of the training data is increased from 300 images to 1200 images. Other augmentations such as changing brightness levels, contrasts, and blurs can also be applied.
- **Step 3: Using Google Colab**
  Create a new notebook on Google Colab as shown in Figure 5.10.
- **Step 4: Folder Structure**
  The folder structure this chapter uses is as given in Figure 5.11:
- **Step 5: Setting Up the Data**
  As seen in the folder structure, the data is split into three ways: training, testing, and validation.

Training: 300 images and their annotations for training

Testing: 60 images and their annotations for testing

Validation: 60 images and their annotations for testing as raw data

Since the number of images for training is only 300, the data is augmented by flipping it into the remaining three directions and changing the augmentations accordingly, which is done in the data_gen.py.

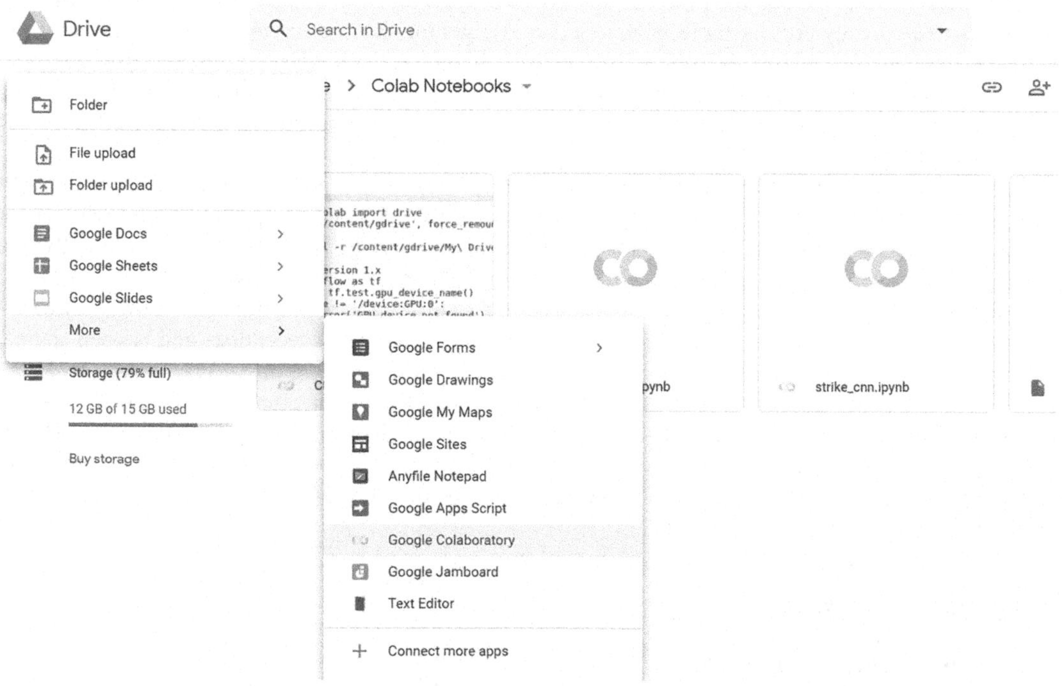

FIGURE 5.10 Creating a new notebook on Google Colab.

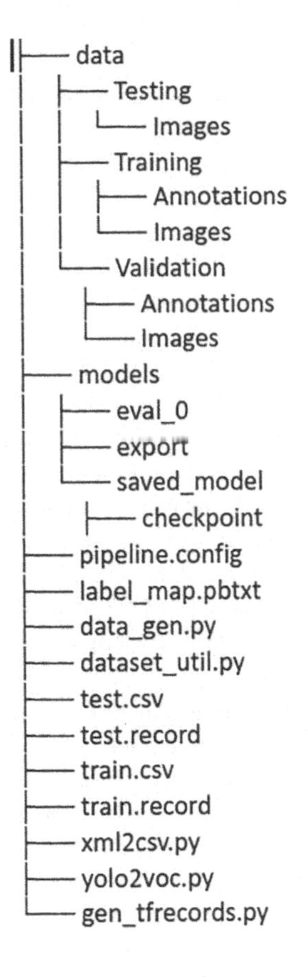

FIGURE 5.11 Directory structure.

This file includes the Python script to flip and mirror the images and then creates corresponding annotations so that there are four images for every single image. Thus, the total number of training images becomes 1200.

The model's folder stores the model checkpoints. It is initially empty.

- **Step 6: Understanding the different scripts and files required by TensorFlow Object Detection API**

  1. **label_map.pbtxt**

     This is a simple text file containing a mapping between classes and integers. It maps every prediction class to an integer in a dictionary format. The API uses this file for both training and detection purposes.

  2. **pipeline.config**

     This file contains the various hyperparameters and paths to various folders and files required for training. It contains hyperparameters such as the initial learning rate, batch_size, and size_of_anchor_boxes and paths to different record files, label_map files, etc. It is also the file that specifies the pre-trained model weights. This file is the soul of the model training, and all the tuning required is done here. The proposed work uses Resnet_101 pre-trained model for training.

  3. **train.csv and test.csv**

     These files contain complete information on the training and testing data. They include the images to be trained and information about each and every labeled bounding box the image has; i.e., it contains the following columns (filename, width, height, class, xmin, ymin, xmax, ymax) as shown in Figure 5.12.

	A	B	C	D	E	F	G	H
1	filename	width	height	class	xmin	ymin	xmax	ymax
2	BloodImage_00258.jpg	640	480	Platelets	257	162	300	202
3	BloodImage_00258.jpg	640	480	WBC	221	223	386	387
4	BloodImage_00258.jpg	640	480	Platelets	541	394	582	435
5	BloodImage_00233.jpg	640	480	RBC	1	164	89	257
6	BloodImage_00233.jpg	640	480	RBC	46	67	171	177
7	BloodImage_00233.jpg	640	480	RBC	289	88	396	196
8	BloodImage_00233.jpg	640	480	RBC	357	185	454	271
9	BloodImage_00233.jpg	640	480	RBC	97	320	230	441
10	BloodImage_00233.jpg	640	480	RBC	13	324	99	395

FIGURE 5.12 Sample CSV files.

These files are generated automatically using the script xml2csv.py, which converts the XML (VOC) files, i.e., the annotation files, into corresponding rows in the CSV. But the data might not always be in XML format. So whatever format the data might be in, it is converted directly to a CSV file, or it is converted to XML format, and then, the discussed script file is used. For example, if the data is in YOLO format, it is first converted to VOC using the script yolo2voc.py and then to CSV using xml2csv.py. These are basic Python scripts and do not require any special knowledge.

4. **train.record and test.record**

   The record files are used to store the data in the CSV files into a format suitable for training by the Object Detection API. It contains the complete dataset in one file, i.e., the images and their annotations all in one place. The only reason for creating the CSV files in this implementation is to convert them to record files using the script in the gen_tfrecords.py file. Thus, two record files are generated, one for training data and the other for testing data. The Object Detection API has several example scripts to generate tfrecord inside the folder "models/research/object_detection/dataset_tools/" which can be used to generate the required tfrecord files.

- **Step 7: Installing TensorFlow Object Detection API on Colab**

  Follow the steps mentioned on the official GitHub page of TensorFlow Object Detection API to install it correctly.

- **Step 8: Setting up TensorBoard (Python library to visualize the results)**

1. **Load TensorBoard**

```
%load_ext tensorboard
```

2. **Give TensorBoard the path where the model's checkpoints are stored to start it in Google Colab's cell** (Figure 5.13).

```
%tensorboard --
logdir "/content/gdrive/My Drive/tf_object_detection/Projects/CRC_Press
/models"
```

   Output:

   This board is a great tool to visualize the results.

   a. The Scalars tab shows different graphs related to the model's performance like precision, recall, and loss.
   b. The Images tab is used for visualizing the predictions on the validation set.
   c. The graphs tab can be used to visualize the whole TensorFlow Graph.
   d. Projector tab in Colab can't be viewed as of now, but it is used to represent high-dimensional embeddings graphically.

   All these results can be seen once the events file for TensorBoard has been created, i.e., once the training starts.

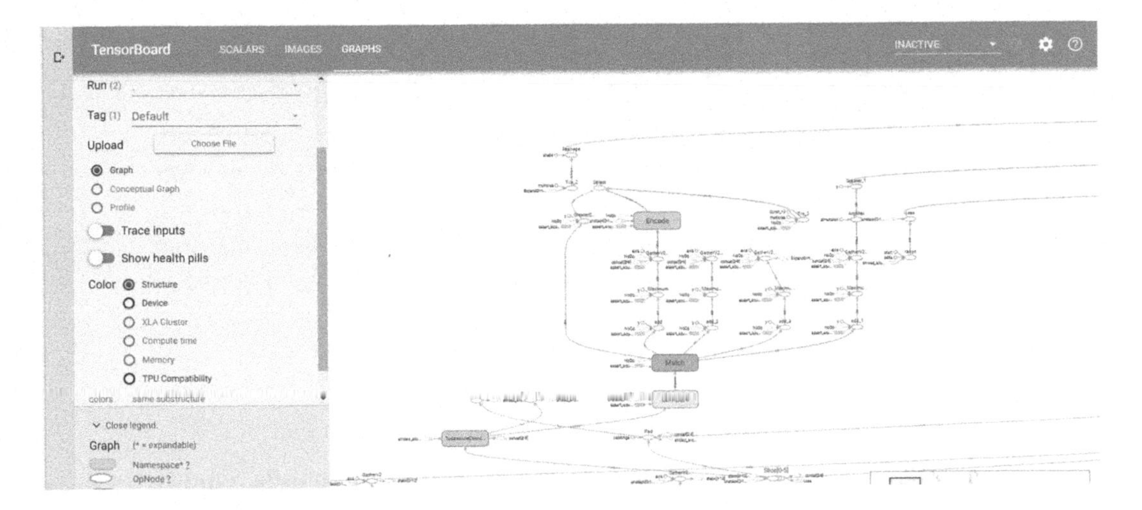

FIGURE 5.13 TensorBoard.

- **Step 9: Actual Training**
  Run the following command to start the actual training

```
!python object_detection/model_main.py \
    --pipeline_config_path=
    /content/gdrive/My\ Drive/tf_object_detection/Projects/CRC_Press/pi
    peline.config \
    --model_dir=
    /content/gdrive/My\ Drive/tf_object_detection/Projects/CRC_Press/mo
    dels \
    --num_train_steps=5000 \
    --sample_1_of_n_eval_examples=1 \
    --alsologtostderr
```

Key Terms:

pieline_config_path: the path to the config file of the model (i.e., pipeline.config file)

model_dir: path where the model checkpoints are to be saved. It is the path to the output directory (i.e., models directory)

num_train_steps: number of epochs for which the model is trained.

sample_1_of_n_eval_examples: As the name suggests, it will sample one of every n input examples. If set to 1, it will sample all the eval examples

alsologtostderr: It simply sends the logs to STDERR standard file

As the training continues, the output of the cell shows many lines. This chapter will go through the important ones here.

1. Output 1 of training cell
   a. In Figure 5.14, the global_step here denotes the number of batches seen by the graph. Therefore, global_step/sec simply refers to the number of batches seen by the graph per second.
   b. Step simply refers to the epoch number
   c. As the name suggests, loss is the total loss of the model and should be as low as possible.

```
INFO:tensorflow:global_step/sec: 2.38884
I0724 07:16:14.443737 139963347396480 basic_session_run_hooks.py:692] global_step/sec: 2.38884
INFO:tensorflow:loss = 0.47458196, step = 3900 (41.861 sec)
I0724 07:16:14.444873 139963347396480 basic_session_run_hooks.py:260] loss = 0.47458196, step = 3900 (41.861 sec)
INFO:tensorflow:global_step/sec: 6.41855
I0724 07:16:30.023600 139963347396480 basic_session_run_hooks.py:692] global_step/sec: 6.41855
INFO:tensorflow:loss = 0.39559072, step = 4000 (15.580 sec)
I0724 07:16:30.024762 139963347396480 basic_session_run_hooks.py:260] loss = 0.39559072, step = 4000 (15.580 sec)
INFO:tensorflow:global_step/sec: 6.43422
I0724 07:16:45.565520 139963347396480 basic_session_run_hooks.py:692] global_step/sec: 6.43422
INFO:tensorflow:loss = 0.56508684, step = 4100 (15.542 sec)
I0724 07:16:45.566784 139963347396480 basic_session_run_hooks.py:260] loss = 0.56508684, step = 4100 (15.542 sec)
INFO:tensorflow:global_step/sec: 6.5895
I0724 07:17:00.741263 139963347396480 basic_session_run_hooks.py:692] global_step/sec: 6.5895
INFO:tensorflow:loss = 0.6033262, step = 4200 (15.176 sec)
I0724 07:17:00.742460 139963347396480 basic_session_run_hooks.py:260] loss = 0.6033262, step = 4200 (15.176 sec)
INFO:tensorflow:global_step/sec: 6.61969
```

FIGURE 5.14 Output 1 of training cell.

2. Output 2 of training cell

   Figure 5.15 helps in the calculation of mean average precision (mAP). Now to calculate the mAP, it is necessary to know a few things. So, let's start with "maxDets." "maxDets" in the above figure simply means the maximum number of detections with the highest scores. The second important term is IoU. IoU simply suggests the correction of each bounding box. As the name suggests, it is the ratio of the area of intersection to the area of union between the ground truth bounding box and the predicted bounding box as shown in Figure 5.16.

   The area of intersection is the common area between the two bounding boxes, whereas the area of union is the total area covered by the two bounding boxes. Now to calculate precision or recall, a few metrics like true positives (TP), false positives (FP), true negatives (TN), and false negatives (FN) are required. A detection is said to be true positive if its IoU is above a specified threshold; else, it is a true negative. The most commonly used threshold is 0.5. True negatives simply refer to every part of the image where there are no bounding boxes, and false negatives are simply those parts of the image where the model misses out the detections. As all the four required variables to calculate precision and recall are available, the following formulae to calculate precision and recall for the detections can be used.

$$\text{Precision} = \frac{\text{TP}}{\text{TP} + \text{FP}}$$

$$\text{Recall} = \frac{\text{TP}}{\text{TP} + \text{FN}}$$

To calculate mAP, as discussed in the paper "The PASCAL Visual Object Classes (VOC) Challenge" (Everingham et al. 2009, 303–338), precision for different model thresholds where the recall normally has the values 0, 0.1, 0.2, … 1.0 is calculated; i.e., 11 precision values for 11 chosen thresholds are calculated. Hence, the mAP is the mean of all the average precision values across all the classes as measured above. In short, mAP is an approximation to the area under the precision–recall curve.

The model has been trained for only 20,000 steps. A significant problem while training the model for a large number of steps is overfitting. Usually, a good idea when to stop the model training is by looking at the different graphs on the TensorBoard, which helps us determine problems, if any, at the earlier steps of training.

The various graphs that can be visualized in TensorBoard are as follows:

1. Graphs of precision vs. steps (Figure 5.17)
2. Graphs of recall vs. steps (Figure 5.18)
3. Graphs for global_norm and global_steps (Figure 5.19)

```
index created!
Running per image evaluation...
Evaluate annotation type *bbox*
DONE (t=1.20s).
Accumulating evaluation results...
DONE (t=0.03s).
 Average Precision  (AP) @[ IoU=0.50:0.95 | area=   all | maxDets=100 ] = 0.453
 Average Precision  (AP) @[ IoU=0.50      | area=   all | maxDets=100 ] = 0.630
 Average Precision  (AP) @[ IoU=0.75      | area=   all | maxDets=100 ] = 0.547
 Average Precision  (AP) @[ IoU=0.50:0.95 | area= small | maxDets=100 ] = 0.000
 Average Precision  (AP) @[ IoU=0.50:0.95 | area=medium | maxDets=100 ] = 0.230
 Average Precision  (AP) @[ IoU=0.50:0.95 | area= large | maxDets=100 ] = 0.701
 Average Recall     (AR) @[ IoU=0.50:0.95 | area=   all | maxDets=  1 ] = 0.280
 Average Recall     (AR) @[ IoU=0.50:0.95 | area=   all | maxDets= 10 ] = 0.443
 Average Recall     (AR) @[ IoU=0.50:0.95 | area=   all | maxDets=100 ] = 0.488
 Average Recall     (AR) @[ IoU=0.50:0.95 | area= small | maxDets=100 ] = 0.000
 Average Recall     (AR) @[ IoU=0.50:0.95 | area=medium | maxDets=100 ] = 0.264
 Average Recall     (AR) @[ IoU=0.50:0.95 | area= large | maxDets=100 ] = 0.754
INFO:tensorflow:Finished evaluation at 2020-07-24-07:16:04
I0724 07:16:04.228512 139963347396480 evaluation.py:275] Finished evaluation at 2020-07-24-07:16:04
INFO:tensorflow:Saving dict for global step 3842: DetectionBoxes_Precision/mAP = 0.45312637, DetectionBoxes_Precision/mAP (large) = 0.7011403, DetectionBoxes_Precision/mAP (medium)
I0724 07:16:04.228791 139963347396480 estimator.py:2049] Saving dict for global step 3842: DetectionBoxes_Precision/mAP = 0.45312637, DetectionBoxes_Precision/mAP (large) = 0.701140
INFO:tensorflow:Saving 'checkpoint_path' summary for global step 3842: /content/gdrive/My Drive/tf_object_detection/Projects/CRC_Press/models/model.ckpt-3842
I0724 07:16:05.427005 139963347396480 estimator.py:2109] Saving 'checkpoint_path' summary for global step 3842: /content/gdrive/My Drive/tf_object_detection/Projects/CRC_Press/model
```

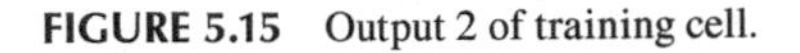

FIGURE 5.15 Output 2 of training cell.

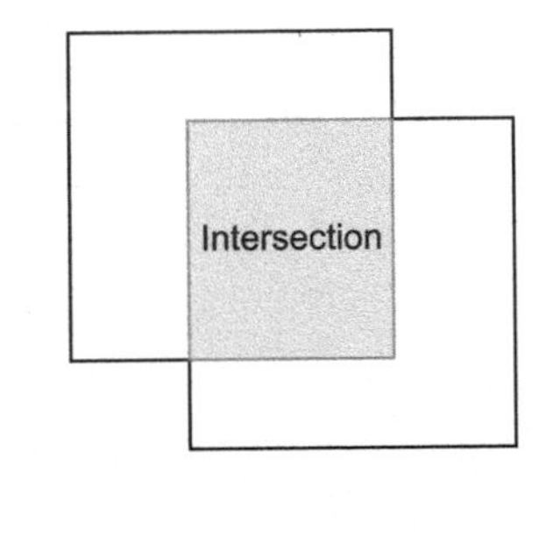

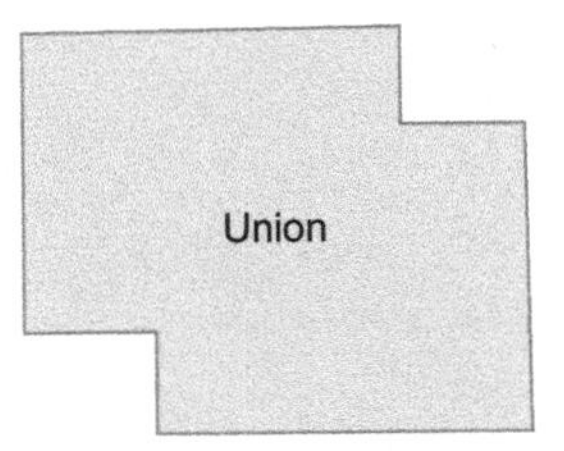

FIGURE 5.16 IoU operation.

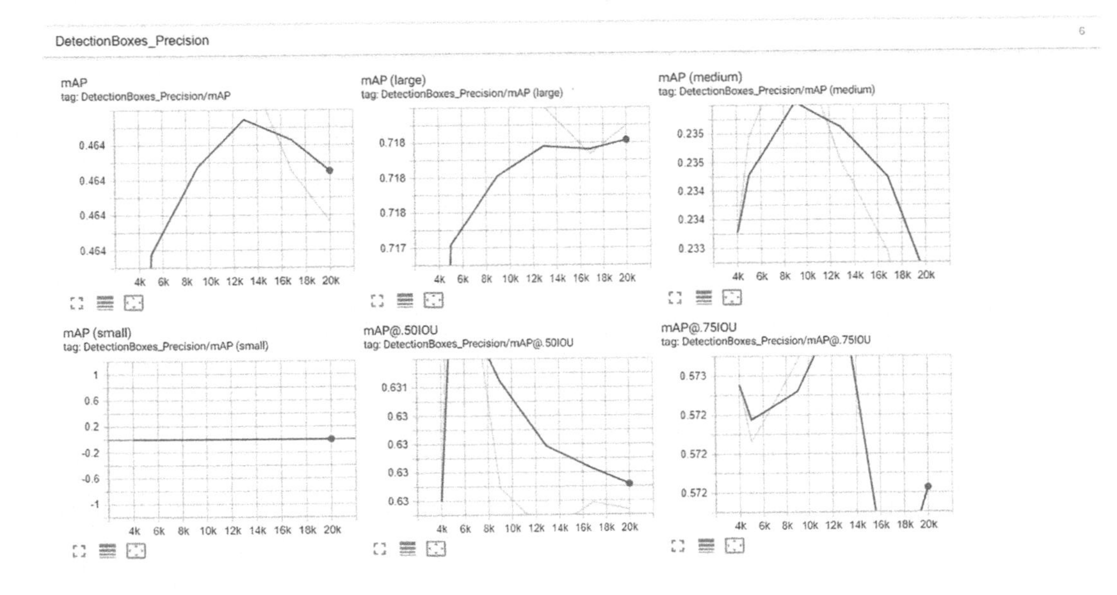

FIGURE 5.17 Precision vs steps.

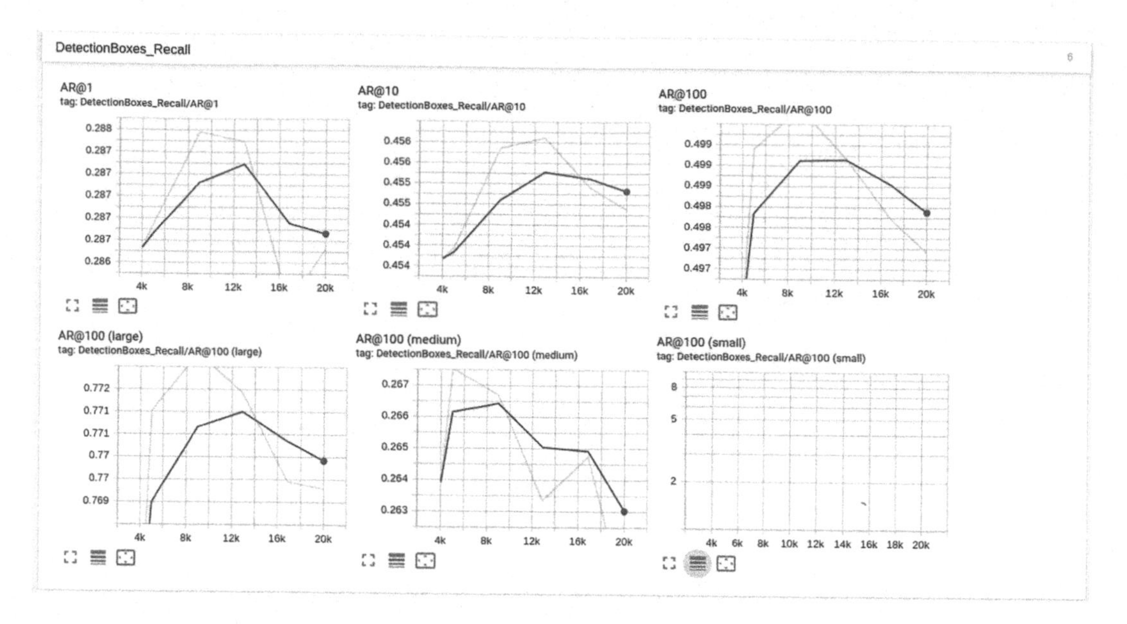

FIGURE 5.18 Recall vs steps.

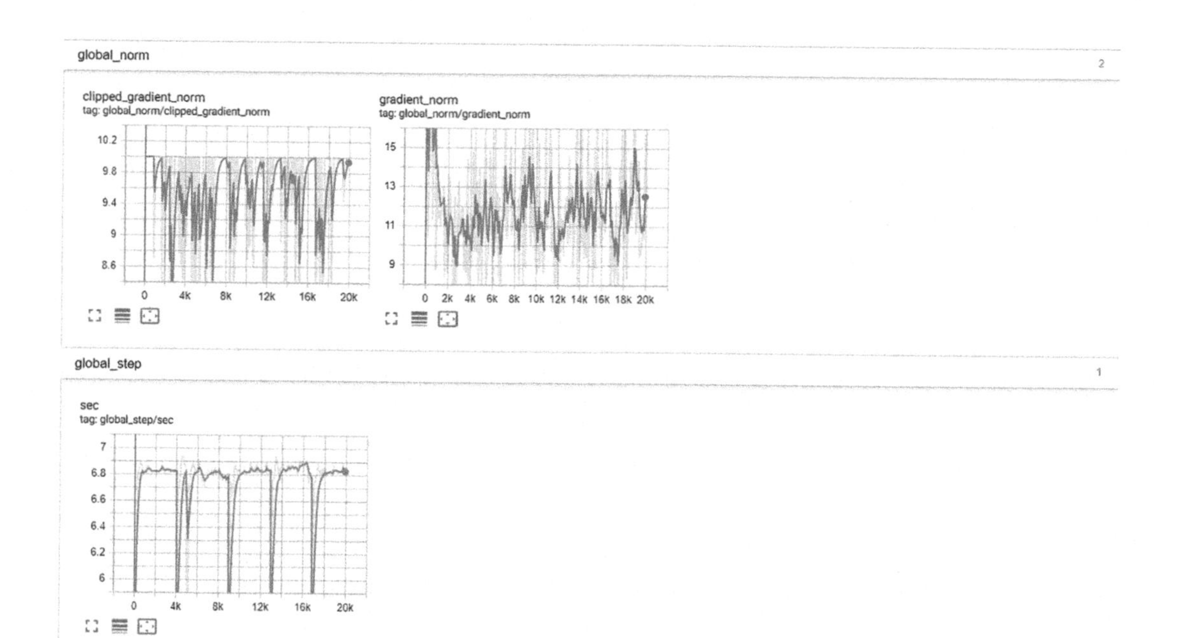

FIGURE 5.19 Global_norm and global_steps.

4. Graph for loss (Figure 5.20)

   When the training is completed, always choose the checkpoint having the least loss. If, while training, the loss saturates or stops decreasing, or increases instead, stop the training and use the checkpoint having the lowest loss till now.

   Apart from graphs, image predictions as shown in Figure 5.21 on the test set can also be seen to check how well the model performs.

- **Step 10: Inference**

  After training the model, comes the part of the evaluation. TensorFlow provides this excellent functionality of exporting a model so that it can be used or served anywhere, i.e., in a Web app or an Android app. The first step to all of this is to create a frozen graph that stores all the values from the model and can then be used for evaluation. To do so, run the following command in a cell.

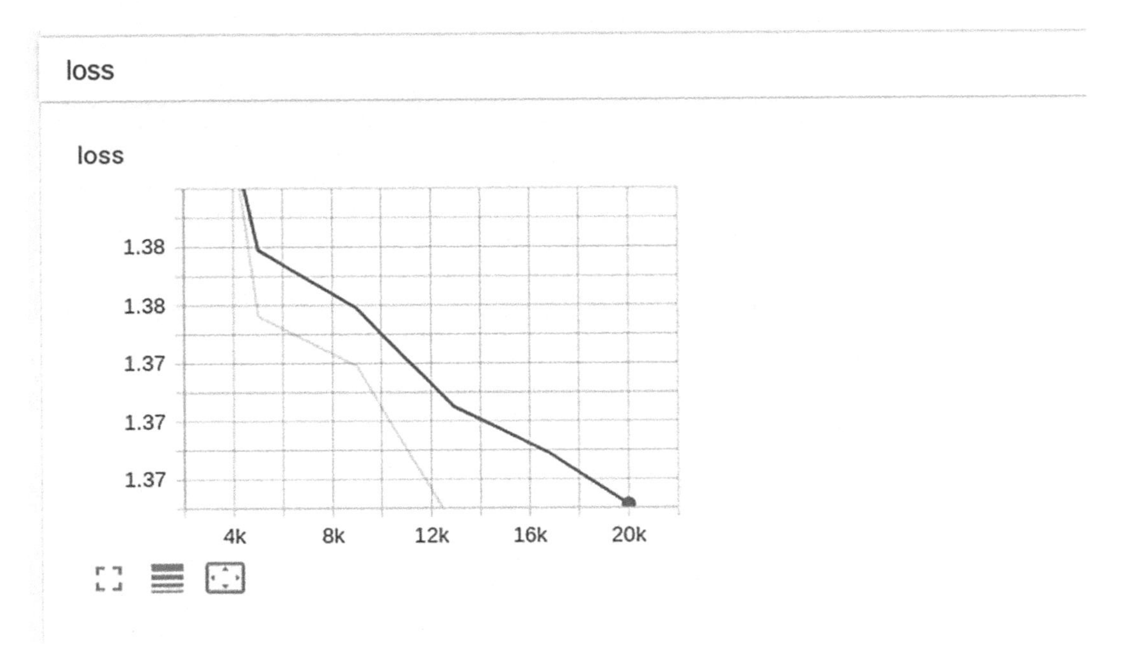

FIGURE 5.20 Loss vs steps.

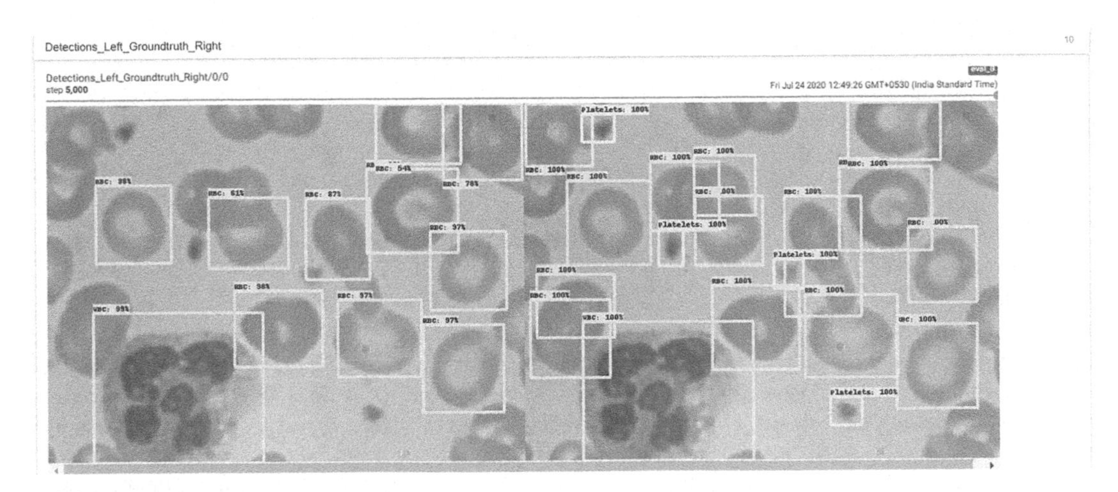

FIGURE 5.21 Results on test set as seen in TensorBoard.

```
!python object_detection/export_inference_graph.py \
      --input_type=image_tensor \
      -- pipeline_config_path=
      /content/gdrive/My\ Drive/tf_object_detection/Projects/CRC_Press/
      pipeline.config \
      --trained_checkpoint_prefix=
      /content/gdrive/My\ Drive/tf_object_detection/Projects/CRC_Press/
      models/model.ckpt-3842 \
      --output_directory=
      /content/gdrive/My\ Drive/tf_object_detection/Projects/CRC_Press/
      models
```

- **Step 11: Evaluation of the model on unseen data:**
  The validation folder contains data that has been used for neither training nor testing. Generally, the meaning of the validation set is different, but here, it's just the name of a folder with raw or unseen images. To evaluate the model, the frozen graph that was created is used. To do so, run the object_detection_tutorial.ipynb file from the TensorFlow Object Detection API folder (complete path from api root folder: /models/research/object_detection/object_detection_tutorial.ipynb)
  The following lines from the file have been modified:

```
# Path to frozen detection graph. This is the actual model that is used
 for the object detection.
PATH_TO_FROZEN_GRAPH = '/content/gdrive/My Drive/tf_object_detection/Pr
ojects/CRC_Press/models/frozen_inference_graph.pb'

# List of the strings that is used to add correct label for each box.
PATH_TO_LABELS = '/content/gdrive/My Drive/tf_object_detection/Projects
/CRC_Press/label_map.pbtxt'


# If you want to test the code with your images, just add path to the i
mages to the TEST_IMAGE_PATHS.
PATH_TO_TEST_IMAGES_DIR = '/content/gdrive/My Drive/tf_object_detection
/Projects/CRC_Press/data/Validation/Images'
TEST_IMAGE_PATHS = [ os.path.join(PATH_TO_TEST_IMAGES_DIR,file ) for fi
le in os.listdir(PATH_TO_TEST_IMAGES_DIR) ]
# Size, in inches, of the output images.
IMAGE_SIZE = (12, 8)
TEST_IMAGE_PATHS
```

After observing the results, they are validated with the original annotations and checked for accuracy, precision, recall, etc.

After running the Jupyter notebook, the predictions can be seen in the output of the last cell as shown in Figures 5.22 and 5.23:

Prediction:

Ground Truth:

Training the model for only 20,000 steps achieves an accuracy of ~92.51%, without any tuning. But accuracy shouldn't be the metric (or the only) that a model should focus on. In the discussed problem statement, predicting more blood cells than a person has cannot be afforded and hence the model should focus on reducing the number of false negatives it gets. Thus, the correct metric for prediction should be recall rather than accuracy. As can be seen from Table 5.1, the model fails to detect blood platelets successfully.

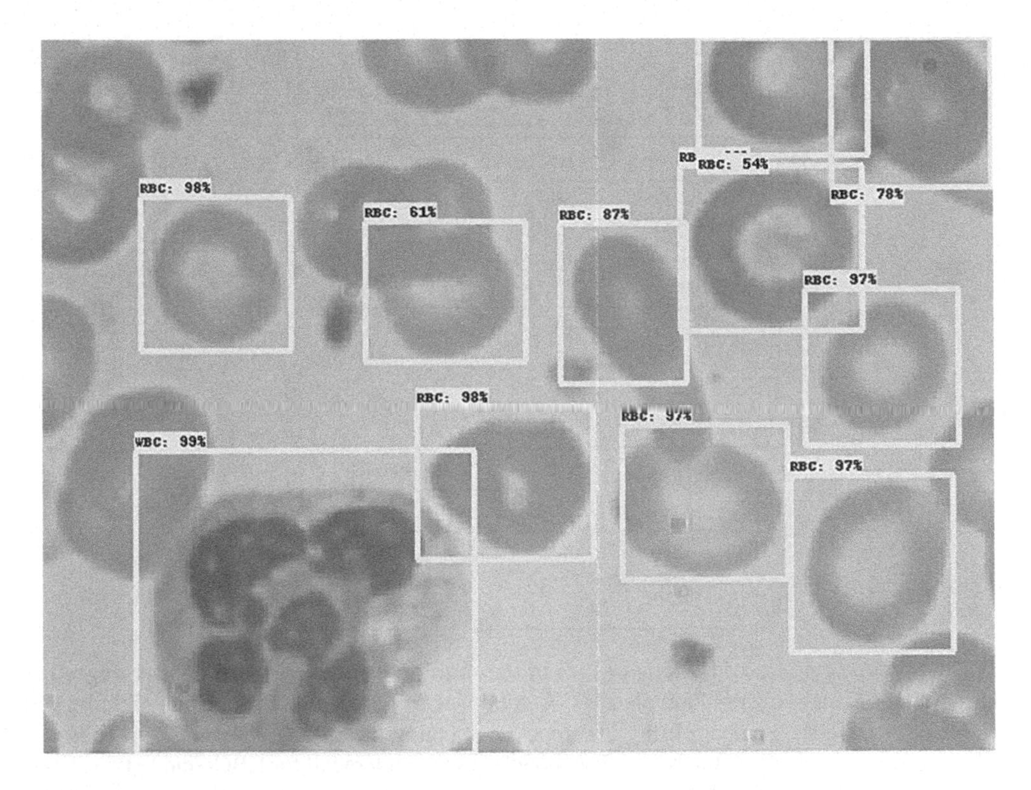

FIGURE 5.22 Result of prediction.

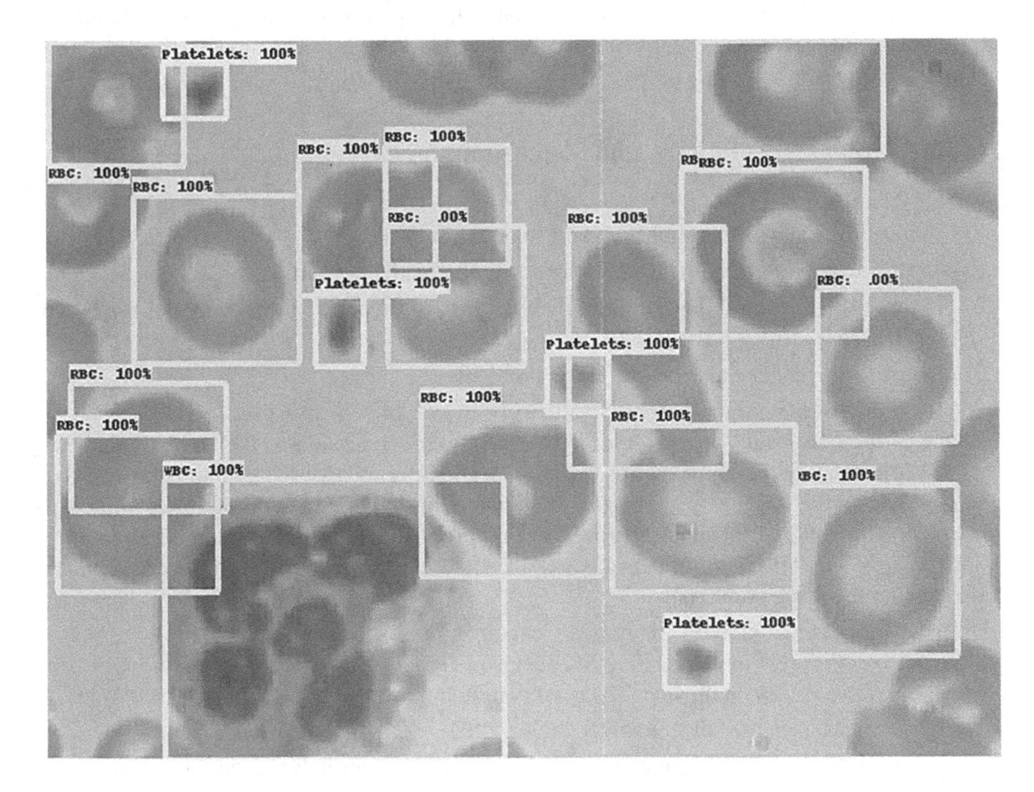

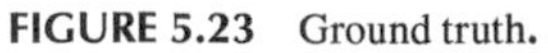
FIGURE 5.23 Ground truth.

TABLE 5.1

Classification Report for WBCs, RBCs and Blood Platelets

	RBCs	WBCs	Platelets
Accuracy	84.78	95.65	57.97
Precision	95.12	90.90	93.75
Recall	88.63	97.56	52.63
F1-Score	91.76	94.11	67.41

TABLE 5.2

Confusion Matrix

		Predicted	
		Positive	**Negative**
Actual	**Positive**	217	70
	Negative	14	113
Recall		75.61	

This can be due to a lack of training or bias in the training data wherein the dataset might have a smaller number of images with blood platelets. A good way to reduce these problems might be getting more data, increasing the number of training steps, or changing the values of various hyperparameters like learning rate and size of anchor boxes. The model shows a high recall for RBCs and WBCs. F1-score is used if we want to maintain a balance between precision and recall. As we can see, the model detects WBCs and RBCs easily but detecting platelets is a difficult task for the model. The trained model also detects a few cells that haven't been labeled in the dataset, which already shows that it's doing a pretty good job! The overall performance of the model can be seen in Table 5.2. The model has an overall recall of 75.61 after 20,000 steps.

5.6.2 Some Problems, Solutions, and Suggestions

1. NaN Loss error

   There can be many reasons for this error to occur, such as

   1. High learning rate, leading to the divergence of a model to infinity.
   2. If the cost function is logarithmic, then it leads to divergence of the model as the cost reaches zero.
   3. Division by zero can also be one of the problems.
   4. Input data contains empty or NaN values.

   One of the reasons this error could occur with the trained model is the manual augmentation. While creating the augmentation script, it should be noted that the bounding boxes should be well within the image when changing the annotations. Having bounding boxes outside the image (not intentionally!) could also lead to this problem. Also, while changing the annotations, keep in mind that none of the values viz. xmin, ymin, xmax, ymax can be zero! This may also lead to the same problem.

2. Graphics processing unit (GPU) cannot allocate memory.

   This problem might most probably not occur on Google Colab, but it occurs a lot of times when the model is trained off-line. The most common reasons for the same are as follows:

   1. Larger batch size than the computer can handle.
   2. Large prefetch_queue_capacity (buffer that gets the images beforehand for training)
   3. Large image dimensions that can be stored for a particular queue size

To overcome this error, simply change their values in the config file.

3. Transport endpoint not connected only on Colab.
   This error occurs because of communication problems between Drive and Google Colab. To get rid of this problem, simply restart the runtime, and the problem will be solved.
4. Problem Generating tfrecords
   The most common reason for this problem is missing annotation for corresponding images in the CSV files. Another thing that people generally forget, which creates problems, is changing the mapping according to the label_map.txt file in the gen_tfrecords.py file.

```
# TO-DO replace this with label map
def class_text_to_int(row_label):
    if row_label == 'RBC':
      return 1
    elif row_label == 'WBC':
      return 2
    elif row_label == 'Platelets':
      return 3
```

There can be many more big and small problems, so to avoid those, follow the following suggestions:

1 Check the data properly before using it for training.
2. Make sure that the data is properly distributed among the different classes to avoid introducing any kind of bias.
3. Have a proper project structure to avoid confusion.
4. Follow the above steps in the given sequence only.
5. If the data is augmented manually, check the data properly after augmenting as it can result in unexpected results or even some painful errors!
6. After the training is completed, always select the checkpoint with the lowest loss for better results. Then, add the checkpoints name in the “checkpoint” file in the “models” directory so that whenever the number of steps for the model is increased, it starts training from the best step till now.
7. Do not increase the number of training steps too much as it can lead to overfitting.
8. Once the training has been completed, try experimenting with the hyperparameters for better results.

These were a few tips, but the list never ends, and as always, the Internet community is always there to help.

5.7 Viability of the Solution

The problem that this chapter solves is based on one of the most common issues from daily life. Everyone in the world, at least once in their lifetime, goes through a CBC test. A general CBC test takes around 24 hours. During these 24 hours, a lot of manual effort goes into completing these tests and getting the results that otherwise could have been used to get more important tasks done. The proposed solution helps in getting the results within a few minutes, which would have otherwise taken hours.

Even though it sounds excellent, the main problem that this kind of solution faces is “trust.” Trusting this type of solution is a major challenge today because these solutions are never accurate, and the kind of problem that the chapter discusses doesn’t allow mistakes. Still, this problem does not primarily affect the solution that this chapter discusses because, in the end, the results are extrapolated in the most modern techniques too. With a bit of tuning and more data, the recall can be easily increased and made >90%. So, after extrapolating the results from the described model predictions, the final results shouldn't be too different from the original ones. It can provide the same results in less time and with a few clicks from your computer. Hence, with the above recall and performance, the above solution can help revolutionize the way medical science has been working to date and assist in easing up the process of CBC tests.

REFERENCES

Alam, M.M. and M.T. Islam. 2019 "Machine learning approach of automatic identification and counting of blood cells." *Healthcare Technology Letters* 6: 103–108.

Cruz, D., C. Jennifer, Valiente, L.C. Castor, C.M.T. Mendoza, B. Jay, L. Jane, and P. Brian. 2017 "Determination of blood components (WBCs, RBCs, and Platelets) count in microscopic images using image processing and analysis." *IEEE 9th International Conference on Humanoid, Nanotechnology, Information Technology, Communication and Control, Environment and Management (HNICEM)*, Manila, pp. 1–7.

Daqqa, K.A., S. Abu, A.Y.A. Maghari, and W.A. Sarraj. 2017 "Prediction and diagnosis of leukemia using classification algorithms." *8th International Conference on Information Technology (ICIT)*, Xi'an, pp. 638–643.

Everingham, M., L. Gool, C.K. Williams, J. Winn, and A. Zisserman. 2009 "The Pascal Visual Object Classes (VOC) challenge." *International Journal of Computer Vision* 88: 303–338.

George- ay, B. and K. Parker. 2003 "Understanding the complete blood count with differential." *Journal of Perianesthesia Nursing: Official Journal of the American Society of PeriAnesthesia Nurses,* 18(2): 96–114.

Girshick, R.B. 2015 "Fast R-CNN." *2015 IEEE International Conference on Computer Vision (ICCV)*, Santiago, pp. 1440–1448.

Habibzadeh, M., A. Krzyżak, and T. Fevens. 2013 "White blood cell differential counts using convolutional neural networks for low resolution images." *ICAISC.*

Kaur, P., V. Sharma, and N. Garg. 2016 "Platelet count using image processing." *2016 3rd International Conference on Computing for Sustainable Global Development (INDIACom)*, pp. 2574–2577.

Krizhevsky, A., I. Sutskever, and G.E. Hinton. 2017 "ImageNet classification with deep convolutional neural networks." *Comunications of the ACM*, 60(6): 1–7.

Meimban, R.J., A.R. Fernando, A. Monsura, J.V. Rañada, and J.C Apduhan. 2018 "Blood cells counting using Python openCV." *14th IEEE International Conference on Signal Processing (ICSP)*, Beijing, pp. 50–53.

Meintker, L., J. Ringwald, M. Rauh, and S. Krause. 2013 "Comparison of automated differential blood cell counts from Abbott Sapphire, Siemens Advia 120, Beckman Coulter DxH 800, and Sysmex XE-2100 in normal and pathologic samples." *American Journal of Clinical Pathology*, 139(5): 641–650.

Ren, S., K. He, R.B. Girshick, and J. Sun. 2015 "Faster R-CNN: Towards real-time object detection with region proposal networks." *IEEE Transactions on Pattern Analysis and Machine Intelligence*, 39: 1137–1149.

Weng, L., 2017 "Object detection for dummies part 3: R-CNN family" https://lilianweng.github.io/lil-log/2017/12/31/object-recognition-for-dummies-part-3.html.

Zhang, A., Z.C. Lipton, M. Li et al., 2020 "Dive into deep learning" https://d2l.ai/chapter_computer-vision/rcnn.html.

6

Application of Neural Network and Machine Learning in Mental Health Diagnosis

Aniruddha Das
Dwarkadas J. Sanghvi College of Engineering

Enakshie Prasad
University of Delhi

Sindhu Nair
Dwarkadas J. Sanghvi College of Engineering, Mumbai

CONTENTS

6.1 Introduction ... 99
6.2 Data Collection Approaches ... 101
6.2.1 *Standard Questionnaire–Based Evaluation* ... 101
6.2.1.1 Hamilton Depression Rating Scale ... 101
6.2.1.2 Beck's Depression Inventory ... 101
6.2.1.3 Patient Health Questionnaire ... 102
6.2.2 Emotion Analysis from Speech Signals Using Convolutional Neural Networks ... 102
6.2.2.1 Working of Convolutional Neural Network Model ... 102
6.2.2.2 Result and Conclusion of the Experiment ... 103
6.2.3 Text-Based Emotion Recognition System ... 103
6.2.3.1 Formal Definition of TBERS ... 104
6.2.3.2 Different Approaches to Implement TBERS ... 104
6.3 Proposed Model ... 106
6.3.1 Model Architecture ... 106
6.3.2 Classifiers Used for Diagnosis ... 106
6.3.2.1 Logistic Model Tree Decision Tree Algorithm ... 108
6.3.2.2 Classification of Stress Recognition Using Artificial Neural Network ... 109
6.4 Therapy Using Intelligent System Games ... 109
6.5 Conclusion and Future Scope ... 111
References ... 112

6.1 Introduction

In the International Health Conference of 1946, the representatives of 61 states decided that health is a state of complete physical, *mental,* and social well-being and not merely the absence of disease or infirmity. This shows that the World Health Organization (WHO) considers mental well-being a crucial aspect of health. According to WHO's 2003 reports, as many as 450 million people suffer from some mental or behavioral disorder. One out of four families is said to have at least one member with a mental disorder. Diagnosis of these underlying mental health–related issues is difficult to assess, quantify, and classify as each individual responds differently and may show a plethora of symptoms. Ignorance toward this critical aspect of an individual's life gives rise to serious consequences, for instance dying by suicide. In cases of kids and teenagers, parents are the primary caregivers but the lack of expertise makes it difficult to classify whether their child needs professional help or not. As awareness rises, several tests are being constructed and some are also available online. However, most of the measures that are available in the public domain are not standardized and, therefore, lack any clinical relevance. The ones that are standardized are usually only accessible to mental health professionals and may help in early diagnosis; however, a number of biases come into play when answering these tests and, thus, the results may not be very accurate.

In 2011, an administrative panel of WHO mentioned that depression will become the primary source of worldwide burden of disease by the year 2030. Mental health refers to the psychological and emotional state of a person. Mental illness is one of the prime issues faced by adolescents. Over years, there have been many therapeutic, pharmacological, and chemotherapeutic approaches to deal with these mental disorders, for example, the use of cognitive behavior therapy, rational emotive behavior therapy, dialectical behavior therapy, journaling, interactive games, and drugs like Prozac and Zoloft among many more. In order to achieve a better success rate in therapies and minimize the number of mental health patients in the longer run, it is crucial to identify these issues as early as possible. According to Spuelman, Caruso, and Goovimsky's diathesis–stress model, mental disorders can have numerous causes—predispositional causes (genetic makeup, biological, social, and family background), precipitating factors (triggering factors like the loss of a loved one), reinforcing factors (maintaining factors like the lack of support, lack of coping skills, and improper mental health care), and mediating factors that buffer between predisposing and precipitating factors (like social support) [1].

Diagnosis is an onerous task, and the consequences of misdiagnosis can be severe. Hence, recognizing and treating mental health issues should be done precisely.

Research in the domain of mental health has grown exponentially, contributing toward the collection of information through various methods, using which one can infer to some extent the status of a patient. Since the onus of the final diagnosis, treatment, and evaluation relies heavily on the perceptivity of psychologists and their personal experiences and opinions regarding a particular matter, it becomes extremely necessary to think of ways that can help in an objective, clear-cut diagnosis. Often a mental disease may not be distinctly related to an underlying biological impairment which makes it very difficult to conduct a diagnosis solely with the help of available neuroscientific instruments. Presently, mental health specialists make use of the pervasively adopted diagnostic categories that have been constructed from expert opinions and enshrined in the DSM-5 and ICD-10 manuals. However, not all symptoms may be shown by a patient and more often than not, the symptoms of different disorders overlap. This makes diagnosis tough and makes it increasingly difficult for the mental health specialist to properly associate a person with a disorder. This makes it evident that the psychopathology that underlies the mental disorders is quite heterogeneous [2,3].

In order to achieve progress in the domain of mental health, it is crucial to overcome two of the main challenges: (i) lack of quality features/parameters which can be used to improve classifications and (ii) unavailability of large supervised datasets which can be used for training. The standard questionnaire approaches, for example, Beck's depression inventory [4] (BDI), are commonly used but work as preliminary tools of diagnosis due to their limitations. Similarly, most of the available measures of disorders are not foolproof and are thus used only as a source of procuring preliminary information. Along with using the multitude of diagnostic tools, professional psychologists invest a lot of emotion and time to study their patients with the intention of identifying and analyzing these behavioral patterns to assist in determining an obvious clinical diagnosis. In addition, it is difficult to assure the veracity of the data presented. In a conscious state, a patient might have a tendency to lie in these questionnaires making the diagnosis process more difficult. With the advancement in technology, the design of intelligent systems can contribute significantly to the betterment of diagnosis as well as therapy. The intervention of artificial intelligence (AI) cannot only help in the collection of different features of a patient but also aid in classifying these disorders to some extent.

This brings to light the application of neural networks (NN) and machine learning (ML) algorithms which can help in the collection of data through different sources and generate a richer dataset, incorporating more elements in addition to the standard questionnaire formats. Currently, the number of cases outnumbers the count of psychologists by a great extent, which engenders several challenges in the way one handles each case. Time being a key constraint, diagnosing all patients accurately becomes an onerous task, creating room for a lot of errors.

ML algorithms (like support vector machines [SVMs], decision trees, and more) focus on identifying underlying patterns in a series of observations. These algorithms can function as unbiased judges and can work without underlying assumptions about any patient. The capability of ML algorithms to process and extract valuable knowledge from extensive data opens the door for improvement in the diagnosis of mental health issues. One of the major challenges in the way information is collected is that a patient is conscious throughout a session and the assumption that the collected data are a true measure of features to be evaluated by the psychologist or an ML model lacks confidence.

Standard approaches of gathering data about a patient (primarily systematic examination, development of questionnaires, and diagnostic interviews) do add a significant value to overcome the lack of confidence parameter, an improvised dataset which uses different approaches like emotion detection through voice notes and sentiment analysis through data in the form of text, and the rate of performance of a patient in a game (one which triggers emotions) can be used as additional features in the input dataset, along with the standard way of collecting data. This would help an AI model in understanding a patient's mental attitude in a particular session and in turn increase the accuracy of a diagnosis. After successful incorporation of these features, one can reach a stage where a modified dataset has been attained for training the ML models which can in turn help in classifying these disorders with more accuracy.

In Section 6.1 of the chapter, different approaches used over time to tackle the dilemma of identifying underlying mental disorders have been discussed. Section 6.1 extends into examining the extant standard questionnaire approach in brief, which is then followed by a subsection, explaining the intervention of NN- and ML-based techniques to elicit a richer set of parameters which can help us overcome the previously mentioned flaws in the standard approach. Section 6.2 of the chapter talks about a proposed model which integrates the above-mentioned approaches, which could give promising results. In Section 6.3 of the chapter, the use of AI-based games as an interactive therapeutic tool was discussed, thus covering all aspects of mental health diagnosis, ranging from identification to therapy using AI intervention. The last section of the chapter summarizes the challenges and discusses the future scope of AI in this domain.

6.2 Data Collection Approaches

6.2.1 Standard Questionnaire–Based Evaluation

Numerous measures are used by mental health specialists to procure preliminary findings that may serve as an indication of depression. Some of the common measures are listed below.

6.2.1.1 Hamilton Depression Rating Scale

Also called Ham-D, this multiple-item questionnaire is designed for adults and is used as an indication of depression in a patient (Hamilton, 1960, 1966, 1967, 1969, and 1980). The questionnaire focuses on symptoms of depression like insomnia, suicidal ideation, feelings of guilt, and somatic symptoms like weight loss. However, the scale has been criticized for focusing heavily on insomnia and burying important diagnostic symptoms such as feelings of worthlessness and anhedonia that are seen as important criteria for depression as per the DSM-IV [5].

6.2.1.2 Beck's Depression Inventory

The BDI has been widely employed as an assessment measure to know about the intensity of depression in patients who are diagnosed with depressive syndromes using the DSM-V (Beck et al., 1988). The reliable and valid measure is used by mental health professionals to assess and document the changes brought about in depressive symptoms and attitudes during therapy. However, the measure falls short at providing a safeguard to combat faking, lying, variable responses, or the tendency of clients to select socially desirable responses [4].

6.2.1.3 Patient Health Questionnaire

The Patient Health Questionnaire-9 (PHQ-9) is a nine-item questionnaire designed to screen for depression and its symptoms in medical and mental healthcare settings [5]. Higher PHQ-9 scores are associated with decreased functional status and increased symptom-related difficulties, sick days, and healthcare utilization. Though a reliable and valid measure of depression, the criteria are not foolproof against response bias.

The major problem with psychological testing is its inability to safeguard against response bias. Very often, the respondents feel the need to give responses that they consider acceptable and socially desirable. This tendency of the clients makes the testing process liable to inaccuracies. Therefore, these tests are used as preliminary criteria at best and not as the sole diagnostic criteria in clinical settings.

6.2.2 Emotion Analysis from Speech Signals Using Convolutional Neural Networks

Diagnosis of mental health–related disorders can be studied by capturing and modeling key behavioral signals [7,8]. From an ML perspective, one can view this study as a regression or classification problem [9,10]. In psychological terms, one can say that mental health disorders are a set of negative emotions. For further studies, let's consider depression as a disorder which the AI model shall classify. During exploration from an AI perspective, rephrasing the problem as detecting depression in a negative-emotion perspective simplifies the context. The goal is to classify depression between nondepressed and depressed categories. (This section of the chapter considers depression an example here, but the study can be extended for other disorders or negative emotion in general too.)

Over time, speech has been considered to be a vital feature in any behavioral-based mental health detection system [11]. It has been studied that patients with depression usually have certain abnormalities in their speech patterns. For example, speech in a patient suffering from depression can be described as lifeless, monotonous with a sad/negative tone in general. Several papers in the past have explored this area, aiming to find ways to detect depression using convolutional neural networks (CNN) [12–14]. In this chapter, another approach toward detecting negative emotions has been explored, i.e., depression using a CNN model.

Deep learning has achieved great feats in the past and outperformed ML models when it comes to ML applications. CNN is one of the most popular areas in deep learning research.

6.2.2.1 Working of Convolutional Neural Network Model

CNN consists of three layers—convolution, pooling, and fully connected. The architecture can be viewed in Figure 6.1. The first layer consists of a convolution core along with an offset in each 2D plane. The network parameters are optimized via weight sharing [15]. During convolution, different kernels are used to extract different attributes of the input image by moving a sliding window over parts of the input image. In the next layer, pooling is done where downsampling occurs. The parameters to be considered as input to the next layer are reduced by selecting only the local optimal values obtained from the convolution layer. Pooling operation helps in effective computation by selecting only the relevant and efficient features which are fed to the model during training. The pooling layer is followed by the fully connected layers which deal with training of the model, learning high-order features, and final classification. This final layer in the architecture is the output stage of CNNs.

CNN is ideally meant for data in the form of images; thus it is essential for us to translate speech in terms of images to carry out binary classification (depressed or neutral). The speech is processed using Fourier transform to produce a spectrogram. This spectrogram image is fed to the CNN. By analyzing the features present in the spectrogram images, the CNN model can learn the association of characteristics to a certain emotion.

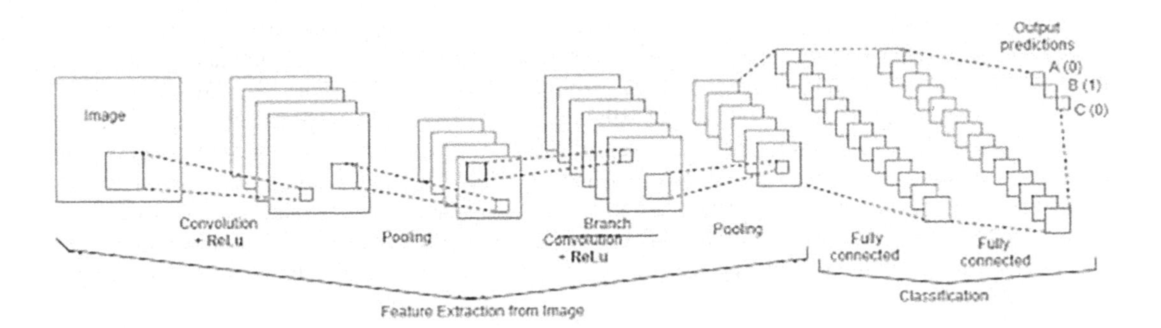

FIGURE 6.1 General CNN architecture [6].

TABLE 6.1
Performance Metrics [6]

F1-Score	Precision	Recall	Accuracy
0.61	0.91	0.46	0.71

After the spectrogram is attained, the next task is to divide the images into training (for training the model) and testing set (for evaluating the final performance). Negative emotions are labeled as 1 and neutral ones as 0. The image is split using a Hamming window of size 4 s and then these segmented speeches are sampled randomly. These spectrums would serve as input to the model.

A six-layer CNN was used in the experiment, consisting of two layers of convolution followed by two layers of max pooling and finally two fully connected layers. Each input image from the segmented spectrogram was of size 513×125 (between 0 and 8 KHz frequencies). The input images were normalized and then provided to the CNN model.

6.2.2.2 Result and Conclusion of the Experiment

The accuracy of the training model came out to be 78.4% and the test set showed 71%. Six students in both categories—negative and neutral—were used in the training set. The total sentences summed up to 396. Three students in both categories were used in the test set, which summed up to 198 sentences. The scores of the final model were (Table 6.1).

$$F_1 = 2 * \frac{\text{Precision} * \text{recall}}{\text{Precision} + \text{recall}} \tag{6.1}$$

Via the results of the above experiment conducted in the paper [15], one can see how CNN method of classifying emotion from speech signals showed significant results and thus shows that speech can be used as a tool of fast emotional recognition. While the above experiment was conducted on a relatively smaller set of samples, one can apply and gain more insight as to how speech signals can play a vital role as an additional feature when it comes to diagnosing mental health. The above experiment deals with a classification scenario, but as mentioned initially, CNNs can be used for regression problems too. Hence, be it a binary classification or a continuous value representing the degree of negativity in an emotion, emotion analysis from speech signals can contribute significantly as a feature in a dataset when being fed for further analysis by other ML algorithms.

6.2.3 Text-Based Emotion Recognition System

One of the ways people deal with the symptoms of mental health–related disorders is via expressing themselves through diary entries and journaling [16]. Writing down helps people in expressing their emotions. While these means are usually considered a let-out, from an AI perspective this can be utilized as another dimension to draw valuable insight into the diagnosis of a patient's mental health status. As mentioned before, behavioral signals carry huge potential in terms of explaining the emotional state of a person. Humans express their emotions not only via verbal cues like speech, but also via nonverbal cues like eye movements, gestures, and facial expressions. Likewise, written texts can serve as an important cue for gaining insight into the minds of individuals. The text-based emotion recognition system (TBERS) is an important dimension which has been discussed heavily in computational linguistics over years [17]. This section of the chapter discusses ways to utilize the capabilities of text-based emotion detection for the purpose of analyzing the emotional state of a patient suffering from mental health–related disorders.

6.2.3.1 Formal Definition of TBERS

Let's convert the current goal into a more structured definition before looking at different approaches toward analyzing the emotion. Let "E" represent the emotion set, "A" represent the author set, "T" represent all possible representations of expressions/emotions in the texts. Given "r" is a function which maps emotion "e" of author "a" from text "t," i.e., r: A×TE, then r would represent the final answer in the classification problem [18].

While E and T may be defined simply, the central issue of emotion recognition systems resides in the fact that the definitions of individual elements, even subsets in both sets of E and T, would be rather confusing. As languages evolve over time, new elements would be added to the set T. In addition to that, the convoluted nature of human language makes it difficult to attach a standard classification of all human emotions, rather one can only view them as "labels" annotated at the end for different purposes.

6.2.3.2 Different Approaches to Implement TBERS

The key toward explicit emotion recognition in text relies on mapping keywords. Understanding the context becomes important while dealing with the task of implicit emotion recognition which is a much more onerous task since emotion is typically hidden within the text. There are several approaches for text-based emotion recognition, some of which are keyword spotting approach, rule-based approach, classical learning–based approach, deep learning approach, and hybrid approaches [19]. This chapter primarily discusses three of the above-mentioned approaches.

6.2.3.2.1 Keyword Spotting Approach

Keyword spotting algorithm consists of five stages (refer to Figure 6.2). In the first stage, text preprocessing occurs where tokens are generated from the input text data, the tokens are then mapped against a predefined set of emotional keywords, and those tokens are passed to the next stage. After the emotional words have been extracted, the intensity analysis of emotional words is performed followed by a negation check. Finally, based on all these stages, an emotion class is assigned as output.

While this method works in cases where emotional words are explicitly mentioned, it fails when it comes to recognizing implicit emotions. For example, the sentence "Today is a happy day" will give

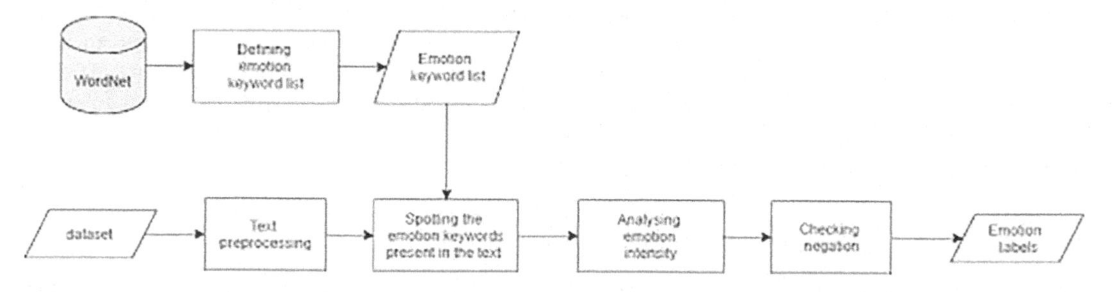

FIGURE 6.2 Keyword spotting approach [19].

good results with this approach since the keyword "happy" can be identified easily and such sentences won't require an understanding of the context. However, in sentences like "I'm not feeling happy," understanding the context becomes important. The context in which the word "happy" is used is in a negative context. Thus, while this approach works for explicit emotion-based texts, it fails while dealing with implicit emotion-based texts.

6.2.3.2.2 Classical Learning–Based Approach

Classical learning–based approach allows a system to learn and improve from experience on its own. Learning-based methods use pretrained classifiers, which use various ML algorithms such as SVMs [20] to classify the text into some category of emotion. SVMs are a supervised ML algorithm which generates an optimal hyperplane in order to predict the class an emotion belongs to in an unseen text (Figure 6.3).

In the first stage, input is preprocessed, which includes substeps like tokenization, stop word removals, Part-Of-Speech tagging, and lemmatization. The next step involves extracting features from the processed text. Given the feature set and emotion labels, the trained SVM classifier is able to assign an emotion label to the unseen text. Apart from SVM, many other classification algorithms like naive Bayes and decision trees can be used as the classifier for such learning-based approaches.

6.2.3.2.3 Deep Learning–Based Approaches

Deep learning methods also learn using experience and attempt to study the input in the form of a hierarchy of concepts. These methods are capable of learning convoluted concepts by using simpler concepts as the foundation. Such methods use long short–term memory (LSTM), an extended form of recurrent neural network which can effectively handle long-term dependencies.

The main steps involved in deep learning–based methods for emotion recognition are given below in Figure 6.4. In the first stage, text preprocessing is performed on the dataset, similar to the preprocessing

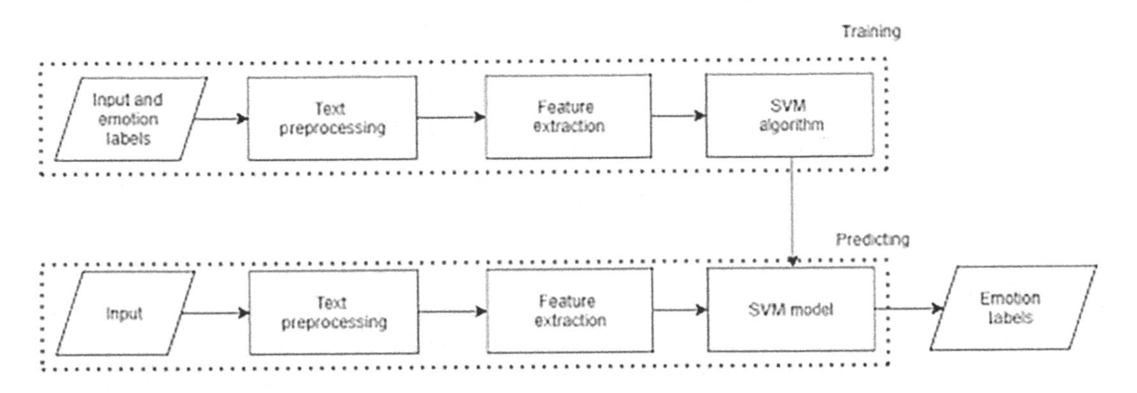

FIGURE 6.3 Classical learning–based approach [19].

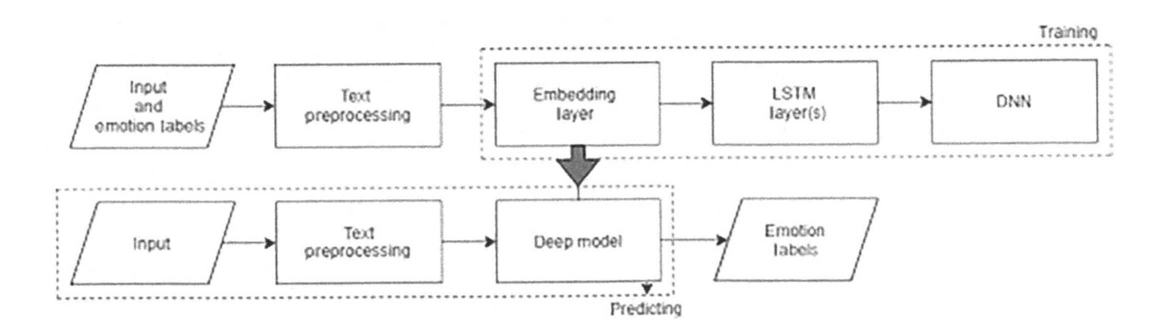

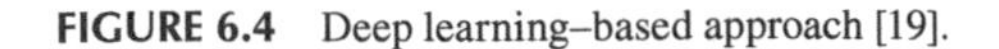
FIGURE 6.4 Deep learning–based approach [19].

in the classical learning–based methods. After this stage, the processed input is fed to the embedded layers, which is then provided as input to the LSTM layers. Finally, the output is passed as an input to a dense neural network, which has an activation function like sigmoid and units equivalent to the number of emotion labels the model can classify the input into. Advanced deep learning methods such as transfer learning by pretraining bidirectional LSTMs, use of transformer-based models like bidirectional encoder representations from transformers.

6.3 Proposed Model

Over years, many approaches as mentioned above have been used to analyze information regarding the status of a patient having mental health–related disorders. While the most common method is the standard questionnaire format, the degree of veracity via that method is questionable because it depends primarily on the patient answering all questions truthfully. The use of AI to analyze emotions from behavioral signals has also proved effective to a great extent as seen above. Eliciting emotions from text has also shown promising results. While all these approaches have proven to be successful to different degrees, no integrated approach has been adopted. In Section 6.1 of the chapter, the dire need of richer features needed to allow ML algorithms to diagnose disorders more effectively was discussed. Currently, the available datasets regarding mental health patients focus on primarily their personal life details, such as sex, age, work-life, marital status, etc. While these parameters do add some value to the dataset, it is difficult to diagnose a patient with such limited information. As seen before, if all these methods to collect data about a patient are used on a patient, they can prove to be effective in diagnosing mental health–related disorders or tracking the progress of any patient under examination.

The performance of ML algorithms depends heavily on the quality of training data they are fed. A dataset with richer and more relevant features can lead to better performances. Although there has not been an integrated approach yet, based on the current available data about patients several attempts have been made to diagnose the mental health of a patient using different ML and neural network (NN) models. While all these algorithms prove effective, the base dataset itself is not that rich to start with; with the integrated approach as proposed above, one can train these classifiers using better features and provide accurate diagnosis in this domain. Such an integrated approach would also open gates for AI-based intelligent systems to prosper in the mental health domain.

6.3.1 Model Architecture

In the architecture of the above-proposed model, there shall be four distinct stages. In the first stage, all different ways of extracting more context about the patient with respect to his or her speech, text, and standardized questionnaires shall be collected. Following this stage, all the results obtained from these individual ways shall be preprocessed which may involve normalization and dealing with categorical data. In the third stage, once the compiled dataset is available, it can be split into training and testing sets, before being fed to the ML classifier. After the training of the model, in the final stage, a class can be assigned, which would, in turn, be the final diagnosis of the status of a mental health patient (Figure 6.5).

6.3.2 Classifiers Used for Diagnosis

One of the key challenges when it comes to mental health diagnosis is the unavailability of enriched datasets [21]. While there are many challenges in terms of the collection as well as the final assessment which would be discussed later in this chapter, there have been many attempts at classifying different mental disorders using various algorithms [22–25]. Based on the currently available datasets, certain attempts at mental health diagnosis have been made via these different approaches. The chapter discusses different approaches used over time along with how they performed on different datasets with limited parameters.

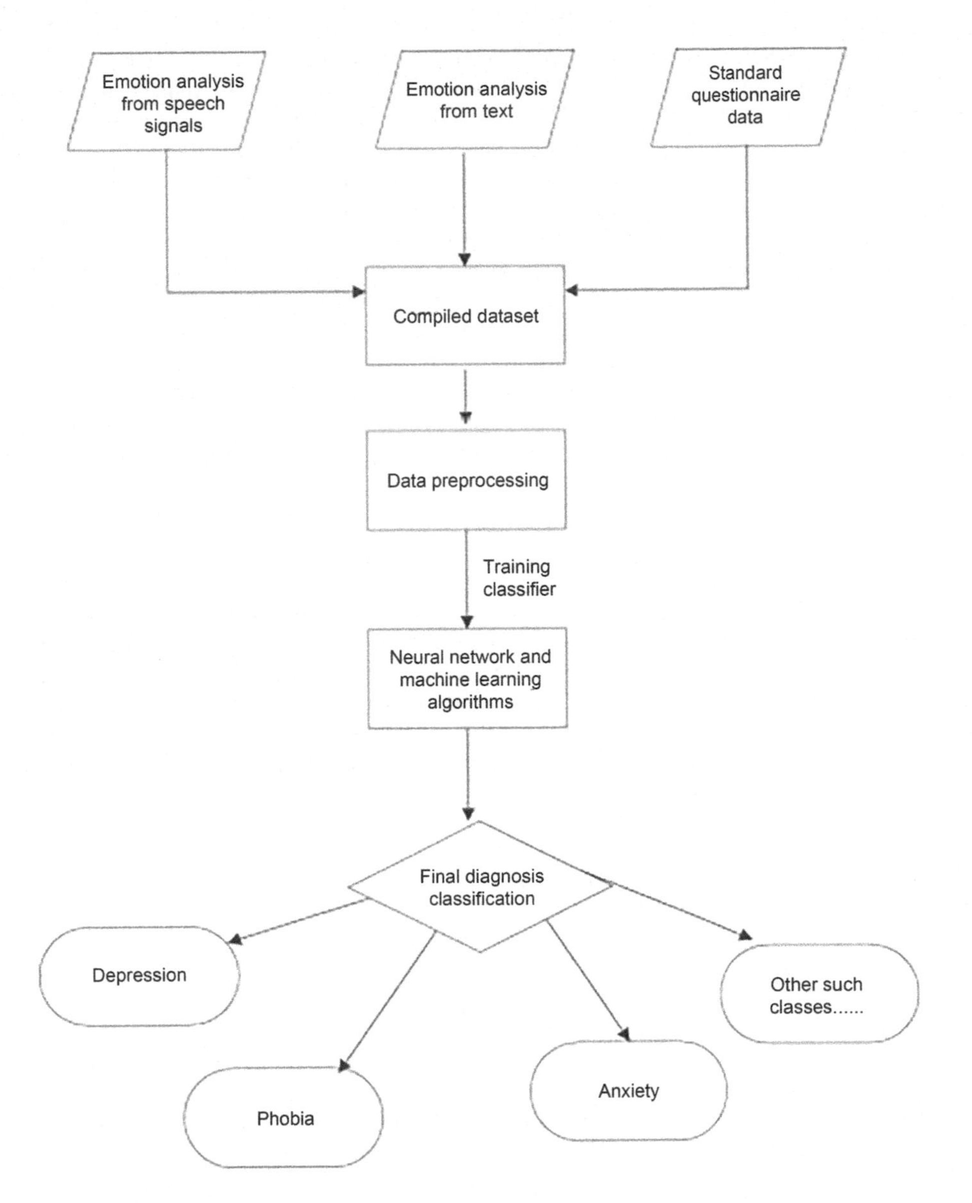

FIGURE 6.5 Proposed model architecture.

In Ref. [22] U. S. Reddy et al. used ML algorithms like logistic regression, decision trees, and random forest, with the aim of finding an algorithm with maximum accuracy and also compared different parameters used in the dataset. In Ref. [23] M. P. Dooshima et al. proceeded with decision trees and naive Bayes algorithms on Weka. Different parameters were selected in the dataset consisting of only 30 patients. These parameters included primarily psychological, demographic, environmental, and biological factors for prediction. These parameters were validated by mental health experts. In Ref. [24] M. Srividya et al. used naive Bayes, K-nearest neighbors (KNN), SVM, decision trees, logistic regression, and random forest algorithms as classifiers. The data collected for this experiment were gathered with the help of questionnaires as mentioned in Section 6.2.1 of this chapter. In this paper [24], the conclusion based on the compiled dataset was that SVM, random forest, and KNN algorithms were the most accurate. To narrow the hypothesis set, the final classes were restricted to depression, stress, and anxiety. In Ref. [25] D. Filip and Jesus used NN as the classifier and restricted the hypothesis set to depression, posttraumatic stress disorder, anxiety, and posteffect of brain injuries on sportspersons.

Most of these approaches rely heavily on the compiled dataset and turn out to be accurate to a certain point; however, with better parameters to evaluate these classifiers can perform much better, resulting in better accuracies. The most standard way of collecting information about patients which is later used to populate datasets is in the questionnaire format. There are some systems which use chatbots to predict mental illness. However, the medium is the only differing point, and this method of garnering knowledge about a patient can also be deemed as a questionnaire format only.

Other approaches such as association rules using data mining techniques have been tried [26,27]. Hybrid classifiers combining naive Bayes and decision trees have been tried out [28], which shows promising results when dealing with small-scaled datasets. In Ref. [29], the author has attempted to use different algorithms utilizing decision trees for classification. The dataset used was from Minnesota Multiphasic Personality Inventory. Out of the different decision tree algorithms like random tree, J48graft, REPTree, BFTree, classification and regression trees, and logistic model trees (LMT), it was seen that LMT outperformed other decision tree algorithms. In the following section, LMT decision tree algorithm is discussed and the results were analyzed. Following which, the chapter presents the study of a stress recognition system using artificial neural networks (ANN).

6.3.2.1 Logistic Model Tree Decision Tree Algorithm

Logistic model tree is a hybrid classifier—mix of decision tree and logistic regression [30]. In Ref. [31], the authors have collected patient data by carrying out a stress monitoring test. Some of the static features included in the dataset used in this experiment were name, occupation, age, treatment history, duration of disease, relationship status, income level, etc. Some of the dynamic factors were location, activity during the test, thoughts, environmental conditions, etc. The proposed model in Ref. [31] is given in Figure 6.6.

To determine the set of rules using the ID3 algorithm, information gain and entropy are calculated where information gain (S, A) is calculated by,

$$E(S)-\sum_{i=1}^{n}\frac{|si|}{S}E(S_i) \tag{6.2}$$

And entropy(S) is calculated using the formula,

$$\sum -p(I)\log_2 p(I) \tag{6.3}$$

Where S comprises all the cases, "A" is the attribute under examination, "n" represents the total partitions. After using the ID3 algorithm, ordinal logistic regression is used for further prediction. The logistic function is given by

$$Y'i = e^n / 1 + e^n \tag{6.4}$$

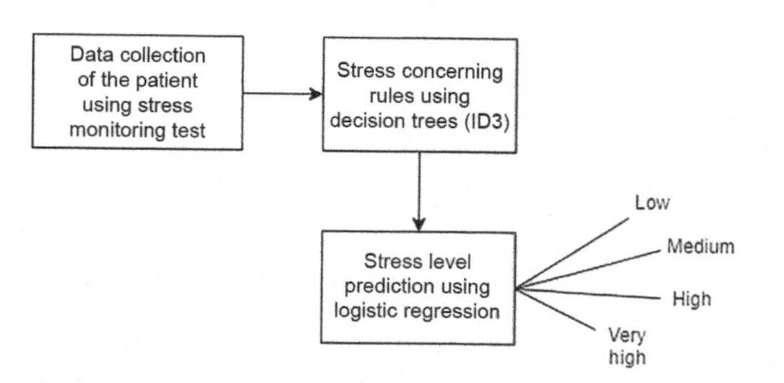

FIGURE 6.6 LMT decision tree block diagram [30].

TABLE 6.2
Results of LMT Decision Tree Algorithm [31]

Stats	LMT Classifier
Precision	63.66%
Recall	61.66%
F-measure	76.66%

Where Y' represents the estimated probability that the ith case is present in a category, the linear regression equation is given by, $n=A+B_1x_1+B_2x_2+\cdots+B_Kx_K$. ("$A$" is the regression constant, B is the slope of regression, and x_1, x_2, and x_3 are the independent variables).

The results of the LMT decision tree algorithm are as follows (Table 6.2).

While in this paper the key aspect being evaluated was stress, the same classifier can be used for evaluation of other disorders like anxiety, depression, phobias, etc.

6.3.2.2 Classification of Stress Recognition Using Artificial Neural Network

In Ref. [31], the authors have considered a different set of parameters in their datasets which revolve more around physiological parameters. These parameters primarily involve galvanic skin response (GSR), electrodermal responses, heart rate, and respiration rate (RR). While these features cannot be grouped under the bracket of behavioral signals completely, to some extent these act as great indicators of stress and other mental health–related disorders too.

In this experiment, a feed-forward back propagation network with two layers has been used for classification. The input layer comprises four network inputs followed by a hidden layer which contains seven neurons. Each neuron undergoes a weighted summation. One neuron is present in the final output layer and the associated value parameter in the output can be used to determine the level of stress. During the training, Levenberg–Marquardt function was utilized and mean squared error was evaluated as a performance index. The linear nature shared between physiological factors and stress recognition supports the use of linear transfer functions in the input and the hidden layers. The parameters provided as input to the ANN classifier were: heart rate, RR, foot galvanic skin response, and hand galvanic skin response. The final output was categorized into stress and no-stress. The threshold used in the classifier is 3.8, above this value, the classifier associates the input tuple to the class "stress" and "no stress" for values below the threshold (Figures 6.7 and 6.8).

The regression analysis throughout the training of neurons helps us determine if there is any linear correlation between the output and the target. During validation testing, it was found that the peak performance was achieved after the third epoch. Data were gathered from 77 people for this experiment and the confusion matrix along with the performance during validation for the same is given below (Figure 6.9 and Table 6.3).

Using ANN as a classifier in the above experiment, classification of a mental health disorder—stress in this case—is achieved. The accuracy achieved in this experiment was up to 99%, sensitivity of 98%, specificity of 100%, and error rate of 1%. Such results show how powerful NN are. While in this case stress was the only class considered, the model can be extended easily for multiclass classification, thus helping in finer classification among the several possible classes of mental health–related disorders like anxiety, phobias, depression, etc.

6.4 Therapy Using Intelligent System Games

Research over the use of games for therapy purposes has been conducted for years now. With much development in the field, there have been many cases where the patient's interaction with these games has led to improved mental health status. The purpose of these games is to elicit a desired emotional/behavioral

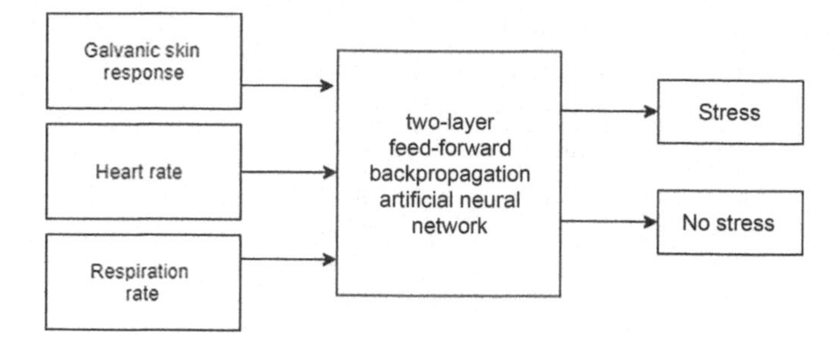

FIGURE 6.7 ANN block diagram for stress classification [31].

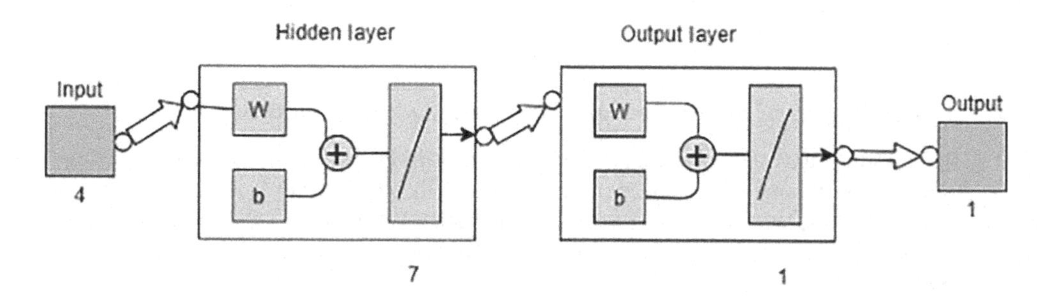

FIGURE 6.8 ANN general architecture [31].

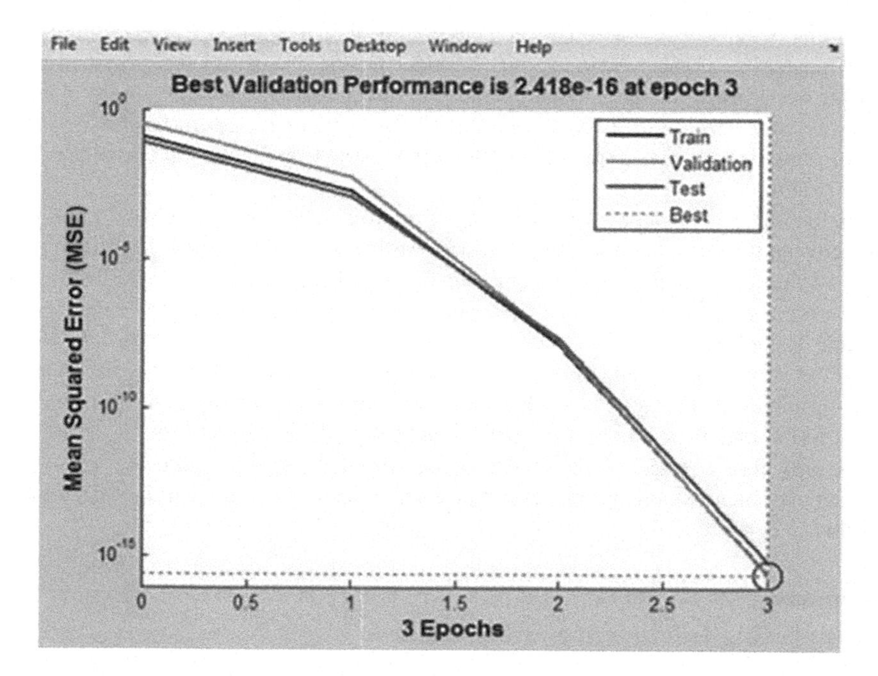

FIGURE 6.9 Performance during validation testing [31].

TABLE 6.3
Confusion Matrix [31]

N=77	"No-Stress" Prediction	"Stress" Prediction
Class "no-stress"	22	0
Class "stress"	1	54

result from the patient. Therapy games usually contain elements like challenge, social relatedness, virtuality, and competition, which usually motivate the user to continue the interaction and achieve the desired training, behavioral, or emotional regulation goals. There have been diverse cases, the case studies of which are described in detail in Ref. [32].

One of the categories of games which proves effective is the use of ReadySetGoals application which is a gamified mobile application aimed at supporting therapeutic goal setting within a substance addiction treatment context. The key stages during the design process were (i) identification of "transfer effect," (ii) examination of "real-world context," (iii) design of a "core gamification loop," and (iv) iterative testing of the prototype. Details about the above design ideologies can be found in Ref. [33].

In some cases, the use of reinforcement learning can help in the development of better games. The principle on which the concept of reinforcement learning works is that AI-based games/products learn over time. It gets rewarded for every correct decision it makes and penalized for the wrong ones, thus making it a better learner gradually. Such games can cater to the specific requirements of every patient by learning from their responses and taking the best course of action suited for a particular patient. Certain products based on the concept of deep reinforcement learning use robotic training to assist mental health patients during their therapy phase [34].

6.5 Conclusion and Future Scope

With the advancement in neuroscience and medical technology, scientists have been able to decipher many aspects of the brain that were once considered a hidden treasure beyond human reach. People have grown to realize the importance of mental faculties and how important their adequate functioning is for a healthy individual. The brain serves as a seat for all cognitions, including memory, speech, thoughts, learnings, and emotions. These along with the dynamism of its functioning makes it the most important organ in the body. Thinking, mood, behavior, and personality are all inevitably dependent on mental health. Mental health includes physical, social, and psychological well-being, and disrupted mental health affects the overall functioning. This makes mental health an important area of study, in need of more and more modern interventions. In the mental health domain, AI-based techniques can bring around significant progress toward the early diagnosis of patients. The advancement in the field of deep learning and ML algorithms helps us discover the underlying information amid a vast pool of data.

As mentioned in the chapter, there are some pitfalls hindering the progress of ML techniques. The lack of datasets available with rich features is a key concern. While there have been many approaches to diagnose mental health–related issues, some of which have been discussed in this chapter, and an integrated approach is needed to first compile a dataset holding better features which would in turn help in better accuracies. Moreover, ML and NN help us overcome the standard drawbacks which come along with the standard questionnaire methods used to elicit information about a patient for diagnosis. A collaboration of these individual approaches can help in providing better data to train NN and ML models for the final diagnosis.

The future scope of research can be the implementation of the proposed model in this chapter. Making a richer dataset available using the mentioned techniques can help AI prosper further in this domain. Some of the key challenges which need to be addressed are the privacy concerns when dealing with such information. Sharing of voice notes or texts publicly for research might not be acknowledged by patients. However, future research as to how one can overcome these common challenges would help more researchers and scientists to evaluate better diagnostic systems. This chapter mentions some of the proposed ideas which can be implemented to ensure effective progress toward early diagnosis of mental health patients.

REFERENCES

1. Spielman, A. J., Caruso, L. S., & Glovinsky, P. B. (1987) A behavioral perspective on insomnia treatment. *Psychiatric Clinics of North America*, 10(4), 541–553.
2. Hyman, S. E. (2007) Can neuroscience be integrated into the DSM-V? *Nature Reviews: Neuroscience*, 8(9), 725–732. doi: 10.1038/nrn2218.
3. Insel, T. R., & Cuthbert, B. N. (2015) Brain disorders? Precisely. *Science (New York)*, 348(6234), 499–500. doi: 10.1126/science.aab2358.
4. Sharp, R. (2015) The hamilton rating scale for depression. *Occupational Medicine*, 65(4). doi: 10.1093/occmed/kqv043.
5. Spitzer, R. L., Williams, J. B. W., Kroenke, K., & Colleagues (1990) Patient Health Questionnaire-9 (PHQ-9). Retrieved from http://www.phqscreeners.com/pdfs/02_PHQ-9/English.pdf.
6. An, H., Lu, X., Shi, D., Yuan, J., Li, R., & Pan, T. (2019) "Mental health detection from speech signal: A convolution neural networks approach", *2019 International Joint Conference on Information, Media and Engineering (IJCIME)*: Lanzhou, China.
7. Gosztolya, G., Grósz, T., Busa-Fekete, R., & Tóth, L. (2014) "Detecting the intensity of cognitive and physical load using adaboost and deep rectifier neural networks", *Proceedings INTERSPEECH 2014 15th Annual Conference of the International Speech Communication Association ISCA*, pp. 452–456: Singapore.
8. Jing, H., Hu, T.-Y., Lee, H.-S., Chen, W.-C. et al., (2014) "Ensemble of machine learning algorithms for cognitive and physical speaker load detection", *Proceedings INTERSPEECH 2014 15th Annual Conference of the International Speech Communication Association ISCA*, pp. 447–451: Singapore.
9. Valstar, M., Schuller, B., Smith, K., Eyben, F. et al., (2013) "AVEC 2013: The continuous audio/visual emotion and depression recognition challenge", *Proceedings of the* 3rd *ACM International Workshop on Audio/Visual Emotion Challenge*, pp. 3–10: Barcelona, Spain.
10. Valstar, M., Schuller, B., Smith, K., Almaev, T. et al., (2014) "AVEC 2014: 3d dimensional affect and depression recognition challenge", *Proceedings of the* 4th *International Workshop on Audio/Visual Emotion Challenge*, pp. 3–10: Orlando Florida.
11. Cummins, N., Joshi, J., Dhall, A. et al., (2013) "Diagnosis of depression by behavioural signals: A multimodal approach", *Proceedings of the 3rd ACM International Workshop on Audio/Visual Emotion Challenge*: Barcelona, Spain.
12. Yang, L., Jiang, D., Xia, X., Pei, E. et al., (2017) "Multimodal measurement of depression using deep learning models", *Proceedings of the 7th Annual Workshop on Audio/Visual Emotion Challenge*, pp. 53–59: Mountain View, California.
13. Yang, L., Sahli, H., Xia, X., Pei, E. et al., (2017) "Hybrid depression classification and estimation from audio video and text information", *Proceedings of the 7th Annual Workshop on Audio/Visual Emotion Challenge AVEC'17*, pp. 45–51: Mountain View, California.
14. Cummins, N., Alice, B., & Schuller, B. W. (2018) "Speech analysis for health: Current state-of-the-art and the increasing impact of deep learning", *Methods*, 151, 41–54.
15. Krizhevsky, A., Sutskever, I., & Hinton, G. (2012) *ImageNet Classification with Deep Convolutional Neural Networks*. Curran Associates Inc: Red Hook, NJ.
16. Shivhare, S. N., & Saritha, S. K. (2014). Emotion detection from text documents. *International Journal of Data Mining and Knowledge Management Process*, 4, 51–57.
17. Shivhare, S. N., Garg, S., & Mishra, A. (2015) "Emotion finder: Detecting emotion from blogs and textual documents," *In International Conference on Computing, Communication and Automation (ICCCA 2015)*: Greater Noida, India.
18. Yang, C., Lin, K. H.-Y., & Chen, H.-H. (2007) "Emotion classification using web blog corpora", *WI'07: Proceedings of the IEEE/WIC/ACM International Conference on Web Intelligence*: Washington, DC.

19. Alswaidan, N. & Menai, M. E. B. (2020). A survey of state-of-the-art approaches for emotion recognition in text. *Knowledge and Information Systems*. doi: 10.1007/s10115-020-01449-0.
20. Amelia, W., & Maulidevi, N. U. (2016) Dominant emotion recognition in short story using keyword spotting technique and learning-based method. *In 2016 International Conference on Advanced Informatics: Concepts, Theory and Application (ICAICTA)*, pp. 1–6: Penang, Malaysia.
21. Laijawala, V., Aachaliya, A., Jatta, H., & Pinjarkar, V. (2020). Classification algorithms based mental health prediction using data mining.
22. Reddy, U. S., Thota, A. V., & Dharun, A. (2018). "Machine learning techniques for stress prediction in working employees", *2018 IEEE International Conference on Computational Intelligence and Computing Research (ICCIC)*, pp. 1–4: Madurai, India.
23. Dooshima, M. P., Chidozie, E. N., Ademola, B. J., Sekoni, O. O., & Adebayo, I. P. (2018). "A predictive model for the risk of mental illness in Nigeria using data mining", *International Journal of Immunology*, 6(1), 5–16.
24. Srividya, M., Subramaniam, M., & Natarajan, B. (2018) "Behavioral modeling for mental health using machine learning algorithms", *Journal of Medical Systems*, 42(5), 88.
25. Filip, D. & Jesus, C. (2015). "A neural network based model for predicting psychological conditions," *International Conference on Brain Informatics and Health*, pp. 252–261: London, UK.
26. Panagiotakopoulos, T. C., Lyras, D. P., Livaditis, M., & Sgarbas, K. N. (2010). A contextual data mining approach towards assisting the treatment of anxiety disorders. *IEEE Transactions on Information Technology in Biomedicine*, 14(3), 567–581.
27. Ordonez, C. (2006). Comparing Association Rules and Decision Trees for Disease Prediction. In *CIKM06: Conference on Information and Knowledge Management*: Arlington, VA.
28. Kohavi, R. (2011). *Scaling up the Accuracy of Naive-Bayes Classifiers: A Decision Tree Hybrid in Data Mining and Visualization*. Silicon Graphics, Inc: Milpitas, CA.
29. Jachyra, D., Pancerz, K., & Gomula, J. (2011). Classification of MMPI profiles using decision trees. *The International Conference on Digital Technologies 2013*, pp. 90–98: Zilina, Slovakia.
30. Landwehr, N., Hall, M., & Frank, E. (2005). Logistic model trees. *Machine learning*, 59(1–2), 161–205.
31. D'monte, S., Tuscano, T., Raut, L., & Sherkhane, S. (2018). Rule generation and prediction of anxiety disorder using logistic model trees. *2018 International Conference on Smart City and Emerging Technology (ICSCET)*, Mumbai, pp. 1–4, doi: 10.1109/ICSCET.2018.8537258.
32. Gao, S., Calhoun, V. D., & Sui, J. (2018). Machine learning in major depression: From classification to treatment outcome prediction. *CNS Neuroscience and Therapeutics* 24, 1037–1052. doi: 10.1111/cns.13048.
33. Siriaraya, P., Visch, V., van Dooren, M., & Spijkerman, R. (2018). Learnings and challenges in designing gamifications for mental healthcare: The case study of the ready set goals application. *2018 10th International Conference on Virtual Worlds and Games for Serious Applications (VS-Games)*, pp. 1–8. doi: 10.1109/VS-Games.2018.8493430.
34. Altameem, T., Amoon, M., & Altameem, A. (2020). A deep reinforcement learning process based on robotic training to assist mental health patients. *Neural Computing and Applications*. doi: 10.1007/s00521-020-04855-1.

7

Application of Machine Learning in Cardiac Arrhythmia

Gresha S. Bhatia, Shefali Athavale, Yogita Bhatia, Tanya Mohanani, and Akanksha Mittal
Vivekanand Education Society's Institute of Technology

CONTENTS

7.1 Introduction 115
7.2 Proposed Machine Learning Implementation 117
7.2.1 Data Preparation 117
7.2.2 Data Preprocessing 117
7.2.3 Development of Model for Detection 118
7.2.4 Training and Experimenting the Model for the Data 119
7.2.5 Detection of Arrhythmia 119
7.3 Experimentation and Results 120
7.4 Evaluation Parameters and Measures 120
7.5 Comparison with Existing Systems 121
7.6 Conclusion 123
References 123

7.1 Introduction

On the global front, many people are suffering from chronic diseases. One such ailment is heart disease, also known as cardiovascular diseases (CVDs), that affects the large size of the population. CVDs include heart attack, stroke, and hypertension caused by disorders of the heart while arrhythmia or irregular heartbeats disturb the electrical system of the heart and cause the heart to abnormally race or skip beats. This further leads to a nonsequential movement of heart signals, causing a major cause of concern. Due to several innovations in healthcare systems and many technological advances, there is a need for an efficient, accurate, early detection, and classification system for arrhythmia. These early detection systems would aid in providing medical assistance for the prevention of heart ailments and in turn save human lives.

Early detection of these short-term, infrequent arrhythmias requires long-term monitoring of electrical activity of the heart. Cardiologists and expert doctors use electrocardiogram (ECG) as a tool to measure and record the electrical activity of the heart and hence diagnose the smooth functioning of the heart. The cardiologists further characterize the heartbeat through extracted sample points from the ECG curve.

Therefore, to provide an effective treatment for arrhythmias, an early diagnosis is important. We can achieve good classification and performances by utilizing computer-assisted techniques and methods. However, many require complex classification mechanisms, long computation times, and a lot of computational power. To add to this further, experts find it strenuous to find minute irregularities from the long-duration ECG recordings which is a manual approach.

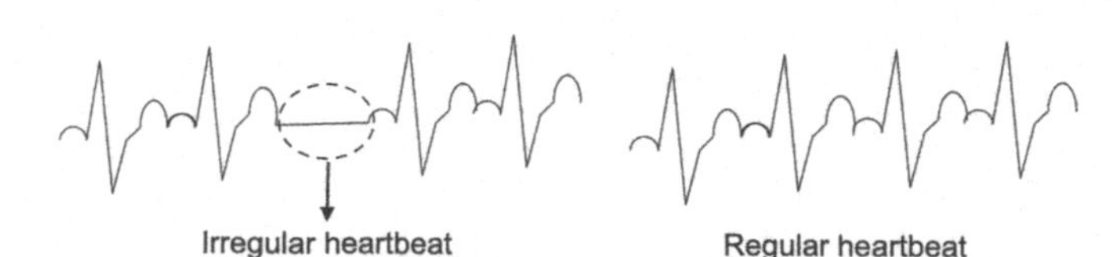

FIGURE 7.1 Regular and Irregular Heartbeat.

Figure 7.1 represents the ECG pattern of a normal heartbeat and arrhythmia beat. It becomes very difficult to differentiate between them and is a manual process. All this causes a lot of misclassified and misunderstood information leading to chaos and affects the ailment diagnosis to a greater extent. Focusing on determining early symptoms and easing the diagnosis process, we aim to utilize machine learning algorithms combined with ECG reports aiding in automating arrhythmia detection and classification. This project will represent an efficient alternative to long-duration ECG recordings and will be useful for doctors and other healthcare professionals to determine the condition of cardiac arrhythmia in patients a priori.

The work in this domain began with a view of establishing a link between the current scenario and the need for analyzing cardiac arrhythmia from a technological perspective. It consisted of gathering information by interacting with domain experts and focusing on various research papers in the medical and technological domains.

Interactions with the experts in the domain revealed that ECG does give an insight into the heart ailment; however, it is all the experience, intuition, and manual process that takes the diagnosis further.

In Ref. [1], Andrew Y. Ng et al. elaborated on an algorithm that detected a wide range of heart arrhythmias using ECG signals through utilizing 34-layer convolution neural networks (CNN). However, it did not detect all types of arrhythmias nor did it compare the accuracy of CNN with different techniques. The authors further indicated that major decision-making was a manual process.

Albert Haque [2] also determined cardiac arrhythmias through the use of neural network as a primary model, and the results are compared with support vector machine (SVM), random forest, and logistic regression. However, it is observed that in the method, the cost of determining false-negative (patient with arrhythmia classified as a nonarrhythmia patient) rate is higher since the algorithm doesn't proceed further to classify them.

The researchers in Ref. [3] elaborated on the use of naive Bayes, SVM, random forest classifier, and neural networks for classification. However, the limitation was there was a need to combine the algorithms in a hierarchical scheme to obtain maximum accuracy.

Himanshu Gothwal et al. explained in Ref. [4] the use of fast Fourier transform to identify features from an ECG signal. Those features generated training as well as testing a dataset for the artificial neural network (ANN) to predict diseases.

The authors of [5] discuss different types of machine learning and deep learning models on the prediction of cardiac arrhythmia. It is inferred that the performance of ANNs is high compared to the usual machine learning models.

The authors in Ref. [6] introduce the use of hybrid classifiers (SVM and ANN). The accuracy obtained using genetic algorithm (GA) and ANN is 99.23% and GA with cuckoo search and ANN with SVM is 99.31%.

Ali Isina and Selen Ozdalilib [7] propose a transferred deep CNN named AlexNet for extracting features from the Massachusetts Institute of Technology–Beth Israel Hospital arrhythmia database. The final classification is carried out by feeding the extracted features to a simple back propagation neural network, which gives a 92% testing accuracy.

A highly efficient algorithm is proposed in Ref. [8] for P-wave detection. The algorithm is based on phasor transform and newly designed rules based on knowledge of heart manifestations and classification (detection of premature ventricular contractions).

In Ref. [9], the authors have trained a deep learning model in which the performance of it is superior for detecting several arrhythmias from ECG records. The main idea that contributes to the performance is the multidimensional representation and multilayer deep learning model. It can recover the structure of 1D data and fit the model based on the training data.

The researchers in Ref. [10] present an echo state network approach that is suitable for processing long-term recordings and large databases as the feature extraction and the algorithm itself both have minimal computational requirements.

In Ref. [11], the author presents the architecture of Xception model, which has a similar parameter count as Inception-v3. The observation showed how convolutions and depth-wise separable convolutions lie at both extremes of a discrete spectrum, with Inception modules being an intermediate point in between.

The authors of this chapter would like to elaborate on the need for applying machine learning techniques toward the detection of cardiac arrhythmia. Moreover, the authors would also like to present a brief summary of all the solutions currently developed and an analysis of various evaluation parameters to determine the best-available framework for the process of detection and increase the test accuracy of this system. This chapter focuses on how machine learning can be used in the detection of cardiac arrhythmia. It further explores the results produced upon the implementation of the proposed methodology followed by evaluation measures which play a significant role in determining the accuracy of the system.

7.2 Proposed Machine Learning Implementation

The implementation of the system which is being discussed in subsequent sections is powered by artificial intelligence (AI). Machine learning, being a subset of AI, has been used to develop the core algorithm of the system. With the help of machine learning models, the system is capable of accurately detecting cardiac arrhythmia.

Figure 7.2 represents the block diagram, consisting of data preprocessing with feature extraction and detection modules. The preprocessing stage is responsible for the conversion of data to a form that can be processed by the detection module. And in the detection stage, a neural network architecture is used to detect arrhythmia from the given input. The detailed methodology is as follows.

7.2.1 Data Preparation

PhysioNet is an online library which has a large collection of clinical datasets for medical and scientific research. For our system, the dataset is generated by downloading the required waveform files from PhysioNet.

7.2.2 Data Preprocessing

The waveform files obtained need to be converted into the form required by the detection model. A spectrogram is a visual representation of the spectrum of frequencies of a signal as it varies with time. A

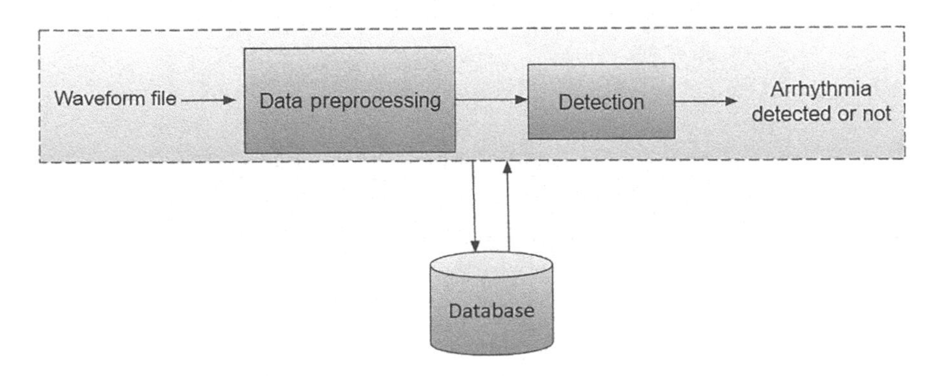

FIGURE 7.2 Block Diagram of the system.

spectrogram is usually depicted as a heat map, i.e., as an image with the intensity shown by varying the color or brightness. These images are augmented before training to improve accuracy.

A function is defined to convert .wav files to spectrogram images. The function first reads the .wav file using python's "SciPy" library function "scipy.io.wavefile.read(filename)" which returns the sample rate and data from .wav file. A spectrogram image of the data (i.e., of .wav file) is plotted using the "specgram()" function in "Matplotlib" library of python.

The spectrogram images obtained are as shown in Figures 7.3 and 7.4.

7.2.3 Development of Model for Detection

The dataset is downloaded from PhysioNet, which consists of 3000 recordings. It consists of .wav files of two classes, namely: abnormal where arrhythmia is present and normal which indicates that arrhythmia is not present.

FIGURE 7.3 Spectrogram Image obtained from the waveform file of a normal patient.

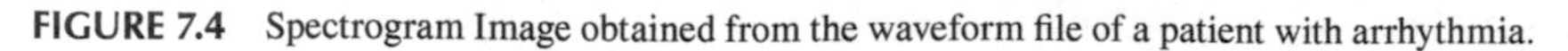

FIGURE 7.4 Spectrogram Image obtained from the waveform file of a patient with arrhythmia.

The model used to detect cardiac arrhythmia is the Xception model. The Xception model just as the name suggests is an extension or extreme version of the inception architecture with a modified depth-wise separable convolution. The model is pretrained on the ImageNet dataset, which contains over 1.2 million images. The Xception model is a CNN, which is 71 layers deep and can classify objects into 1000 categories.

7.2.4 Training and Experimenting the Model for the Data

The dataset is already converted into two-dimensional (2D) spectrogram images. This dataset is then further divided into the ratio of 80:20 into training and validation, respectively. The training part of the dataset is then used to train the model since the Xception model is not pretrained for detecting arrhythmia. Once the training is done, the remaining 20% of the dataset is used to validate and test the model trained. Since the testing dataset is new for the model, the accurate detection of arrhythmia will make the model more efficient and useful for deployment.

7.2.5 Detection of Arrhythmia

The working of the detection module is as shown in Figure 7.5 and represented as:

7.2.5.1. The detection module takes the ECG signals of the patient as input.

7.2.5.2. These input signals are converted into 2D spectrogram images using a SciPy library.

7.2.5.3. The obtained spectrogram image is then given to the Xception model, which employs the transfer learning algorithm for detecting the class of arrhythmia. Transfer learning is one of the approaches used in machine learning where a model makes use of knowledge acquired for one task to solve the related tasks. The use cases of transfer learning include feature extraction and fine-tuning of the model. The Xception model was previously trained on the ImageNet dataset.

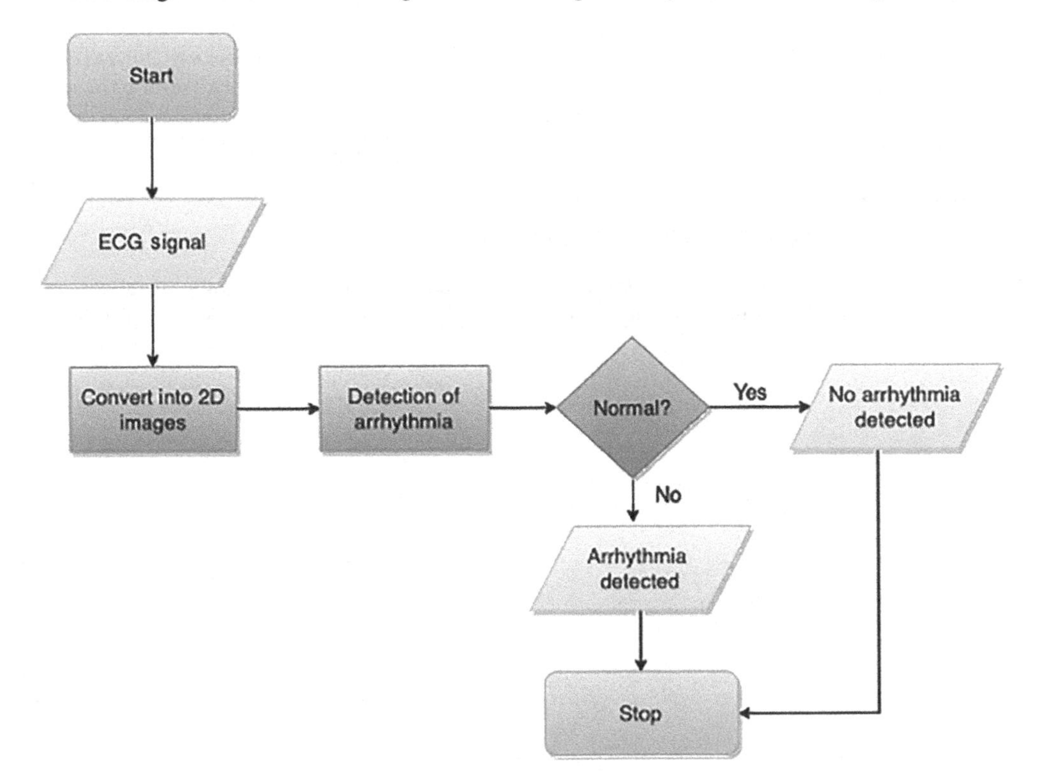

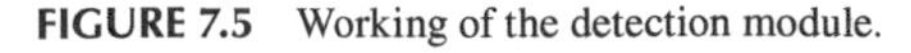
FIGURE 7.5 Working of the detection module.

Its last fully connected layers were replaced with new fully connected layers based on the number of classes the dataset has. The model is making use of previous weights and features such as the most prominent pixels in an image to help detect the new classes more accurately.

7.2.5.4. The model then generates the results which can be: arrhythmia detected or arrhythmia not detected.

Deployment of the Model and Testing it on Real-World Scenarios

While deploying this trained and tested model in real life, we can make use of a website to which the waveform will be given as input. This waveform will then be converted into a spectrogram image. This spectrogram is then processed by applying a CNN to detect the class of cardiac arrhythmia. The model can be leveraged for further improvement in the methodology.

7.3 Experimentation and Results

As mentioned previously, the Xception model has been trained using the training dataset (80% of the main dataset) to detect arrhythmia. The model was then tested on the remaining 20% of the dataset, i.e., the test dataset, and the performance was evaluated and is as shown in Figure 7.6. Since the model could not provide satisfactory accuracy, some of the hyperparameters like the learning rate, the number of epochs, and batch size were changed. After tweaking the hyperparameters, the model was again trained on the training dataset, and its performance is as shown in Figure 7.7.

The graphs (Figures 7.6 and 7.7) generated when the model was trained and tested are as shown below.

An epoch is a hyperparameter which essentially means one full round of training of a model. The graphs (Figures 7.7 and 7.8) show the plot of epoch vs loss and the plot of epoch vs accuracy respectively obtained after training the model for 10 epochs. The light grey line here represents the learning curve of the model obtained after training it on the training dataset and the dark grey line represents the learning curve obtained from the testing dataset. Our dataset has been split into an 80:20 ratio for training and testing.

Xception model is a deep CNN and just like any other CNN, the model updates its weights after every epoch. The number of epochs is never fixed and it has to be kept sufficiently large, allowing the learning algorithm to run until the error has been minimized. As the number of epochs increases, more times the weights get updated in the neural net and the underlying learning algorithm learns better.

7.4 Evaluation Parameters and Measures

In this section, the parameters which should be used for determining the accuracy of the system are described. The authors also discuss the accuracy of the system and present samples to substantiate the information.

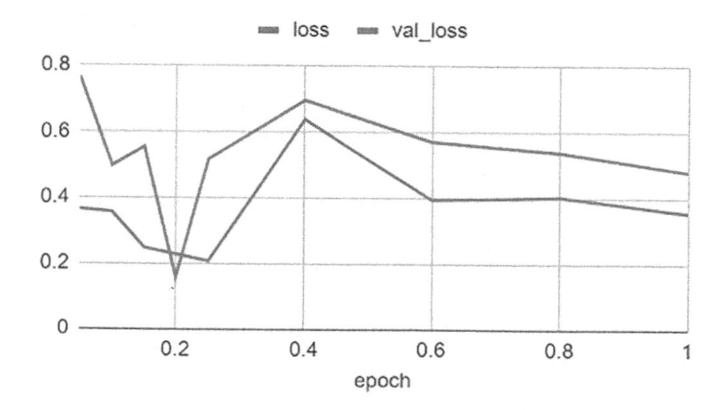

FIGURE 7.6 Training and Validation loss curve when trained the Xception model.

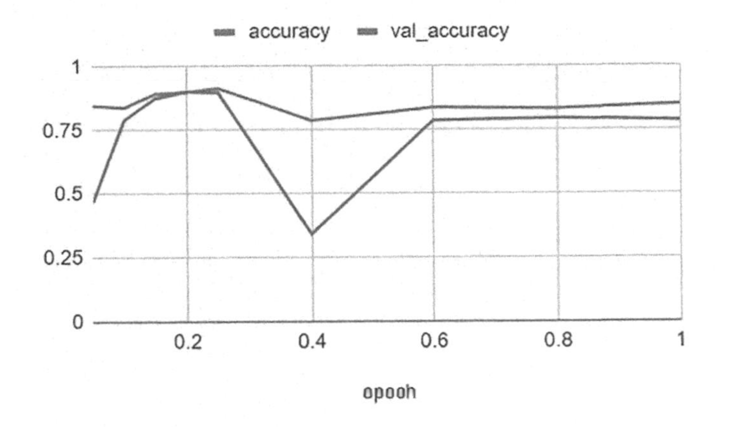

FIGURE 7.7 Training and Validation accuracy curve when trained the Xception model.

The accuracy of the system is measured as the percentage of correctly classified instances.

$$\text{Accuracy} = (\text{TN} + \text{TP})/(\text{TN} + \text{TP} + \text{FN} + \text{FP})$$

where,

TP = Number of true positives
FN = Number of false negatives
FP = Number of false positives
TN = Number of true negatives
In our system:

True positive: Case where a patient has arrhythmia is correctly detected by the system
True negative: Case where a patient is normal is correctly detected by the system
False positive: Case where a patient is normal but the system detects an arrhythmia
False negative: Case where a patient has arrhythmia but the system detects as normal

Test Cases:

The following test cases consist of spectrogram images which are converted from the ECG recordings of the patient. Two cases are being presented below in order to show the behavior of the system.

Test case 1: The spectrogram image (Figure 7.8) and its output (Figure 7.9) denote the abnormal class of arrhythmia.

For the previous image, the system detected the probability of the abnormal class as 98.47% and the probability of normal class as 1.53%. Since the probability of abnormal class is much higher than the normal class, the system classifies this patient to be of abnormal class, i.e., the patient has arrhythmia. Hence, we can conclude that the system correctly classified the testing data.

Test case 2: The spectrogram image as represented in Figure 7.10 and its output represented in Figure 7.11 denote the normal class of arrhythmia.

The system detected that for the previous image the probability of the normal class is 100% and the probability of abnormal class is 0%. The system classifies this patient to be normal class, i.e., the patient has no arrhythmia. Hence, the system correctly classified the testing data.

7.5 Comparison with Existing Systems

Table 7.1 shown below gives a comparison of the existing system namely HeartToGo system and iBoSen system with the proposed detection system that detects arrhythmia using .wav files.

FIGURE 7.8 Spectrogram Image containing the abnormal class of arrhythmia.

```
Using TensorFlow backend.
2020-04-04 17:42:20.583786: I tensorflow/core/platform/cpu_feature_guard.cc:142] Your CPU supports instru
ctions that this TensorFlow binary was not compiled to use: AVX2
Top 1 ===================
Class name: abnormal
Probability: 98.47%
Top 2 ===================
Class name: normal
Probability: 1.53%
```

FIGURE 7.9 Output of the model predicting the class probability of the image.

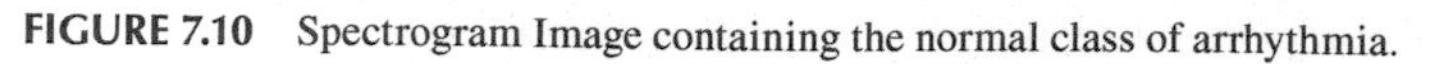

FIGURE 7.10 Spectrogram Image containing the normal class of arrhythmia.

```
Using TensorFlow backend.
2020-04-04 17:48:34.797225: I tensorflow/core/platform/cpu_feature_guard.cc:142] Your CPU supports instru
ctions that this TensorFlow binary was not compiled to use: AVX2
Top 1 ====================
Class name: normal
Probability: 100.00%
Top 2 ====================
Class name: abnormal
Probability: 0.00%
```

FIGURE 7.11 Output of the model predicting the class probability of the image.

TABLE 7.1
Comparison of Existing Systems

Characteristics	HeartToGo System	iBoSen System	Proposed System
Platform	Cell phone–based wearable device	Wearable ECG monitoring system	Software-oriented system
Technique	Artificial neural network	Hidden Markov model	Transfer learning
Output	Generates cardiac health summary reports	Displays ECG waveform	Detects whether cardiac arrhythmia is present or not
Evaluation measures	Physical movement intensity and summary of heart rate (bpm)	Accuracy, sensitivity, and positive predictivity	Accuracy, precision, recall, and F1-score

7.6 Conclusion

This chapter focuses upon the purpose behind making use of technologies like machine learning in the healthcare sector. Along with that, the chapter highlights an effective system for the detection of arrhythmia from ECG signal data obtained from ".wav" files. The rising amounts of heart-related cases in the past few years and increasing technological developments have motivated the authors to employ AI. According to a survey conducted across India, a majority of the respondents who suffered from heart ailments were of the age-group 60 and above. Patients of this age-group are comparatively more vulnerable to deaths than patients of younger age-groups. Moreover, individuals from poor financial backgrounds are unable to afford the expenses incurred by the optimal therapy. Hence, it becomes extremely important to diagnose patients at an early stage and provide them access to quality healthcare services in order to avoid any serious issues in a later period of time. AI is largely being employed in the field of healthcare and a quick early diagnosis of CVDs can be made possible by providing the doctors with highly accurate AI-driven systems. This chapter elaborates upon a potential system which can be used by doctors to facilitate early diagnosis of heart-related diseases. The system proposed in this chapter works by taking the ECG recording of the patient as an input and converts it into a 2D image. The machine learning model, which has been used in this system, is trained to detect cardiac arrhythmia by utilizing more than 3000 recordings, of which 80% of the data is used for training and 20% to validate the detection. Evaluation of the system is further done through the parameters of accuracy, precision, recall, and comparison with the existing systems.

REFERENCES

1. Andrew Y. Ng, Pranav Rajpurkar, Awni Y. Hannun, et al. "Cardiologist-level arrhythmia detection with convolutional neural networks", 6 July 2017.
2. Albert Haque, "Cardiac dysrhythmia detection with GPU-accelerated neural networks", 2015. https://pdfs.semanticscholar.org/7bd7/581cbfa9d40e872d3759d7c0a026b94662d9.pdf?_ga=2.243844557.992428240.1598638039-359226557.1591635536.
3. Vasu Gupta, Sharan Srinivasan, and Sneha S. Kudli, "Prediction and classification of cardiac arrhythmia", 2013. https://pdfs.semanticscholar.org/8d5c/cd84af537948c1227d6b1b504f34b433a58c.pdf?_ga=2.243279181.992428240.1598638039-359226557.1591635536.

4. Himanshu Gothwal, Silky Kedawat, and Rajesh Kumar, "Cardiac arrhythmias detection in an ECG beat signal using FFT and ANN", *Journal of Biomedical Science and Engineering*, 2011. https://www.researchgate.net/publication/215446357_Cardiac_Arrhythmias_Detection_In_An_ECG_Beat_Signal_Using_Fast_Fourier_Transforms_and_Artificial_Neural_Network.
5. V. Sai Krishna and A. Nithya Kalyani "Prediction of cardiac arrhythmia using artificial neural network", *International Journal of Recent Technology and Engineering (IJRTE)*, 2019, 8(1S4). https://www.ijrte.org/wp-content/uploads/papers/v8i1s4/A10860681S419.pdf.
6. Pooja Sharma, D. V. Gupta, and Surender Jangra "ECG Signal based arrhythmia detection system using optimized hybrid classifier" *International Journal of Innovative Technology and Exploring Engineering (IJITEE)*, 2019, 8(9), 1–6.
7. Ali Isina and Selen Ozdalilib "Cardiac arrhythmia detection using deep learning" *In 9th International Conference on Theory and Application of Soft Computing, Computing with Words and Perception, ICSCCW 2017*, Budapest, Hungary, August 2017.
8. Lucie Maršánová, Andrea Němcová, Radovan Smíšek, et al. "Advanced P wave detection in ECG signals during pathology: Evaluation in different arrhythmia contexts", December 2019. https://www.nature.com/articles/s41598-019-55323-3.
9. K. S. Rajput, S. Wibowo, C. Hao, et al. "On arrhythmia detection by deep learning and multidimensional representation", March 2019. https://www.researchgate.net/publication/332140572_On_Arrhythmia_Detection_by_Deep_Learning_and_Multidimensional_Representation.
10. Miquel Alfaras, Miquel C. Soriano, Silvia Ortin "A fast machine learning model for ECG-based heart rate classification and arrhythmia detection" *Interdisciplinary Physics, Frontiers in Physics Journal*, 2019. https://www.frontiersin.org/articles/10.3389/fphy.2019.00103/full.
11. François Chollet, Xception: Deep learning with depth wise separable convolutions" *2017 IEEE Conference on Computer Vision and Pattern Recognition (CVPR)*, Honolulu, HI, 2016. https://ieeexplore.ieee.org/document/8099678.

8

Advances in Machine Learning and Deep Learning Approaches for Mammographic Breast Density Measurement for Breast Cancer Risk Prediction: An Overview

Shivaji D. Pawar and Kamal Kr. Sharma
Lovely Professional University

Suhas G. Sapate
Annasaheb Dange College of Engineering and Technology

CONTENTS

8.1 Introduction 126
8.2 Machine Learning Approach 127
8.2.1 Preprocessing 127
8.2.2 Breast Border Detection Algorithms 128
8.2.2.1 Gray-Level Thresholding 129
8.2.2.2 Iterative Optimal Thresholding 129
8.2.2.3 Otsu's Optimal Thresholding 130
8.2.2.4 Minimum Cross-Entropy Thresholding 130
8.2.3 Pectoral Muscle Removal Algorithms 130
8.2.3.1 Image-Based Approach 131
8.2.3.2 Model-Based Approach 132
8.2.4 Segmentation 133
8.2.4.1 Region-Growing Algorithm 133
8.2.4.2 Graph-Cut Algorithm 134
8.2.4.3 Fuzzy C-Means 134
8.2.4.4 Watershed Algorithm 134
8.2.4.5 Otsu's Optimal Thresholding 135
8.2.4.6 Fusion of K-Means and Region-Growing Algorithms 135
8.2.5 Feature Extraction and Classification 135
8.2.5.1 Statistical Feature Extraction 135
8.2.5.2 Feature Reduction 137
8.2.5.3 Classification 137
8.3 Deep Learning Approach 137
8.3.1 Preprocessing 138
8.3.2 Design of Convolutional Neural Network Algorithm 138
8.3.2.1 Input Layer 138
8.3.2.2 Convolutional Layer 138
8.3.2.3 Max Pooling Layers 138
8.3.2.4 Activation Function 138
8.3.2.5 Classification Layer 138

8.3.2.6 Dropout Layer 138
8.3.2.7 Kernel Selection 139
8.3.2.8 Development of Baseline (Density Map) for Breast Density Prediction 139
8.3.3 Validation and Testing of Machine and Deep Learning Model 139
8.3.4 Related Work 139
8.4 Findings and Discussion 140
8.5 Conclusion 141
References 141

8.1 Introduction

Breast cancer is one of the leading cancers affecting women health and it is the second foremost cause of cancer deaths in the world. "The International Agency for Research on Cancer" performed a study of 20 different world areas in 2018 and highlighted "18.1 million" new cancer patients and "9.6 million" cancer deaths (Bray et al. 2018, 394). Mammographic views represent fibroglandular tissues which can be measured in terms of mammographic breast density (MBD) or percentage mammographic density appears white in color on mammogram projecting a fibroglandular tissue in the breast as compared to the whole breast (Science Daily, 2016). Breast density is a significant parameter in medical diagnosis, especially for breast cancer risk prediction. A Medio lateral oblique (MLO) view of MBD is depicted in Figure 8.1.

In the case of dense breasts, the mammogram's sensitivity decreases by 48% as compared to 98% in fatty breasts; hence every second or third cancer may undergo misprediction from a prediction which will cause a threat to life and an increase in treatment cost (Science Daily, 2016). Breast density calculation from mammography is a complex task in breast cancer prevention and treatment.

The mammographic density pattern was firstly described qualitatively by Wolfe (1976) into four classes. The author performed density measurements with the help of a polar planimeter by transforming radiographic densities on a continuous scale. To improve screening technology and remove the confusion about mammographic findings, "The breast imaging-reporting and data system (BIRADS)" was the standardized mammographic reporting, suggesting four major types for classifying breast density on a pattern-based approach known as "BIRADS 4th edition." This standard is further modified as "BIRADS 5th edition," which not only depends upon the percentage distribution but also calculates breast density according to the incidence of a region of fibroglandular tissue that may hide underlying cancer (Liberman and Menell, 2002). Nowadays, "BI-RADS 5th edition" is the most commonly used method of measuring breast density in clinical radiology practice.

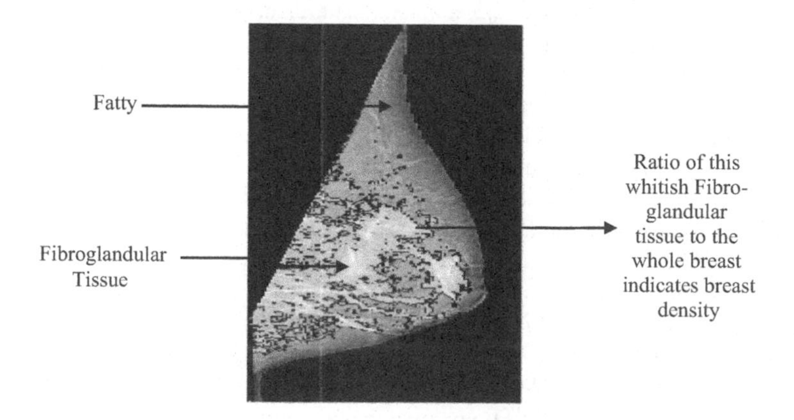

FIGURE 8.1 Concept of mammographic breast density.

In the last two decades, different researchers have proposed many semiautomatic and automatic methods for the measurement of breast density to support the radiologists as an additional opinion for breast density measurement and to reduce the possible humanoid errors. However, due to accuracy variations, no method is practically suitable to classify breast density quantitatively. The success story of breast density measurement lies in the basic blocks of an image processing pipeline. Performance parameters of breast density measurement can be reached at their optimal level if a more advanced version of algorithms is used in the image processing pipeline.

The basic inclination behind this chapter is to provide a study and analysis of different approaches used by various researchers for automatic breast density measurement. The major focus is to provide an overview of the current status, development, and future scope of machine learning and deep learning approaches for breast density measurement and subsequently compare them in terms of computational complexity and accuracy.

This chapter is based on the selected machine and deep learning–based research articles on breast density measurement from IEEE Xplore, Science Direct, Springer, Elsevier, and Google Scholar database. The organization of the chapter hereafter is as follows. Section 8.1 highlights the comparative analysis of selected machine learning algorithms used for breast density measurement. Section 8.2 provides an analysis of major deep learning approaches. Section 8.3 presents the comparative analysis of machine learning and deep learning algorithms in terms of accuracy and computational complexity as well as merits and demerits. The important observations and research findings are discussed in Sections 8.4, and 8.5 concludes the chapter.

8.2 Machine Learning Approach

If there is fast-forward to 10 years of our life, then machine learning is part of every industrial application like high-tech products, speech recognition, image processing, and many more. Machine learning has the ability to learn handcrafted features from the objects and classify them as per the design requirement. In the last decade, many researchers attempted to develop machine learning approaches for the measurement of breast density which is discussed in this section. The machine learning approach follows the pipeline of an image processing system, in which the entire model is divided into four stages such as preprocessing, segmentation, feature extraction, and classification. This image processing pipeline is illustrated in Figure 8.2.

8.2.1 Preprocessing

This is the most significant stage of breast density measurement. Input images for breast density measurement are MLO and craniocaudal views of digital mammograms. These views typically show different areas such as pectoral muscles, breast tissues consisting of adipose and fibroglandular tissues, and the background air region outside the breasts which consists of some artifacts and labels. These areas of input digital mammograms are depicted in Figure 8.3.

Mammographic breast density is directly proportional to the area occupied by the fibroglandular tissues in the breasts which appear whitish on the mammogram. Thus, a small presence of pectoral muscles, artifacts, labels, and background noise which appear whitish can affect the performance of breast density measurement and increase false-positive rates. Therefore, the successful removal

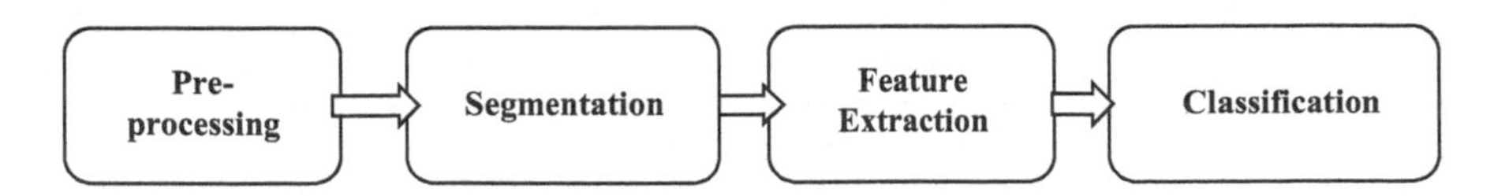

FIGURE 8.2 Machine learning approach for breast density measurement.

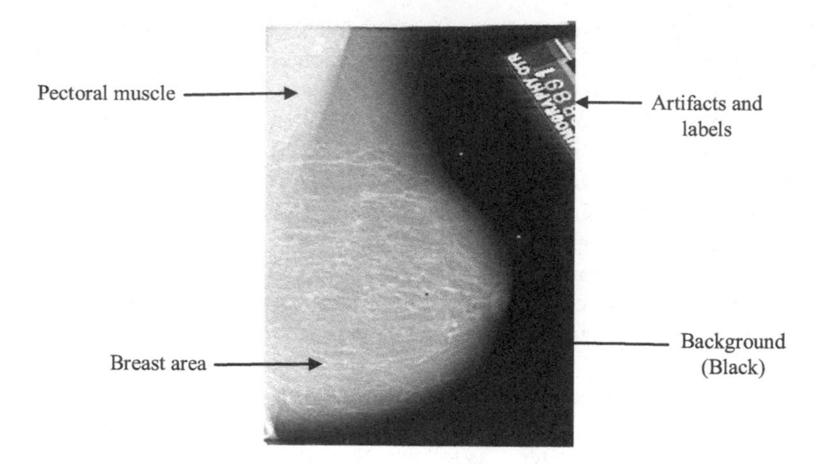

FIGURE 8.3 Different breast regions on MLO view of a mammogram. Courtesy: DDSM dataset Rebecca et al. (2017).

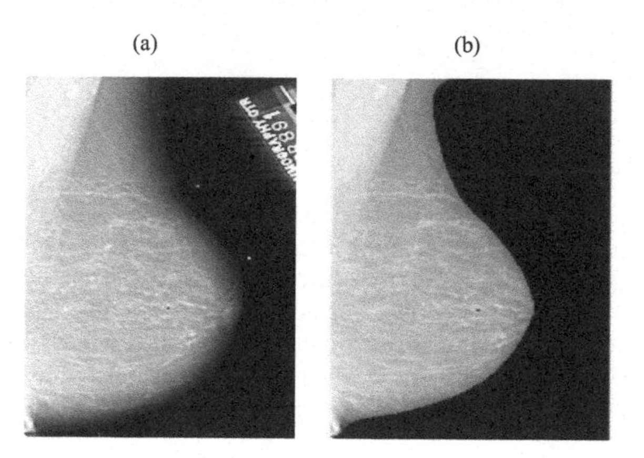

FIGURE 8.4 Concept of pectoral muscle removal. (a) MLO view of a mammogram with artifacts and labels, and (b) mammogram after removal of artifacts and labels.

of all these areas from an input digital mammogram is the basic objective of preprocessing of digital mammograms. Preprocessing of a digital mammogram is performed in two parts: one is breast border detection and the other is pectoral muscle removal. The concept of preprocessing is shown in Figure 8.4.

8.2.2 Breast Border Detection Algorithms

The basic objective behind breast border detection is to isolate the breast part from its background on digital mammograms. The basic focus here is to identify the tissue–air interface in the breast area and remove any noise or artifacts present in digital mammograms. Thresholding acts as a perfect tool to perform this task due to the simplicity in coding and good computational speed. The basic design parameter in thresholding is the selection of a threshold value that segments the image into two or more than two parts. As per the literature survey, there are different thresholding methods used by different researchers for the detection of breast border which solely depends upon the manual or automatic threshold selection. The major thresholding techniques used for breast border detection are depicted in Figure 8.5.

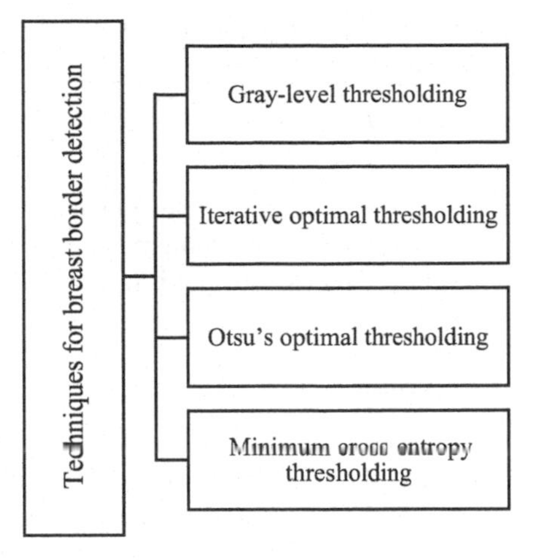

FIGURE 8.5 Major breast border detection techniques from the literature.

8.2.2.1 Gray-Level Thresholding

This is one of the simple methods used for image segmentation, in which variation of pixel intensity of different sections of a digital mammogram is used to select a threshold value. Proper thresholds allow segmenting the mammogram into breast border and background. As per the literature survey, different researchers (Subashini et al. 2010; Liu et al. 2010, 11; Keller et al. 2012) used global thresholding where the basic principle is described mathematically by Equation 8.1.

$$g(x,y)=\begin{cases} 1, & f(x,y)>T \\ 0, & f(x,y)\leq T \end{cases} \tag{8.1}$$

As mentioned earlier, this method is simple to implement but most of the time the output of this algorithm is affected by Gaussian noise of zero mean when the standard deviation of intensity variation is 10–50. This will cause unclear boundaries between the breast border and background air region; hence it needs further post processing. To overcome this problem, two approaches are investigated known as iterative optimal thresholding and Otsu's optimal thresholding which are discussed below.

8.2.2.2 Iterative Optimal Thresholding

The basic focus of iterative optimal thresholding is to reduce a weighted sum of squared error of the breast border and select the optimal threshold to segment the breast border and background. This approach is used by Devi and Vidivelli (2018) by selecting two different background and breast border pixels which are B and b and then with the help of observation determined the mean gray levels $\mu B^{(t)}$ and $\mu b^{(t)}$ for B and b respectively with the help of Equations 8.2 and 8.3.

$$\mu B^{(t)}=\frac{\sum_{(i,j)\in B} f(i,j)}{|B|} \tag{8.2}$$

$$\mu b^{(t)}=\frac{\sum_{(i,j)\in b} f(i,j)}{|b|} \tag{8.3}$$

And finally, a new threshold is calculated with the help of Equation 8.4.

$$T^{t+1} = \frac{\mu B^{(t)} + \mu b^{(t)}}{2} \tag{8.4}$$

If, $T^t = T^{t+1}$ then exit, otherwise use T^{t+1} to segment the image into B and b and continue until threshold T is stable. This algorithm works well as compared to gray-level thresholding and also produces comparable results concerning Otsu's thresholding but consists of a drawback of fuzzy nature output.

8.2.2.3 Otsu's Optimal Thresholding

To enhance segmentation accuracy and avoid the fuzzy nature of the output, Tortajada et al. (2012) proposed the option of Otsu's optimal thresholding with a bimodal distribution. Sharp and deep valleys in-between two peaks are used to separate the breast border. This method performs a complete search to find the intensity difference between and within the group to identify the threshold which separates the breast border from another part of a mammogram. This algorithm provides good quality segmentation but due to complete search requires exponential computation as well as larger memory requirement.

8.2.2.4 Minimum Cross-Entropy Thresholding

To reduce computational complexity, Tzikopoulos et al. (2011) proposed another alternative to find minimum cross-entropy between different parts of the mammogram. In this algorithm, the author performed pixel-to-pixel level cross-entropy between the background and breast border to create a histogram. This concept reduces the computational complexity but is prone to sensitivity to the noise. Comparative analysis of breast border detection algorithms is highlighted in Table 8.1.

All the above-listed algorithms show variations in segmentation accuracy from low to optimal level, but as the accuracy of segmentation increases and then computational complexity increases, there is a need for a more refined version of algorithms that will reduce computational complexity with good segmentation accuracy.

8.2.3 Pectoral Muscle Removal Algorithms

Fibroglandular tissues and pectoral muscles have the same intensity; hence the small incidence of pectoral muscles in the breast area may cause a reduction in classification accuracy. Accurate removal of pectoral muscles from the breast area is a more thought-provoking and challenging job in front of researchers due to its change in size, shape, and position in each mammogram. The concept of pectoral muscle removal in digital mammograms is illustrated in Figure 8.6.

TABLE 8.1

Merits and Demerits of Different Breast Border Detection Methods

Methods	Computational Complexity	Accuracy	Merits	Demerits
Gray-level thresholding	Low	Less than optimal level	Easy for coding and less computational complexity	A less-clear boundary between background and foreground and narrow in applications
Iterative optimal thresholding	Average	Good	Average computational complexity	Image output details are fuzzy in nature.
Otsu's optimal thresholding	High	Good	Simple in coding	Requires uniform illumination. The histogram should be bimodal
Minimum cross-entropy thresholding	Low	Less than optimal level	Less computational complexity	Sensitive to noise

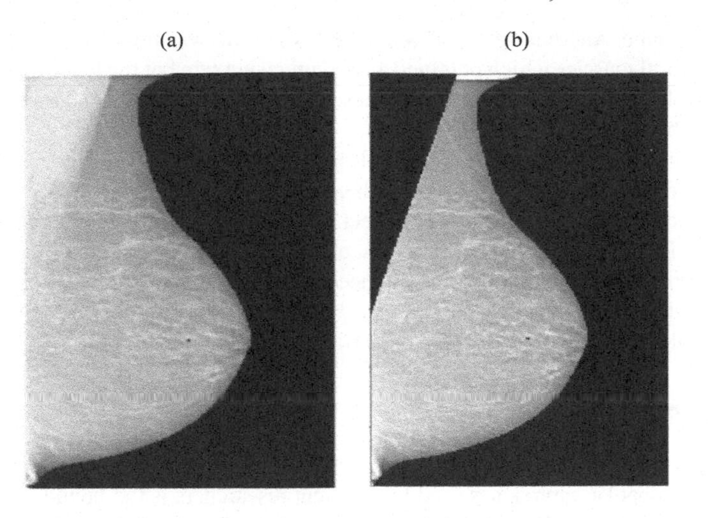

FIGURE 8.6 Concept of pectoral muscle removal. (a) MLO view of a mammogram with pectoral muscle and (b) mammogram after removal of pectoral muscles.

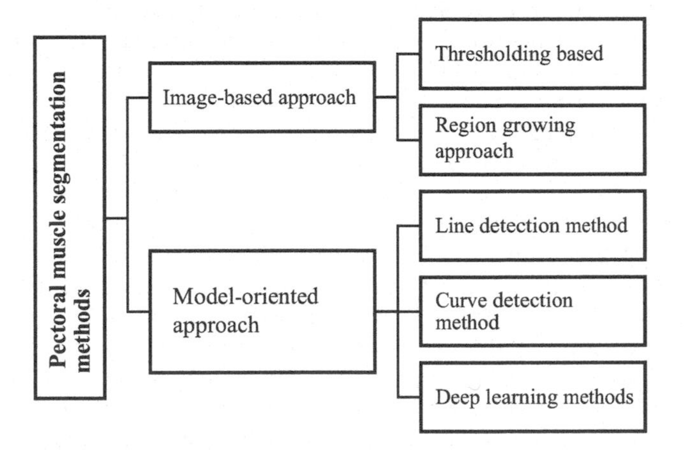

FIGURE 8.7 Pectoral muscle removal algorithms from the literature.

Different authors have proposed either model-based or image-based approaches for pectoral muscle removal as shown in Figure 8.7.

8.2.3.1 Image-Based Approach

Region growing and thresholding are the major algorithms utilized in this approach for the removal of pectoral muscles. A gray-level thresholding algorithm with an enhancement filter was used by Vikhe and Thool et al. (2016) on 322 digital mammograms. This algorithm consists of a low acceptance rate due to unclear boundaries of the pectoral muscles and breast area and also the problem of the selection of accurate threshold. Instead of thresholding region, the growing algorithm is used by different authors due to high segmentation accuracy and only the design of initial seed selection is the major concern with this algorithm Sapate et al. (2018).

Authors in Maitra et al. 2013 proposed dynamic seed selection to identify boundaries of the pectoral muscles. In this algorithm, a triangle is extracted to suppress the pectoral muscles and for initial seed selection and then the Cartesian slope-intercept equation is used to remove the pectoral muscles. The basic merit of this algorithm is automatic seed selection but fails to remove a pectoral muscle which is not in a triangular shape. Another option of watershed algorithm was proposed by Taifi et al. (2018) for removal of the pectoral muscles which obtained a good precision rate but the basic limitation of this algorithm is the cause of over-segmentation if noise present in mammograms. This drawback is overcome if we use the marker-based watershed algorithm.

The region merging algorithm was proposed by Rahman et al. (2019) and tested on 200 images. Output images of this algorithm consist of local abrupt changes and discontinuities, hence require further post processing with Gabor filter along with edge connect. This post processing provides segmentation results at an optimal level. Image-based algorithms consist of different limitations like threshold selection, automatic seed selection, over-segmentation, and the need for further post processing (Kotia et al. 2021). To overcome this drawback, research is diverted toward a model-based approach which is a combination of image processing algorithms with different image transformation techniques.

8.2.3.2 Model-Based Approach

In this approach, the pectoral muscle is modeled as a straight line or curve or learned features by deep learning. One of the popular approaches used by different researchers is the boundary of pectoral muscle is modeled as a straight line with the help of Hough transform and Randon transform. A combination of Hough transform with canny edge detection with different edge features can be applied to the image to detect a particular angle by which the pectoral muscle is removed from the image. This approach was proposed by Sreedevi and Sherly et al. (2015) in which they used a global thresholding approach to detect the pectoral muscles from the mammogram with the combination of Hough transform and connected component labeling and obtained an accuracy of 90.06% on 161 images. Canny edge detection with five different edge features and Hough transform this combination is used by Rampun et al. (2019) and they obtained "dice similarity coefficients" of 98.8% and 97.8% mini-Mammographic Image Analysis Society (MIAS) dataset. The straight line–based approximation works well if canny edge detection can identify the pectoral muscle edge but sometimes if it fails to identify the edge, then this technique is unable to remove the pectoral muscles.

Ability to construct the original density from projection data and computational simplicity, Randon transform is another technique used by Liu et al. (2010, 11). This algorithm is used to forecast the objects in the image by applying a set of angles, and resulting projections are used to add intensities of the pixels in all directions which are known as line integral. In this method, the local variance of neighbors is calculated on all the pixels of region of interest, and the resulted variance image Randon transform is implemented to identify the most significant line which represents the edge of muscles. The contents detected on the left side of the detected line are removed as pectoral muscles. Morphological operation with multilevel wavelet transforms as another alternative is performed by Mughal et al. (2017) with a segmentation accuracy of 92.2%. Despite many advantages like the ability to force certain geometric properties, good handling of noisy data, and easy adjustment of noisy data, this method is having certain disadvantages like larger storage space and domain restrictions and useful to detect only one object at one time.

Polynomial curve fitting is another approach used by many researchers to remove the pectoral muscles from the breast area. In this algorithm, the curve is estimated from the straight line which will be the best fit to a series of input samples at fixed pixel intervals and develop a smooth curve for pectoral muscle removal. This approach was implemented by Subashini et al. (2010) and Devi and Vidivelli (2018) which is basically a regression analysis, having flexibility for different shapes and simple implementation. A similar approach was also proposed by Mustra et al. (2009), in which a combination of polynomial curvature estimation and adaptive histogram equalization is used for pectoral muscle removal. This technique has limitations as the degree of polynomial curve increases then this algorithm is unstable; hence there must be a proper trade-off between shape and degree.

Due to composite variations of the pectoral muscles, handcrafted methods have certain limitations and to overcome this nowadays concept deep learning is used by different authors. Wang et al. (2019) used a novel approach with the help of deep learning by utilizing U-Net to find an impact on preprocessing technique such as image normalization, zero and extrapolated padding techniques, and different contrast between background and breast with the help of training U-Net on 2000 regularized images and obtained "median dice similarity coefficients" of 0.8879 and 0.9919 respectively for pectoral and breast segmentations from 825 testing images. Comparative analysis of pectoral muscle removal algorithm is highlighted in Table 8.2.

TABLE 8.2

Comparative Analysis of Pectoral Muscle Removal Algorithms

Comparison Parameters	Image-Based Approach	Model-Based Approach
Operating principle	Clustering or thresholding	Mathematical analysis like regression, line integral, or learning from data
Computational complexity	Low	Moderate to high
Accuracy	82%–95.65%	90%–96.65%
Merits	Simple to implement. Effective storage and fast retrieval	Ability to force certain geometric properties from the data Capable to learn complex hierarchical features Provides better accuracy than handcrafted technique
Demerits	Difficulty in automating seed generation	Large storage space required Domain restrictions

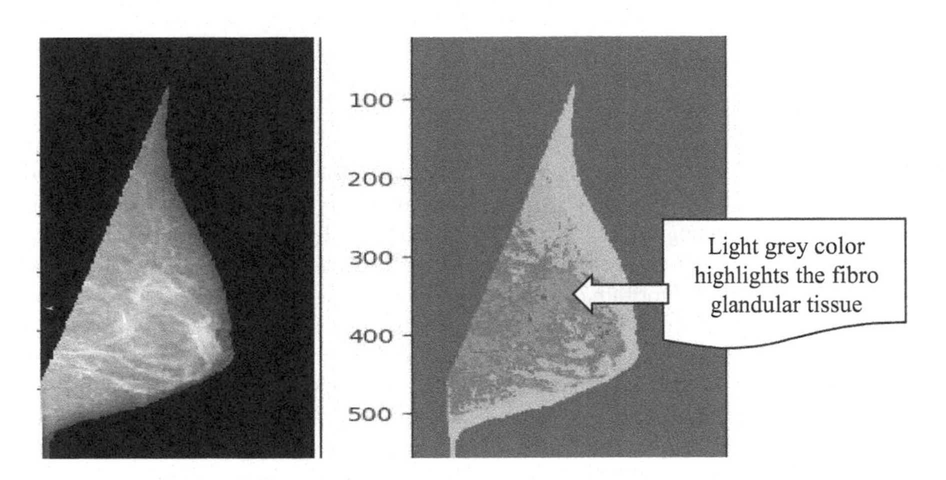

FIGURE 8.8 Concept of segmentation algorithm.

8.2.4 Segmentation

This is the most significant activity in breast density measurement in which fibroglandular tissue is separated from the whole breast area. The entire concept of fibroglandular tissue segmentation is shown in Figure 8.8.

During our literature survey, we came across different methods of segmentation which is shown in Figure 8.9.

8.2.4.1 Region-Growing Algorithm

This is one of the most utilized techniques in which pixels in fibroglandular tissues are labeled with a unique label that is different than other regions. This algorithm starts the process by selecting a seed pixel within the image and identifies the neighbor pixel recursively using similarity measures or homogeneous property. This process is repeated until no more pixels remain unlabeled and to select neighbor either a four-connected or eight-connected or zero-connected window is used. This concept is used by Liu et al. (2011) on 88 full-field digital mammography (FFDM) and provides the total rate of agreement of 86.4%. Another similar approach, with certain modifications, was proposed by Devi and Vidivelli (2018), and the best accuracy was obtained at 128×128 window. The overall classification accuracy obtained was 89.7%. The accuracy of this method is depending upon neighbor selection, window size, and classifier selection. To enhance the segmentation accuracy, all these points should be integrated to design a modified algorithm.

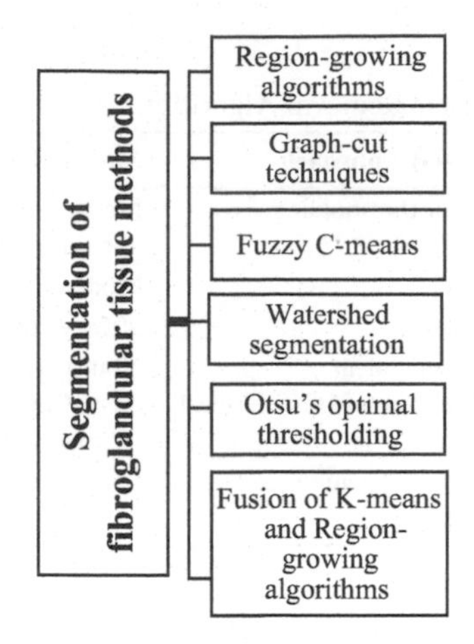

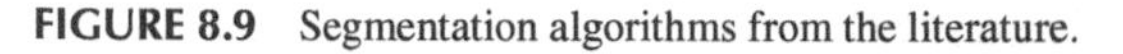

FIGURE 8.9 Segmentation algorithms from the literature.

8.2.4.2 Graph-Cut Algorithm

During the last two decades, the graph cut technique has shown good results in the segmentation of different medical images such as MRI and CT, but it is less utilized in the segmentation of fibroglandular tissues with the help of mammograms. Only two research articles by the same author (Saidin et al. 2009, 2012) explore the use of graph-cut techniques to find the boundary of different breast tissue regions in the mammograms. In this algorithm, the image is divided into two parts, known as "objects" and "background." The minimum cut of the graph will decide to determine the energy function which is minimized either locally or globally. To get better segmentation, a version of max-flow algorithm by Boykov and Kolmogorov was used by the author. In this algorithm, the first mammogram image is represented as a graph with the number of edges and nodes. After defining data cost in terms of boundary term and smooth cost by region terms, these terms are optimized by energy optimization. Over-segmentation is one of the major issues to be tackled in the future for better segmentation accuracy.

8.2.4.3 Fuzzy C-Means

This is one of the popular approaches used for the segmentation of medical images that is classified into hard or fuzzy clustering depending upon whether the pattern of data belongs exclusively to a single cluster or several clusters with a different degree. Fuzzy clustering is considered to be superior to hard upon clustering as it represents the relationship between input pattern data and clusters more naturally. Fuzzy clustering tries to minimize a heuristic global cost function by exploiting the fact that each pattern has some graded membership in each cluster. Keller et al. (2012) proposed adaptive multi-cluster fuzzy C-means algorithms for automatic segmentation of fibroglandular tissues from the breasts. In this algorithm, after the removal of the pectoral muscles from the breasts, an optimal number of clusters was found with the help of an adaptive histogram of the image. Basic innovation in this algorithm was the adaptive nature for the determination of optimal numbers of the clusters which were based on breast tissue properties. The selection of an accurate value of C is the major issue in this algorithm.

8.2.4.4 Watershed Algorithm

This segmentation concept was taken from topography in which a gray-level image represents the altitude and the region with a constant gray level constitutes the flat zone of an image. Region edges correspond to high watershed and low-gradient region interiors correspond to catchment basins. Catchment basins of the topographic surface are homogeneous in the sense that all pixels belonging to the same catchment basin are connected with the basin region. Watershed is defined as lines separating catchment basins, which belong to different minima. Regions that separate watersheds are known as catchment basins. Due to its effectiveness in intensity-based segmentation, Chang et al. (2006) used this method for fibroglandular tissues from the whole breasts. In this algorithm, depending upon intensity gradient value of the pixels image is classified into catchment area as low gradient pixels and catchment baseline as a high gradient pixels.

Watershed algorithms work well if each local minimum corresponds to an object of interest. In such cases, the watershed line represents the boundaries of each object. If there are many more minima than the object of interest in the image, then the image will be over-segmented and, to avoid over-segmentation, the watershed algorithms are constructed by an appropriate marking function.

8.2.4.5 Otsu's Optimal Thresholding

Generally, thresholding is a statistical decision theory in which the basic focus is to minimize the average error incurred in assigning the pixels to two or more than two groups which are also called a class. In this regard, Otsu's method is an attractive approach as maximizes the variance between the classes which is a well-known statistical discrimination. The basic idea behind this approach is to identify threshold classes which should be distinct in terms of the intensity values of pixels. This concept will provide the best separation in terms of intensity values. Olsen and Mukhdoomi et al. (2007) utilized pixel intensity-based segmentation in which automatic selection of thresholding was performed by Otsu's optimal thresholding. In this algorithm, a discriminant analysis was performed to divide the foreground and background by maximizing the discriminant measure function. Pixel is divided into two classes, i.e., fibroglandular and breast tissue. The probability of each class is calculated to decide optimal thresholding and, after implementation of the algorithm, the image is converted into binary to calculate breast density. In this approach, the author claimed 83.3% accuracy on the digital mammogram.

8.2.4.6 Fusion of K-Means and Region-Growing Algorithms

To overcome the disadvantages of the region-growing algorithm, input parameters, such as seed point and threshold values, which are selected manually and hamper the segmentation accuracy are required. Hence, to increase segmentation accuracy and enhance automation, Elmoufidi et al. (2015) proposed a fusion of K-means and region-growing algorithms for automatic segmentation and detection of the boundary of different disjoints of the breasts. In this proposed work, the authors performed preprocessing on the mini-MIAS dataset to enhance the contact of breast image, remove the noise, and separate breasts from the background. After preprocessing, K-means algorithm automatically generates seed points and threshold values for each region which is given as input to the region-growing algorithm to divide the mammogram into homogeneous regions according to the intensity of the pixel. In this proposed method, the mean precision percentage segmentation accuracy of all cases reached 92.87%. Comparative analysis of segmentation methods is highlighted in Table 8.3.

8.2.5 Feature Extraction and Classification

The basic objective of medical image processing is to identify unique features for classification. For pattern recognition, texture features and statistical features are more significant than other features. Feature extraction, feature reduction, and classification are discussed in this section.

8.2.5.1 Statistical Feature Extraction

Most review articles used common statistical features such as mean, variance, dispersion, average energy, entropy, skewness, and kurtosis features of the image with a combination of higher-order statistical parameters which are calculated from co-occurrence or run-length matrices and a wavelet approach. There is software available to calculate such features like Mazda and Pyradiomics; the output of such software is depicted in Figure 8.10.

TABLE 8.3

Comparative Analysis of Different Fibroglandular Tissue Segmentation Methods

Segmentation Methods	Datasets	Classification Accuracy	Implementation Accuracy	Merits and Demerits
Region-growing algorithms	88 FFDM	Moderate	87%	The accuracy of this method was affected by empirically selected wavelets, classifier, and sub regions
Graph-cut techniques	Mini-MIAS dataset	Moderate	Completeness CM=0.435	Initial seeds are marked by the user which will affect the classification accuracy. Need to use other algorithms to mark the initial seed automatically
Fuzzy C-means	160 FFDM	High	Jaccard indices of $J=0.62\pm0.22$	Values of k generally found sensitive to the change in peaks or changes in histogram construction, hence shows an average agreement with the radiologists' prediction
Watershed segmentation	Mini-MIAS dataset	Moderate	87.5%	Chances of over-segmentation, hence requires the support of marking functions
Otsu's optimal thresholding	Mini-MIAS dataset	High	83.3%	The histogram should be bimodal. Do not use any object structure or spatial coherence
The fusion of K-means and region-growing algorithms	Mini-MIAS dataset	Moderate	92.87%	Automatic selection of k causes enhancement inaccuracy

Images			
Area	177590	189980	198240
Mean	81.39719	87.95794	87.26706
Variance	10904.45	9869.961	9702.966
Skewness	0.722403	0.59211	0.626045
Kurtosis	-1.21181	-1.1828	-1.11876
Perc.01%	1	1	1

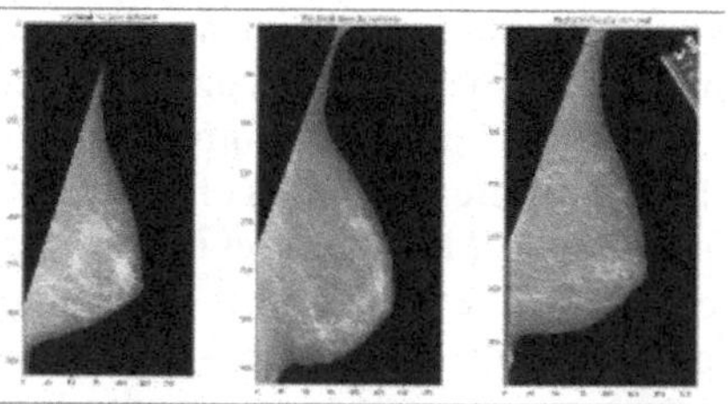

FIGURE 8.10 Feature extraction techniques. Courtesy: Mazda software Szczypiński et al. (2009).

8.2.5.2 Feature Reduction

We can extract nearly about 300 different first-order and second-order statistical features to calculate breast density from fibroglandular tissues. After feature extraction, the major concern is how to select dominant features by reducing the dimensions of the features. Different methods are used to reduce the dimensions of the features such as principal component analysis, linear discriminant analysis, chi-square probability, and gain ratio. With the help of this technique, we can make a proper trade-off between classification accuracy and the dimension of the features.

8.2.5.3 Classification

During our literature survey, we came across two machine learning algorithms which are support vector machine and K-nearest neighbor. The basic reason behind this is most of the research is performed on a medium-scale dataset that can easily fit into the memory size of the desktop, and it is easy to achieve comparatively better classification accuracy in both classifier training and feature extraction. In the future, there is a need to extract more features on a larger dataset to enhance classification accuracy with different classifiers (Sharma et al. 2020). Comparative analysis of different surveyed classification algorithms is shown in Table 8.4.

8.3 Deep Learning Approach

Due to different shortcomings of handcrafted techniques and the availability of larger datasets, the recent research focus is diverted toward the learned feature ability of convolutional neural network (CNN). For the last 2 years, many researchers have used a deep learning approach to classify MBD as per BI-RADS. As compared to the machine learning approach, this approach does not require any segmentation or feature extraction steps. The generalized approach of this approach is shown in Figure 8.11.

During the literature survey, all the authors used a generalized approach which is discussed in this section.

TABLE 8.4

Comparative Analysis of Feature Extraction and Classification Algorithms

Classification Approach	Classification Accuracy
Support vector machine (SVM)	77.3%–95.44%
K-nearest neighbors (KNN)	78%
Semiautomatic approach	73.91%–84%

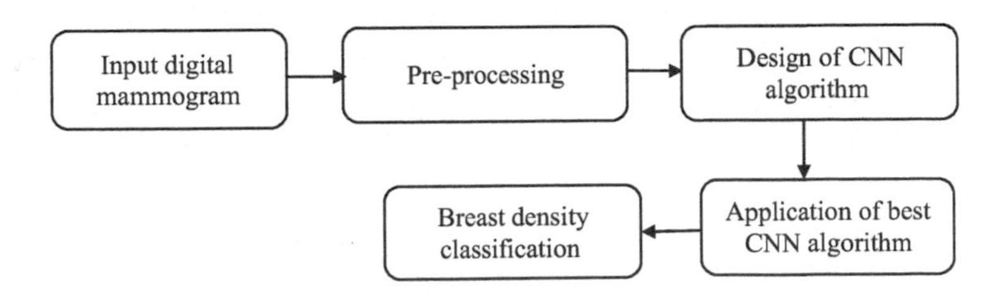

FIGURE 8.11 Generalized approach of CNN algorithm.

8.3.1 Preprocessing

Input mammographic images consist of some noisy and potentially inaccurately labeled images. Hence, the need to perform some preprocessing to achieve the following tasks:

1. Remove air and chest muscle from each image.
2. Histogram equalization for the adjustment of intensity distributions.
3. Resizing of images into a smaller size to increase computational complexity.
4. The mean image of training data has to be generated and subtracted from each image to ensure that each pixel has zero mean.

8.3.2 Design of Convolutional Neural Network Algorithm

This section describes the basic building blocks of CNN architectures used for the measurement of mammographic density.

8.3.2.1 Input Layer

The input layer of the neural network consists of an input neuron to take preprocessed data into the system. In most of the research articles, data size of 512×512 is used, and it may also vary depending upon the resize requirement of the system.

8.3.2.2 Convolutional Layer

This is the fundamental building block of a CNN architecture consisting of multiple neurons with learnable weights, responsible for extracting discriminating features from an image, and generates an output feature map. Many convolutional layers practically depend upon learning accuracy. In review articles, nearly about three to six convolutional layers are used and the number of CNN layers is variable during the training of CNN.

8.3.2.3 Max Pooling Layers

This layer is specialized to select maximum activation from a nonoverlapping rectangular region of size $R_x \times R_y$, thus downsampling an image by a factor of R_x and R_y in x-y direction respectively. Thus, this layer selects features which are superior and position invariant which will help in faster convergences and improves the generalization performance of the model.

8.3.2.4 Activation Function

Hyperbolic tangent (tanh) activation function works significantly better than the sigmoid function for different convolutional layers due to stronger gradient and more robust, easier, and faster optimization of the loss function.

8.3.2.5 Classification Layer

This is the output layer which consists of four neurons corresponding to the four-class classification problem (BIRADS I, II, III, and IV), which is activated by a SoftMax activation function.

8.3.2.6 Dropout Layer

Dropout layers prevent over-fitting and enhance the performance of a CNN classifier. Generally, dropout layers have been used between the fully connected networks to avoid over-fitting.

8.3.2.7 Kernel Selection

The choice of kernels and their respective sizes determines the number of trainable parameters which in turn affects the performance of CNN architecture. A study suggests that if the number of feature maps or trainable parameters is equal to or greater than the number of samples, then there exists a risk of over-fitting. To avoid over-fitting, the number of features should be much less than the number of samples. Thus, the number of kernels in each layer and their respective sizes were chosen such that the number of trainable parameters is less than the number of samples to avoid over-fitting.

8.3.2.8 Development of Baseline (Density Map) for Breast Density Prediction

The most generalized method to perform the task of breast density classification as per the literature review is to train the classifier with a feature based histogram of pixel intensity in the image. The difference in pixel intensity occurs as fat appears darker than the fibroglandular tissues as it absorbs much of the radiation whereas the adipose tissues allow the radiation to get through more easily, hence the use of pixel intensity histograms as a feature. Hence the need to develop a threshold-based density map and classify them into four classes as per BI-RADS.

8.3.3 Validation and Testing of Machine and Deep Learning Model

After the successful design of trained CNN, we will validate the model with validation dataset and try to make certain modifications depending upon the available accuracy and sensitivity to develop the final model for testing. Figure 8.12 shows the validation and testing phase of the CNN model.

And division of the dataset for the training of CNN is shown in Table 8.5.

Most recent research articles used the area under the receiver operating characteristic curve, confusion matrix, and kappa statistics to analyze the performance analysis of breast density measurement.

8.3.4 Related Work

In the last 2 years, many authors have used the deep learning approach which is discussed in the earlier section for the measurement of breast density. Ciritsis et al. (2019) performed breast density estimation with the help of a deep CNN on 20,578 single-view mammograms of different patients with the help of 11 convolutional layers and three fully connected layers and obtained a validation accuracy of 90.9%. Yasar et al. (2018) proposed the concept of complex wavelet transforms to separate the sub matrix in terms of real and imaginary feature coefficients, generate three image features which are applied to an artificial neural network (ANN), and provide a satisfactory classification of breast density.

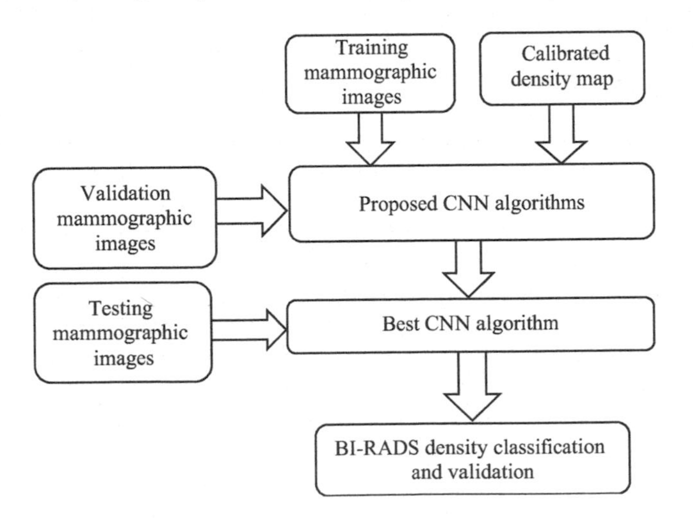

FIGURE 8.12 Validation and testing phase of the CNN model.

TABLE 8.5

Division of Dataset for Different Phases of CNN

% of Dataset	CNN Phases	Proposed Accuracy
70%	Training	Up to 85%
20%	Validation	Up to 90%
10%	Testing	Up to 90%–95%

Combination of CNN and (sparse) auto encoder on 493 mammographic images applied by Kallenberg et al. (2016), in which multistage image patches that capture both detailed and contextual region with multilayer convolutional architecture and sparse auto encoder, obtained an average area under the curve (AUC) of 0.61. As per a research article by Benjamin et.al. (2019), all the mammograms between 2006 and 2015 were reviewed with transfer learning of deep learning networks in which weight is utilized from Image Net. Hyper parameter optimization was performed to calculate prediction loss and accuracy. As per results, the ability to correctly classify interval cancers was moderate (AUC=0.65), and contingency table analysis showed that network correct classification accuracy was 75.2%.

Ha et al. (2019) proposed a novel 3×3 convolutional network based on the mammographic dataset to predict breast cancer risk stratification in which model is trained with 1474 mammographic datasets. Input image size was 256×256, and this model achieved an overall accuracy of 72% and independent of MBD. Lizzi et al. (2019) performed a residual convolutional network for breast density classification. Two approaches for classification were used: one is two-class dense and nonsense breast tissue and the second classification was according to BIRADS four standard classes. Maximum accuracy obtained for images 650 x 650 pixel size is equal to 78%. And two-class accuracy was 89.4%. Another approach to classifying breast density into four types by using an image processing pipeline such as image enhancement, segmentation, and feature extraction to increase the sensitivity of detecting breast cancer.

Bovis et.al. (2002) adopted a variation on bootstrap aggregation ("bagging") to meet the assumption of input dataset which is necessary for classifier combination. Feed-forward ANN is trained with different input data space using 10-fold cross-validation. In a dataset of 377 patients, MLO mammogram is used to classify breast density into two-class and four-class classifications and obtained average recognition rates on these tests are 96.7% and 71.4%.

In the future, there is a need to increase recognition rates for four-class classification algorithms. A further increase in results is possible by increasing computing power, increase in the dataset, and increasing the quality of ground truth (Sharma et al. 2020). Comparative analysis of the deep learning approach is highlighted in Table 8.6.

8.4 Findings and Discussion

MBD measurement is an interest-growing, challenging, and complex area of research, hence the basic motivation behind writing this book chapter is to find the future research direction. During the survey, we have recorded some observations which are discussed in this section.

In preprocessing, high-intensity tags and artifacts on the background of mammograms can hamper the classification accuracy, hence the need to consider all these points while designing a preprocessing algorithm. In every mammogram, the size and shape of the pectoral muscles change, hence the design of a single algorithm is not suitable for the removal of the pectoral muscles. The small presence of pectoral muscles in the breast area causes an error in density measurement, hence the integration of model and image processing algorithm is the research need of pectoral muscle removal algorithm.

All surveyed preprocessing algorithms show variation in segmentation accuracy from low to optimal level but as the accuracy of segmentation increases then computational complexity increases, hence there must be a proper trade-off between computational complexity and segmentation accuracy. As per pawar et al.(2020), segmentation algorithm provides moderate accuracy due to over-segmentation and under-segmentation, hence the fusion of two algorithms can be a better option to enhance the classification

TABLE 8.6

Comparative Analysis of Deep Learning Approaches

Author	Deep Learning Model	Technical Specification if any	Classification Accuracy
Ciritsis et al.	Deep CNN	11 convolutional layers and 3 fully connected layers on 20,578 single-view mammograms	90.9%
Yasar et al.	Deep CNN	Complex wavelet transforms for feature extraction	-
Kallenberg et al.	Combination of CNN and (sparse) auto encoder	493 mammograms	AUC of 0.61
Benjamin et al.	Transfer learning of deep learning network of Image Net	Hyper parameter optimization	AUC of 0.65 and classification accuracy was 75.2%
Ha et al.	Deep CNN	Novel 3×3 convolutional networks	Classification accuracy of 72%
Lizzi et al.	Residual convolutional network		Two-class accuracy of 89.4% and four-class of 78%
Bovis et.al.	Feed-forward ANN.	Bootstrap aggregation ("bagging") on 377 MLO	Two-class accuracy of 96.4% and four-class of 71.4%

accuracy. Classification algorithms can provide a better accuracy if there is no proper trade-off between feature extraction and feature selection. The dataset used for testing by different authors is from different manufacturing units of mammograms, hence there is no standard ground truth for comparison.

Due to the availability of huge datasets and an increase in computational power, deep learning is one of the significant research options used recently by different authors with the help of convolutional networks with different novel approaches like transfer learning, a feed-forward neural network with hyper parameter optimization, and the concept of bootstrap aggregation. These convolutional networks used variable hidden and classification layers from 3 to 11. Overall classification accuracy for four-class classification was found in-between 70% and 90%. Hence, there is still a need of refining classification accuracy in terms of an increase in the dataset, feature extraction techniques, and the development of quality ground truth.

8.5 Conclusion

In this book chapter, we have provided an extensive overview of different machine and deep learning approaches used for breast density measurement and classification. Although there are many research challenges discussed in the existing techniques, there is a positive hope for the solution of accurate and precise breast density measurement due to rapid development in computer vision, machine learning, and deep learning algorithms. Future research directions for machine learning approaches will be a certain modification in preprocessing and segmentation algorithms such as fusion of two or more algorithms and dynamic seed selection, and the proper trade-off between feature extraction and feature selection may act as a key to success in accurate classification. In the deep learning approach, the availability of a larger size of validated dataset and good quality ground truth for comparison is a major concern. The observations and findings mentioned in various sections will surely help develop a novel, accurate, precise, and automated breast density measurement tool for breast cancer prediction, which will be clinically useful for medical practitioners.

REFERENCES

Bovis, K. 2002. "Classification of mammographic breast density using a combined classifier paradigm". *Medical Image Understanding and Analysis*, 2002, 177–180.

Bray, F., Farley, J., Soerjomataram, I., Siegel, R. L., Torre, L. A., & Jemal, A. 2018. "Global cancer statistics 2018: GLOBOCAN estimates of incidence and mortality worldwide for 36 cancers in 185 countries." *A Cancer Journal for Clinicians*, 68(6), 394–424. https://doi: .org/10.3322/caac.21492.

Chang, R. F., Chang-Chien, K. C., Takada, E., Suri, J. S., Moon, W. K., Wu, J. H. K., Chen, D. R. 2006. "Three comparative approaches for breast density estimation in digital and screen-film mammograms". *In Annual International Conference of the IEEE Engineering in Medicine and Biology – Proceedings* (pp. 4853–4856). doi: 10.1109/IEMBS.2006.260218.

Ciritsis, A., Rossi, C., De Martini, I. V., Eberhard, M., Marcon, M., Becker, A. S. Boss, A. 2019. "Determination of mammographic breast density using a deep convolutional neural network". *British Journal of Radiology*. doi: 10.1259/bjr.20180691.

Devi, S. S., & Vidivelli, S. 2018. "Classification of breast tissue density in digital mammograms". *In Proceedings of 2017 International Conference on Innovations in Information, Embedded and Communication Systems, ICIIECS 2017* (Vol. 2018-January, pp. 1–7). doi: 10.1109/ICIIECS.2017.8276139.

Elmoufidi, A., El Fahssi, K., Jai-Andaloussi, S., & Sekkaki, A. 2015. "Automatically density-based breast segmentation for mammograms by using dynamic K-means algorithm and Seed Based Region Growing". *In Conference Record - IEEE Instrumentation and Measurement Technology Conference* (Vol. 2015-July, pp. 533–538). doi: 10.1109/I2MTC.2015.7151324.

Ha, R., Chang, P., Karcich, J., Mutasa, S., Pascual Van Sant, E., Liu, M. Z., & Jambawalikar, S. 2019. "Convolutional neural network based breast cancer risk stratification using a mammographic dataset". *Academic Radiology*, 26(4), 544–549. doi: 10.1016/j.acra.2018.06.020.

Kallenberg, M., Petersen, K., Nielsen, M., Ng, A. Y., Diao, P., Igel, C., Lillholm, M. 2016. Unsupervised deep learning applied to breast density segmentation and mammographic risk scoring. *IEEE Transactions on Medical Imaging*, 35(5), 1322–1331. doi: 10.1109/TMI.2016.2532122.

Keller, B. M., Nathan, D. L., Wang, Y., Zheng, Y., Gee, J. C., Conant, E. F., & Kontos, D. 2012. "Estimation of breast percent density in raw and processed full-field digital mammography images via adaptive fuzzy c-means clustering and support vector machine segmentation". *Medical Physics*, 39(8), 4903–4917. doi: 10.1118/1.4736530.

Kotia J., Kotwal, A., Bharti, R., & Mangrulkar, R. 2021. "Few shot learning for medical imaging". In: Das S., Das S., Dey N., & Hassanien AE. (eds) *Machine Learning Algorithms for Industrial Applications*. Studies in Computational Intelligence. vol. 907. Springer: Cham. doi: 10.1007/978-3-030-50641-4_7.

Liberman, L., & Menell, J. H. 2002. "Breast imaging reporting and data system (BI-RADS)". *Radiologic Clinics of North America*. doi: 10.1016/S0033--8389.

Liu, L., Wang, J., & He, K. 2010. "Breast density classification using histogram moments of multiple resolution mammograms". *In Proceedings -2010 3rd International Conference on Biomedical Engineering and Informatics, BMEI 2010* (Vol. 1, pp. 146–149). doi: 10.1109/BMEI.2010.5639662.

Liu, Q., Liu, L., Tan, Y., Wang, J., Ma, X., & Ni, H. 2011. "Mammogram density estimation using sub-region classification". *In Proceedings -2011 4th International Conference on Biomedical Engineering and Informatics, BMEI 2011* (Vol. 1, pp. 356–359). doi: 10.1109/BMEI.2011.6098327.

Lizzi, F., Atzori, S., Aringhieri, G., Bosco, P., Marini, C., Retico, A., Fantacci, M. E. 2019. "Residual convolutional neural networks for breast density classification". *In BIOINFORMATICS 2019-10th International Conference on Bioinformatics Models, Methods and Algorithms, Proceedings; Part of 12th International Joint Conference on Biomedical Engineering Systems and Technologies, BIOSTEC 2019*, Prague, Czech Republic, (pp. 258–263).

Maitra, I. K., Nag, S., & Bandyopadhyay, S. K. 2013. "Mammographic density estimation and classification using segmentation and progressive elimination method". *International Journal of Image and Graphics*, 13(3), 1350013. doi: 10.1142/s0219467813500137.

Mughal, B., Muhammad, N., Sharif, M., Saba, T., & Rehman, A. 2017. "Extraction of breast border and removal of pectoral muscle in wavelet domain". *Biomedical Research (India)*, 28(11), 5041–5043.

Mustra, M., Bozek, J., & Grgic, M. 2009. "Breast border extraction and pectoral muscle detection using wavelet decomposition". *IEEE EUROCON 2009*. doi: 10.1109/eurcon.2009.5167827.

Olsen, C., & Mukhdoomi, A. 2007. "Automatic segmentation of fibro glandular tissue". *In Lecture Notes in Computer Science (Including Subseries Lecture Notes in Artificial Intelligence and Lecture Notes in Bioinformatics)*, (Vol. 4522 LNCS, pp. 679–688). doi: 10.1007/978-3-540-73040-8_69.

Pawar, S., Sapate, S., and Sharma, K. 2020. "Machine learning approach towards mammographic breast density measurement for breast cancer risk prediction: An overview". *Proceedings of the 3rd International Conference on Advances in Science & Technology (ICAST)*. doi: 10.2139/ssrn.3599187.

Taifi, K., Ahdid, R., Fakir, M., Elbalaoui, A., Safi, S., & Taifi, N. 2018. "Automatic breast pectoral muscle segmentation on digital mammograms using morphological watersheds". *In Proceedings - 2017 14th International Conference on Computer Graphics, Imaging, and Visualization* (pp. 126–131). doi: 10.1109/CGiV.2017.24.

Rahman, M. A., Jha, R. K., & Gupta, A. K. 2019. "Gabor phase response-based scheme for accurate pectoral muscle boundary detection". *IET Image Processing*, 13(5), 771–778. doi: 10.1049/iet-ipr.2018.5290.

Rampun, A., López-Linares, K., Morrow, P. J., Scotney, B. W., Wang, H., Ocaña, I. G., Macía, I. 2019. "Breast pectoral muscle segmentation in mammograms using a modified holistically-nested edge detection network". *Medical Image Analysis*, 57, 1–17. doi: 10.1016/j.media.2019.06.007.

Lee, R. S., Gimenez, F., Hoogi, A., Miyake, K. K., Gorovoy, M., & Rubin, D. L. 2017. A curated mammography data set for use in computer-aided detection and diagnosis research. *Scientific Data*, 4, Article number: 170177. doi: https://doi.org/10.1038/sdata.2017.177.

Saidin, N., Ngah, U. K., Sakim, H. A. M., Siong, D. N., & Hoe, M. K. 2009. "Density-based breast segmentation for mammograms using graph cut techniques". *In IEEE Region 10Annual International Conference, Proceedings*. doi: 10.1109/TENCON.2009.5396042.

Saidin, N., Sakim, H. A. M., Ngah, U. K., & Shuaib, I. L. 2012. "The graph cuts technique, breast density, and abnormality detection". *In Proceedings - IEEE-EMBS International Conference on Biomedical and Health Informatics: Global Grand Challenge of Health Informatics* (pp. 361–364). doi: 10.1109/BHI.2012.6211588.

Sapate, S. G., Mahajan, A., Talbar, S. N., Sable, N., Desai, S., & Thakur, M. 2018. "Radiomics based detection and characterization of suspicious lesions on full-field digital mammograms". *Computer Methods and Programs in Biomedicine*, 163, 1–20. doi: 10.1016/j.cmpb.2018.05.017.

Sharma, K.K., Pawar S.D., & Bali B. 2020. "Proactive preventive and evidence-based artificial intelligence future healthcare." *International Conference on Intelligent Computing and Smart communication 2019: Algorithms for Intelligent systems*. doi: 10.1007/978-981-15-0633-8_44.

Sreedevi, S., & Sherly, E. 2015. "A novel approach for removal of pectoral muscles in digital mammogram". *Procedia Computer Science*, 46, 1724–1731. doi: 10.1016/j.procs.2015.02.117.

Subashini, T. S., Ramalingam, V., & Palanivel, S. 2010. "Automated assessment of breast tissue density in digital mammograms". *Computer Vision and Image Understanding*, 114(1), 33–43. doi: 10.1016/j.cviu.2009.09.009.

Szczypiński, P. M., Strzelecki, M., Materka, A., & Klepaczko, A. (2009). MaZda—A software package for image texture analysis. Computer Methods and Programs in Biomedicine, 94(1), 66–76. doi: 10.1016/j.cmpb.2008.08.005.

The U iversity of Malaya. "Breast density in quantifying breast cancer risk." *Science Daily,* December 9th, 2016. http.2016/12/161209111849.

Tortajada, M., Oliver, A., Martí, R., Vilagran, M., Ganau, S., Tortajada, L., Freixenet, J. 2012. "Adapting breast density classification from digitized to full-field digital mammograms". *In Lecture Notes in Computer Science (Including Subseries Lecture Notes in Artificial Intelligence and Lecture Notes in Bioinformatics)* (Vol. 7361 LNCS, pp. 561–568). doi: 10.1007/978-3-642-31271-7_72.

Tzikopoulos, S. D., Mavroforakis, M. E., Georgiou, H. V., Dimitropoulos, N., & Theodoridis, S. 2011. "A fully automated scheme for mammographic segmentation and classification based on breast density and asymmetry". *Computer Methods and Programs in Biomedicine*, 102(1), 47–63. doi: 10.1016/j.cmpb.2010.11.016.

Vikhe, P. S., & Thool, V. R. 2016. "Intensity based automatic boundary identification of pectoral muscle in mammograms". *Procedia Computer Science*, 79, 262–269. doi: 10.1016/j.procs.2016.03.034.

Wang, K., Khan, N., Chan, A., Dunne, J., & Highnam, R. 2019. "Deep learning for breast region and pectoral muscle segmentation in digital mammography". *PSIVT 2019: Image and Video Technology* (pp. 78–91). doi: 10.1007/978-3-030-34879-3_7.

Wolfe, J. N. 1976. "Risk for breast cancer development determined by the mammographic parenchymal pattern". *Cancer,* 37(5), 2486–2492. doi: 10.1002/1097–-0142.

Yasar, H., Kutbay, U., & Hardalac, F. 2018. "A new combined system using ANN and complex wavelet transform for tissue density classification in mammography images". *4th International Conference on Computer and Technology Applications, ICCTA* (pp. 179–183). doi: 10.1109/CATA.2018.8398679.

9

Applications of Machine Learning in Psychology and the Lifestyle Disease Diabetes Mellitus

Ruhina Karani, Dharmik Patel, Akshay Chudasama, Dharmil Chhadva, and Gaurang Oza
Dwarkadas J. Sanghvi College of Engineering

CONTENTS

9.1 Introduction....145
9.1.1 Application of Machine Learning in Psychology....145
9.1.1.1 Detecting Depression Using Machine Learning....145
9.1.2 Application of Machine Learning in Detecting Lifestyle Disease....146
9.1.2.1 Detecting Diabetic Retinopathy Using Machine Learning....146
9.2 Machine Learning for Depression Detection....146
9.2.1 Preprocessing....147
9.2.2 Algorithms Used....148
9.2.3 Score Generation....148
9.3 Machine Learning for Diabetic Retinopathy Detection....150
9.3.1 Preprocessing....150
9.3.2 Algorithms Used....152
9.3.2.1 Training Phase....152
9.3.2.2 Testing Phase....152
9.4 Conclusion....153
References....153

9.1 Introduction

Owing to the advancement in medical technologies, it is now possible to diagnose some diseases efficiently and cost-effectively. One of the technologies that open the door for various possibilities is machine learning. The aim of this chapter is to apply machine learning algorithms to explore the possibility of detecting severe health problems pertaining to psychology and diabetes mellitus.

9.1.1 Application of Machine Learning in Psychology

One of the common and serious medical illnesses that negatively affect how a human being feels is depression. Depression is a state of mental illness that can change an individual's thinking, behavior, and physical well-being. According to the World Health Organization (WHO), one in four people in the world will be affected by mental disorders at some point in their lives.

9.1.1.1 Detecting Depression Using Machine Learning

In India, the National Mental Health Survey 2015–2016 reveals that nearly 15% of Indians are suffering from mental health issues, in the age-group of 15–49 years. It has been observed that majority of mobile users are in the age-group of 15–25 years. Therefore, data collected from mobile phones can act as an aid in identifying levels of depression within this age-group. Mobiles that passively collect and analyze behavioral sensor data 24/7 can help in detecting the depression of a person. These passively sensed data are then used to generate an overall depression score based on Patient Health Questionnaire(PHQ)-8 survey.[1–5] K-means algorithm is used to generate the depression score. If the generated depression score is higher than the threshold value, then the score is shared with the trusted individuals provided by the user.

9.1.2 Application of Machine Learning in Detecting Lifestyle Disease

Lifestyle diseases like diabetes, asthma, cardiovascular diseases, hypertension, etc., are increasing day by day. According to WHO, India has the highest number of diabetics. One of the major health concerns related to diabetes mellitus is diabetic retinopathy (DR).

9.1.2.1 Detecting Diabetic Retinopathy Using Machine Learning

A human eye has tiny blood vessels responsible for providing nourishment to the retina. In severe diabetes, these vessels get blocked due to excessive sugar, and in response to this, new blood vessels are generated to keep the nourishment supply intact. This damage is called DR. DR is one of the significant causes of blindness, and millions of people in the world are blinded by it. DR can be averted if detected in time; however, symptoms are often visible too late to provide appropriate treatment. DR can be categorized as nonproliferative diabetic retinopathy (NPDR) and proliferative diabetic retinopathy (PDR). The classification of an eye into one of these categories depends on the presence of features. Machine learning can be used to eliminate the manual process in the detection of DR and the need for an expert by making the entire process a fully automated system.[6–9] This system will allow people to upload the fundus images and obtain the results in a shorter period. Also, the cost of examining and evaluating will decrease drastically.

9.2 Machine Learning for Depression Detection

This book chapter discusses an android application designed to detect levels of depression among mobile users using a machine learning algorithm by continuously recording user physical activities and tracking sleep duration and sociability using data collected from smartphones. The app collects physical activity records (e.g., still, walking, running, in a vehicle, and cycling), sleep duration, and sociability through a number of missed calls, incoming and outgoing calls, duration of incoming and outgoing calls, etc. The app also collects phone unlock events, screen on time, number of apps used and their usage period, notifications received, and number of notifications that are clicked or ignored. User physical activities are collected using Activity Recognition API that reads bursts of data from mobile sensors and the data are stored in the database every hour. Light sensor data are collected from mobile light sensors every 2 minutes and after 30 minutes the average of the light sensor data collected is stored locally in a mobile. Call logs are recorded for each day using the react-native-call-log library. The phone unlock event and screen on time are recorded using a broadcast receiver. App usage is collected using android UsageStatsManager.

The app creates a structured build of the data collected from the mobile phone every day (i.e., after midnight) and sends it to the server. The app continuously checks for the presence of the build and the active Wi-Fi connection every 10 minutes. If the build and active Wi-Fi connection are present, the app will send the data to the secured server. Once the build is successfully sent to the server, the build gets deleted locally. Figure 9.1 depicts the architecture of the system. It includes an android device to passively sense the data with the help of various sensors embedded in it. These sensed data items are regarded as features for evaluating questions in the eight-item PHQ-8 depression scale. Additionally, the score generated is also sent back to the android device.

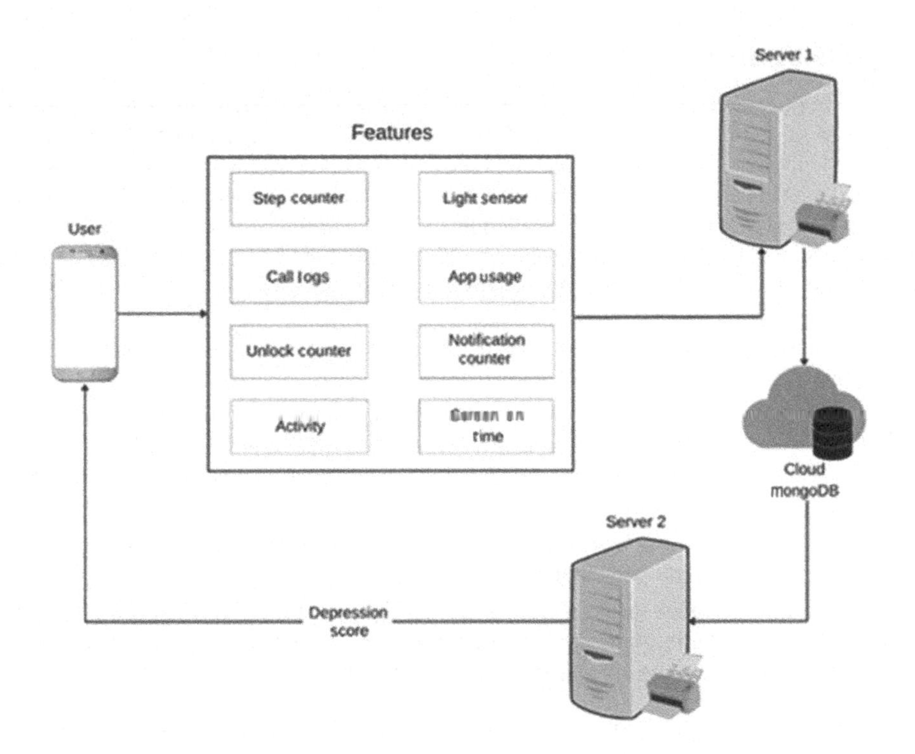

FIGURE 9.1 Architecture of depression detection using machine learning on a smartphone.

9.2.1 Preprocessing

The application uses activities, sleep time, usage stats, call stats, step counts, number of notifications, and screen time as features from the dataset. Data are requested from the server to perform preprocessing on it. Preprocessing involves two phases viz cleaning and transformation of data.

1. **Cleaning:** Cleaning involves eradicating duplicate values and filling the missing values with average values in each feature.
2. **Transformation:** Data transformation converts three features (notification data, call stats, and usage stats) into corresponding one numerical value. Sleep time, screen on time, activities, and step count are taken as it is, no transformation done. The applications installed in smartphones are categorized into five types: social media, game, video player, browser, and others. Each type is assigned some weight needed for converting the values of data. Features are converted to one value as follows:

$$\text{Call stats} = \text{Duration of incoming and outgoing calls} * w_1$$

$$+\text{number of missed calls } * w_2 + \text{unique calls} * w_3$$

$$\text{Usage stats} = \text{No. of social media apps} * w_1 + \text{no. of gaming apps} * w_2$$

$$+\text{No. of browsing apps} * w_3 + \text{no.of video player apps} * w_4 + \text{other apps} * w_5$$

$$\text{Total notification of apps} = \text{No. of notifications}\left(w_1 + w_2 + w_3 + w_4 + w_5\right)$$

$$\text{Notification} = \text{Total notification} * w_5 + \text{total notifications clicked} * w_1$$

$$+\text{total notification of apps} * w_3 + \text{total clicked notification of apps} * w_2$$

$$*\text{Note-}w_1 = 0.8,\ \ w_2 = 0.6,\ \ w_3 = 0.4,\ \ w_4 = 0.3,\ \ w_5 = 0.2\ \left(\text{weights}\right)$$

9.2.2 Algorithms Used

After performing preprocessing steps as specified in Section 9.2.1, K-means clustering algorithm is used to identify levels of depression among users. K-means is an iterative algorithm which tries to partition data into K-predefined clusters. Clusters are subgroups which collect data similar to each other. The less variation found within clusters, the more homogeneous (similar) data points are within the same cluster.

Data are organized into two types; the total collected data and the new week data. For each feature, it is necessary to calculate the range of values for each level, i.e., level 0, 1, 2, and 3. In order to get this range of values for each feature, the following steps are performed:

1. Use K-means algorithm to find centroids pertaining to each of the four levels from the total collected data.
2. Calculate the distance between each centroid and take the minimum distance from the calculated distances. For example, if the centroids are 20, 25, 35, and 50, then the distances calculated between the centroids are 5, 10, and 15. Out of all the distances calculated, 5 is the minimum distance.
3. In order to find the range for level 0, add and subtract the minimum distance calculated in Step 2 from the average of total collected data.
4. Similarly, to find the range for levels 1, 2, and 3, multiply the minimum distance with values 2, 3, and 4 (i.e., values for level number) respectively. Then add and subtract the newly calculated minimum distance from the average of total collected data.

Consider a feature Number of Unlocks as an example. Suppose for person 1 and person 2, the average of total collected data of Number of Unlocks feature is 30 and 50 respectively. Along with this, the minimum distance for person 1 and person 2 is 5 and 10 respectively. The calculated range of values for each level is shown in Table 9.1.

Once the range in which the average weekly data falls in is identified, the corresponding level, i.e., level 0, 1, 2, and 3 will be assigned to that feature for the new week's data. Figures 9.2 and 9.3 show K-means clustering graphs for PHQ questions of "poor appetite or overeating" and "trouble concentrating on things, such as reading a newspaper or watching television" respectively.

9.2.3 Score Generation

Once the levels are calculated for each feature for a week, these levels are used in the PHQ-8 questionnaire to identify the levels of depression. There are eight questions in the PHQ-8 questionnaire and, for each question, a set of features are shortlisted. The shortlisted features for each question in the PHQ-8 questionnaire are shown in Table 9.2. Each feature is assigned a weight according to its importance in the given question. These weights are multiplied with the levels identified. The results of multiplication are added and divided by 100 to generate the score for the respective question (i.e., between 0 and 3). The same procedure is followed for each question and, at the end, the scores of each question are added to get the final depression score. Each feature is associated with weights according to its relevance to the PHQ-8 questions. For example, question 1 of PHQ-8 from Table 9.2 has a weight of 17 assigned to each screen on time, unlocks, call logs, and app usage respectively, 12 to sleep and step counter, and 8 to notification.

TABLE 9.1

Unlocks Range of Values for Each Level

Person 1	**Person 2**
Level 0: 25–30 and 30–35	Level 0: 40–50 and 50–60
Level 1: 20–25 and 35–40	Level 1: 30–40 and 60–70
Level 2: 15–20 and 40–45	Level 2: 20–30 and 70–80
Level 3: >15 and <45	Level 3: >20 and <80

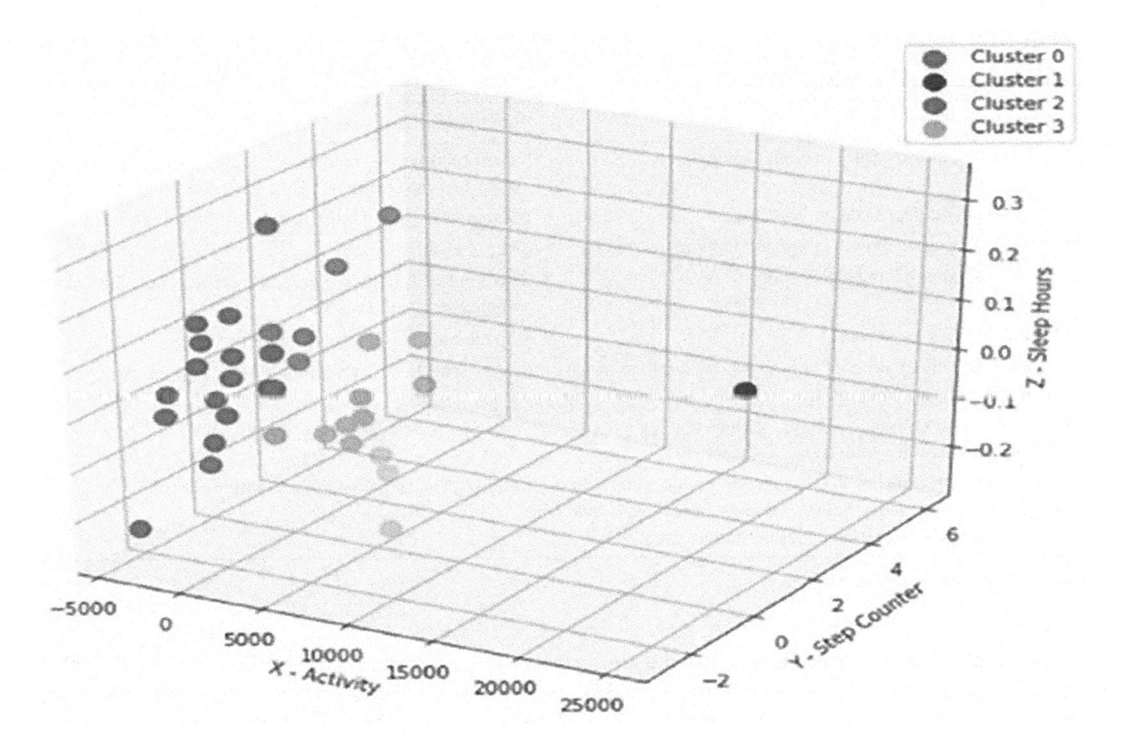

FIGURE 9.2 K-means clustering graph for PHQ-8 question "poor appetite or overeating".

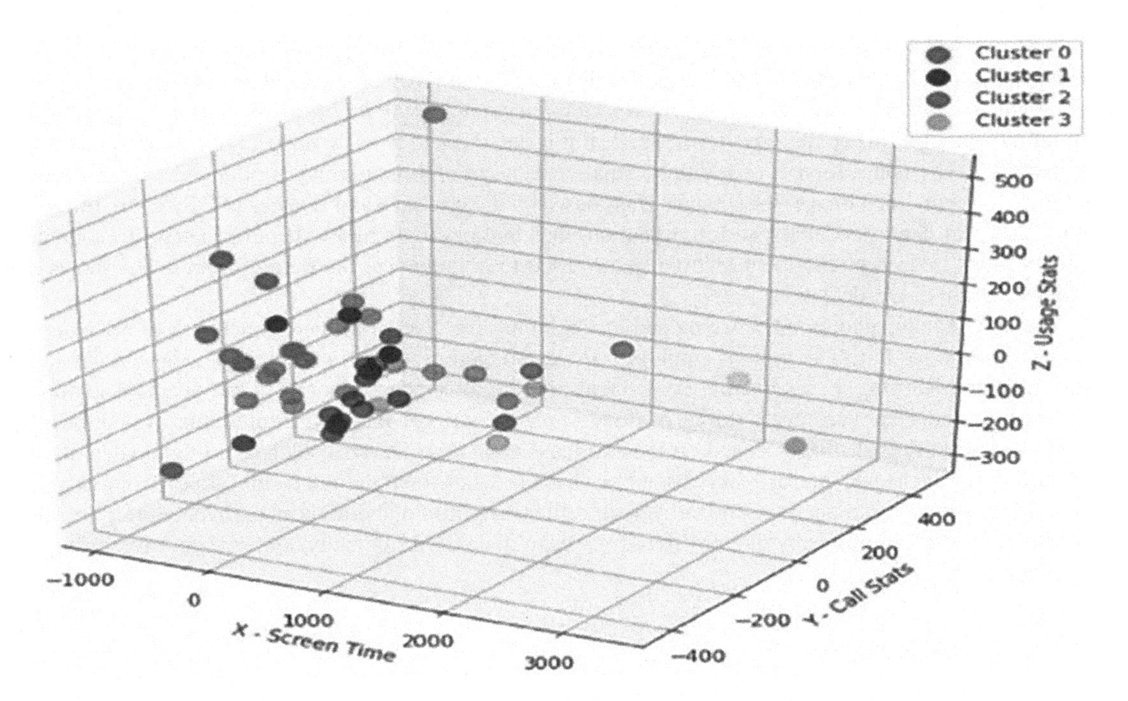

FIGURE 9.3 K-means clustering graph for PHQ-8 question "trouble concentrating on things such as reading a newspaper or watching television".

TABLE 9.2

Questions and Their Corresponding Features

Questions	Shortlisted Features
Q1: Little interest or pleasure in doing things	Screen on time ↑ ↓, unlocks ↑ ↓, call logs, app usage ↑ ↓, sleep ↑ ↓, step counter ↑ ↓, notification ↑ ↓
Q2: Feeling down, depressed, or hopeless	Average of the remaining seven-question score
Q3: Trouble falling or staying asleep or sleeping too much	Sleep ↑ ↓
Q4: Feeling tired or having little energy	Step counter ↓, activity value, screen on time ↓, unlocks ↓, app usage ↓, call logs ↓
Q5: Poor appetite or overeating	Step counter ↑ ↓, sleep ↑ ↓, activity value
Q6: Feeling bad about yourself or that you are a failure or have let yourself or your family down	Sleep ↓, call logs ↓, screen on time ↓, app usage ↓
Q7: Trouble concentrating on things, such as reading a newspaper or watching television	Screen on time ↑, app usage ↓, call logs ↓
Q8: Moving or speaking so slowly that other people could have noticed. Or the opposite—being so fidgety or restless—that you have been moving around a lot more than usual	Call logs ↑ ↓, step counter ↑ ↓, activity value

Note: ↑ - increasing, ↓ - decreasing.

Once the depression score is generated as described above, it is compared with the threshold value. The depression score higher than the specified threshold value implies that the user is suffering from depression and needs help. The generated score is then shared with the trusted individuals provided by the user.

9.3 Machine Learning for Diabetic Retinopathy Detection

DR is an ocular disorder caused by aftereffects of poststage diabetes mellitus that affects the retina. The increased sugar level in diabetics leads to blocking of tiny blood vessels present around the retina. Since these vessels are responsible for providing nourishment to the retina, their blockage is attempted to be countered by the creation of new blood vessels. Complications during proliferation of blood vessels lead to retinal damage; this condition is termed DR. If it is not checked at early stages, it can lead to lifelong blindness.[10] The application uses multiple, smaller datasets of fundus images obtained from various sources as input. Each image is assigned a class based on the presence and severity of DR. These images have different degrees of noise and illumination, thus making them highly heterogeneous. To counter this, various image enhancement techniques are used to homogenize the dataset. Figure 9.4 shows a block diagram of DR detection.

The color fundus photographs are passed through multiple layers of a neural network to get a definite prognosis for DR. The system employs a three-pronged algorithm combining image processing and machine learning, i.e., principle component analysis of machine learning and adaptive gamma correction, and Gaussian thresholding in image processing, that facilitates automatic evaluation of high-resolution retinal images based on the extracted unhealthy features, such as soft exudates, hard exudates, hemorrhages, microaneurysms, etc. It then classifies the obtained result into different stages of severity such as no diabetic retinopathy, nonproliferative mild diabetic retinopathy, nonproliferative moderate diabetic retinopathy, nonproliferative severe diabetic retinopathy, and proliferative diabetic retinopathy.[10–13]

9.3.1 Preprocessing

Figure 9.5 shows the color fundus image that is used as an input.

Several steps as described below are performed for preprocessing of input images.

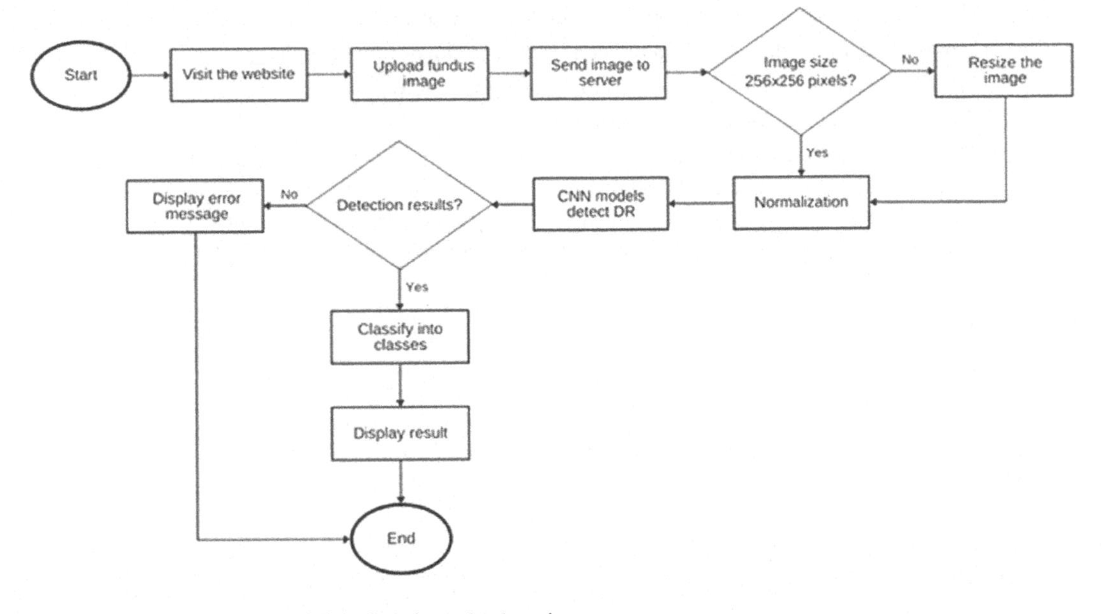

FIGURE 9.4 Block diagram of diabetic retinopathy detection.

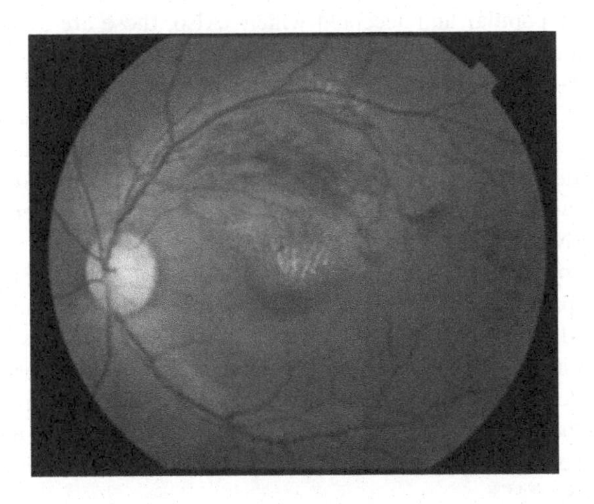

FIGURE 9.5 Color fundus image. (Reprinted from "Branch retinal vein occlusion associated with quetiapine fumarate" "by Ku C. Yong, Tan A. Kah, Yeap T. Ghee, Lim C. Siang, and Mae-Lynn C. Bastion, Department of Ophthalmology, Universiti Kebangsaan Malaysia Medical Centre (UKMMC) and Universiti Malaysia Sarawak (UNIMAS)" is licensed under CC BY 2.0.)

Step 1: The images in the EyePACS dataset are cropped and resized to a standard size of 256 * 256 pixels. It was found that this size was the most appropriate to retain the features while reducing the image size. Also, cropping and resizing around 50% of the images preserved the aspect ratio, but the remaining images were stretched.[6] To tackle this issue, the following steps were taken:

1. Resize the image height to 256 pixels.
2. Calculate the center (C) of the image.
3. Move to the $(C-128)^{th}$ pixel.
4. Extract the image from $(C-128)^{th}$ pixel to $(C+128)^{th}$ pixel.

Step 2: Images are normalized by subtracting the local average color.

1. **Normalization**: The normalization can be broken down into two parts:
   i. Calculating the local average color by using Gaussian blur.
   ii. Subtracting the local average color.
2. **Data augmentation**: The dataset used suffers from the data imbalance problem because 75% of the dataset was class 0 and the rest of the 25% was shared by the remaining classes. Therefore, to counter this problem, data augmentation is done. For this, the processed images were rotated at 90° and 270°, resulting in an increase in the class size.[6]

9.3.2 Algorithms Used

9.3.2.1 Training Phase

In this phase, the processed images are used to train the machine learning models. Convolutional neural networks (CNN) are used to classify these images. For training, three pretrained CNN models are used. They are:

- VGG16
- VGG19
- Inception-v3

All these models are very popular and accepted widely. Also, these are some of the leading models used for image classification. For training and classification purposes, Keras deep learning framework on TensorFlow backend is used. While training the machine learning models on the dataset, it is very crucial to go through the hyperparameter selection phase. The hyperparameter selection phase helps us to better the model by tuning certain parameters, such as learning rate, activation functions, number of epochs, etc. In order to find a perfect spot where the models depicted superior performance, a trial-and-error approach was used to find the values of these hyperparameters. Furthermore, all three models behaved differently on the same data for different hyperparameters.

In order to find a better optimization technique, various techniques such as Stochastic gradient descent (SGD), root mean square propagation (RMSprop), and adoptive moment estimation (Adam) were experimented on. It was realized that SGD provided consistent and desirable results and hence all the models use SGD optimization. Initially, the entire dataset was divided into training, validation, and testing sets in a ratio of 60:20:20 respectively. Consequently, 32,164 images were used for training, 7026 for validation, and 7026 for testing purposes, with labels. During the training phase, the models were trained using epochs equal to the multiples of 5 and hence after every execution the results were analyzed to find an optimal number of epochs.[6,10–13]

9.3.2.2 Testing Phase

After the completion of training phase, the testing stage was initiated to measure the performance of the models on unseen data values. For this, the images from the testing set were used and the results were analyzed. The images were passed in batches related to a particular class so that the performance metrics and insights of a particular model and a specific class can be obtained.

CNN has proven itself as the most effective neural network for image classification. It also performs the best in detecting the presence of DR. For detection, three renowned and largely accepted pretrained networks such as Inception-v3, VGG16, and VGG19 were used. The topmost layers were replaced with something that was needed for this problem. Two fully connected layers at the top of the architecture were added, with the first layer consisting of 4096 neurons and the second with 1024 neurons. The activation function used was rectified linear unit (ReLU). The topmost layer had five neurons, where one neuron will be fired for each of the classes. This layer used Softmax as the activation function. All the models were implemented on Keras deep learning framework using TensorFlow as the backend.[6]

- **VGG16:** This model was proposed by Karen Simonyan and Andrew Zisserman.[14] It made the use of 3×3 convolutional kernels and 2×2 pooling kernels. Max pooling was used in the pooling layers. The two fully connected layers at the top have 4096 neurons. The topmost layer is a prediction layer which consists of 1000 nodes to predict 1000 objects.[14] This layer was altered to five nodes to predict five predefined classes of DR.
- **VGG19:** This model was again proposed by Karen Simonyan and Andrew Zisserman. It differed from VGG16 in terms of network architecture, the additional convolutional layer in the third, fourth, and fifth blocks. The topmost layer was again altered to five nodes for predicting five levels of severity.[14]
- **Inception-v3:** It is a newer CNN architecture with a depth of 42 layers.[8] The computation of Inception-v3 is much more efficient than any VGG networks. The convolutional layers are 3×3 and the max pooling layers are 3×3 as well as 8×8.[6]

9.4 Conclusion

It is said that a smartphone knows more about its user than the user itself. Therefore, it is possible to detect psychological illnesses like depression using the data collected from smartphone sensors. Early-stage depression detection is crucial in helping the fast recovery of patients. Section 9.2 of this chapter shows the method used for detecting depression among mobile users. The results obtained using this method are quite satisfactory but they can be improved further by conducting the test for a longer duration and for a large number of users. New methods can also be explored further to improve the accuracy of score prediction and also to help the practitioners diagnose depression.

Section 9.3 of this chapter describes the method used for detecting DR. Applying machine learning algorithm as a tool in technology to detect lifestyle diseases like DR can help diabetic patients to detect it at an early stage and seek consultations with a dialectologist in time.[15] The results obtained by tuning hyperparameters of pretrained models such as VGG16, VGG19, and Inception-v3 to detect DR are quite satisfactory. But, Inception-v3 model outperformed VGG16 and VGG19 in terms of overall performance. New methods can also be explored further to improve accuracy and precision.

REFERENCES

1. R. Wang, W. Wang, A. Dasilva, J. F. Huckins, W. M. Kelley, T. F. Heatherton, and A. T. Campbell, "Tracking depression dynamics in college students using mobile phone and wearable sensing," *In Proceedings of the ACM on Interactive, Mobile, Wearable and Ubiquitous Technologies*, Vol. 2, No. 1, New York, March 2018.
2. A. Mehrotra, R. Hendley, and M. Musolesi, "Towards multi-modal anticipatory monitoring of depressive states through the analysis of human-smartphone interaction," *In Proceedings of the 2016 ACM International Joint Conference on Pervasive and Ubiquitous Computing: Adjunct*, ACM, New York, pp. 1132–1138, September 2016.
3. K. Demirci, M. Akgonul, and A. Akpinar, "Relationship of smartphone use severity with sleep quality, depression, and anxiety in university students," *Journal of Behavioral Addictions* 4(2), 85–92, 2015.
4. A. A. Farhan, C. Yue, R. Morillo, S. Ware, J. Lu, J. Bi, J. Kamath, A. Russell, A. Bamis, and B. Wang, "Behavior vs. introspection: Refining prediction of clinical depression via smartphone sensing data," *In IEEE Wireless Health (WH)*, Bethesda, MD, pp. 1–8, 2016.
5. F. Wahle, T. Kowatsch, E. Fleisch, M. Rufer, and S. Weidt, "Mobile sensing and support for people with depression: A pilot trial in the wild," *JMIR mHealth and uHealth* 4(3), 2–4, 2016.

6. A. Jain, A. Jalui, J. Jasani, Y. Lahoti, and R. Karani, "Deep learning for detection and severity classification of diabetic retinopathy," *In 1st International Conference on Innovations in Information and Communication Technology (ICIICT)*, Chennai, India, pp. 1–6, 2019.
7. M. M. Engelgau, L. S. Geiss, J. B. Saaddine, et al., "The evolving diabetes burden in the United States," *Annals of Internal Medicine*, 140(11), 945–950, 2004.
8. B. Zhang, F. Karray, Q. Li, and L. Zhang, "Sparse representation classifier for micro aneurysms detection and retinal blood vessel extraction," *Information Sciences*, 200, 78–90, 2012.
9. P. Massin, A. Erginay, T. Walter, and J. Klein, "A contribution of image processing to the diagnosis of diabetic retinopathy detection of exudates in color fundus images of the human retina," *IEEE Transactions on Medical Imaging*, 21(10), 1236–1243, 2002.
10. K. Ng, J. Suri, O. Faust, and R. Acharya, "Algorithms for the automated detection of diabetic retinopathy using digital fundus images: A review," *Science Journal of Medical Systems*, 36, 145–157, 2010.
11. E. Benetti, F. Massignan, E. Pilotto, et al., "Screening for diabetic retinopathy: 1 and 3 nonmydriatic 45-degree digital fundus photographs vs 7 standard early treatment diabetic retinopathy study fields," *American Journal of Ophthalmology*, 148(1), 111–118, 2009.
12. E. Y. Ng, C. Chee, T. Tamura, U. Acharya, and C. M. Lim, "Computer-based detection of diabetic retinopathy stages using digital fundus images," *In Proceedings of the Institute of Mechanical Engineers, Part H: Journal of Engineering in Medicine*, pp. 545–553, 2009.
13. P. S. Bhat, U. R. Acharya, C. M. Lim, J. Nayak, and M. Kagathi, "Automated identification of different stages of diabetic retinopathy using digital fundus images," *Journal of Medical Systems*, 32(2), 107–115, 2008.
14. K. Simonyan and A. Zisserman, "Very deep convolutional networks for large-scale image recognition" arXiv:1409.1556, 2014.
15. R. S. Mangrulkar, "Retinal image classification technique for diabetes identification," *In Proceedings of 2017 International Conference on Intelligent Computing and Control, I2C2 2017, Institute of Electrical and Electronics Engineers Inc.*, pp. 1–6, 2018. doi: 10.1109/I2C2.2017.8321873.

10

Application of Machine Learning and Deep Learning in Thyroid Disease Prediction

Aditi Vora, Ramchandra S. Mangrulkar, Narendra M. Shekokar, and Meera Narvekar
Dwarkadas J. Sanghvi College of Engineering

CONTENTS

10.1 Chapter Flow 155
10.2 Introduction 155
10.3 Related Work with Thyroid Prediction 156
10.4 Machine Learning Model for Thyroid Prediction 157
 10.4.1 Supervised Learning 157
 10.4.2 Unsupervised Learning 159
10.5 Implementation of Model for Thyroid Prediction 161
10.6 Impact/Case Study of Work 162
10.7 Advantages 162
10.8 Conclusion 163
10.9 Future Scope 163
References 163

10.1 Chapter Flow

The proposed chapter begins with the introduction of machine learning and deep learning and exactly what it is and how it has become so important in this digital era to know about these technologies. We, later on, discuss how deep learning and machine learning algorithms help us in predicting thyroid disease. Later on, this chapter further focuses on some solutions for the prediction of thyroid disease.

10.2 Introduction

The thyroid gland is an important gland in our human body. It is a butterfly-shaped gland, which is located at the base of our neck and is small in size. It is part of a detailed network of glands called the endocrine system. The endocrine system coordinates many activities of the human body. The thyroid gland performs the task of manufacturing hormones that regulate our body's metabolism. The thyroid hormone also controls many other activities such as how fast the calories are burned in the human body and the speed of our heartbeats. The study reveals that women are more affected by thyroid than men. Out of every eight women at least one develops a problem of thyroid disease. Due to thyroid disease, women develop problems with their menstrual cycles because it can make the menstrual cycle irregular, and other problems that women face are in their pregnancy or they don't even get pregnant. There are some types of thyroid problems that affect women which are hypothyroidism, hyperthyroidism, or goiter, or even they can be affected by thyroid cancer. Some women also go to doctors to get tested for thyroid disease if they have:

1. Some thyroid problems
2. Undergone some surgery or radiotherapy that affects the thyroid gland
3. Some conditions such as anemia, goiter, or type 1 diabetes

There are two types of thyroid disease that we focus on here:

1. Hypothyroidism
2. Hyperthyroidism

Hypothyroidism is the problem when the thyroid gland does not produce sufficient thyroid hormones that are necessary for our human body and is also called an underactive thyroid. Hypothyroidism affects our body's functions like metabolism, etc. Hypothyroidism is also caused by Hashimoto's disease. The symptoms of hypothyroidism include feeling cold, constipation problems, weakness in the muscles, weight gain, feeling sad, depressed, pain in the muscles or joints, thinning of hair, pale or dry skin, less sweating, etc. Also, bad cholesterol can occur and it may increase the risk of heart disease.

Hypothyroidism can be cured with medicine that makes our thyroid hormone function normally. With the help of thyroid hormone pills, hypothyroidism can be treated. Hyperthyroidism, also called an overactive thyroid, causes the thyroid gland to make more thyroid hormones than our usual body requires. The common cause of hyperthyroidism can be Graves' disease. It is a problem with the immune system. Symptoms of Graves' disease are not noticed in the early stages. They usually begin slowly. But, gradually, an increase in metabolism can cause symptoms such as weight loss even if we eat the same or more amount of food. Other symptoms include eating more than the usual amount of food, feeling nervous, anxious, or irritated, sleeping problems, irregular heartbeat, shaking of hands and fingers due to anxiety or stress, more sweating and feeling hotter than usual, weakness in the muscles, and diarrhea. Hyperthyroidism also increases our risk to come in contact with osteoporosis. Osteoporosis is a condition in which our weak bones break.

The parathyroid glands are usually paired and are very small and they are so-called because they lie in a close relationship with the thyroid gland. Two glands are present: one inferior and the other superior on either side. There is also the trachea and larynx. The trachea, also called awindpipe, is the main gateway to the lungs. It is divided into left and right bronchi. The larynx, also called the voice box, is the passageway for air between the pharynx and the trachea. The larynx plays a pivotal role in human speech.

10.3 Related Work with Thyroid Prediction

Shaik Razia and M. R. Narasinga Rao [1] proposed a method for thyroid prediction using the density based spatial clustering of applications with noise (DBSCAN)algorithm for predicting thyroid disease. It classifies the dataset using hierarchical multiple classifiers. Also, the hypothyroid disease was predicted using the linear discriminant analysis algorithm. Other classification algorithms used were support vector machines (SVMs), decision trees, and radial basis function (RBF). Multilayer perceptron (MLP) was used in this study. Also, progressive learning vector quantization neural network was used for automatic thyroid segmentation. From the study, it was found that RBF provides the best accuracy and is best suitable for the prediction of thyroid disease with the least number of iterations and the highest peak signal-to-noise ratio values.

Ankita Tyagi et al. [2] proposed a system for thyroid prediction using various machine learning algorithms; SVMs and decision trees were used to predict the estimated risk on a patient's chance of obtaining thyroid disease. Age, gender, thyroid-stimulating hormone (TSH), and other attributes were used for examination. The datasets were possessed from the University of California, Irvine (UCI) machine learning repository. The work was done in two stages. The first stage was subset selection and the next stage was the prediction of disease. The model can further be enhanced to any desired level by increasing the inputs. SVMs and decision trees outperformed in this study.

Liyong Ma et al. [3] proposed a system to diagnose thyroid from SPECT images using convolutional neural networks (CNN). SPECT images use gamma-ray cameras to collect image data. It is a nuclear medicine imaging method. SPECT can show changes in blood flow, function, and metabolism of organs or lesions, which is beneficial for the early diagnosis of the disease. It is mainly important for the clinical diagnosis of the disease. SPECT imaging makes a final diagnosis of thyroid disease. Machine learning is used to diagnose a disease using SPECT images. The SPECT imaging method had superior performance than other methods.

David Dov et al. [4] proposed a method for thyroid malignancy prediction from whole slide images (WSI) and multiple instance learning approaches typically used for the analysis of WSI divide the image (bag) into instances which are used to predict a single-bag level. The five-fold cross-validation approach was used for training, validation, and testing. The deep learning–based algorithm identifies informative instances and assigns them local malignancy scores that are incorporated into a global malignancy prediction [5,6]. The deep learning algorithm was also compared to human-level performance. The algorithm proved to be better than other algorithms used.

Arvind Selwal and Ifrah Raoof [7] proposed a thyroid disease prediction system using MLP machine learning model. The system uses 7–11 features of humans for examining the disease. The system uses a gradient descent backpropagation algorithm for training the machine learning model. The preprocessing of the dataset is done to remove the ambiguities if any. The error backpropagation algorithm is used to train the pattern classifier. Using MLP, the system achieved almost 100% accuracy.

Edward Kim et. al [8] proposed a deep learning-based application for thyroid cytopathology. In the proposed system the disease was studied at the cellular level and which can help predicting thyroid accurately. The system uses custom thyroid ontology which is combined with multimedia data and deep convolutional network is used.

Peiling Tsou et. al. [9] proposed a thyroid prediction system where a deep learning convolutional neural network is trained on images based on histopathology. The model was also tested on different types of tissue samples. The performance of this model was also good. The accuracy of this model was 95.2%.

10.4 Machine Learning Model for Thyroid Prediction

Machine learning is a part of artificial intelligence which performs data analysis. The systems are built in such a way that they can learn from the data, analyze the results, and also predict the output, and a more accurate model can be obtained. The model is improved with experience and the model itself learns from the experiences and the output obtained is much clearer. Nowadays, machine learning is used in many sectors such as healthcare, finance, automation, entertainment, transportation, retail, and many more. Machine learning technology requires minimal human intervention and that is why it is widely in use today. Amazon, Netflix, and many other applications are based on machine learning. More complex and vast amounts of data can now be analyzed using machine learning.

Thyroid diseases can benefit a lot when using machine learning algorithms. Already systems are built that classify hypothyroidism and hyperthyroidism and predict diseases. The system uses many algorithms like decision trees, random forest, CNN, and clustering. Algorithms such as SVMs, iterative dichotomiser (ID3), and many such machine learning algorithms can be used to predict the disease. SVMs classify the disease by constructing a hyperplane which gives positive and negative classes and depending on that the classification is done.

Various machine learning models are proposed as follows:

10.4.1 Supervised Learning

Supervised learning is one type of machine learning task that always maps input to output. The data is always labeled which consists of many training examples. Supervised learning has an input and the desired output. The model can also learn from its experience in the case of supervised learning.

1. **K-Nearest Neighbor**: K-nearest neighbor (KNN) is also a type of supervised learning algorithm. It is simple and easy to implement machine learning algorithm. It calculates distances from a given dataset and is based on a similarity or proximity measure. The classification is done based on the majority and the class is assigned to the most frequent value occurring. KNN uses a K-fold cross-validation approach using ten-folds to evaluate the output. When given a training tuple, KNN algorithm simply stores the value and simply waits for the test tuple, and hence this algorithm is called a lazy learner algorithm because it waits for some more information. Therefore, KNN algorithm is not used much in the prediction of thyroid disease because more time is spent waiting to get the instances (Figure 10.1).

   The medical dataset is given, then the missing values are checked, then the training is performed on the classifier with data having no missing values, and the output is the predicted value. The output predicted value is hence imputed in the missing fields that are given to the KNN classifier and then we get the combined data. The process is repeated for all attributes.
2. **Decision Trees:** A decision tree is a graph-like structure which has nodes and edges, and it is helpful in the prediction of thyroid disease. ID3 is the best-suited algorithm for decision tree classification. Decision trees are useful mainly for large datasets, and it becomes easy for classification when we use decision trees. Decision trees are helpful to classify hypothyroidism, hyperthyroidism, and euthyroidism based on the symptoms and other factors. The classification and regression trees are also used for the prediction depending on the probabilities. The decision tree model gives the highest accuracy of prediction (Figure 10.2).

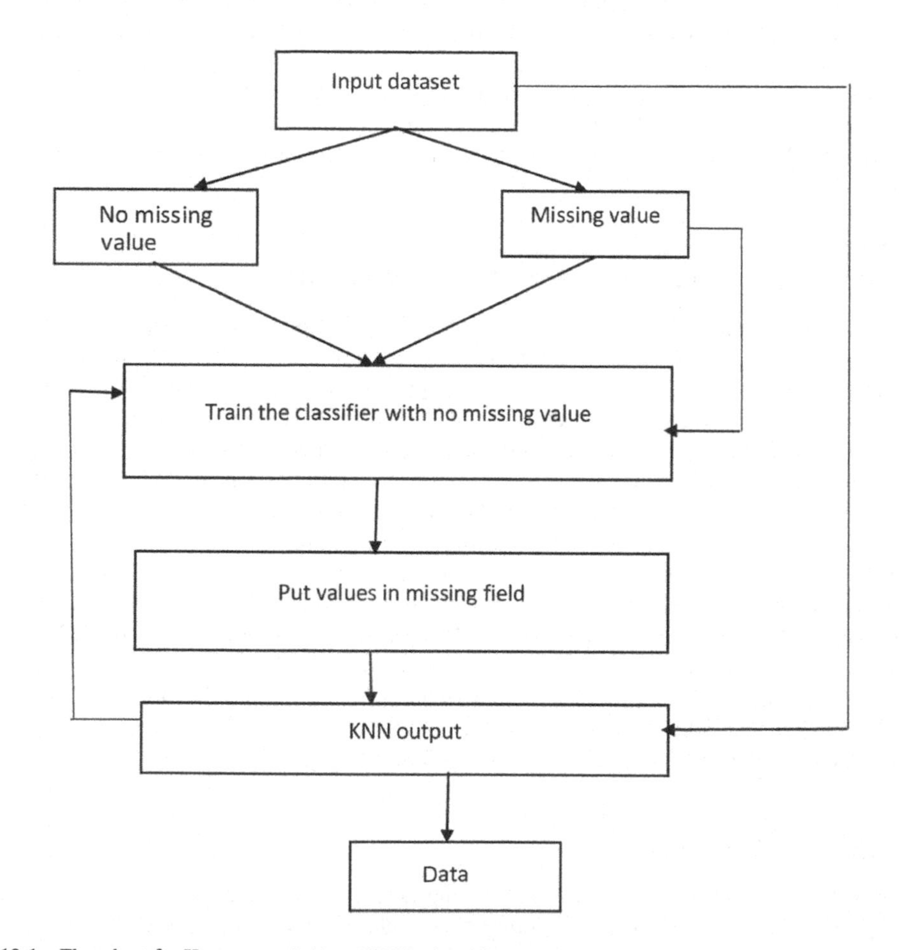

FIGURE 10.1 Flowchart for K-nearest neighbor (KNN) algorithm.

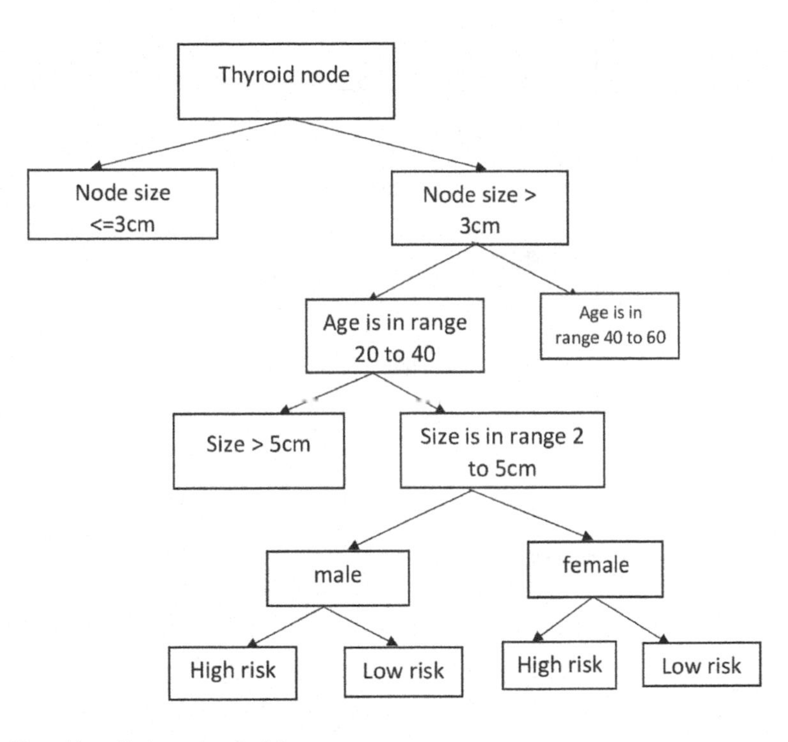

FIGURE 10.2 Thyroid prediction using decision trees.

The above decision tree represents thyroid prediction with steps 1&2. Step 1 has a thyroid nodule and it is classified into size ≤2 and size >2. It is classified further depending on the age-group, i.e., ≤30 age ≥60 and age<30 or age>60. Again, the age-groups are classified according to the thyroid nodule sizes, and accordingly gender is classified. In step 2, we have hypoechogenicity and isoechogenicity. If there is hypoechogenicity, then the output is the correlation with the clinical picture, repeat fine needle aspiration (FNA) and then refer for surgery. If the output is isoechogenicity, then the output is that we have to correlate with the clinical picture and, if the history is negative and no alarm sign is present, then the output is followed up.

3. **Support Vector Machines**: SVMs are a supervised learning algorithm in which we always want to maximize the margin between the data points and the hyperplane. The main aim of SVMs is to find a hyperplane in an N-dimensional space (N—the number of features) that distinctly classifies the data points given to us.

The optimal hyperplane is obtained which maximizes the margin and separates the data points (Figure 10.3).

SVMs are very useful in the classification of thyroid disease, and they always give 99% accuracy in most cases. They help in the classification of hypothyroidism and hyperthyroidism. They classify the data points on the positive and negative sides of the hyperplane. Positive means there is thyroid and the output is 1, and negative means there is no thyroid and the output is 0. The objective always will remain the same that is to minimize the error as much as possible.

10.4.2 Unsupervised Learning

Unsupervised learning is a machine learning technique in which our model is not supervised. Instead, we need to allow the model to work on its own and improve upon the output, and it does not have anyone for supervision of the model. The data is always unlabeled and the model learns on its own. Unsupervised

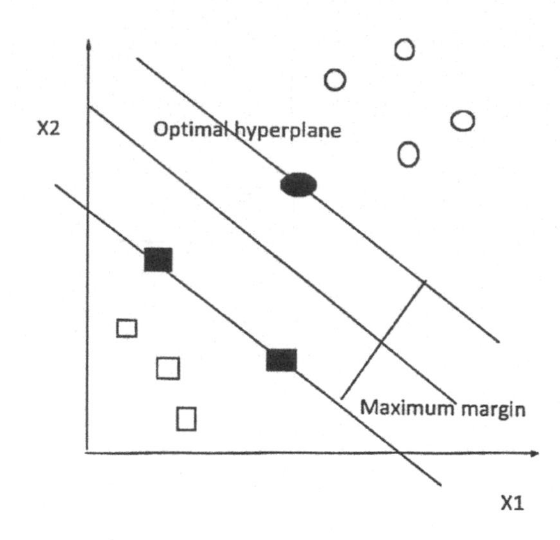

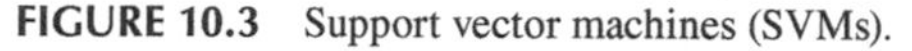
FIGURE 10.3 Support vector machines (SVMs).

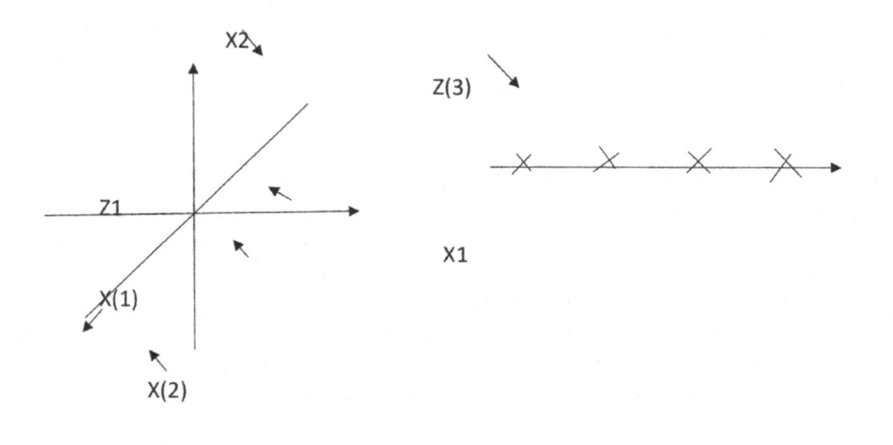

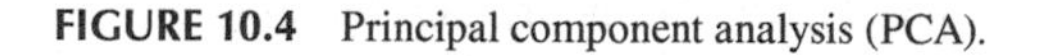
FIGURE 10.4 Principal component analysis (PCA).

learning allows us to perform more complex processing tasks. Unsupervised learning models are more unpredictable than supervised learning models.

1. **Principal Component Analysis**: Principal component analysis (PCA) is mainly used in predictive models and it is used mainly for dimensionality reduction. Dimensionality reduction is mainly used for obtaining a precise output and it removes unnecessary features. It is a statistical procedure and uses an orthogonal transformation to convert different observations of possibly correlated variables into a set of values of uncorrelated variables which are linear and are called principal components (PC). The output of the PCA algorithm is a set of PC. The thyroid dataset is reduced using PCA. The reduced hypothyroid dataset is then fitted with various classification algorithms and then the prediction is done (Figure 10.4).

The dataset is reduced from 2D to 1D.

10.5 Implementation of Model for Thyroid Prediction

The model is implemented using various machine learning algorithms like decision trees, SVMs, and random forest and also can be done using CNN. In this system [5] different types of thyroid are predicted, i.e., hypothyroidism and hyperthyroidism, and also proper treatment is given to patients. SVMs and decision trees give proper accuracy and output as compared to other algorithms in use. SVMs build a hyperplane which makes it precise for classifying the output, and the accuracy is almost 99%. Decision trees also give proper output based on the ID3 algorithm. The input dataset is taken from the clinical data. All types of diseases such as Graves' disease and Hashimoto's disease are being compared and all the outputs are compared.

CNN is used here in this system to predict the output. It basically belongs to a class of deep neural networks and is usually used for data visualization. It has an input and output layer and there are multiple hidden layers. The hidden layers of CNN mainly consist of a set of convolutional layers that perform the operation of convolution with dot product or simple multiplication.

The activation function is a rectified linear unit (ReLU) function and is usually followed by additional convolutions such as pooling layers, fully connected layers, and normalization layers, and they are the hidden layers because their inputs and outputs are masked by activation functions and final convolution function. The network architecture that is used is DenseNet because it gives a more precise output. It directly connects all the layers so that there is no need for backpropagation to again check for the values. DenseNet is also less parametric and therefore the training becomes easier. In DenseNet architecture, the features of the previous layer are concatenated with the same weights in every cross-layer (Figure 10.5).

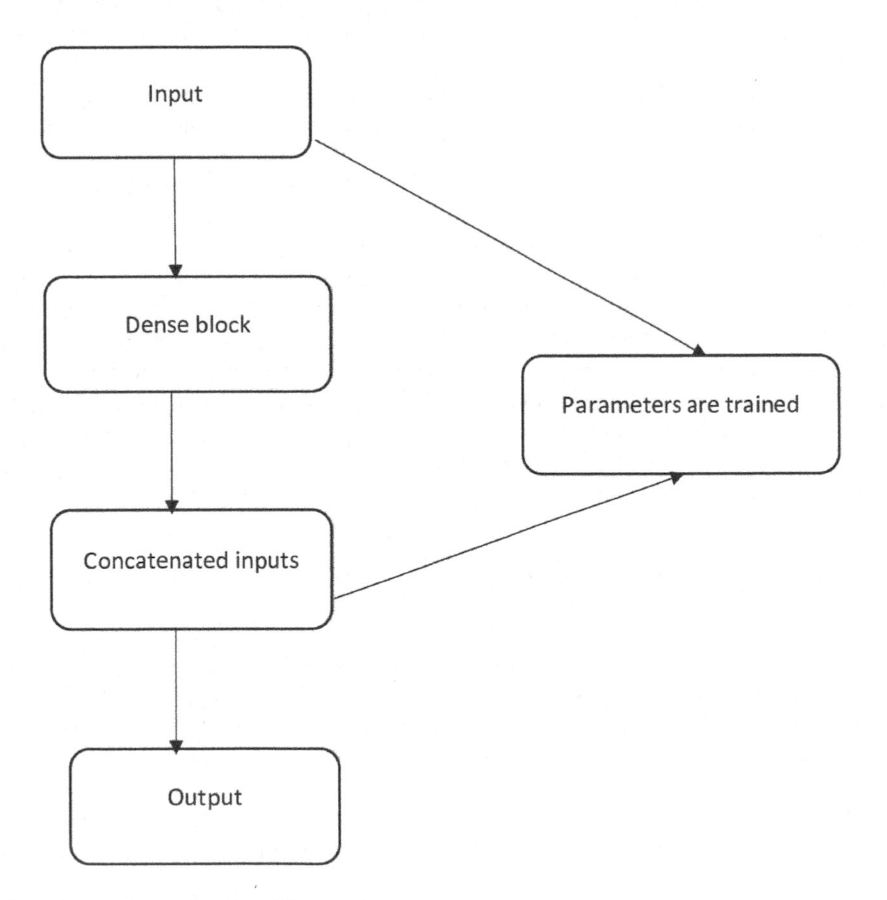

FIGURE 10.5 Dense block using DenseNet algorithm.

The input is given to the dense layer and the inputs are concatenated to get the output, and in the improved dense layer, trainable parameters are present. There are different layers such as normalization, ReLU, and convolutional, then features are extracted, and then the input parameters and the trainable parameters are concatenated to get output features. The output size of the convolutional layer is 110×110; the pooling layer size is 54×54; the dense block layer (1) size is 54×54 and the transition layer (1) size is also 54×54 and 27×27, the dense block layer (2) size is 27×27 and the transition layer (2) size is 27×27 and 13×13, and the dense block layer (3) size is 13×13 and the transition layer (3) size is 13×13 and 6×6; and the classification layer size is 1×1.

10.6 Impact/Case Study of Work

Thyroid disease is that kind of disease which mostly affects women. There are many changes in the human body when a thyroid disease occurs and it becomes difficult to detect in the early stages. Therefore, the proposed system will detect the thyroid in the very early stages. The impact of this model for predicting thyroid disease will prove very useful because more and more men and women who are suffering from thyroid will be able to use this system. This system also uses CNN, so the accuracy will also be very good. The system was never made using CNN so it becomes a challenging task whenever we start using or propose a new technology that is not being used previously. Deep learning algorithms have nowadays proven to be efficient in predicting various diseases in the healthcare sector. In this system, prediction of all types of thyroid, i.e., hypothyroidism and hyperthyroidism is done, and we are also trying to recommend medicines to patients so that it becomes easily curable. The main problem with the thyroid is that it does not easily get detected because of its symptoms. The human body does not produce enough thyroid hormones for proper functioning. Different cases are being considered as mentioned below.

The first case is a young woman with abnormal thyroid function who went to the physician for a test. She had a past medical history of fibromyalgia and does regular checkups. She also informed the doctor that recently her mother got hypothyroidism, her checkups were normal but she had hair loss problems, but that also improved because she was taking supplements. There was no evidence of thyroid eye disease and her blood was also properly functioning but the TSH was low. She has to then do other tests regarding thyroid diagnosis. With the deep learning model, she can undergo various treatments for thyroid.

The other case was an 11-year-old girl suffering from hyperthyroidism. There was no past medical history of thyroid disease but still her reports showed that she had hyperthyroidism and she had symptoms such as weight loss and was not able to tolerate the heat. She also experienced a downfall in her grades at school due to her thyroid. The diagnosis of Graves' disease was also made due to the laboratory results. So, our system can help the girl to fight against hyperthyroidism. The treatment can accordingly be done according to the results given by the proposed system.

Graves' disease is mostly caused in children.

10.7 Advantages

The advantages of using this system are mainly the prediction of disease. It accurately classifies the disease and gives higher accuracy. This system has no negative impact on patients. This system is also helpful for doctors to use in hospitals because manually checking for thyroid becomes difficult for detection. So, with the help of this system, their work becomes easier. This system has no side effects in the detection of thyroid disease. By using a deep learning algorithm and also neural networks in this system, the system, therefore, becomes more accurate and suitable for predicting the disease. For both clinical and statistical classification, thyroid detection is difficult as well as important. Deep learning makes this task easier.

10.8 Conclusion

It is seen in the chapter that a deep learning algorithm is used to successfully and accurately classify thyroid disease. Thyroid detection is a difficult and challenging task to be done manually. Therefore, this system helps to predict whether a patient has thyroid disease or not depending on the symptoms and also suggests proper medication. For statistical data also, thyroid prediction is difficult.

Due to extremely unbalanced and vast amounts of data with many different types of attributes and features, the traditional models give poor performance. Therefore, deep learning and neural networks are good choice for prediction. Deep learning is a flexible modeling technique which correctly predicts the data and it is also meant for the prediction of large amounts of data.

Diagnosis of thyroid using deep learning is a robust task with respect to different sampling variations. The system proposed is robust and it is also highly reliable for the prediction.

10.9 Future Scope

This system will efficiently help the doctors in diagnosing the disease and predicting appropriate and accurate results.

The future scope is that more work needs to be done on optimization and the accuracy of the system. The other scope is that more features can be added to the system for optimization purposes and also to make the system give accurate suggestions regarding the surgeries and other treatments that need to be done for curing this disease. The reliability of the model also needs to be improved given different features and make the model more robust in nature.

REFERENCES

1. Shaik Razia and Manda Rama Narasinga Rao "Machine learning techniques for thyroid disease diagnosis: A review" *Indian journal of Science and Technology* (2016), 9(28), 1–9. DOI: 10.17485/ijst/2016/v9i28/93705.
2. Ankita Tyagi, Ritika Mehra, and Aditya Saxena "Interactive thyroid disease prediction system using machine learning technique" *IEEE International Conference* (2019), IEEE. DOI: 10.1109/PDGC.2018.8745910.
3. Liyong Ma, Chengkuan Ma, Yuejun Liu, and Xuguang Wang "Thyroid diagnosis from SPECT images using convolutional neural network with optimization" *US National Library of Medicine National Institutes of Health* (2019), NCBI. DOI: 10.1155/2019/6212759.
4. David Dov, Shahar Z. Kovalsky, Jonathan Cohen, Danielle Elliott Range, Ricardo Henao, and Lawrence Carin Fellow "A deep learning algorithm for thyroid malignancy prediction from whole slide cytopathology images" *IEEE International Conference* of Machine Learning for Healthcare (2019), IEEE.
5. M. Shyamala and P. S. S. Akilashri "Thyroid disease prediction by machine learning technique from healthcare communities" *International Journal of Computer Science and Engineering* (2018) 6(11), 237–242.
6. https://www.womenshealth.gov/a-z-topics/thyroid-disease.
7. Arvind Selwal and Ifrah Raoof "A multilayer perceptron based improved thyroid prediction system" *Indonesian Journal of Electrical Engineering and Computer Science* (2020), 17, 524–533. DOI: 10.11591/ijeecs.v17.i1.
8. Edward Kim, Miguel Corte-Real, and Zubair W. Baloch "A deep semantic mobile application for thyroid cytopathology" *Proceedings of SPIE 9789, Medical Imaging 2016: PACS and Imaging Informatics: Next Generation and Innovations* (2016). DOI: 10.1117/12.2216468.
9. Peiling Tsou, and Chang-Jiun Wu "Mapping driver mutations to histopathological subtypes in papillary thyroid carcinoma: Applying a deep convolutional neural network" *Journal of Clinical Medicine* (2019), 8(10), 1675. DOI: 10.3390/jcm8101675.

11

Application of Machine Learning in Fake News Detection

Smita Bhoir, Jyoti Kundale, and Smita Bharne
Ramrao Adik Institute of Technology

CONTENTS

11.1 Introduction 166
11.1.1 Fake News: What It Is? 167
11.1.2 Essential Concept 167
11.1.2.1 Writing Mode–Based Fake News Analysis 167
11.1.2.2 Propagation-Based Fake News Analysis 168
11.1.2.3 User-Based Fake News Analysis 168
11.1.3 Types of Fake News 168
11.1.4 Overview of Chapter 168
11.2 Traditional Approaches for Fake News Detection 169
11.2.1 Physical Fact-Checking 169
11.2.2 Automatic Fact-Checking 169
11.3 Writing Mode–Based Approaches for Fake News Detection 170
11.3.1 Falsehood Identification and Analysis 170
11.3.2 Deception in NEWS 171
11.4 Dissemination-Based Approaches for Fake News Detection 172
11.4.1 Fake News Propagation Patterns 172
11.4.2 Models Based on Fake News Propagation 172
11.4.3 Dissemination-Based Fake News Detection 172
11.5 Integrity-Based Fake News Detection 173
11.5.1 Determining the Accuracy of Headline News 173
11.5.2 Determining the Accuracy of News Outlets 173
11.5.3 Determining the Accuracy of News Comments 173
11.5.4 Determining the Accuracy of Spreader News 173
11.6 Application of Machine Learning in Fake News Detection 173
11.6.1 Proposed Work 174
11.6.2 *Proposed Methodology* 174
11.6.3 *Design and Workflow Diagram of Proposed System* 175
11.6.4 Experimental Results and Analysis 176
11.7 Open Research Challenges in Fake News Detection 176
11.7.1 Redetection of Fake News 176
11.7.2 Identification of Trustworthy Contents 181
11.7.3 Fake News in Cross-Domain 181
11.7.4 Deep Learning and Fake News 181
11.7.5 Fake News and Social Media 181
11.8 Conclusion 181
References 181

11.1 Introduction

In the last two decades, digital media is continuously growing in every part of communication. Due to the ever-increasing availability of internet sources, nowadays people are reading news via websites and online newspapers. The media shared on social networking websites have a great impact on human life as growing numbers of people prefer to learn the news from sources like Facebook and YouTube (Tanvir et al. 2019). But the various news articles are unverified while posting through the social network. Hence, the authenticity of these materials must be checked. Extensive spreading of fake news through social media is increasing widely over time. It makes a negative impact on civilization and people also. The exponential rise of fake news and its impact on politics, equality, and civic confidence have increased the need for fake news detection and analysis.

Recently in 2020, various fake news and articles, rumors have spread about the COVID-19 virus in various countries through digital communication. Government officials from various countries are keeping an eye on fake news related to COVID-19 viruses. During this pandemic situation, false news spreads even quicker than coronavirus. Fake messages sometimes trigger community tensions, lynching cases, and negative stereotypes of individuals, specific groups, and communities. However, false news becomes a major threat to a life-threatening pandemic. Thus, there is a need to stop false information generation through various resources. Twitter assures that virus results end in factual posts, and Instagram diverts people looking for bug data to a special message with accurate data (Marr 2020).

In the year 2016, the importance of fake news detection was increased during the U.S. presidential election campaign (Guess et al. 2019). Many case studies in economics, politics, and psychology illustrate that the dissemination of fake news has emerged as a world issue and caught greater public attention. The leading cause is that as opposed to mainstream news sources such as newspapers and television, fake news can be produced and reported online quicker with minimum cost. However, as an ideal platform for accelerating the dissemination of fake news, social networks eliminate the actual distance gap among people, providing rich platforms for sharing, forwarding, voting, and reviewing, and inspire people to discuss any news (Orso et al. 2020). This kind of interaction around online media can increase serious implications and provide potential political and economic benefits to individuals. It became easier to spread false news among people with a greater impact on reality. Political and psychological factors play a significant role in increasing public trust in false news and further encouraging the dissemination of fake news. Nowadays, various political parties and corporate companies are trying to reach the crowd by creating fake accounts on social media. To achieve more financial profit, small-scale commercial start-ups are also using this platform.

The existing research studies indicated that the detection of fake news reached an average accuracy of 50%–55%. As compared to other types of media, this situation is more important for fake news, as the news, a symbol of trustworthiness and objectivity, is comparatively easier to gain public confidence. The need for an in-depth study is inspired by many viewpoints on the origin of fake news, what is to produce it, the way of dissemination, and its detection. This chapter aims to establish a structured approach for the thorough analysis of fake news. Since fake news is not known and current fake news studies are limited, further research into the relevant areas can assist as a basis to extend false news investigation.

The chapter is divided into various sections, which includes Section 11.1 an introduction about fake news and describes various definitions of fake news, fundamental concepts regarding fake news and its detection, an overview of existing fake news detection methods, and their different types. As mentioned earlier, a study of this chapter emphasizes four perspectives. Traditional approaches for fake news detection are given in Section 11.2, which mainly focuses on fact-checking. It highlights physical fact-checking and automated fact-checking techniques used for fake news detection. Section 11.3 covers writing mode–based approaches for fake news detection into which its detection and deception analysis are given. Dissemination-based approaches for fake news detection are given in Section 11.4. It covers fake news dissemination patterns, models of fake news dissemination, and dissemination-based fake news detection. Section 11.5 highlights integrity-based approaches for fake news detection which further gives details about assessing news headline integrity, assessing news source integrity, and assessing news spreader integrity. An application of machine learning in fake news detection and the proposed model for the same is given in Section 11.6. In Section 11.7, various open research challenges in fake news detection are discussed. Finally, this chapter summarizes the need for fake news detection and the application of machine learning (ML) approaches for the same.

11.1.1 Fake News: What It Is?

The meaning of news refers to the common man as the authenticated source of information. So, what makes the news "fake"? The word "fake" is often connected to them with words such as "false," "inauthentic," and "misleading" in the context of "fake news." No universal definition has been given for fake news. Based on several terminologies used for interpreting news as "fake news," some definitions are found in the literature. Different studies link this concept with the words like "maliciously false news," "false news," "satire news," "disinformation," "rumor," etc. According to Zhou and Zafarani (2018) based on the authenticity, intention, and whether the news is true or not, the fake news can be defined in two ways.

Definition I: "Fake news is false news." This definition covers the broad aspect of fake news which includes the statements, posts given by the organization, or public figures.

Definition II: "Fake news is purposely and verifiably corn news from a news source." The given term covers the specific facet of fake news in terms of false news purposely featured. These types of news are more harmful. Instead of telling truth to the public, it's better to mislead the people by providing disinformation.

11.1.2 Essential Concept

Core human perception and psychological hypotheses developed through various fields provide the new idea for qualitative and quantitative analysis toward analyzing the massive fake news data. The life cycle of fake news is shown in Figure 11.1. Different kinds of theories have been identified to study fake news analysis. These can be broadly categorized as: (i) writing mode–based fake news analysis; (ii) dissemination-based fake news analysis; and (iii) user-based fake news analysis.

11.1.2.1 Writing Mode–Based Fake News Analysis

It is an attempt to spot false news by capturing the news material manipulators in writing form. The writing mode–based approach also emphasizes the need to analyze news material. The writing style of these articles or posts is different from true news articles. There are two traditional style–based categories based on deception and objectivity (Tandoc et al. 2018). Falsehood-oriented lexicon-based procedures are used to capture the tricky arguments from the news body. The study of deception aims at studying

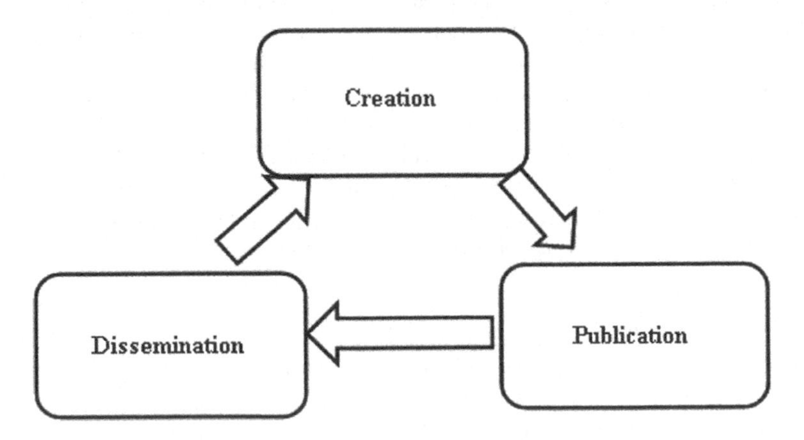

FIGURE 11.1 The life cycle of fake news.

the patterns of misleading material through various types of information like short text, multimedia contents, online news articles, and reviews given by people. The majority of deception studies had been identified with how to analyze the different styles of design to detect fake news and other forms of misleading content. The style-based feature and pattern can be grouped together based on the attributes and linguistic structural features of the news content. Objectivity oriented focus on styles which misguides the users easily. This may suggest diminished impartiality of the news material which leads potentially to deceive consumers easily. Hyperbiased style represents the behavior of producing false news under the influence of any political pressure. Yellow press articles or news does not contain sound news rather than trusting the eye-catching headlines (Shu et al. 2017). In this type, a bag-of-words approach also can be considered as it considers the frequency of each word n-gram or bi-gram (Thota et al. 2018).

11.1.2.2 Propagation-Based Fake News Analysis

It is useful to predict credibility in the news. This study is also useful for how different users take to spread and propagate misinformation through various sources. This type of analysis is also based on the parameters like the difference between spreading the regular news versus false news. Is there any pattern followed when misinformation is spread from various domains like political, economic, education, and social networking websites? Does the news from a variety of domains affect the propagation of false news? All these factors are important while analyzing the propagation-based fake news analysis.

11.1.2.3 User-Based Fake News Analysis

This kind of hypothesis analyzes fake news from a user's viewpoint. It also examines how users are interacting with false news. What kind of role users are playing in producing, propagating, and intervening in false news? Some of the users can be categorized as malicious users and instinctive users. Malicious users are those kinds of users who create and spare false information for certain benefits. Instinctive users are those users whose involvement is unplanned but they are obsessed with social encouragement.

11.1.3 Types of Fake News

Before the frequent use of social media, newsprint to radio/television were the main sources of news. Due to the tremendous use of the internet, nowadays people prefer to listen and watch news through online media. The human itself cannot differentiate between information and misinformation. There are many theories in psychology and perception, based on which different types of fake news are identified. These are mainly categorized as "satire news," "parody news," "fabricated news," and "manipulated news." "Satire news" is the most common type of fake news, in which the content has been created in a comic way to attract an audience with news updates. Examples include different types of mimic content circulated on the famous person through social media or any article containing dark humor on any person or situation, etc. "Parody news" is another form of fake news type which differs from satire news as it uses nonrealistic information to inject humor. "Fabricated news" is a type of fake news in which contents are not having any truthful basis. They are false contents used intentionally for spreading misinformation. It can be propagated easily through websites or any online social media network. Manipulated news is used intentionally to hide true information or alter true information or pictures. These types of manipulation can be done on real images or videos to create incorrect descriptions about a person, event, or place. Due to the easy availability of photo editing software, it is easily possible.

11.1.4 Overview of Chapter

This chapter summarizes each perspective with different analyzable and functional knowledge given by news and its spreaders with flexible tactics and structures and its relevant techniques. The purpose of this chapter is to study, summarize, evaluate, and analyze the latest research on fake news widely and extensively. It includes analysis, detection, and intervention strategies for fake information. Different materials related to news such as title, skin text, creator, distributor, and social-based information can be used to continue fake news research. This also addresses feature-based and relationship-based methods to research fake news and access resources such as different websites, as well as various forms of tools to support fake news studies.

Various ML techniques are used in the literature to detect fake news detection. ML-based processes like neural networks, natural language processing, and deep learning can be used. In this chapter, an efficient FAKE NEWS detector model is proposed using ML approaches. The experimental results have shown that the better accuracy of fake news detection in earlier phases from various sources over the internet like a post, news, and social media can be achieved by applying appropriate natural language processing techniques. The open issues and challenges such as fake news detection at early stages, identifying trustworthy contents, and fake news intermediation contents are also presented in this survey. Cross-domain fake news analysis is also one of the main challenges while detecting fake news. Through analyzing the features of fake news and problems identified in fake news studies, here several possible research challenges are highlighted that can facilitate further development.

11.2 Traditional Approaches for Fake News Detection

Traditional approaches for fake news detection can be studied using a different perspective. These can be from the creation of news to the circulation of news throughout different phases of fake news detection. The fake news information can be news-related which covers major headlines, the textual content of news, creator of news, and originator of news. It can be relevant to the social context which includes different opinions, reactions, and spreader networks. The most common approach for the detection of fake news detection is based on information. The information-based method is focusing on the detection of false information present in fake news. This process is known as fact-checking. This method aims to determine the validity of news by matching information derived from news material to be checked (e.g., its assertions or statements) with established true knowledge. The information-based approach is further classified into physical fact-checking and automated fact-checking, which are discussed in detail in Sections 11.2.1 and 11.2.2.

11.2.1 Physical Fact-Checking

Physical fact-checking is done by a person or a small group of experts according to their domain expertise. These are the fact inspectors that check the authenticity of the news. This kind of verification consists of checking the source of data, date, and time mentioned along with place. These persons are experts who can detect accurate results highly, but for the large volume of data, this process might be time-consuming. Hence, various websites provide this physical fact-checking service for the users. Examples of these websites are "factcheck.org," "factchecker.in," "politifact.com," "snopes.com," "punditfact," etc. Some websites provide fact-checking on articles and news based on different types of content like news items or non-news items to determine the content's credibility. Nowadays, many social networking sites like Facebook and Twitter have used this mechanism as they know the importance of identifying false information contents. To recognize and avoid dissemination of false information across fake stories, likes, and shares, Twitter has established a potential solution (Atodiresei et al. 2018).

11.2.2 Automatic Fact-Checking

Automatic fact-checking is heavily dependent on automatic tools and techniques used based on ML. ML techniques like information retrieval and natural language processing techniques provide the auto-processed information to detect the real facts within the articles or news and have been widely accepted. Automatic fact-checking is based on extraction from fact-based data and information evaluation. The authenticity of the news reports to be checked is calculated by comparing the news content details with the evidence available on the knowledge base. ML techniques often extract raw knowledge and clean up the data using different processing techniques. Automated fact-checking eliminates the information which is redundant, conflicting, and from unreliable sources. There are various traditional ML-based models, and deep learning–based models are present to check false data. In traditional methods, features are extracted at image or text level. It is based on a support vector classifier, random forest algorithm, and XGBoost (Pérez-Rosas et al. 2017, Zhou and Zafarani 2018).

11.3 Writing Mode–Based Approaches for Fake News Detection

Writing mode–based approach is one of the important aspects of fake news detection. The writing style is focused on the exploring contents of the news. Writing-based approach gives a clear idea about the intention of accessing the news in public. Based on the writing style, one can identify whether the given news is fake or not depending on the nature of it, which can be easily detected by the ML algorithm. Deception analysis and detection is a booming area for research in the case of fake news detection with spreading false contents across a variety of information. Such studies of deception usually do not find the intrinsic features of the type of news articles. Such a type of approach will be helpful for the reader to better understand and facilitate the widespread style of delusion inside different types of information.

11.3.1 Falsehood Identification and Analysis

The study of falsehood detection and analysis aims at studying the patterns of misleading material through various types of information in statements tweeted in social media, reviews given on e-commerce sites, images, and videos— a text which is widely used for communication over various software applications.

I. **Style of Falsehood Models***:* The material type of false knowledge (e.g., fake news) aimed at deceiving audience would be very diverse from the fact one while forensic psychological tests have shown that claims originating from concrete observations vary from that focus on imagination in meaning and accuracy (Siering et al. 2016). Such intuitions and basic assumptions have inspired and achieved great style-based studies of manipulation, whether for tweets, online messages, review sites, or news reports. The performance and result of the deception studies have been discussed in the later section by experimenting using online reviews, news, and statements.

II. **Style-Based Characteristics and Patterns***:* A collection of quantifiable characteristics, mostly ML attributes, is commonly described in the content types.

   a) **Language Features Based on Characteristics/Attributes**: These features are based on theory and are also called theory-oriented features. The "sensory ratio" function captures the concept described by the fact-detailed analysis that fake events convey lower degrees of sensory data equivalent to actual events. There are various parallel dimensions that can be considered which can be grouped together to define characteristic-based features such as extent, difficulty, ambiguity, subjectivity, nonimmediacy, feeling, multiplicity, informality, specificity, and readability. It's very important to consider an accurate feature while doing characteristic-based deception. It is a time-consuming process and requires an additional level of quantification process. There is a need to categorize features across different domains to obtain the most accurate deception detection. More systematic work is urgently needed to identify the most insightful and theoretical features which can better catch material deception. The following Table 11.1 shows attribute-based features proposed by various researchers.

   b) **Feature Based on Language Structure**: The evolving natural language processing techniques are applied to extract features from given content. At the lexicon-level frequency of words, characters are checked using the n-gram model. The syntax is checked using the part-of-speech tagger. Although these attributes are less important, observable, and predictable in deception studies, calculating them in comparison with characteristic-based

TABLE 11.1
Features Based on the Characteristics

Year	Author	Attribute Type	Feature Present
2012	Afroz, Brennan, and Greenstadt (2012)	Quality, complexity, uncertainty	Character count, sentence count, average number of sentences, percentage of modal verb
2013	Shojaee et al. (2013)	Quality, nonimmediacy, sentiment	Character count, number of exclamations, number of quotations
2015	Hauch et al. (2015)	Sentiment, quality, specificity	Positive words, verb count, temporal ratio
2015	Pak and Zhou (2015)	Specificity, quality, uncertainty, sentiment	Word count, exclusive term, number of positive words, tentative terms
2016	Siering, Koch, and Deokar (2016)	Sentiment, quality, complexity	Positive words, sentence count, the average number of punctuations
2016	Yang et al. (2018)	Complexity, quality diversity	The average number of words, verb and noun count, lexical diversity
2017	Braud and Søgaard (2017)	Diversity, informality, subjectivity	Redundancy, misspelled words, number of subjunctive verbs
2017	Potthast et al. (2017)	Quality, complexity, nonimmediacy	Paragraph count, number of words, number of quotations
2017	Volkova et al. (2017)	Sentiment, uncertainty, nonimmediacy	Number of positive words, tentative terms, rhetorical questions

features is fairly easy. The characteristics inside or across a variety of languages for each study level achieve maximum performance and highlight the best value features that can be accomplished at a single-language level. The following conclusions can be drawn:

1. In the case of a single language, the lexicon level gives better results compared to syntax, semantics, and discourse level.
2. It is always better to combine features across different languages rather than considering a single language.

c) **Analyzing Fake News Based on Content Style**: Evaluation based on the consideration of false information may greatly differ from evaluating the deception of many other kinds of knowledge. To detect dissatisfaction in style-based news stories, more explicit hints and trends should be found, assisted by highly associated theories, especially as regards journalism. Some of the individuals responsible for the dissemination of false news have power as to what the topic is about, then through the nature of their vocabulary they can be revealed (Yang et al. 2018). Generally, the news covers various topics like health, education, politics, and economics in different languages across various domains. It is necessary to explore cross-domain and cross-language analysis. The origin of fake news takes place through an important public event. Fake news creators take financial advantages from breaking news. Deep learning and ML models can be used to detect fake news (Gogate et al. 2017, Shu et al. 2017).

III. **Deception Detection Strategies***:* The feature vector approach is commonly used for writing mode–based deception. The style of content is given as input to the ML model to determine the classification or regression problem. Most of the studies have been performed using a supervised learning approach where data is labeled.

11.3.2 Deception in NEWS

The method proposed to date for the analysis of deception doesn't differentiate between news and other kinds of data. So, the system will concentrate on how to evaluate fake news content types, a different form of the analysis of deception in certain knowledge types.

The general deception-based technique can be used for fake news detection based on style. The supervised and semisupervised approaches can be used. The reason is that the labeled dataset is limited. It is very difficult for a human being to create such types of datasets having manual labeling.

11.4 Dissemination-Based Approaches for Fake News Detection

Other than style and knowledge base in fake news detection, how the news spreads and how it propagates by different users is also important to know. In this Section 11.4, brief ideas related to fake news propagation patterns, models based on fake news propagation, and propagation-based fake news detection are given. Several factors can be considered: How users represent the news? How measures are taken under consideration for propagating news? How does fake news spread other than regular news? And how various platforms are used to spread news like Twitter, WhatsApp, etc.? For such types of queries, the answer is unclear.

The fake news can be spread in cascading ways. It consists of a root and other nodes. Links represent how the news has traveled from one user to another. Root node indicates the origin of fake news who first publish the news and nodes below root node represent a user that propagates that fake news via links. The cascaded propagation for fake news can be analyzed qualitatively and quantitatively. In terms of qualitative analysis, propagation patterns are considered, and in the case of a quantitative analysis, a mathematical model that represents how a propagation takes place is considered. There are few questions related to this like how users describe the news? difference between spreading real versus fake news? how can one recognize fake news? and what is the variation found in fake news according to a different domain like social media, education, and politics? Many questions are still unanswered. Cascaded fake news is also divided into two types: hop-based and time-based. In hop-based, depth is measured in terms of the maximum number of steps and breadth is nothing but the number of users. The size is depending on the users involved in spreading. Time-based cascade analysis gives a trees-like structure.

11.4.1 Fake News Propagation Patterns

It is further divided into two parts: one is particular patterns that describe spreading fake news and the other compares it with the spreading of real news. The measurement is done using a hop- and time-based model. Propagation of fake news varies according to language, domain, and social website.

11.4.2 Models Based on Fake News Propagation

A qualitative analysis of fake news is done by a mathematical model. Such types of models are helpful in understanding, quantifying, and predicting fake news. Time is taken to spread fake news represented through regression analysis (Najar et al. 2012, Du et al. 2014). Propagation dynamics are represented by classical models like epidemics and economics. Epidemics represent an overall user that spreads the news. The economic-related model represents individual decisions of when to forward the news or delete the fake news.

11.4.3 Dissemination-Based Fake News Detection

Dissemination-based fake news detection model is further classified into two categories: (i) cascade-based fake news detection and (ii) network-based fake news detection methods.

Cascade-based techniques give a detailed path depending on the similarity of cascading news to the other—true or false. Informative representation is used for determining fake news. The network-based model is homogeneous, heterogeneous, or hierarchical (Vishwanathan et al. 2010).

11.5 Integrity-Based Fake News Detection

While studying fake news from an integrity perspective, there is a need to identify a knowledge base related to news and social issues. In this context, they differentiate a trustworthiness-based analysis of fake news from one focused on dissemination since it is possible to identify fake news at times using only secondary data and without taking into account news content or social ties. Such findings suggest that in some instances fake news identification can be simplified to identify an inaccurate source of websites.

11.5.1 Determining the Accuracy of Headline News

Assessing the trustworthiness of news headlines also decreases the identification of clickbait headlines whose main aim is to draw visitors' attention and to divert them to click on links that will take them to false pages. Examples of such clickbait include "tips to lose weight in 7 days" and "10 Bollywood actresses who got married to old-age people." The clickbait is often combined with fake news articles as they are effective tools for fake news to gain a high click rate and public confidence. The current clickbait identification studies focus on linguistic feature identification and use deep learning techniques to avoid user spread of such news.

11.5.2 Determining the Accuracy of News Outlets

A study of fake news indicated that most fake news stories are published by false news websites (Mourao and Robertson 2019) that view themselves as legitimate news publishers. Such findings suggest that to some degree, the content, reliability, and political ideology of the source websites influence the accuracy and legitimacy of the news. Old style web ranking algorithms measure website reputation to improve user search query responses by search engines. Web spam detection algorithms can thus be graded into: (i) content-based algorithms that evaluate web content features such as word counts and repetition of content and (ii) link-based algorithms that detect web spam using graph information.

11.5.3 Determining the Accuracy of News Comments

Posts from users on news outlets and social media often hold useful information about views and opinions; however, posts are mostly overlooked. In the field of e-commerce, the review of the product plays a vital role, thus negative comments, i.e., a false review about the product creates a greater impact on business (Zhou and Zafarani 2018). Models for determining how plausible a statement is divided into: (i) content-based, (ii) behavior-based, and (iii) model-based on a graph (network).

11.5.4 Determining the Accuracy of Spreader News

Users play a significant role in the dissemination of fake news. They are responsible for the distribution of fake news in various ways. User reputation has been used to research false news, especially from a dissemination perspective, either directly or indirectly through their posts and interconnections. Nevertheless, as the difference between malicious and natural remains ambiguous, only a few studies have suggested user vulnerability—innocent users may also engage regularly and inadvertently in false news activities. Reclassification of users of fake news operations is participants and nonparticipants. Participants are further divided into nasty users and inexperienced users based on their actions, each group being defined by different techniques for the identification and intervention of false news.

11.6 Application of Machine Learning in Fake News Detection

A ML strategy called the structure for rumor detection has been developed to legitimize signals from vague articles so that a user can understand falsehoods (Sivasangari et al. 2018). In the above sections,

TABLE 11.2

Various Techniques for Fake News Detection Concerning the Performance Measure Parameter

Techniques/Performance	Complexity	Toughness	Extendibility	Interpretability
Decision trees	Moderate	Moderate	Moderate	High
Neural network	High/low	Moderate/low	Moderate	Low
Bayesian	High/low	Moderate/low/high	High	High
Kernel-based	Low/how	High/moderate	Moderate	Low
Rule-based	Moderate/high	Moderate	Low	Better
Instance-based	High/low/moderate	Moderate	Low	Better

a detailed study of different types of fake news detection approaches is given. The study indicates that many research challenges in the area of fake news detection can be solved by applying appropriate ML and natural language processing techniques. The ML techniques that can be applied in various approaches are summarized in Table 11.2.

11.6.1 Proposed Work

Distributing information was more expensive before the internet, and there were much clearer definitions of what constituted news and media making it easier to control or self-regulate. Yet, the rise of social media has broken down many of the barriers that have stopped the dissemination of false or misleading news in democracies. It has allowed everyone to create and disseminate knowledge, particularly for those who have shown themselves to be most skilled at how social network functions. The objective of the proposed work is to build a model using the best ML technique/s, which can detect whether the news is fake or not based on its contents. This proposal requires a natural language processing perspective (Mourao and Robertson 2019). A significant part of the aim is to compare and evaluate the outcomes from several different implementations of models and provide an overview of the results. The main advantageous feature of the proposed model is that it classifies news with better accuracy.

Major objectives of this project are:

1. To construct a knowledge base from posts or articles.
2. To build a supervised learning model to classify the news whether it's trustworthy or fake news.
3. To improve the accuracy of prediction whether a news article or piece is real or fake.

11.6.2 Proposed Methodology

A significant part of the aim is to compare and evaluate the outcomes from several different implementations of models and provide an overview of the results. It uses advantageous features of naïve Bayes and the random forest classifier and proposes a hybrid model for classification. The process includes preprocessing of data and eliminating stop words and refining data, then separating features from the data to be categorized and then classifying the data as fake or real news.

The proposed work is summarized as follows, also given in Figure 11.2.

1. Data preprocessing and refining
2. Feature extraction
3. Classification using classifiers as fake news or real news
    1. **Data Preprocessing and Refining**: The data received from the post or any article needs to be preprocessed and refined using data preprocessing techniques to improve efficiency. Enhancements like removal of stop words, tokenization, sentence segmentation, and punctuation removal are used.

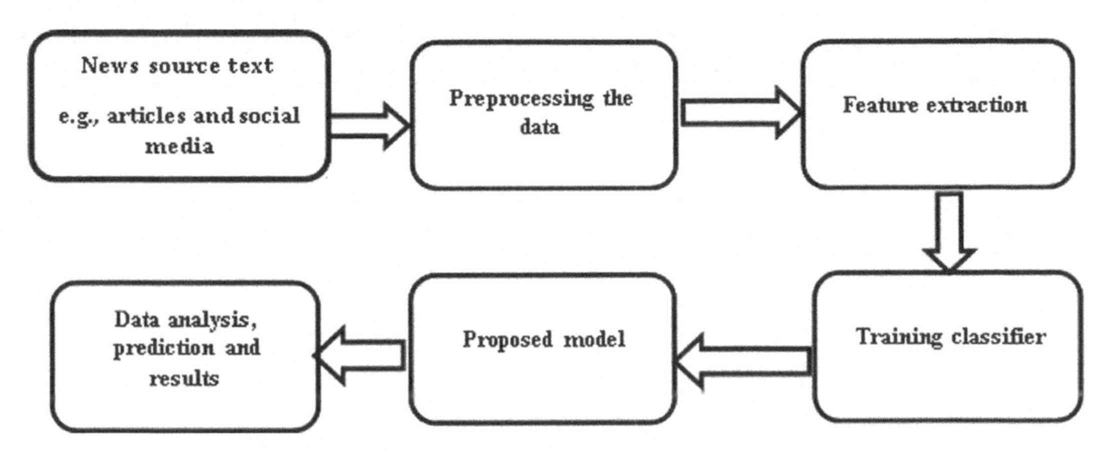

FIGURE 11.2 Fake news detection methodology.

2. **Feature Extraction**: The feature extraction step plays a vital role in the detection of fake news in which features can be selected based on the:
    - i) word count
    - ii) authenticity
    - iii) clout
    - iv) length
    - v) tone
3. **Classification Using Classifiers as Fake News or Real News**: The survey of standard classifiers indicates that the accuracy cannot be improved using the single standard classifier. Thus, the model is proposed as shown in above Figure 11.2 to improve the accuracy of the classification of news fake or real. A hybrid model using advantageous features of naïve Bayes classifier and random forest classifier (Reddy et al. 2020) is implemented.
    - a) In naïve Bayes classification technique, it assumes that feature words are independent of each other which is almost impossible. The classification accuracy is less compared to other ML classifiers.
    - b) Random forest, on the other hand, has many advantages like tolerance to noisy data and speed, and accuracy is much better than other ML classifiers.
    - c) But it is inappropriate to say an individual classifier to be the best or standard model. Hence, in the proposed system, naïve Bayes classifier is combined with random forest by utilizing advantageous features and minimizing disadvantages of both to enhance the performance of the classifier.

11.6.3 Design and Workflow Diagram of Proposed System

The steps involved in the prediction of fake news through the proposed system are given as follows, also given in Figure 11.3.

1. A user registers (if he is a new user) or logs into the system.
2. A user enters news to be detected as FAKE or REAL.
3. The proposed model fetches data from knowledge base built initially (FAKE NEWS knowledge base) and generates results.
4. Prediction results are generated, which generate results as FAKE or REAL NEWS.
5. Then, the user's decision plays a role in whether to accept or reject the results.

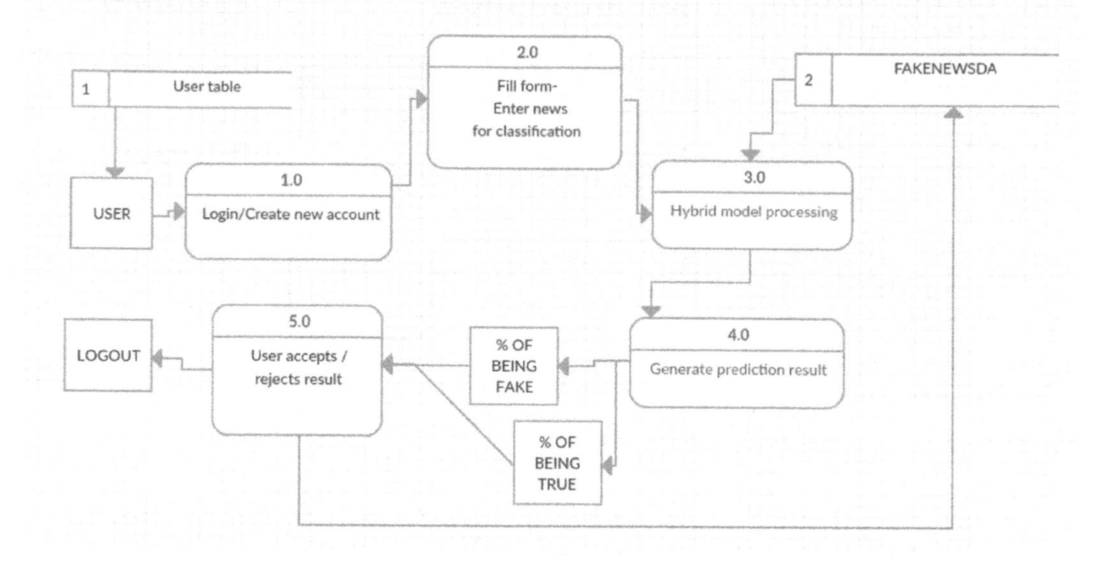

FIGURE 11.3 Workflow diagram.

11.6.4 Experimental Results and Analysis

The model is trained for around 50,000 posts and articles and tested for results. The results are analyzed based on the classification accuracy, recall, precision, and F1 score.

A user has to register to the system. After successful registration, the user can log into the system. A graphical user interface (GUI) for user login as shown in Figures 11.4–11.7 gives GUI for users to enter news to be classified and enter categories of the news either social, political, sports, etc., and the source of the news either Facebook, Twitter, news article, etc.

The comparative analysis of performance measures of existing classification ML techniques with the proposed method is given in Table 11.3, which showed that the proposed model accuracy of fake news detection is improved. The graphical representation of the comparison of the proposed model with the individual model is given in Figures 11.8–11.11.

11.7 Open Research Challenges in Fake News Detection

Based on the current state of fake news research, the following research challenges are given, which give deeper insights for the understanding of fake news. Various research challenges in the detection of fake news are given as:

a) Predetection of fake news
b) *Identification of trustworthy contents*
c) Fake news in cross-domain
d) Deep learning and fake news
e) Fake news and social media

11.7.1 Redetection of Fake News

Detection of false news at an earlier stage is helpful to avoid the spread of fake news. If a person is unaware of false news and he or she is a trustworthy person, then such news can be easily spread. Once

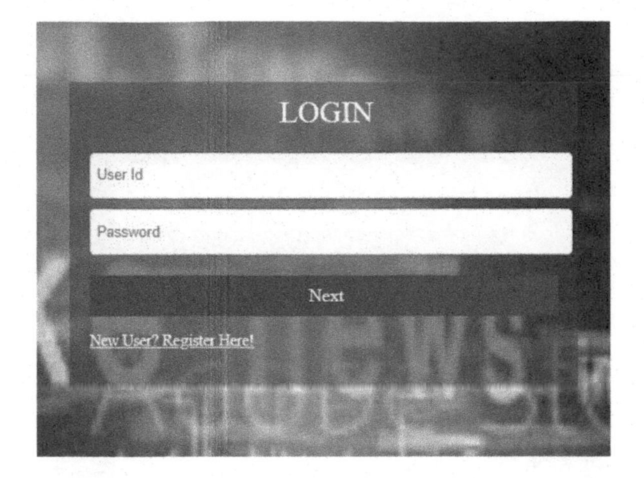

FIGURE 11.4 GUI: user login.

FIGURE 11.5 GUI: search page.

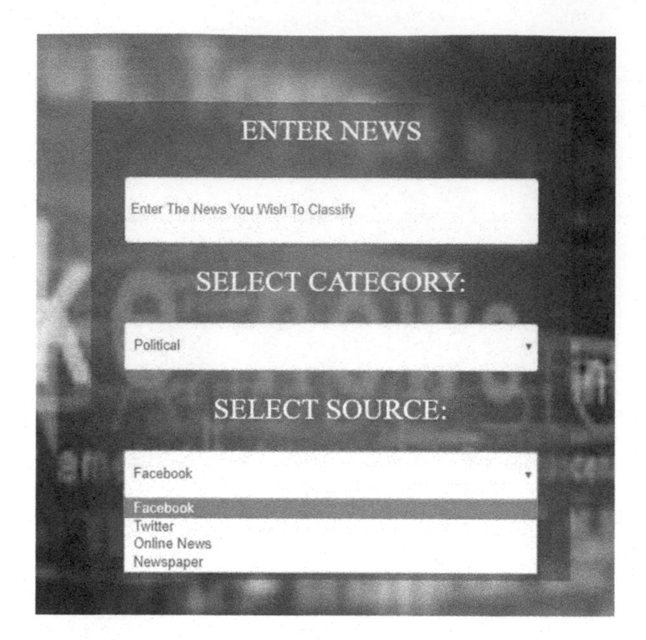

FIGURE 11.6 GUI: category dropdown.

FIGURE 11.7 GUI: source dropdown.

TABLE 11.3
Comparative Analysis of Performance Measures

Classifier	Parameter	Accuracy	Precision	Recall	F1 score
Naïve Bayes	False	0.58	0.58	0.49	0.53
	True	0.62	0.62	0.70	0.66
	Avg/total	0.60	0.60	0.61	0.60
SVM	False	0.53	0.53	0.52	0.53
	True	0.60	0.60	0.62	0.61
	Avg/total	0.57	0.57	0.57	0.57
Random forest	False	0.63	0.63	0.43	0.51
	True	0.62	0.62	0.79	0.69
	Avg/total	0.63	0.63	0.62	0.61
Proposed model	False	0.88	0.88	0.82	0.81
	True	0.82	0.82	0.86	0.85
	Avg/total	0.85	0.85	0.84	0.83

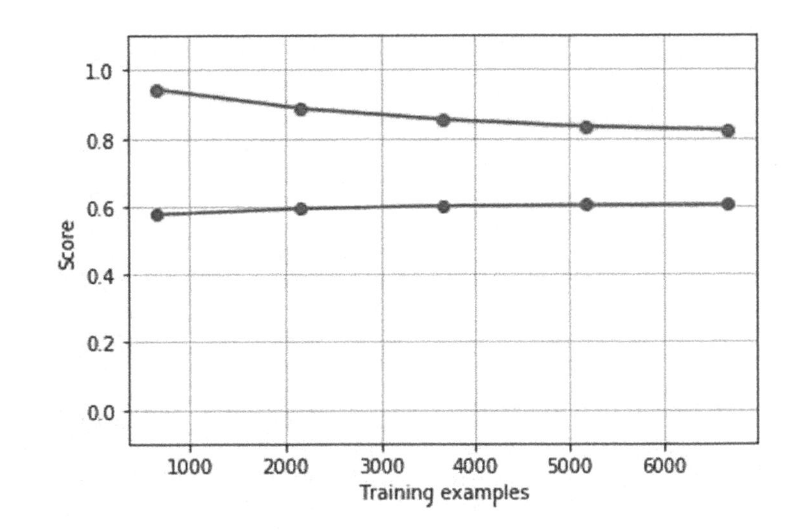

FIGURE 11.8 Naïve Bayes vs. proposed model.

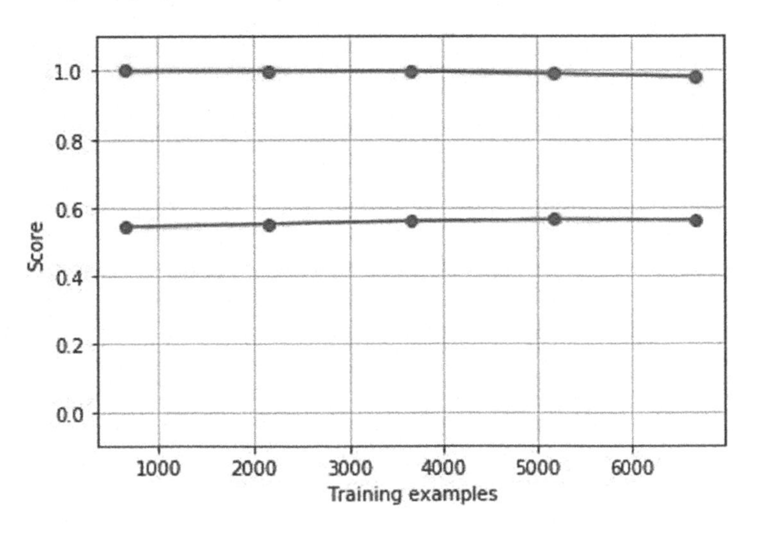

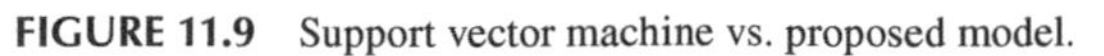

FIGURE 11.9 Support vector machine vs. proposed model.

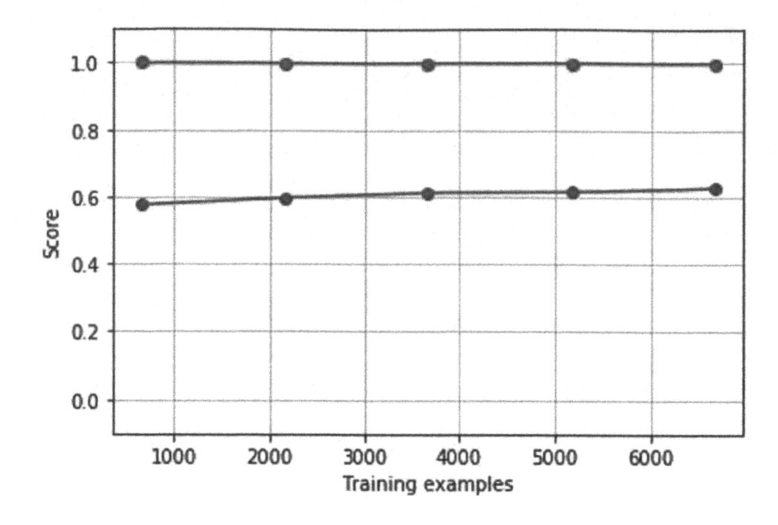

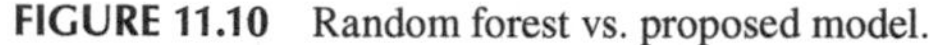
FIGURE 11.10 Random forest vs. proposed model.

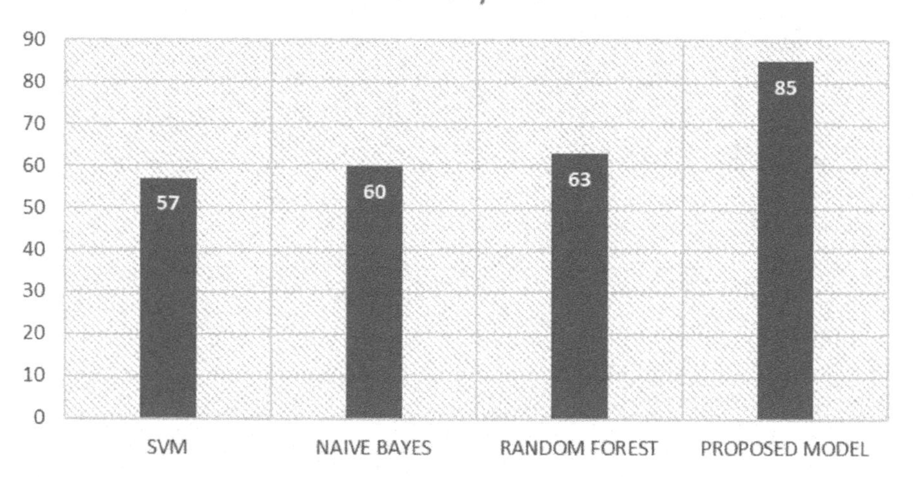

FIGURE 11.11 Comparative analysis indicating improved accuracy using proposed model.

fake news gains their trust, it becomes difficult to rectify them. Finally, the output of ML techniques may be adversely affected by restricted knowledge.

To tackle such research challenges and perform early fake news detection, one can concentrate on:

1. the construction of a dynamic knowledge base, the early detection of fake news becomes difficult whose knowledge base does not exist. The research challenge identified indicates the need for technologies related to constructing a real-time knowledge base to realize timely ground-based reality updates;
2. features of consistency, in particular, features capable of capturing the generality of deceptive writing styles across subjects, contexts, languages;
3. and defining verifiable information and topics one can boost the efficiency of the identification of false news.

11.7.2 Identification of Trustworthy Contents

With new knowledge being generated and shared online at an exponential pace, the identification of verifiable material may increase the effectiveness of fake news detection and interference through ordering a check-worthy text.

The trustworthiness can be evaluated by considering various factors like the impact of news on society, nation, origin, etc. There is a need to create web applications to check the trustworthiness of a piece of news.

11.7.3 Fake News in Cross-Domain

Fake news created an impact in almost any area and false people making use of the spread of false news for their benefit. Thus, the detection of false news in almost any domain requires analysis of fake news concerning context, topic, and language used (Guess et al. 2019).

11.7.4 Deep Learning and Fake News

Deep learning studies support false news analysis and gain benefits on their own (Surani et al. 2021). Current fake news research highlighted that to improve efficiency and accuracy of detection of fake news deep learning and neural networks can be used.

11.7.5 Fake News and Social Media

Social media has created a greater impact on society. Many people across the world got connected with the help of various social networking websites. The reviews and opinions shared on social media are spread over the network, thus it becomes easier to spread fake news with greater speed. The research challenges exist in the identification of types of users who generate news and build an intelligent system to identify fake news or post before it is available publicly.

11.8 Conclusion

In the last two decades, digital media is continuously growing in every part of communication. Fake news detection is an emergent field of study that gained attention but poses some difficulties because of the limited number of resources available. The exponential growth of fake news and its degradation of democracy, justice, and public confidence have increased the need for its study, identification, and intervention. In this chapter, the need for fake news detection and its life cycle are explained. The research work in this field is discussed thoroughly with various approaches that can be used for the detection of fake news such as traditional approaches, writing mode–based approaches, dissemination-based approaches, and integrity-based approaches. By analyzing the features of fake information and real problems in fake news studies, this chapter highlighted some possible research tasks and solutions that can be provided through the application of various ML approaches. In this chapter, an efficient FAKE NEWS detector model is proposed using ML techniques. The experimental results show that 85% of accuracy has been achieved on fake news detection in earlier phases from various sources on the internet like the post, news, and social media by applying appropriate natural language processing techniques. The open issues and challenges are also discussed with probable examination tasks which can encourage additional growth of fake news research and effective methodologies for the detection and analysis of fake news.

REFERENCES

Afroz, Sadia, Michael Brennan, and Rachel Greenstadt. 2012. "Detecting hoaxes, frauds, and deception in writing style online." *In 2012 IEEE Symposium on Security and Privacy,* pp. 461–475, 33rd IEEE Symposium on Security and Privacy, S and P 2012-San Francisco, CA.

Atodiresei, C.-S., A. Tănăselea, and A. Iftene. 2018. "Identifying fake news and fake users on Twitter." *Procedia Computer Science*, 126, 451–461. doi: 10.1016/j.procs.2018.07.279.

Braud, Chloé, and Anders Søgaard. 2017. "Is writing style predictive of scientific fraud?" *Proceedings of the Workshop on Stylistic Variation*, 37–42. doi: 10.18653/v1/W17-4905.

Du, Nan, Yingyu Liang, Maria Balcan, and Le Song. 2014. "Influence function learning in information diffusion networks." *In International Conference on Machine Learning*, pp. 2016–2024.

Gogate, Mandar, Ahsan Adeel, and Amir Hussain. 2017. "Deep learning-driven multimodal fusion for automated deception detection." *In 2017 IEEE Symposium Series on Computational Intelligence (SSCI)*, pp. 1–6.

Guess, Andrew, Jonathan Nagler, and Joshua Tucker. 2019. "Less than you think: Prevalence and predictors of fake news dissemination on Facebook." *Asian-Australasian Journal of Animal Sciences*, 32(2), 1–9. doi: 10.1126/sciadv.aau4586.

Hauch, Valerie, Iris Blandón-Gitlin, Jaume Masip, and Siegfried L. Sporer. 2015. "Are computers effective lie detectors? A meta-analysis of linguistic cues to deception." *Personality and Social Psychology Review*, 19(4), 307–342.

Marr, B. 2020. "Coronavirus fake news: How Facebook, Twitter, and Instagram are tackling the problem." *Forbes*. https://www.forbes.com/sites/bernardmarr/2020/03/27/finding-thetruth-about-covid-19-how-facebook-twitter-and-instagram-are-tackling-fake-news/.

Mourao, Rachel, and Craig Robertson. 2019. "Fake news as discursive integration: An analysis of sites that publish false, misleading, hyperpartisan, and sensational information." *Journalism Studies*, 1–19. doi: 10.1080/1461670X.2019.1566871.

Najar, Anis, Ludovic Denoyer, and Patrick Gallinari. 2012. "Predicting information diffusion on social networks with partial knowledge." *In Proceedings of the 21st International Conference on World Wide Web*, pp. 1197–1204.

Orso, Daniele, Nicola Federici, Roberto Copetti, Luigi Vetrugno, and Tiziana Bove. 2020. "Infodemic and the spread of fake news in the COVID-19-era." *European Journal of Emergency Medicine*. doi: 10.1097/mej.0000000000000713.

Pak, Jinie, and Lina Zhou. 2015. "A comparison of features for automatic deception detection in synchronous computer-mediated communication." *In 2015 IEEE International Conference on Intelligence and Security Informatics (ISI)*, pp. 141–143.

Pérez-Rosas, Verónica, Bennett Kleinberg, Alexandra Lefevre, and Rada Mihalcea. 2017. Automatic detection of fake news. ArXiv preprint. ArXiv:1708.07104.

Potthast, Martin, Johannes Kiesel, Kevin Reinartz, Janek Bevendorff, and Benno Stein. 2017. "A stylometric inquiry into hyperpartisan and fake news." ArXiv Preprint, ArXiv:1702.05638.

Reddy, Harita, Namratha Raj, Manali Gala, and Annappa Basava. 2020. "Text-mining-based fake news detection using ensemble methods." *International Journal of Automation and Computing*, 17(2): 210–221. doi: 10.1007/s11633-019-1216-5.

Shojaee, Somayeh, Masrah Azrifah Azmi Murad, Azreen Bin Azman, Nurfadhlina Mohd Sharef, and Samaneh Nadali. 2013. "Detecting deceptive reviews using lexical and syntactic features." *In 2013 13th International Conference on Intelligent Systems Design and Applications*, Salangor, Malaysia, pp. 53–58.

Shu, Kai, Amy Sliva, Suhang Wang, Jiliang Tang, and Huan Liu. 2017. "Fake news detection on social media: A data mining perspective." *ACM SIGKDD Explorations Newsletter* 19(1): 22–36.

Siering, Michael, Jascha-Alexander Koch, and Amit V. Deokar. 2016. "Detecting fraudulent behavior on crowdfunding platforms: The role of linguistic and content-based cues in static and dynamic contexts." *Journal of Management Information Systems*, 33(2), 421–455.

Sivasangari, V., P. V. Anand, and R. Santhya. 2018. "A modern approach to identify the fake news using machine learning." *Indian Journal of Pure and Applied Mathematics*, 118(20), 10.

Surani, Mehdi, and Ramchandra Mangrulkar. 2021. "Online public shaming approach using deep learning techniques." *Journal of University of Shanghai for Science and Technology*. doi: 10.51201/jusst12675.

Tandoc Jr, Edson C., Zheng Wei Lim, and Richard Ling. 2018. "Defining 'fake news' a typology of scholarly definitions." *Digital Journalism* 6(2), 137–153.

Tanvir, Abdullah-All, Ehesas Mia Mahir, Saima Akhter, and Mohammad Rezwanul Huq. 2019. "Detecting fake news using machine learning and deep learning algorithms." *2019 7th International Conference on Smart Computing and Communications, ICSCC 2019*, pp. 1–5. doi: 10.1109/ICSCC.2019.8843612.

Thota, A., P. Tilak, S. Ahluwalia, and N. Lohia. 2018. "Fake news detection: A deep learning approach." *SMU Data Science Review,* 1(3), 21.

Vishwanathan, Vichy N., Nicol N. Schraudolph, Risi Kondor, and Karsten M. Borgwardt. 2010. "Graph kernels." *The Journal of Machine Learning Research*, 11, 1201–1242.

Volkova, Svitlana, Kyle Shaffer, Jin Yea Jang, and Nathan Hodas. 2017. "Separating facts from fiction: Linguistic models to classify suspicious and trusted news posts on Twitter." *In Proceedings of the 55th Annual Meeting of the Association for Computational Linguistics,* Vancouver, Canada (Vol. 2, Short Papers), pp. 647–53.

Yang, Yang, Lei Zheng, Jiawei Zhang, Qingcai Cui, Zhoujun Li, and Philip S. Yu. 2018. "TI-CNN: Convolutional neural networks for fake news detection." ArXiv Preprint, ArXiv:1806.00749.

Zhou, Xinyi, and Reza Zafarani. 2018. "Fake news: A survey of research, detection methods, and opportunities." ArXiv Preprint ArXiv:1812.00315.

12

Authentication of Broadcast News on Social Media Using Machine Learning

Smita Sanjay Ambarkar
Lokmanya Tilak College of Engineering

Narendra M. Shekokar
Dwarkadas J. Sanghavi College of Engineering

Monika Mangla and Rakhi Akhare
Lokmanya Tilak College of Engineering

CONTENTS

12.1 Introduction 185
12.2 Literature Survey 186
12.3 Fake News Detection Using Machine Learning 187
12.3.1 Data Retrieval 187
12.3.2 Data Pre-Processing 187
12.3.3 Data Visualization 188
12.3.4 Tokenization 188
12.3.5 Feature Extraction 188
12.3.6 Machine Learning Algorithms for Fake New Detection 188
12.3.6.1 Support Vector Machine (SVM) 189
12.3.6.2 Naive Bayes Classifier 189
12.3.6.3 Decision Tree 190
12.3.6.4 ANN 190
12.3.7 Training and Testing Model 190
12.4 Comparative Study 191
12.5 Conclusion 193
References 193

12.1 Introduction

News is considered to be a fake if it is factually incorrect, misinterprets the facts, or spreads through any unauthenticated media. Nowadays, social media is a prominent tool to perform various social activities. Unfortunately, this tool is being equally used for some unethical purposes by some antisocial elements. Resultantly, sometimes havoc and panic situations are created in society. Also, political parties take advantage of social media to circulate fake facts about oppositions to gain the confidence of the general public. This fact was realized during the 2016 US presidential election. Fake news (false stories) is primarily circulated through social media compared to any other mediums of communication (TV, Radio, etc) owing to cost and ease of communication. Despite the widespread usage of social media, fake news detection over social media is a challenging task as the detector model needs to identify the correlation between news creators, the news subject, and the credibility of the news to identify the impact of news on the society.

Machine learning (ML) is a competent tool to design a fake news detector. There are numerous supervised and unsupervised machine learning algorithms in order to design the training machine. This chapter also puts forth the training model that identifies the fake news, based on a certain degree of accuracy.

Social media is an excellent source of communication. It is one of the most popular technologies among all age groups of the community. The prime reason for its popularity and demand is its cost-effectiveness that encourages multiple users to communicate interactively with their peers. However, cost and connectivity attract malicious users for creating social media accounts and thus it becomes home to cyber users, trollers, social bots, and cybercriminals. Cybercriminals perform social engineering using social bots. Social bot [1] is a malicious computer application created by an attacker to communicate with the people on social media, interact with other bots, and spread fake news. The survey reveals that millions of social bot accounts were created to distort the US 2016 presidential election [2], and these bots tweeted in favor of Trump or Clinton. Usually, trolls are malicious emotional activators who disturb online activities by spreading fake news on social media. However, evidence suggests that ~1000 Russian trolls get paid for committing 'Anti-social activity'. Trolling enables easy and fast circulation of fake news as trolling provokes the negative behavior of the people like fear, anger, irrationalism, distrust, etc. [3]. Cyborg users [4] are antisocial humans with social media accounts; they also connect their accounts with bots to spread fake news. In a nutshell, all these partisan malicious accounts are prominent sources of fake news proliferation in social media. Social media created a new way to explore the news all over the world, but the spreading of fake news depraves this paradigm. This motivates us to impede the flow of fake news by proposing a generalized solution for this problem. The studies reveal that the genuine community of social media believes in fake news because of several human characteristics factors [5]. Moreover, the frequency of fake news occurrence is an influencing factor that implies the principle "the more you hear/read, the more you believe." Hence, the fake news will also be considered authentic by the community at large if its exposure increases [6,7]. Social credibility is another influencing factor as there is no way in social media to check the credibility of the source. If a person gets the news from a trusted friend, they believe the news. If the dissemination of fake news is not restricted, then the society will be hampered at large, and hence this study is motivated to curtail the consequences of fake news.

The chapter has been organized into various sections. The importance of fake news detection and the motivation is discussed in this section – Section 12.1. Related work is presented in Section 12.2. The employment of various ML algorithms for fake news detection is compared in Section 12.3. Further, all these algorithms are compared in Section 12.4. Finally, the conclusion is presented in Section 12.5.

12.2 Literature Survey

This year the entire world is under the severe impact of COVID-19 and that certainly has witnessed a huge surge in the fake news related to healthcare. This fake news directly affects human lives; hence it is of utmost importance to curtail its circulation. The authors have surveyed the diverse literature to get insight into the system.

Gurav et al. [8] proposed a novel ML-based fake news detection method and implemented it in the Facebook Messenger chatbot with 81.7% accuracy. They used the naïve Bayes classifier in the BuzzFeed news data set and increased accuracy of 74% on the test set. The authors also used this approach for automating fake news detection on Twitter. Further, Thota et al. [9] used stance detection techniques for fake news detection. Stance detection is the process of automatically finding the relationship between two pieces of text. In Ref. [9], authors predicted the stance using a news article and news headline pair. The stances were classified as "agree," "disagree," "discuss," or "unrelated,", based on similarity among them. The authors explored different ML algorithms to find the best results and compared them to classify the stance between the article and the headline.

Zhou and Zafarani [10] studied and evaluated fake news based on four perspectives: the false knowledge it carries, its writing style, its propagation patterns, and the credibility of its source. The authors highlighted different research work based on reviews. Authors identified different fundamental theories across various disciplines to encourage interdisciplinary research on fake news. The survey is useful for experts in computer and information sciences, social sciences, political science, and journalism to research fake news.

Zhang et al. [11] proposed a novel automatic inference model for fake news detection named fake detector. This model is based on textual feature extraction to build a deep diffusive network model for identification for news articles, creators, and subjects. The authors experimented on different datasets and compared them with the fake detector model. This proposed model provided better results as compared to others. Similarly, Ahmed et al. [12] proposed a model that used different classification models like decision tree, SVM, logistic regression, K nearest neighbor to process features extracted from the text. Emergent is a real-time data-driven rumor identification approach. It automatically identifies the rumors, but where human input is required, it is not automated.

O'Brien [13] used a simple CNN model for pre-processing of the small dataset, which extracts useful features that humans may or may not be able to detect. In this model, generalizations, colloquialisms, and exaggeration techniques are used to detect fake news. The proposed CNN model could handle complex relationships between the detection of these patterns and the decision for classification, which humans could not handle.

Ambarkar and Akhare [14] analyze the impact of social media on society. The authors particularly analyze the emotional health of society hampered by social media messages, which may include fake messages as well. The authors put forth the analysis using the WhatsApp messaging system.

The literature survey reveals that machine learning techniques provide significant support for detecting fake news. The authors of this chapter hence proposed the generalized architecture in the next section.

12.3 Fake News Detection Using Machine Learning

In today's era, social media is a great source of information. The news is broadcasted and spread at an exponential rate, which is creating a huge impact on the public. Recently, the entire world is witnessing the pandemic outbreak of COVID-19 which creates havoc in the entire world [15]. Fake news regarding COVID-19 will further create panic among the people. So, this chapter proposes a model to detect the forged news from the social media platform. The model consists of several steps as shown in Figure 12.1.

1. Data Abstraction
2. Data Pre-processing
3. Data Visualization
4. Tokenization
5. Feature Extraction
6. Machine Learning Algorithms
7. Training and Testing Model
8. Evaluation Metrics

The steps for detection of fake news are elaborated below.

12.3.1 Data Retrieval

Data retrieval is an important process that aims to retrieve data from various sources. Among various sources, Twitter is an important source of data. For the same purpose, users create a Twitter account that is used for tweeting. It is agreed that a tweet from a verified account is believed to be more reliable as compared to a tweet from a non-verified account. Thus, Twitter takes a long process to verify an account before stamping it as a verified account.

12.3.2 Data Pre-Processing

Once the data is downloaded from social media, it is pre-processed to remove any punctuation symbols and emoticons. Further, the whole data is also converted to lower case letters to maintain consistency in the text. Additionally, repeating characters like URL tags and stop words are also removed out of the dataset during pre-processing.

12.3.3 Data Visualization

Once the data is pre-processed, it is fed to the model so as to understand the data through pictorial representation. For the same, data is represented pictorially in the form of graphs and charts.

12.3.4 Tokenization

Tokenization refers to the splitting of the text into words or small chunks of words. There are various inbuilt functions available in the NLTK (Natural Language Toolkit) python module itself. Here, tokenization returns an array of strings that can also be empty.

12.3.5 Feature Extraction

Feature extraction is a process of dimensionality reduction that aims to reduce the raw data into more manageable groups for processing. The prime motive of feature extraction is to reduce the processing time of the compiler. It also leads to efficiency enhancement for detecting the value of the word. Here, we use the technique Term Frequency – Inverse Document Frequency (TF-IDF) to represent the frequency of the word in the document.

TF-IDF is an efficient method for extracting string, based on the occurrence. In this proposed architecture, we assume that if the word is repeated more number of times in the text, then it is an important word and it should be considered for extraction. Further, we calculate the term frequency using normalization of the occurrence of the word according to the size of the text.

$$\text{So, TF} = \text{Document count(word)/Total number of words in the document.}$$

IDF is used to filter the extraction process further.TF provides equal importance to every repeating word in the text. However, some repetitive words like "a,", "the" will be of less importance and should not be considered in the extraction process. IDF method reduces the effect of such repetitive words. IDF is calculated using the following formula.

$$\text{IDF}\ (w) = \log\left(\text{Total number of documents/Number of documents containing word}\ w\right)$$

Further, to extract the more important word, the product of TF-IDF is calculated so that important words will get a higher score.

$$\text{TF-IDF(word)} = \text{TF(word)} * \text{IDF(word)}$$

Here, the TF-IDF extracts each important word in the document; hence, once the frequency of the important words is extracted, the entire dataset can be dropped to further enhance the processing speed.

12.3.6 Machine Learning Algorithms for Fake New Detection

Machine learning algorithms are powerful algorithms used for detecting the similarities of the system [15]. Therefore, it is a challenging task to use machine learning algorithms for detecting fake or wrong news. This section presents the extensively used machine learning algorithms for fake news detection along with their features and characteristics.

The ML algorithms have been broadly classified into supervised and unsupervised learning. The broad classification of these algorithms has been given in the following Figure 12.2.

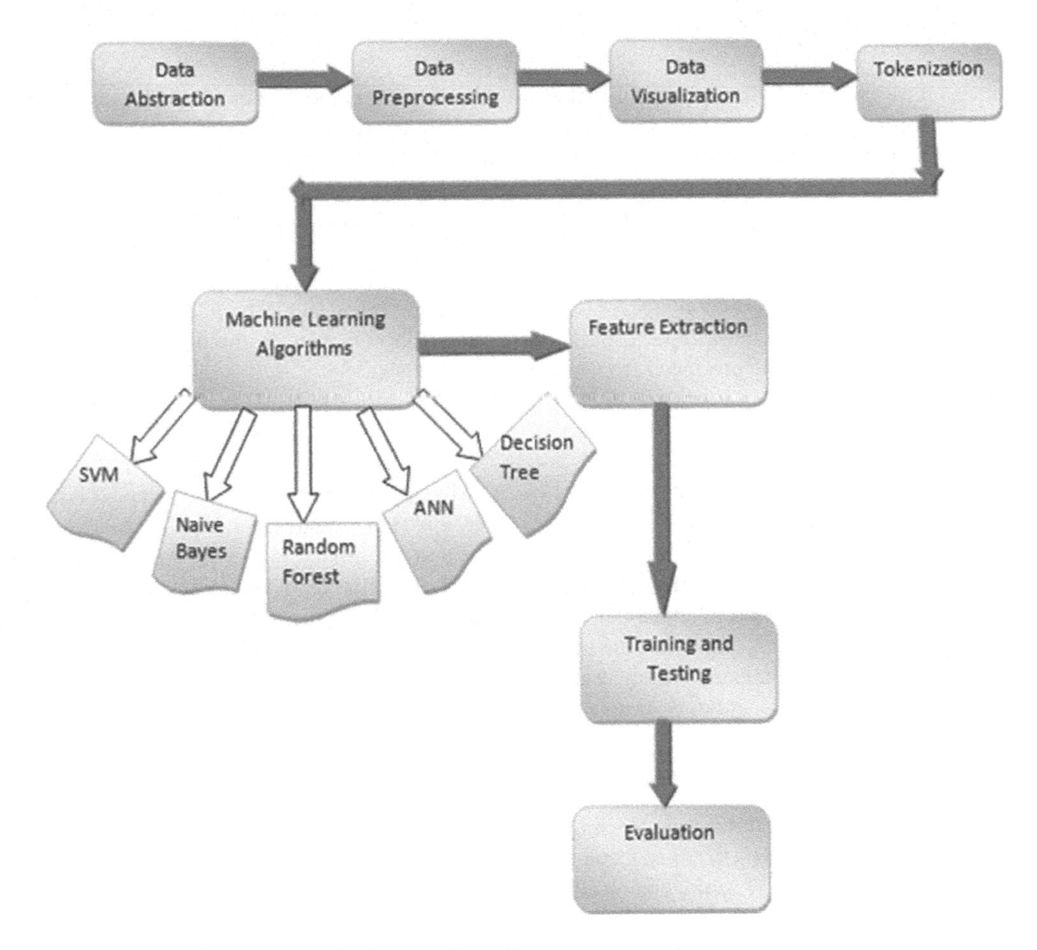

FIGURE 12.1 General Architecture for fake news detection.

Training and testing are the two basic aspects of each ML algorithm as shown in Figure 12.2. The statistical method is used to train the input data set, and resultantly, it helps to understand the network performance. Afterward, flaws in the system will be detected using the testing method.

12.3.6.1 Support Vector Machine (SVM)

SVM is a supervised learning algorithm that classifies the multidimensional labeled data using a margin hyperplane. SVM is broadly classified into linear and nonlinear. Here, linear SVM forms a linear kernel using linear algebra whereas Non-Linear SVM is used for nonlinearly separated data. Further, one-class and multi-class SVMs are also available to implement signature-based and anomaly-based detection. Here, multi-class and one class are the supervised and unsupervised machine learning algorithms [16].

12.3.6.2 Naive Bayes Classifier

Naive Bayes classifier consists of a series of classifiers based on probability estimation, and the classifiers follow Bayesian network structure. Naive-based classifiers used the target functions having discrete value as output. Moreover, independent attribute values were used in the naïve-based classifiers. Primarily, three naïve Bayes algorithms are available, viz. Gaussian naïve Bayes, multinomial naive Bayes, and Bernoulli naive Bayes. However, naive Bayes is a simple classifier, and it fails to detect various kinds of attacks as the underlying assumption is not true for IoT [17] dataset.

12.3.6.3 Decision Tree

The decision tree is an ML algorithm that makes use of a tree-like model to conclude a decision. It is represented using If-else rules. It can be employed for discrete as well as continuous values. A decision tree is a powerful algorithm against an error-prone dataset. The speciality of the algorithm is that even if some attributes are missing from the dataset, still the algorithm will work fine and provide the result. An algorithm is efficient for the attribute-value pair and the discrete output value target functions.

Random Forests:

It is a classification and regression algorithm training decision trees. Here, the final decision depends on the decision of individual trees. It performs better in case of low correlation amongst individual trees. More accurate predictions are produced from uncorrelated models [18].

12.3.6.4 ANN

ANN is an Artificial Neural Network inspired by the working of the human nervous system. ANN consists of three basic layers (i) Input layer, (ii) Hidden layer, and (iii) Output layer. The input layer is fed with the input dataset. Further, the processing is performed at the hidden layer, and finally, the output layer generates the output. In ANN, the calculated error fed back to the input layer known as backpropagation. This error could be used to adjust the network to minimize the error rate.

The following Section 12.3.7 analyzes the fake news detection models and also puts forth the performance of various above-mentioned widely used machine learning algorithms. The model will never be complete without the below-mentioned training and testing phase.

12.3.7 Training and Testing Model

Once the network is developed, it is trained using the training dataset using supervised learning. The algorithm makes use of training data to adjust features that are later used to evaluate various performance metrics, viz. accuracy, precision f1 score, and recall. The model is tested to minimize various error metrics.

The proposed model shows the detailed steps used for fake news detection. This fake news detection model can be implemented in the form of the structure of the different architectural models shown in Figure 12.3. The authors [19] suggest the classifications based on data, feature, model, and application.

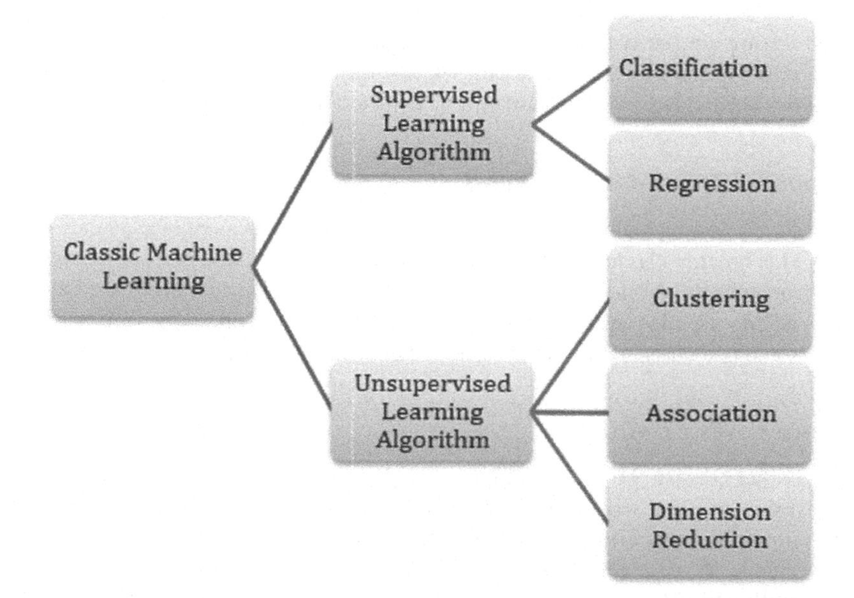

FIGURE 12.2 Classification of machine learning algorithms.

The data-oriented model is based on a benchmark dataset. However, there is no benchmark dataset available from which all the relevant features could be extracted; hence, it is a challenging issue to prepare such a benchmark dataset.

The data-oriented model aims at detecting fake news in the early stage. The main source for the feature-based model is the news and social media. However, the existing feature-based model concentrates on general user profile features than user-specific profile features. The model-based method is more promising. It extracts the available features and uses the machine learning approach as supervised or unsupervised to classify the news as fake or genuine. This paper is used in the model-based approach to propose the generalized method. Application-oriented models extend its scope as it not only detects the fake news but also explores the dimensions or ways used to diffusion and intervention in social media.

There are multiple ways of fake news detection techniques as shown in Figure 12.3. Our proposed work concentrates on social media datasets and uses various machine learning algorithms for the detection of fake news.

12.4 Comparative Study

This section proposes a comparative analysis of the research work for the detection of fake news. This section considers the diverse research and analyzes the research of various authors. The authors of this chapter outlined some of the challenges in real-time fake news detection techniques.

Size of Content: With the escalation of social networks, the users are also growing exponentially. Every second, huge content is being generated that floods the internet with numerous messages. It will

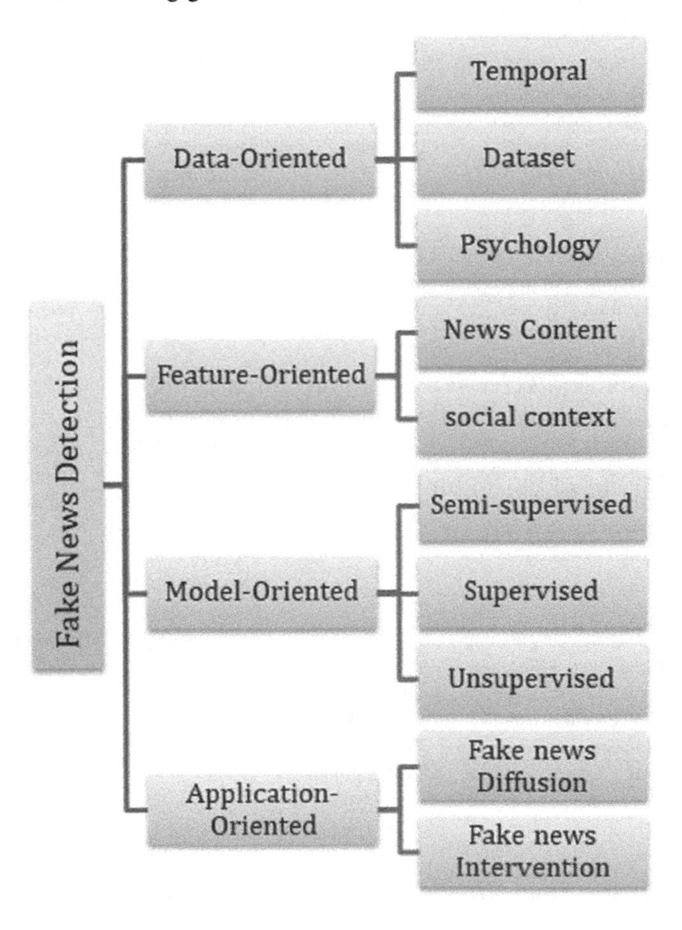

FIGURE 12.3 Fake news detection models.

TABLE 12.1

Comparative Study of recent work by various researchers

Authors	Proposed Approach	Model	Features	Performance Analysis
Stein and ZuEissen [21]	This paper proposes a Bayesian approach for automatic text classification using class-specific features	LapPLSI	Document and topic-based modeling	LapPLSI 76.4%
Mishu and Rafiuddin [22]	This paper uses various machine learning algorithms to classify the text. It also put forth the comparative analysis of all the algorithms	Multidimensional NB, Logistic Regression, SVC, Linear SVC	Features are extracted using document frequency, TF-IDF (term frequency inverse document frequency), count vectorizer	Document frequency Multinomial NB—72% LR—73.5% SGD—76% SVC—78% linear SVC—83.3% voting—89% TF-IDF Linear SVC—96.3% Voting—96.4% SGD—96.2%
Kaur et al. [23] for news articles	This paper proposes the multilevel voting model for the detection of fake articles using articles' textual data	SGD, PA, Multinomial NB, Gradient Boosting DT, AdaBoost	TF-IDF, hashing-vectorizer count-vectorizer	TF-IDF 97.2% Count-vectorizer 96.5% Multilevel voting 90.2%
Qazvinian et al. [24] for Tweeter	The model proposed in this paper will identify the tweets containing rumors	Naïve Bayes	Content-based, Twitter-specific memes network-based	Random 50.1% Uniform 43.9% Tag 58.9% URL 66.4% TAG+URL 68.9%
Gupta et al. [25] for Tweeter dataset and also corresponding user information	This paper considers the Boston Marathon case study to analyze the Tweeter contents	Logistic regression	Credibility global engagement topic engagement, likability, social reputation	
Caetano et al. [26] for news posts and WhatsApp database	This paper proposes the predictive model to predict four different types of news 1. hoaxes, 2. clickbait, 3. suspicious (Satire) 4. propaganda	Linguistic models	TF-IDF, Doc2Vec	It analyzes the WhatsApp chat groups and puts forth results according to three layers – message layer, user layer, and group layer
Benevenuto et al. [27] for real YouTube user information	This paper analyzes the video and identifies the spammers in social video networks. The video response crawler was proposed in this paper	SVM	Individual characteristics of user behavior, video attributes, the social relation between users via video response interactions	The video response crawler classifies the YouTube uses as spammers, promoters, and legitimates
Chhabra et al. [28] for Phishtank, which is a malicious URL dataset [29]	This paper detects phishing websites using URL static features. The proposed method uses messages from Facebook, Orkut, and Twitter	Naïve Bayes, logistic regression, DT, SVM-RBF, SVM-linear, SVM-sigmoid	Static, grammar, lexical, and vectors	The drawback of the shortened URL is that it hides the actual URL and promotes the phishing activities. Authors put forth that space gain for half of the phishing URLs in their dataset was 37%, which was significantly less than space gain in general URLs on the Internet

be a real challenge to propose an algorithm that is scalable enough to process on such a huge million and billions of datasets [20].

Prevention: Malicious activities on the internet such as fake news, phishing scams cause tremendous harm to users in a very short time. The real challenge is to provide a solution that will detect such malicious activities in real-time. Our study reveals that the algorithms put forth by the various author will detect malicious activities after the attack.

Anonymity: Anonymity is tightly assigned to the internet as the internet medium itself is anonymous. Moreover, this medium is worse because of the lack of privacy regulations and tracing technologies. Cybercriminals hide their identities by creating multiple and fake profiles with a very easy verification process. Imposing authentication and validation in such huge cyberspace with diverse technology is an extremely challenging task.

To overcome the above challenges, the authors of this paper outlined below the comparative analysis of different techniques based on various parameters (Table 12.1).

12.5 Conclusion

It has been witnessed from various instances in history that the news over social media has a huge impact on the emotional and psychological health of humankind. However, unfortunately, this platform is also used to spread fake news by wrongdoers. Hence, considering the impact of news on social media, it is necessary to verify the authenticity of this news. As a result, several researchers have suggested various models that use ML algorithms to authenticate news on social media. The chapter discusses the importance of fake news detection. The chapter also discusses the significant work of various researchers that has been proposed recently in this direction. The generalized machine learning-based approach is put forth to give the implementation direction. At last, it analyzes the various research for fake news detection and summarizes the efficiency of various machine learning algorithms. Fake news detection can also be extended in the direction of detecting fake images as images play a major role in affecting the thinking of the public.

REFERENCES

1. Ferrara, Emilio, Onur Varol, Clayton Davis, Filippo Menczer, and Alessandro Flammini. "The rise of social bots." *Communications of the ACM* 59, no. 7 (2016): 96–104.
2. Bessi, Alessandro, and Emilio Ferrara. "Social bots distort the 2016 US Presidential election online discussion." *First Monday* 21, no. 11–7 (2016): 1–14.
3. Cheng, Justin, Michael Bernstein, Cristian Danescu- Niculescu-Mizil, and Jure Leskovec. "Anyone can become a troll: Causes of trolling behavior in online discussions." *In Proceedings of the 2017 ACM conference on computer supported cooperative work and social computing*, Portland, Oregon, USA, pp. 1217–1230 (2017).
4. Chu, Zi, Steven Gianvecchio, Haining Wang, and Sushil Jajodia. "Detecting automation of twitter accounts: Are you a human, bot, or cyborg?" *IEEE Transactions on Dependable and Secure Computing* 9, no. 6 (2012): 811–824.
5. Paul, Christopher, and Miriam Matthews. "The Russian "firehose of falsehood" propaganda model." RAND Corporation (2016): 2–7.
6. Zajonc, Robert B. "Attitudinal effects of mere exposure." *Journal of Personality and Social Psychology* 9, no. 2 (1968): 1.
7. Zajonc, Robert B. "Mere exposure: A gateway to the subliminal." *Current Directions in Psychological Science* 10, no. 6 (2001): 224–228.
8. Gurav, Subhadra, Swati Sase, Supriya Shinde, Prachi Wabale, and Sumit Hirve. "Survey on automated system for fake news detection using NLP & machine learning approach." *International Research Journal of Engineering and Technology (IRJET)* 6, no. 1 (2019): 308–309.
9. Thota, Aswini, Priyanka Tilak, Simrat Ahluwalia, and Nibrat Lohia. "Fake news detection: A deep learning approach." *SMU Data Science Review* 1, no. 3 (2018): 10.

10. Zhou, Xinyi, and Reza Zafarani. "Fake news: A survey of research, detection methods, and opportunities." arXiv preprint, arXiv:1812.00315 2 (2018).
11. Zhang, Jiawei, Bowen Dong, and S. Yu Philip. "Fakedetector: Effective fake news detection with deep diffusive neural network." *In 2020 IEEE 36th International Conference on Data Engineering (ICDE)*, Dallas ,Texas, pp. 1826–1829, IEEE (2020).
12. Ahmed, Sajjad, Knut Hinkelmann, and Flavio Corradini. "Combining machine learning with knowledge engineering to detect fake news in social networks-a survey." *In Proceedings of the AAAI 2019 Spring Symposium*, Stanford University Computer Science Department, Palo Alto, California, vol. 12 (2019).
13. O'Brien, Nicole. "Machine learning for detection of fake news." PhD dissertations, Massachusetts Institute of Technology, (2018).
14. Ambarkar, Smita, and Rakhi Akhare. "A study to analyze impact of social media on society: WhatsApp in particular." *International Journal of Education and Management Engineering* 10, no. 1 (2020): 1.
15. Ambarkar, Smita Sanjay, and Narendra M. Shekokar. "Improving security of IoT networks using machine learning-based intrusion detection system." In Vasudevan H., Michalas A., Shekokar N., Narvekar M. (eds), *Advanced Computing Technologies and Applications*, pp. 199–210. Springer, Singapore (2020).
16. Schölkopf, Bernhard, Robert C. Williamson, Alex J. Smola, John Shawe-Taylor, and John C. Platt. "Support vector method for novelty detection." *Advances in Neural Information Processing Systems* 12 (2000): 582–588.
17. Deokar, Sanjivani, Monika Mangla, and Rakhi Akhare. "A secure fog computing architecture for continuous health monitoring." In *Fog Computing for Healthcare 4.0 Environments*, pp. 269–290. Springer, Cham (2021).
18. Cao, Qimin, Yinrong Qiao, and Zhong Lyu. "Machine learning to detect anomalies in web log analysis." *In 2017 3rd IEEE International Conference on Computer and Communications (ICCC)*, Chengdu, China, pp. 519–523. IEEE (2017).
19. Mangla, Monika, Rakhi Akhare, and Smita Ambarkar. "Context-aware automation based energy conservation techniques for IoT ecosystem." In *Energy Conservation for IoT Devices*, pp. 129–153. Springer, Singapore (2019).
20. Akhare, Rakhi, Monika Mangla, Sanjivani Deokar, and Vaishali Wadhwa. "Proposed Framework for Fog Computing to Improve Quality-of-Service in IoT Applications." In *Fog Data Analytics for IoT Applications*, pp. 123–143. Springer, Singapore, 2020.
21. Stein, Benno, and Sven Meyer zuEissen. "Retrieval models for genre classification." *Scandinavian Journal of Information Systems* 20, no. 1 (2008): 3.
22. Mishu, Sadia Zaman, and S. M. Rafiuddin. "Performance analysis of supervised machine learning algorithms for text classification." *In 2016 19th International Conference on Computer and Information Technology (ICCIT)*, Dhaka, Bangladesh, pp. 409–413, IEEE (2016).
23. Kaur, Sawinder, Parteek Kumar, and Ponnurangam Kumaraguru. "Automating fake news detection system using multi-level voting model." *Soft Computing* 24, no. 12 (2020): 9049–9069.
24. Qazvinian, Vahed, Emily Rosengren, Dragomir Radev, and Qiaozhu Mei. "Rumor has it: Identifying misinformation in microblogs." *In Proceedings of the 2011 Conference on Empirical Methods in Natural Language Processing*, Edinburgh, Scotland, UK. pp. 1589–1599 (2011).
25. Dewan, Prateek, Mayank Gupta, Kanika Goyal, and Ponnurangam Kumaraguru. "Multiosn: Realtime monitoring of real world events on multiple online social media." *In Proceedings of the 5th IBM Collaborative Academia Research Exchange Workshop*, New York NY, pp. 1–4 (2013).
26. Caetano, Josemar Alves, Jaqueline Faria de Oliveira, Hélder Seixas Lima, Humberto T. Marques-Neto, Gabriel Magno, Wagner MeiraJr, and Virgílio A. F. Almeida. "Analyzing and characterizing political discussions in WhatsApp public groups." arXiv preprint arXiv:1804.00397 (2018).
27. Benevenuto, Fabricio, Tiago Rodrigues, Virgilio Almeida, Jussara Almeida, and Marcos Goncalves. "Detecting spammers and content promoters in online video social networks." *In Proceedings of the 32nd International ACM SIGIR Conference on Research and Development in Information Retrieval*, Boston MA, USA, pp. 620–627 (2009).
28. Chhabra, Sidharth, Anupama Aggarwal, Fabricio Benevenuto, and Ponnurangam Kumaraguru. "Phi.sh/$ocial: The phishing landscape through short URLs." *In Proceedings of the 8th Annual Collaboration, Electronic Messaging, Anti- Abuse and Spam Conference*, Perth, Australia, pp. 92– 101 (2011).
29. Shu, Kai, Amy Sliva, Suhang Wang, Jiliang Tang, and Huan Liu. "Fake news detection on social media: A data mining perspective." *ACM SIGKDD Explorations Newsletter* 19, no. 1 (2017): 22–36.

13

Application of Deep Learning in Facial Recognition

Jimit Gandhi, Aditya Jeswani, Fenil Doshi, Parth Doshi, and Ramchandra S. Mangrulkar
Dwarkadas J Sanghvi College of Engineering

CONTENTS

13.1 Introduction to Facial Recognition 195
13.2 Introduction to Deep Learning (DL) 197
13.3 Deep Learning Models for Facial Recognition 199
13.3.1 DeepFace 199
13.3.2 FaceNet 200
13.3.3 ArcFace 201
13.3.4 Baidu 201
13.3.5 Face Recognition Datasets 202
13.3.6 Comparison 203
13.3.7 Loss Functions Used to Improve the Network 203
13.3.7.1 Euclidean-Distance Based Loss 203
13.3.7.2 Softmax Loss and Its Variants 204
13.4 Scope and Challenges 204
13.5 Conclusion 205
References 205

13.1 Introduction to Facial Recognition

Facial recognition has been at the forefront of technological advancements in the fields of image processing and computer vision in recent times. While several state-of-the-art techniques for facial recognition are available in today's time, it is interesting to first understand and analyze how this technology first came into being and how it has developed over the years. This will provide a true understanding of the intuitiveness behind the technologies currently used to perform the task.

At this point, the authors first look at why facial recognition technology has become such a lucrative field of research over the years. Several other unique identification systems have been in place over the years. But facial recognition, unlike fingerprint recognition, retinal scanning, and any other such technology, provides the option to perform recognition without the conscious participation of the user. This provides an opportunity for several government and private organizations to identify potential threats or assets without alerting the individual, thus proving to be immensely useful. Apart from this, unlike fingerprint scanning, the task of facial recognition can be relatively easily undertaken from recorded footage also. While results might not be as assuring as fingerprints and retinal scans, the aforementioned reasons really make facial recognition an important characteristic of modern technology.

Probably the first facial recognition was developed when Woodrow Wilson Bledsoe used a RAND tablet [1] to manually calculate the dimensions between various features within a face. These values were then stored in a database for comparison. Clearly, while this system was pretty archaic, it laid the foundation for several modern techniques. Following this, A. J. Goldstein et al. [2] came up with a set of

22 features that could provide relevant, distinctive, relatively independent measures which could be used for reliable judgment. Their work developed a robust foundation and provided objective metrics to make comparisons between a set of defined features.

One of the most revolutionary works in the field was the development of Eigenfaces [3]. Turk and Pentland treated the task of facial recognition as a two-dimensional recognition problem. They used eigenvectors called Eigenfaces to store the feature spaces of 2-D faces that were assumed to be in an upright position. Unlike previous methods, their feature space did not relate to any specific feature such as length of nose, the wideness of ears, etc. This enabled them to use an unsupervised approach to learn these feature spaces. They not only use motion detection to track the faces, but also provide a mechanism to use these eigenvectors to achieve more precision while tracking the head, which is usually constantly moving. Using this concept also enabled them to learn new faces. This can be done in an unsupervised manner. A new image is initially stored as "unknown," and if patterns cluster to this "unknown," a new face can be identified.

The Eigenfaces method was an important step forward in the field of facial recognition; however, it suffered greatly in terms of illumination. The performance of the system was poor if the data to be tested consisted of faces under different illumination conditions. A newer approach called Fisherfaces was proposed to reduce the impact of lighting conditions on the facial recognition system. This technique uses the fact that a labeled dataset is provided, in order to build a class-specific method termed as Fisher's linear discriminant [4]. The transformation matrix used is class-specific, which enables the technique to avoid illumination being captured in terms of the extent to which it is captured in the Eigenface method. The dimensionality reduction method used works better than principal component analysis used in Eigenface because it looks to maximize the determinant of the ratio of the between-class scatter plot to the within-class scatter plot. However, the performance of this algorithm is influenced by the kind of data on which it is trained. For example, if all the facial images in the training data are under bad illumination and the images on which it is tested are very well illuminated, then the method is likely to find the wrong components, thereby misclassifying the image.

Soon after, the Defense Advanced Research Projects Agency (DARPA) along with the National Institute of Standards and Technology started realizing the importance of facial recognition technology. They developed the Face Recognition Technology (FERET). The goal of their program was to encourage research in this field, and to do so, they developed a common database of faces. There were three phases to this program. In this phase, the program collected 673 sets of images, hence gathering over 5000 total images. They used the large gallery test, false-alarm test, and rotation test to evaluate automatically. In Phase 2, the dataset was improved. It now contained 1109 sets of images with 8525 total images for 884 individuals. Finally, in the last phase, 456 other sets were added. The evaluation in this phase assessed the accuracy of the algorithms used before. Overall, this program provided the necessary baseline for researchers to compare their work and analyze if they have generated meaningful results.

More recently, tech giants started to emerge. While focusing on several other areas as well, most of them had a dedicated team of researchers trying to develop the most innovative, accurate, and lightning-fast facial recognition software. With the advent of powerful processors, ultra-fast storage devices, and high-resolution cameras on cell phones, the immense compute power that limited the processing of facial recognition in real time now seemed to turn into reality. One such method used was 3D recognition. Apple's FaceID [5] is a prime example of this technology. They use the depth in eye sockets, shape of the nose, and other such vital features to distinguish data. The major reason for using this system was that they were unaffected by lighting conditions. In FaceID, a mesh of dots was projected on the face. The depth of each dot was calculated, which then helped in distinguishing one person from another. However, using 3D analysis means that the system will be susceptible to changes in expressions, which is a natural human tendency.

Some other techniques use the analysis of skin texture [6] and combinations of all of these methods. They use periocular skin texture for biometric modality. They extract skin texture features using local binary patterns and then perform matching using City Block (a measure of distance for calculating similarity). Gurton et al. [7] have developed a technique for thermal imagery using polarimetric imaging. They use long-wave-infrared images to perform the task. Using polarimetric information, they extracted subtle surface features from faces for subject identification.

All these technologies have mainly focussed on accuracy along with speed. In general, all these recognition techniques can be roughly divided into two categories: feature based and view based. While feature-based algorithms deal with the geometric features of a face, view-based algorithms focus on the actual view of the image of the face. While these feature-based techniques follow a sort of a divide and conquer approach where each aspect of the face is considered separately for analysis before clubbing them together, some other techniques use a holistic approach, which considers the recognition of the face as a whole.

13.2 Introduction to Deep Learning (DL)

Deep learning is a representation learning technique [8] i.e., there are multiple levels of nonlinear modules which are used to transform the representation at one level into a representation at another, more abstract level. In representation learning, the machine is fed with the raw data for the task to be performed, allowing it to learn the importance of the different features in the data in order to efficiently complete the objective of pattern detection or classification.

Deep learning techniques make use of layers to understand the different features present in the input data, and these layers are arranged in the form of a network. Every layer consists of a varying number of nodes in them, which receive input from the previous layer or the raw data stream in the case of the first layer. The arrangement of the nodes is inspired by the neurons in the human brain, where every neuron transmits a signal to the next neuron through electrical impulses across a synapse, thus calling the architecture an artificial neural network (ANN). The ANNs consist of three types of layers:

1. **Input Layer**: It is the first layer of the neural network that receives the raw input data which is fed to the system for achieving the objective.
2. **Hidden Layer:** These are the intermediate layers, present between the input and output layers, which are used for feature extraction and understanding the relations between the features present in the input data. The neural network architectures generally have two to three hidden layers; however, the networks where the number of hidden layers is very large are called deep neural networks.
3. **Output Layer**: It is the final layer of the neural network that provides the results as per the task at hand, such as the class to which the input belongs in case of classification tasks, a real value for predictions, etc. (Figure 13.1).

The neurons present in every layer of the network apply a nonlinear function to the data received from the previous layer to produce a value that is passed on to the next set of neurons. This nonlinear function used by the neuron is called an activation function (Figure 13.2).

Depending on the requirement, the different activation functions that can be used by a neuron are:

1. **Sigmoid Function**: It is mostly used in the feed-forward neural networks for binary classification problems since the output lies between 0 and 1. The sigmoid function is generally used in the output layers.
2. **Hyperbolic Tangent Function**: The *tan h* function is another zero-centered function whose output lies in between -1 and +1. It is smoother as compared to the sigmoid function and hence preferred in certain applications. In addition to this, it has also been shown to give better performance in multi-layer neural networks.

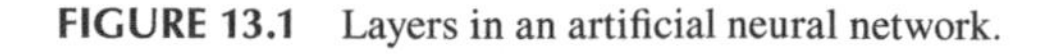

FIGURE 13.1 Layers in an artificial neural network.

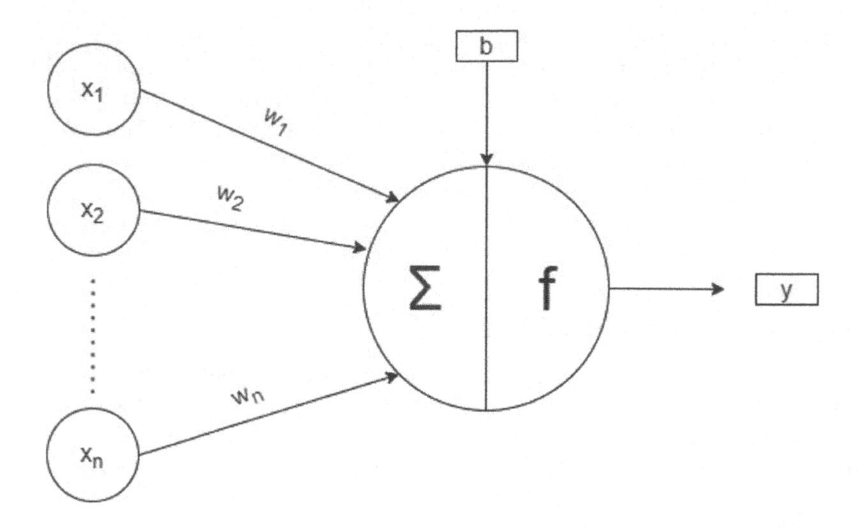

FIGURE 13.2 A simple neuron.

3. **Softmax Function**: The softmax function is another AF that is used for classification problems, specializing in multi-class classification problems.
4. **Rectified Linear Unit (ReLU)**: The ReLU activation function performs corrections on the input values which are <0, thereby eliminating the vanishing gradient problem. It achieves this by performing thresholding operation as:

$$f(x) = \max(0, x) \tag{13.1}$$

In order to obtain improved performances, many variations of the activation functions have been produced, which are used depending on the specific objective to be completed.

A specific type of deep learning network, called convolutional neural network (CNN), has over time proved to be extremely useful for image-based tasks, identifying the different aspects of images and helping differentiate between features.

CNNs are a specific kind of network architecture used for image processing applications. An image is nothing but a matrix of values, representing the pixels present in the image. While these pixel values can be treated as numbers and used in traditional approaches of classification, there is a loss of some of the spatial and temporal data that is present in the image. This is where CNNs gain advantage. The CNNs are capable of retaining these aspects of the image, thereby better understanding the feature space and the characteristics of the input image. The CNN architecture consists of three different types of layers:

1. **Convolutional Layer**: It is the first layer of the network architecture that has the same depth dimension as the input image while the width and height dimensions can be arbitrarily chosen. It performs the convolution operation on the input image matrix, i.e. element-wise multiplication [8], sums up the values obtained, and generates the new element for the output matrix. The process begins by considering the upper left corner of the matrix and ends when the entire matrix has been covered.
2. **Pooling Layer**: Its function is to progressively reduce the spatial size of the representation to reduce the amount of parameters and computation in the network, and hence to also control overfitting. The pooling layer operates independently on every depth slice of the input and resizes it spatially, using the MAX operation.
3. **Fully Connected Layer**: It is the last layer in the CNN architecture. Every neuron in this layer is connected to every neuron in the previous layer, i.e. there are full connections to activations in the previous layer. The output is computed as matrix multiplications carried out normally in the neural networks with a certain bias offset.

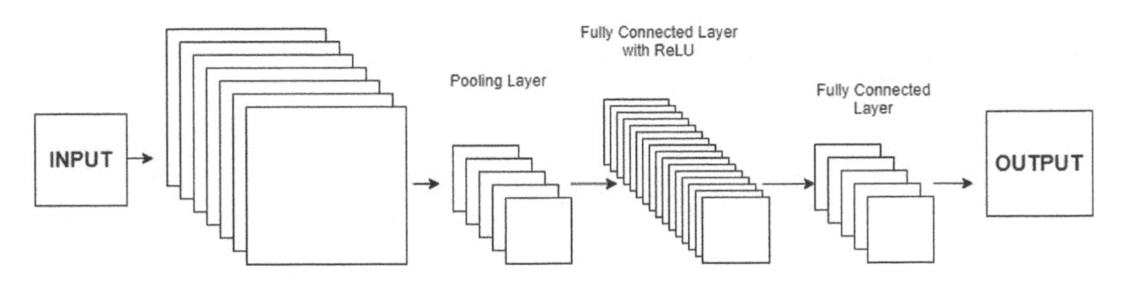

FIGURE 13.3 Architecture of a Convolutional Neural Network (CNN).

The following image shows a basic CNN architecture with the different layers involved in the processing of data given as input (Figure 13.3).

The use of deep learning techniques has shown to provide a high degree of accuracy of tasks in a number of domains, ranging from finance, medicine, and image analysis to even tasks related to understanding text and speech. The improved performance over the traditional machine learning algorithms has caused the widespread deployment of these algorithms. However, although the performance of these algorithms produces the desired outcome with precision, they require sufficient computational power and resources, which may not be feasible at times. There is thus a trade-off between the complexity of the problem and the deep learning technique used. Hence, before implementing a neural network to solve the task at hand, it is necessary to validate if the problem is large enough to require a neural network solution for the same.

One of the most famous architectures for CNNs comes from the work on the task of classification as carried out by Alex Krizhevsky et al. [9] at the University of Toronto. Their work on using graphics processing units (GPUs) for training large convolutional neural networks sparked the modern-day revolution of deep learning in computer vision tasks, as researchers realized that the computational power had finally caught up to make large neural networks feasible. The innovations from this work include the use of multiple GPUs for training and the ReLU activation function which helps in the prevention of overfitting. Most importantly, AlexNet, as their network is popularly referred to, led to an influx of different CNN architectures which could be used as image feature extractors for the task of facial recognition, among other computer vision tasks. This will be elaborated upon in Section 13.2, where the authors discuss the use of deep learning models in facial recognition.

13.3 Deep Learning Models for Facial Recognition

Due to the emergence of deep learning models, the focus on facial recognition techniques shifted from the traditional methods such as Eigenfaces [3] toward neural-network-based techniques, which are usually more accurate. The shift has been caused due to several underlying reasons, chief among which are the availability of computational power due to the advent of GPUs and the presence of large datasets which are a prerequisite for DL-based facial recognition models. In contrast to the traditional facial recognition methods which used hand-crafted features, these models directly worked on the raw image pixels. The following sections give a brief overview of the evolution of facial recognition techniques and the large datasets required for these models.

13.3.1 DeepFace

The research by Taigman et al. [10] was amongst the first works that focused on using deep learning models such as CNNs. Their model achieved near human-level performance on the benchmark Labeled Faces in the Wild dataset [11].

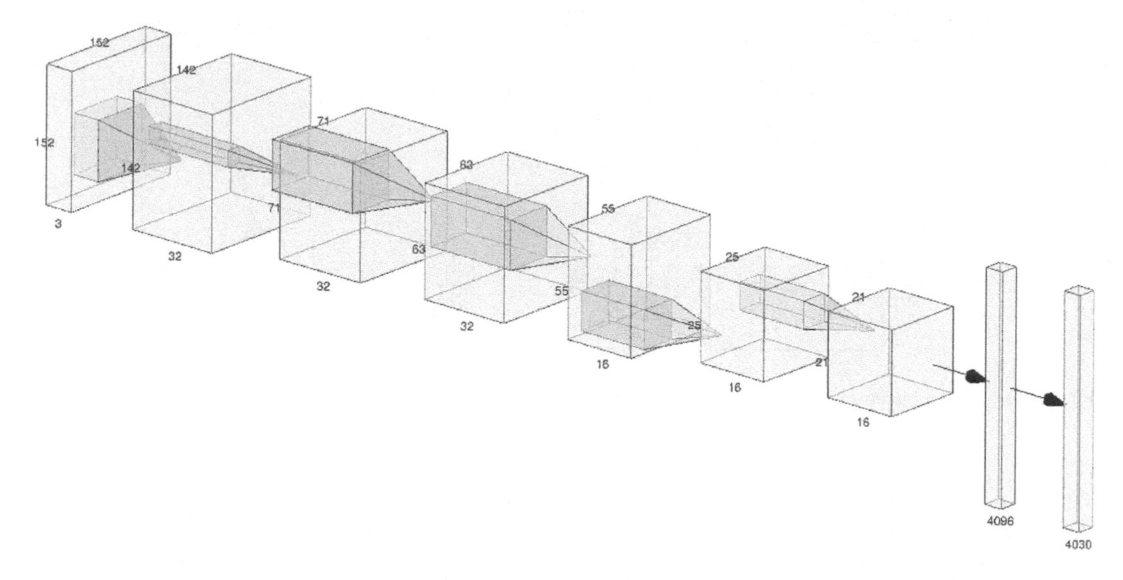

FIGURE 13.4 DeepFace architecture.

The core idea of DeepFace and other facial recognition methods based on deep learning models is that a CNN, or other similar models, can be used to extract a feature vector for a particular face. This vector would be unique for each face, thus allowing facial verification and recognition. In more traditional methods for facial recognition, the features were handcrafted, while in the newer models, the neural networks are assigned the task of generating the necessary feature vector.

DeepFace achieved the creation of the necessary feature vector by training their model using the cross-entropy loss on the Social Face Recognition (SFC) dataset [10], which contains facial images of 4030 people. At test time, feature vectors are then compared using a weighted χ^2 loss function. To measure the similarity between the two vectors, other measures such as Euclidean distance and cosine distance can be used as well, in place of χ^2.

As shown in Figure 13.4, the DeepFace model was based on AlexNET [9]. They used multiple convolutional and max pooling layers to extract the features from an RGB image of size 152×152 pixels. These features were then passed on to fully connected layers, ultimately ending up with a 4096-dimensional feature vector, which is used to compare different images.

13.3.2 FaceNet

FaceNet [12] was a major step forward from DeepFace [10] as it led to an evolution in the types of loss functions used in training a facial recognition network. While DeepFace started the use of deep learning methods in place of the more traditional feature engineering systems, it was FaceNet that led to the improvement and creation of novel loss functions that were specifically tailored to the task of face recognition, rather than being modified from other computer vision tasks.

Similar to DeepFace, FaceNet employed a CNN for feature extraction, but rather than using AlexNet, FaceNet used the GoogleNet-24 [13], popularly known as the Inception network. FaceNet was more efficient in terms of image representation since the feature embedding generated is of 128 dimensions, as compared to the 4096-dimensional vector used in DeepFace.

The main novel contribution of FaceNet was the use of the triplet loss for training the neural network. The triplet loss, as elaborated on in the section, is specifically designed for the task of facial recognition, which is in contrast to the general softmax loss used in DeepFace. Due to its specialized nature, training using the triplet loss ensures that the feature representations for each face are more accurate and efficient. The triplet loss involves training on three images at the same time, where two images are of the same person while the third image is of a different person. The goal is to maximize the Euclidean distance between the images of different people while minimizing the Euclidean distance between images of the same person.

To ensure proper training, it is necessary to carefully choose the triplets such that it is a hard example for the network to learn from. If the images are chosen randomly without any thought, most triplets would be very easy examples for the network, and there will not be any learning due to backpropagation. Schroff et al., the authors of FaceNet, came up with an online selection technique for choosing the hardest examples from each mini-batch generated during training.

During testing, the 128-dimensional feature vectors generated for different images are compared using Euclidean distance to find the similarity between the two images.

13.3.3 ArcFace

ArcFace [14] or Additive Angular Margin Loss for Deep Face Recognition focuses on the problem of large-scale facial recognition using deep convolutional neural networks. In large-scale facial recognition, usually, the features are spread across a large space, and hence, the boundaries between different classes tend to blur out. As such, it gets tougher to estimate class boundaries and, in turn, make accurate predictions. As a solution, Arcface uses a unique loss function that brings together features of the same individual and at the same time pushes features of other individuals farther away. This new loss function is elaborately tested on over ten facial recognition benchmarks, and it is found that ArcFace regularly performs better than most other state-of-the-art techniques.

While softmax loss and triplet loss are popular approaches for large-scale facial recognition, they have certain cons associated with them. Softmax loss works well for close-set classification problems but not as well for open-set face classification problems. Additionally, the size of the transformation matrix also linearly keeps increasing. On the other hand, in triplet loss, the face triplets increase drastically for large datasets, and it is also tough to use semi-hard sample mining. To offset these, Sphereface [15] introduced angular margin to enforce intr-class compactness. However, to make computations, they used an integer-based multiplicative angular margin which made the training unstable. This was stabilized by a hybrid approach using softmax, but because of the use of integers, Softmax seemed to dominate the training. Cosface [16] improves on this problem by using a cosine penalty.

Arcface improves on these loss functions. A cosine function is used to calculate the distance between different features, an additive angular margin is used instead of a multiplicative angular margin, and the cosine function is used again to get the target logits. After this, rescaling is used, similar to softmax. The final loss function is defined as:

$$L = -\frac{1}{m}\sum_{i=1}^{m}\log\left(\frac{e^{s\left(\cos\left(\theta_{yi}\right)+m\right)}}{e^{s\left(\cos\left(\theta_{yi}\right)+m\right)}+\sum_{j=1,j\neq y_i}^{n} e^{s\cos\left(\theta_j\right)}}\right) \tag{13.2}$$

where L is the loss function, m is the angular margin, y_i is the ith class, s is the radius of the hypersphere of the learned embedding features, θ_j is the angle between the weight W_j and the feature x_i.

Arcface was trained on several datasets like LFW [11], CFP-FP [17], AgeDB-30 [18], etc, and in each, Arcface performed better than all other loss functions.

13.3.4 Baidu

Baidu Research's Institute of Deep Learning came up with a novel two-staged approach to extract meaningful features for the task of facial recognition. Using a data-driven approach, they use a multi-patch deep convolutional neural network as well as deep metric learning to extract features. Using this, very discriminative, yet low-level features, can be extracted [19].

In the first step, that is multi-patch deep CNNs, they use nine CNNs and feed them an input 2D image. Each layer has a softmax layer at the end for multiclass classification. In the end, all these outputs are concatenated together to achieve higher dimension features.

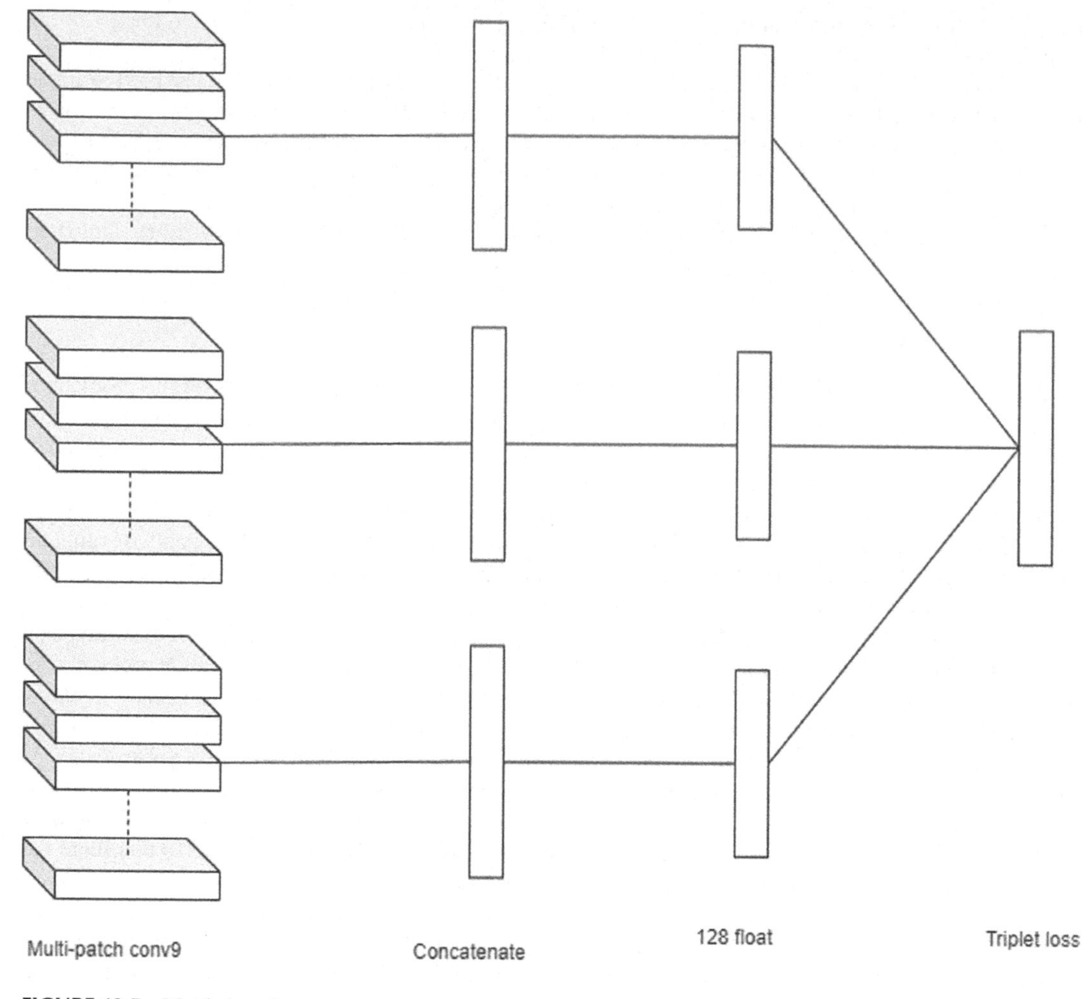

FIGURE 13.5 Metric learning.

In the second step, that is metric learning, a triplet loss is used to shorten the L2 distance of the samples. The high-dimensional features are fed to it, and a low-dimensional output is received which is discriminative enough for retrieval and verification (Figure 13.5).

The deep embedding approach is experimented on the labeled faces in the Wild [11] dataset and achieves a remarkable accuracy of 99.77% under a pair-wise approach.

13.3.5 Face Recognition Datasets

There are a number of different datasets used in the face recognition task. Most datasets are of a huge size, since face recognition using deep learning models requires a massive amount of data. The following sections describe some of the numerous datasets used in the face recognition task.

- **Labeled Faces in the Wild (LFW) [11]**

  The LFW dataset, created by Huang et al., is one of the oldest datasets used in face recognition. However, it is still relevant due to the quality of images used and is widely used as a benchmark to test the performance of different face recognition models in an unconstrained environment. It is the basis for the comparison between different algorithms as shown in Table 13.1.

It has 13,223 images of 5749 people. Out of the 5749 different people, 1680 people have two or more images in the dataset. By modern standards, this is quite a small dataset, so it is usually used for testing purposes or as a benchmark to compare the performance of different algorithms.

However, the LFW dataset has some flaws. It does not contain many images of women or different ethnicities. It also ignores extreme conditions such as pose or illumination which play a major role in facial recognition.

- **Social Face Classification (SFC) [10]**

  This dataset was created by Taigman et al. in order to train the DeepFace models. It was created through a large collection of photos from Facebook. It contains 4.4 million faces from 4030 people. Each person has between 800 and 1200 images in the dataset. Since it has a large number of images, it is suitable for being used for training large deep learning models such as those used in DeepFace and FaceNet.

- **YouTube Faces (YTF) [20]**

  This dataset by Wolf et al. is similar to LFW but focuses on studying the problem of face recognition in videos. It contains 3425 videos of over 1500 celebrities, which forms a subset of the LFW dataset. It is a more specialized dataset than LFW, as its focus lies in the specific domain of face recognition and verification at the level of videos, rather than singular images.

13.3.6 Comparison

TABLE 13.1

Accuracy of the Techniques on LFW Dataset

Technique	Base Architecture	Loss Function	Accuracy
FaceNet [12]	GoogleNet-24	Triplet-loss	99.63
DeepFace [10]	AlexNet	Softmax	97.35
DeepID2 [21]	AlexNet	Contrastive loss	99.15
DeepID3 [22]	VGGNet-10	Contrastive loss	99.53
Baidu [23]	CNN-9	Triplet-loss	99.77
VGGface [24]	VGGNet-16	Triplet-loss	98.95
Arcface [14]	ResNet-100	Arcface	99.83
Cosface [16]	ResNet-64	Cosface	99.33
Ring loss [25]	ResNet-64	Ring loss	99.50

13.3.7 Loss Functions Used to Improve the Network

The base network for face recognition is trained on more than 10^6–10^7 images. Training a network on such a huge dataset requires high-end resources that are generally not available. Pretrained models and architectures are available that are trained on this huge dataset. Publicly available smaller datasets in the range of 10^3 images are used for improving the base network. The accuracy is improved by experimenting with various loss functions and utilizing it for improving the predictions on smaller datasets. Some of the traditionally and widely used loss functions are explained below.

13.3.7.1 Euclidean-Distance Based Loss

The images are projected into a Euclidean space, and the Euclidean distance between the two image representations is used to train the network. This distance between them is mainly along with -

13.3.7.1.1 Using Contrast Loss

$$L = (1 - y_{ij})\max\left(0, M^- - \left\|f(x_i) - f(x_j)\right\|_2\right) + y_{ij}\max\left(0, \left\|f(x_i) - f(x_j)\right\|_2 - M^+\right) \tag{13.3}$$

where x_i and x_j are the images; $y_{ij}=1$ means x_i and x_j are matching samples, and $y_{ij}=0$ means nonmatching samples. $f(\cdot)$ is the feature embedding of the corresponding image, where f is generally some form of neural network. M^+ and M^- control the margin by which the two images should be distinct or similar.

This loss was introduced in Ref. [20] where authors trained a Siamese-like network with this loss function. It considers the absolute difference between the matching and nonmatching pairs.

13.3.7.1.2 Using Triplet Loss

Triplet loss was introduced along with FaceNet [12]. It considers the relative difference between the matching and nonmatching pairs. Here, the data is in the form of triplets of anchor, positive, and negative images. An anchor is the subject image. A positive image is the one that matches with the anchor and a negative image forms a nonmatching pair with the anchor image. The loss is then calculated as -

$$L = \max\left(0, \|f(A)-f(P)\|_2 - \|f(A)-f(N)\|_2 + \alpha\right) \tag{13.4}$$

Where A, P, N are anchor, positive, and negative images, respectively. α is the margin and $f(.)$ is some deep neural network representing a nonlinear function.

Some of the models use both Softmax loss and Triplet loss concurrently by training the network first on Softmax and then fine-tuning it on Triplet Loss. Both of these losses may sometimes face instability while training. Hence, recently, Center-loss [26] was developed in order to stabilize optimization. Center loss computes the distance from center, where center is learnt for each class in multiclass classification and penalizes higher distances.

13.3.7.2 Softmax Loss and Its Variants

One of the most used loss functions for multiclass classification is softmax Loss in the deep learning literature. Many efforts were taken in order to improve the softmax loss by adding new components that increased the decision boundary, improved the convergence properties of the network, etc. In Ref. [27], the authors used large-margin softmax, L-Softmax, a variant of softmax that increased the margin between the matching and non-matching components that increased the inter-class separability between the classes and also helped in avoiding overfitting.

In Ref. [28], the authors devised a new loss function, angular softmax (A-Softmax) for their architecture SphereFace. A-Softmax helps in distinguishing "angularly" features of the face that lie on the hypersphere manifold. Angular margin parameter "m" controls the margin required for the discrimination.

There have been multiple other losses like CoCoLoss [29], RingLoss [25] that normalize the weights and features in order to improve the performance on face recognition tasks.

13.4 Scope and Challenges

Face recognition is a technology that is no longer of a purely academic nature due to its widespread use in real-world applications. This unique position at the intersection of academic research and practical implications creates a novel set of challenges for any researcher working on this topic. Some of the most prominent challenges are detailed below, such as security issues and the extreme levels of accuracy required.

1. **Security**: This is the most important challenge faced by any face recognition system since the practical implications of any failure by a face recognition model can have disastrous consequences. For example, a face recognition system working as part of airport security may fail to identify a terrorist entering under a fake identity. Such a failure can lead to a massive loss of lives. One of the active areas in face recognition, and computer vision in general, is the use of adversarial attacks to fool deep learning models. These attacks, while invisible to the human eye, can confuse a deep network into misclassifying a face. Another security issue that has to

be dealt with by face recognition systems pertains to the use of generative adversarial networks for creating or modifying realistic faces.

2. **Comprehensibility and Interpretability**: There exist, despite the widespread usage of deep networks, several open questions regarding the fundamentals of deep neural networks used in face recognition. One of the questions is the prominent effect of adversarial attacks on face recognition models while being completely invisible to the human eye. Another fundamental question pertains to the data capacity of a face recognition model. Researchers are actively working on improving the interpretability of face recognition models.
3. **Extreme Accuracy**: In most applications, an accuracy of 95% would be considered exceptional, but such accuracy would not be good enough to be used with face recognition systems, considering the implications of a misclassification. Hence, researchers have to continuously work on these models to extract the maximum possible accuracy and performance from these models. This leads to the creation of massive datasets, rather than newer architectures.

13.5 Conclusion

The task of facial recognition has been under study for many years, and with every advancement in technology and computational systems, there has been a marked improvement in the accuracy achieved for the task. With the Eigenface method being the first major technique for facial recognition, the development of feature-based, view-based, and holistic approaches laid the foundation for modern-day development of facial recognition systems. Consequently, the field received a significant boost through the use of neural networks that are trained on varied data sources described in the chapter.

This chapter serves to highlight how the use of deep learning systems has contributed to the development of effective facial recognition systems. The availability of GPUs and frameworks to build neural network models has made it easier for researchers to test out different hypotheses quicker and validate their results with the various databases containing facial images to be recognized by the models. While the models described in the chapter have achieved great accuracy, the evolving neural network systems and resources available are helping researchers to develop more effective systems, with the machines coming close to human accuracy in identifying people.

Facial recognition is an active research field with numerous methods being tested frequently. Accuracy is of significant importance in facial recognition since it has application in the fields of security, accessibility, identity verification, and surveillance among others, where the margin for error is extremely small.

REFERENCES

1. Davis, Malcolm R., and Tom O. Ellis. "The RAND tablet: A man-machine graphical communication device." *In Proceedings of the October 27–29, 1964, Fall Joint Computer Conference, Part I*, pp. 325–331, 1964, San Francisco, CA.
2. Goldstein, A. Jay, Leon D. Harmon, and Ann B. Lesk. "Identification of human faces." *Proceedings of the IEEE* 59(5), 748–760, 1971.
3. Turk, Matthew A., and Alex P. Pentland. "Face recognition using eigenfaces." *In Proceedings. 1991 IEEE Computer Society Conference on Computer Vision and Pattern Recognition*, pp. 586–587, IEEE Computer Society, 1991.
4. Belhumeur, Peter N., João P. Hespanha, and David J. Kriegman. "Eigenfaces vs. fisherfaces: Recognition using class specific linear projection." *IEEE Transactions on Pattern Analysis and Machine Intelligence* 19(7), 711–720, 1997.
5. Apple Inc. "About face ID advanced technology." Apple Support, February 26, 2020. https://support.apple.com/en-in/HT208108.

6. Miller, Philip E., Allen W. Rawls, Shrinivas J. Pundlik, and Damon L. Woodard. "Personal identification using periocular skin texture." *In Proceedings of the 2010 ACM Symposium on Applied Computing*, pp. 1496–1500, 2010, Sierre, Switzerland.
7. Gurton, Kristan P., Alex J. Yuffa, and Gorden W. Videen. "Enhanced facial recognition for thermal imagery using polarimetric imaging." *Optics Letters* 39(13), 3857–3859, 2014.
8. LeCun, Yann, Yoshua Bengio, and Geoffrey Hinton. "Deep learning." *Nature* 521(7553), 436–444, 2015.
9. Krizhevsky, Alex, Ilya Sutskever, and Geoffrey E. Hinton. "Imagenet classification with deep convolutional neural networks." *In Advances in Neural Information Processing Systems*, pp. 1097–1105, 2012, Lake Tahoe, Nevada.
10. Taigman, Yaniv, Ming Yang, Marc'Aurelio Ranzato, and Lior Wolf. "Deepface: Closing the gap to human-level performance in face verification." *In Proceedings of the IEEE Conference on Computer Vision and Pattern Recognition*, pp. 1701–1708, 2014, Columbus, OH.
11. Huang, Gary B., Manu Ramesh, Tamara Berg, and Erik Learned-Miller. Labeled faces in the wild: A database for studying face recognition in unconstrained environments. Technical report, 2007.
12. Schroff, Florian, Dmitry Kalenichenko, and James Philbin. "Facenet: A unified embedding for face recognition and clustering." *In Proceedings of the IEEE Conference on Computer Vision and Pattern Recognition*, pp. 815–823, 2015, Boston, MA.
13. Szegedy, Christian, Wei Liu, Yangqing Jia, Pierre Sermanet, Scott Reed, Dragomir Anguelov, Dumitru Erhan, Vincent Vanhoucke, and Andrew Rabinovich. "Going deeper with convolutions." *In Proceedings of the IEEE Conference on Computer Vision and Pattern Recognition*, pp. 1–9, 2015, Boston, MA.
14. Deng, Jiankang, Jia Guo, Niannan Xue, and Stefanos Zafeiriou. "Arcface: Additive angular margin loss for deep face recognition." *In Proceedings of the IEEE Conference on Computer Vision and Pattern Recognition*, pp. 4690–4699, 2019, Long Beach, CA.
15. Liu, Weiyang, Yandong Wen, Zhiding Yu, Ming Li, Bhiksha Raj, and Le Song. "Sphereface: Deep hypersphere embedding for face recognition." *In Proceedings of the IEEE Conference on Computer Vision and Pattern Recognition*, pp. 212–220, 2017, Honolulu, Hawaii.
16. Wang, Hao, Yitong Wang, Zheng Zhou, Xing Ji, Dihong Gong, Jingchao Zhou, Zhifeng Li, and Wei Liu. "Cosface: Large margin cosine loss for deep face recognition." *In Proceedings of the IEEE Conference on Computer Vision and Pattern Recognition*, pp. 5265–5274, 2018, Salt Lake City, UT.
17. Sengupta, Soumyadip, Jun-Cheng Chen, Carlos Castillo, Vishal M. Patel, Rama Chellappa, and David W. Jacobs. "Frontal to profile face verification in the wild." *In 2016 IEEE Winter Conference on Applications of Computer Vision (WACV)*, pp. 1–9, IEEE, 2016, Lake Placid, NY.
18. Moschoglou, Stylianos, Athanasios Papaioannou, Christos Sagonas, Jiankang Deng, Irene Kotsia, and Stefanos Zafeiriou. "Agedb: The first manually collected, in-the-wild age database." *In Proceedings of the IEEE Conference on Computer Vision and Pattern Recognition Workshops*, pp. 51–59. 2017, Honolulu, Hawaii.
19. Liu, Jingtuo, Yafeng Deng, Tao Bai, and Chang Huang. Targeting ultimate accuracy: Face recognition via deep embedding, 2015.
20. Hadsell, Raia, Sumit Chopra, and Yann LeCun. "Dimensionality reduction by learning an invariant mapping." *In 2006 IEEE Computer Society Conference on Computer Vision and Pattern Recognition (CVPR'06)*, vol. 2, pp. 1735–1742, IEEE, 2006, New York, NY.
21. Sun, Yi, Yuheng Chen, Xiaogang Wang, and Xiaoou Tang. "Deep learning face representation by joint identification-verification." *In Advances in Neural Information Processing Systems*, pp. 1988–1996, 2014, Montreal, Canada.
22. Sun, Yi, Ding Liang, Xiaogang Wang, and Xiaoou Tang. "Deepid3: Face recognition with very deep neural networks." arXiv preprint arXiv:1502.00873, 2015.
23. Liu, Jingtuo, Yafeng Deng, Tao Bai, Zhengping Wei, and Chang Huang. "Targeting ultimate accuracy: Face recognition via deep embedding." arXiv preprint arXiv:1506.07310, 2015.
24. Parkhi, Omkar M., Andrea Vedaldi, and Andrew Zisserman. "Deep face recognition." In Xianghua Xie, Mark W. Jones, and Gary K. L. Tam, editors, *Proceedings of the British Machine Vision Conference (BMVC)*, pp. 41.1–41.12. BMVA Press, London, 2015.
25. Zheng, Yutong, Dipan K. Pal, and Marios Savvides. "Ring loss: Convex feature normalization for face recognition." *In Proceedings of the IEEE Conference on Computer Vision and Pattern Recognition*, pp. 5089–5097, 2018, Salt Lake City, UT.

26. Wen, Yandong, Kaipeng Zhang, Zhifeng Li, and Yu Qiao. "A discriminative feature learning approach for deep face recognition." *In European Conference on Computer Vision*, pp. 499–515, Springer, Cham, 2016.
27. Liu, Weiyang, Yandong Wen, Zhiding Yu, and Meng Yang. "Large-margin softmax loss for convolutional neural networks." *ICML,* 2(3), 7, 2016.
28. Liu, Weiyang, Yandong Wen, Zhiding Yu, Ming Li, Bhiksha Raj, and Le Song. "Sphereface: Deep hypersphere embedding for face recognition." *In Proceedings of the IEEE Conference on Computer Vision and Pattern Recognition*, pp. 212–220, 2017, Honolulu, Hawaii.
29. Liu, Yu, Hongyang Li, and Xiaogang Wang. "Rethinking feature discrimination and polymerization for large-scale recognition." arXiv preprint arXiv:1710.00870 (2017).

14

Application of Deep Learning in Deforestation Control and Prediction of Forest Fire Calamities

Muskan Goenka and Ramchandra S. Mangrulkar
Dwarkadas J Sanghvi College of Engineering

CONTENTS

14.1 Introduction 209
14.2 Problems and Relevance to Today's Society/Environmental Need 210
14.2.1 Deforestation 210
14.2.2 Animal, Insect Attack 211
14.2.3 National Security on Borders and Illegal Smuggling of Goods 211
14.2.4 Forest Fires 211
14.3 Brief Solution for the above Problems 211
14.4 Reasons and Causes of Wildfires 211
14.5 Methods to Detect, Predict and Control the Density of Forests 212
14.5.1 Methods to Detect Forest Fires 212
14.5.2 Methods to Predict Forest Fires 215
14.5.3 Deforestation Control 216
14.5.3.1 Dataset Required 216
14.5.3.2 Processing and Analysis of Images 216
14.5.3.3 Model Creation Using Logistic Regression 216
14.6 Conclusion 220
Bibliography 221
Journal Article 221
News Article 221

14.1 Introduction

The year 2020 has been one of the most challenging years in world history. With Australia's wildfires, also known as Black Summer, Philippines' volcanic eruption, the recent California wildfires caused by lightning, the tension between Iran and the USA, and, above all, the Coronavirus has let the world into astonishment and fear of the near future. Coronavirus outbreak is getting out of hand even after having such advanced technologies. What one needs to understand from this is that if "NOT NOW, THEN NEVER." It's already high time humans need to get responsible and aware of their actions. At this point in time, the environmental and surrounding situation has paralyzed humans, resulting in economic and various other losses.

There is no need to describe the adverse effects of climatic changes, environment degradation, ozone layer depletion, the drastic increase in the level of carbon dioxide, Amazon fires, and melting of glaciers in Antarctica, have on humankind. Among these, deforestation tops the list as people are still not serious and are selfish; they cannot see their dark future. The aim of this paper is to bring technology-driven innovative solutions for such problems. Taking about forest fires prediction and deforestation control, here are a few statistics to give an idea about the loss caused by these fires: **Amazon wildfires**—the fire started around January 2019 and extended up to October 2019, burning 906,000 ha of land, releasing 140 million metric tons of carbon dioxide that is equivalent to the annual emission of 30 million cars. **Australian bush fires**—started around August 2019 and lasted for several months, destroying 18,636,079 ha (12.35 million acres) of land, killing millions of animals. These bush fires cause changes in weather and cause more fire through a thunderstorm which results in lightning. To add on to that, recent California fires were majorly caused by lightning.

The primary reason for these fires is caused by human activities like clearing of land, deforestation. Forest fires and deforestation are correlated to each other. Thus, it would be beneficial and efficient if the control and prediction of both the issues are addressed in one. In order to control deforestation, one of the solutions can be the monitoring of satellite images of the forest areas. Today, we have such powerful satellites like Geofen 4 that have a color resolution of slightly <50 m (which can track aircraft carriers by their wake at sea) along with a thermal image resolution of 400 meters (good for forest fire identification).

Thus, the live data of the satellite image format is needed. For this, one can approach a government body or an organization, explaining the need for the satellite images. One can get high-resolution images that will help classify and keep control over the density of the forest. At times, LiDAR technology, which measures the distance to the target by illuminating the target with laser light, thus regulating the reflected light with sensors, is also used. Similarly, in order to predict forest fires, various machine algorithms can be applied to the data. For example, regression and classification can be used to predict a day's temperature. There need to be many factors for examining a particular land. With the help of machine learning and deep learning algorithms, one can figure out the stats from the data.

K-Means clustering algorithm, logistic regression, Support vector machine (SVM), random forest, etc. can be used with or without Principal Component Analysis (PCA) for forest fire prediction. After training the model with considerable datasets, deforestation control and fire prediction models merged into one as both work in parallel. The final model will be tested by giving data of a particular forest region, for example, take data of the forest area of Assam in India. By using the satellite image, the same can be compared to the present image of the region with the previous images (few days or a few months back images) that will intimate deforestation practice taking place in a particular area; similarly, by considering all the parameters of the region like temperature, smoke, geographical conditions, natural calamities, etc., future prediction and per cent chances of fire taking place in that region can be analyzed. Thus, any region can be monitored by its satellite image and geographical conditions and can save humankind from a disaster.

The author of this proposed chapter would like to present a detailed exploration of the topics covered, as mentioned above, along with a complete and detailed preview of the model. Moreover, it will also describe the use of the model at other places as well. Information and research going on in the discussed field will also be shared so that the readers could take some potential steps toward saving the environment.

The rest of the chapter is organized as: problem and relevance today's society/environmental need and the general solution resolving the mentioned problem; next, reasons and causes of forest fires, followed by methods to detect them; methods to predict fires using machine learning; and finally, density control using deforestation. Each method is discussed in detail along with the algorithm. Then, the chapter is concluded with the conclusion and references.

The following features all the typical social or environmental problems where, due to lack of proper monitoring, humans are facing extreme loss which is irreversible. It elaborates the significant issues in the government, public, and environmental sectors.

14.2 Problems and Relevance to Today's Society/Environmental Need

14.2.1 Deforestation

Forest ecosystems are a critical component of the world's biodiversity as the living world is directly dependent on them than other ecosystems. About 31% of global land is forested. Approximately half the forest area is relatively intact. More than one-third is primary forest (i.e. naturally regenerated forests of native species, where there are no visible indications of human activities and the ecological processes are not significantly disturbed). The total forest area is 4.06 billion hectares. Deforestation and forest fires taking place at alarming rates have led to immense biodiversity loss. In the 1990s, it is estimated 420 million hectares of forest have been lost for conversion into other land uses by humans. And, from 2015 to 2020, the rate of deforestation is increased to 10 million hectares per year. According to FAO, the GlobalTreeSearch database reports the existence of 60,082 tree species. Due to deforestation and climatic change, humans have already lost hundreds of species.

14.2.2 Animal, Insect Attack

Recently, locust attack resulted in tremendous loss of crops. Talking about India, when rabi crops were recently harvested, the locust attack could have taken a heavy toll on India's kharif produce, if not controlled in time. Such similar swarms are formed due to species behavior, habits, and migration. However, these are no new enemies. Many countries like India face such attacks.

14.2.3 National Security on Borders and Illegal Smuggling of Goods

Security and safety is the top priority when it comes to one's nation. Many intruders cross national borders illegally, where monitoring becomes difficult. Continuous control over army bases and outposts is necessary. Similarly, in isolated, vast areas like deserts, smuggling of commodities takes place on a large scale. Theft done through sea route requires high-level monitoring/ contact with ships.

To check the reach and supply of electricity in the rural parts. There is no alert or prediction for forest fires either by natural or human-made means.

14.2.4 Forest Fires

In recent days, forest fires are taking place so frequently, converting a giant portion of the dense forest into ashes.

14.3 Brief Solution for the above Problems

For the abovementioned problems, there is a need for a system that will monitor and predict abnormal activities (forest fire) with the help of satellite and machine learning to predict future forest-fires, where cameras can only monitor a small portion. Also, they can't be used to monitor open and wide areas like deserts, oceans, dense forests, acres of farming land. As illegal things even occur at night, it becomes extremely difficult to track them by human patrolling. So, in such cases, as in the Pulwama attack too, if the entry and movement of the terrorists could be tracked or noted as suspicious, it could have helped Indian soldiers. Satellite communication is changing the world scenario by communicating securely over the channel.

Consider two significant problems deforestation and forest fires: the following deals with the major causes and reasons behind such incidence to understand and create a model based on these parameters that could help us pinpoint and control the problem effectively.

14.4 Reasons and Causes of Wildfires

Fires are a recurring process of nature in forests. They are either caused naturally or by man. Below are few major reasons for such fires: high-latitude forests often experience wildfires caused by lightning, volcanic eruptions that eject hot rocks and lava, climatic conditions, and the rustling of leaves during extreme temperatures. The majority of fires across the world are caused by humans either accidentally or due to negligence or on purpose, where the use of controlled fire is made to clear part of the land for farmlands, for building houses, etc. They use slash-and-burn agriculture to speed up the process of clearing forest land.

Not all fires are bad as they clear dried and dead under-bush which helps the forest to restore its ecosystem. But, NOT ALL FIRES ARE GOOD. Fires release an immense amount of smoke and carbon into the air, affecting the carbon cycle and consuming/destroying trees that absorb this carbon during photosynthesis. These fires even release greenhouse gases, destroying natural resources and human structure, making human life difficult. And, in turn, these poisonous gases lead to chronic respiratory diseases.

Humans burn an average of 175 million acres of forest and grassland according to statistics estimated by scientists.

14.5 Methods to Detect, Predict and Control the Density of Forests

Following up, one could understand the reasons and causes of forest fires and density reduction of extensive forests. Now, the following will introduce the methods to detect and predict forest fires and how deforestation will be controlled using satellite image.

14.5.1 Methods to Detect Forest Fires

Many fire detection sensor systems exist, but these systems are difficult to apply on large open spaces like dense forests and spacious grounds, on high altitudes, and even at places of extreme temperature. These systems, because of their response delay, maintenance cost, and dysfunction due to wild animal attack and other problems, make it more difficult to implement.

Below is the **proposed methodology** with which forest fire detection can be done.

So, to detect forest fires initially, there are various motivating factors for the use of image processing-based method over other measures in the era of rapid development in digital camera technologies and CCD or CMOS digital cameras, which resulted in increasing image quality with high resolution and decreased the cost of the cameras. The second factor is the broad coverage areas with digital cameras. With such embedded satellite cameras, one can get expected images of any region. And lastly, the response time of image processing models is far better than those of sensor-based systems. Using the satellite image processing technique, the image will be processed as per the pixels. A detailed explanation of the process is as follows: to get fire pixels, background subtraction is applied for regional detection.

Further, segmented moving regions are converted from RGB to YCbCr color space where five fire detection rules for separate fire pixels are undertaken. YCbCr algorithm effectively separates the luminance from chrominance and can also separate high-temperature fire pixel centers, as the fire at the high-temperature region is white. Later, an essential measure of the temporal variation is applied to differentiate between fire and fire-like objects. Using this, an automatic forest fire system can be formed (Figures 14.1–14.4).

Experimentation and Results:

Conversion from RGB to YCbC

```
% RGB to YCbCr with Matlab
I = imread('rgb.png');
figure(1), imshow(I);
```

A color image has three channels (red, green, blue). Below is a code to access each component of the image:

```
% RGB to YCbCr with Matlab
R = I(:, :, 1);
G = I(:, :, 2);
B = I(:, :, 3);
figure(2), imshow(R), figure(3), imshow(G), figure(4), imshow(B);
```

```
% RGB to YCbCr with Matlab
I2 = rgb2ycbcr(I);
Y = I2(:, :, 1);
Cb = I2(:, :, 2);
Cr = I2(:, :, 3);
figure(5), imshow(I2), figure(6), imshow(Y), figure(7), imshow(Cb),
figure(8), imshow(Cb);
```

Below is the applied code for fire detection:

```
import cv2
import numpy as np
import matplotlib.pyplot as plt

frame = cv2.imread("test/2.jpg")
hsv = cv2.cvtColor(frame, cv2.COLOR_BGR2GRAY)
img = cv2.GaussianBlur(hsv, (3,3), 0)

# Detecting edges
sobelx = cv2.Sobel(img, cv2.CV_64F, 1,0,ksize=5)
sobely = cv2.Sobel(img, cv2.CV_64F, 0,1,ksize=5)
```

Original image:

FIGURE 14.1 Forest fire detection original image.

Vertical edges:

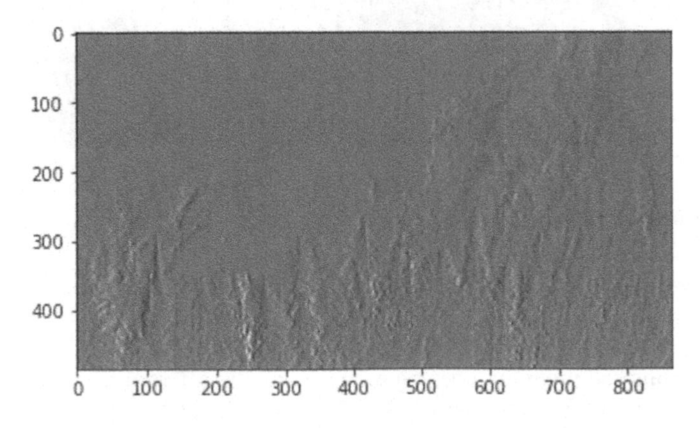

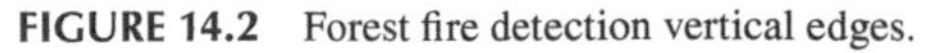

FIGURE 14.2 Forest fire detection vertical edges.

Horizontal edges:

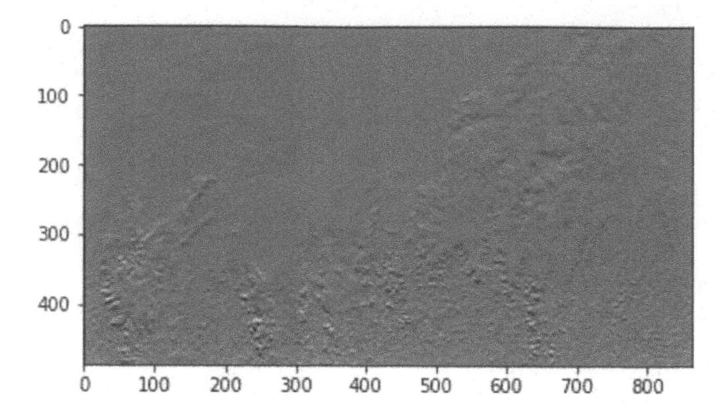

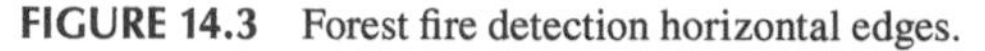

FIGURE 14.3 Forest fire detection horizontal edges.

```
# Detecting red fire
height, width, _ = frame.shape
thresh = 175
for i in range(height):
  for j in range(width):
    b, g, r = frame[i, j]
    if not r > thresh and r > g > b:
      frame[i, j] = 0
    elif not r > g + 50:
      frame[i,j] = 0
    else:
      frame[i,j] = 255
plt.imshow(frame)
```

Threshed image

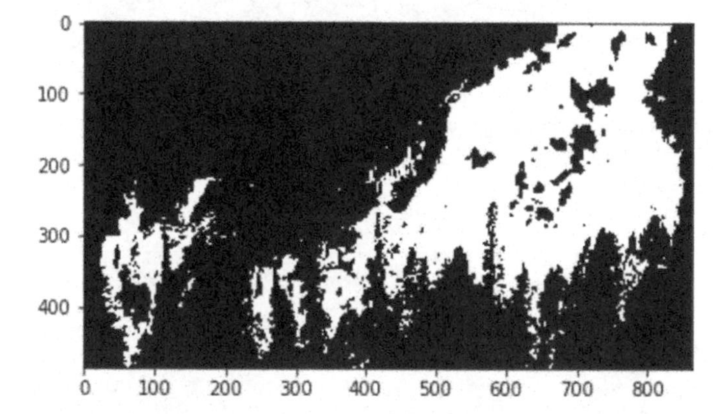

FIGURE 14.4 Detecting red fire-threshed image.

Working: Taking into consideration the vertical edges and horizontal edges values which are deduced from the original image, these values compared to the thresh value assigned will give an output image showing the lighted area and forest area by different colors.

Further, by extending the program, one can even display the affected region by analyzing various color representations according to the fire intensity. These colors can be seen, as the hottest region will be seen as white, while the medium intensity with red, orange, and so on. The intensities can be given a numerical range; for example, red represents (~200–500). This is one of the ways implemented on small data; other approaches can be used to deal with big data as well.

Limitations of This Method: There are no such major limitations as this approach is one of the popular ways of image processing.

14.5.2 Methods to Predict Forest Fires

Researchers used various other ways: for example, neural networks (NN) for the prediction of fires caused by humans. Infrared scanners and NN, when used together, could reduce forest fire false alarms, gaining 90% accuracy. Spatial clustering (FASTCiD) was adopted by a few who could detect forest fire spots in satellite images. In 2005, SVM in satellite images from North America obtained 75% success at finding smoke at 1.1 km pixel level.

Methods like logistic regression, decision trees (DT), and random forest (RF) are used for detection of fires in the Slovenian forests, using satellite data and meteorological data.

Forest meteorological conditions like temperature and wind speed majorly influence forest fires, and several fire indexes like Forest fire Weather Index (FWI) use such data to create a model that can predict fires from previous similar datasets with the help of a data mining approach. This will be useful to predict the burned area of forest fires. There are various data mining techniques to implement: support vector machine (SVM) and random forest. Random forest algorithm is suited for multiclass and works well with numerical and categorical features, while SVM is suited for two-class problems. Selection of a four distinct feature setup includes using spatial, temporal, and FWI components like fuel moisture codes, drought code, duff moisture code, fire behavior index, build-up index, and fire weather index.

Also, the best configuration uses an SVM and four meteorological inputs (i.e. temperature, relative humidity, rain, and wind) that are also capable of predicting the burned area of small fires which occur frequently. This knowledge can be useful to improve firefighting measures and to manage resources (e.g. prioritizing targets for air tankers and ground crews) (Figures 14.5–14.9).

Proposed Methodology: Working on with such attributes, month, day, temp, wind, rain, etc. the following algorithm (SVR) can be applied to predict fires.

```
from sklearn.svm import SVR
from sklearn.model_selection import GridSearchCV
svr = SVR()
parameters = {"kernel": ("linear", "poly", "rbf"), "C": (1, 5, 10),
"epsilon": (0.1, 0.2, 0.5)}
clf = GridSearchCV(svr, parameters)
clf.fit(X_train, y_train)

from sklearn.metrics import mean_squared_error as mse
y_preds = clf.predict(X_test)
mse(y_test, y_preds)
```

Result: 969.1022948044374 (just an example of how this value is received by working on a particular set of data).

This program uses one of the machine learning algorithms for the prediction. SVR is a regression algorithm that supports both linear and nonlinear regressions. It is based on SVM's principle. SVR differs from SVM in the way that SVM is a classifier used for the prediction of discrete categorical labels, while SVR is used for the prediction of a continuous ordered variable. Here, the results show the mean squared error value that will help for predicting the burned area of small fires. This was just a small implementation of the proposed algorithm.

Limitations of This Method: In order to take into consideration all the attributes contributing to these fires, we will need other algorithms as well which can handle all these attributes and big data. SVR does not perform well if the datasets are very large and contain more noise.

14.5.3 Deforestation Control

In order to analyze a particular region properly, one needs to know the location and area of deforestation, the rate and speed at which deforestation is taking place, and the reasons and causes behind deforestation. To answer the above requirements, Geographic Information System (GIS) and remote sensing can be used, as remote sensing can provide fast and inexpensive datasets and GIS for analyzing the location, types, and rates of changes. Studying the elevation of the working forest area and considering necessary latitudes and longitudes, the diversity of elevation, slope, population, rivers, etc. can be understood. Below is the proposed methodology.

14.5.3.1 Dataset Required

The dataset to be taken into consideration should be similar to satellite images: e.g. Landsat TM image and Landsat ETM+ image with a general resolution of about 28.5 m or the images with high resolution which can be used to extract data from pixel according to density. Topographic maps could be used for the geo-referencing of two images.

14.5.3.2 Processing and Analysis of Images

Usually, three major errors are encountered when a satellite image is generated from satellite sensors: (i) Sensor error: this is when there are some errors in the geometry and the measured brightness values of the pixels. (ii) The second error can be caused because of atmospheric parameters that affect the sensors by the amount of radiation received. (iii) The third one is the geometric errors related to the earth's surface like the earth's curvature, the rotation of the earth, elevation differences, and satellite positioning. Either of the processes should resolve these errors.

Radiometric Correction: In order to use division and subtraction of images (background images), a correction needs to be made for the Haxe, Sunagle, and skylight errors.

Geometrical Correction: In image registration (geo-referencing), the most crucial task is the selection of control points, especially at the time when there is a long period between the map and the image. For geo-referencing the images, the first-order polynomial equation can be used, which could help in removing the errors related to rotation and scaling of the image. The given equation is:

$$X = a_0 + a_1x + a_2y \; Y = b_0 + b_1x + b_2y \tag{14.1}$$

where x and y are the coordinates of a point in the first coordinate system (old system), and X and Y are its new coordinates in the new coordinate system.

14.5.3.3 Model Creation Using Logistic Regression

It is a special type of regression used when one wants to study the probability of two different classes. For example, here, the forest area can be either stable or destroyed. The logistic regression model can be displayed as:

$$\log\log \text{it}(p) = a + b_1x_1 + b_2x_2 + b_3x_3 \tag{14.2}$$

where p is a dependent variable and gives the probability of any of the two conditions. The variables x_1, x_2, and x_3 are dependent variables defining the phenomenon, and b_1, b_2 and b_3 are their coefficients: a is the additive coefficient.

The following algorithm with the code can be used to find the density change from the images.

Experimentation:
Algorithm:

1. Read Image.
2. Normalize Values.
3. Find median of image pixels.
4. Find deviation in median and subtract it from median (values greater than this threshold are used for density).
5. Calculate the density of the image.
6. Perform steps 1–5 for images at different time stamps, and subtract the density (use global threshold value to compare images).
7. Increase in density=Afforestation, decrease=deforestation.

```
def threshold_median(self, std=False, median=None):
  if median is None:
    median = np.median(self.image)
    if std:
      median -= np.std(self.image)
  self._mean = np.mean(self.image)
  self._median = np.median(self.image)

self.image[self.image < median] = 0
self.image[self.image >= median] = 1

# calculate_diff("Bareilyoutput.tif", "Gondaloutput.tif")
image_list = ["overtime/"+file for file in os.listdir("overtime/")]
_med, _den = None, None
thresh = 0.1
for image in image_list:
  __den = _den
  _med, _den = main(image, False, _med)
  _den *= _med
  if __den:
    diff = __den - _den
    if diff < 0:
      print("Deforested by {:.2f}".format(diff))
      if abs(diff) > thresh:
        print("ALERT: This is above the threshold value, an abnormal
        activity is detected!")
    else:
      print("Afforested by ", diff)
      print("Keep planting more trees for a better future!")
  else:
    print("Initializing variables")
```

Original image

FIGURE 14.5 Deforestation 2001 original image.

Initializing Variables

Original image

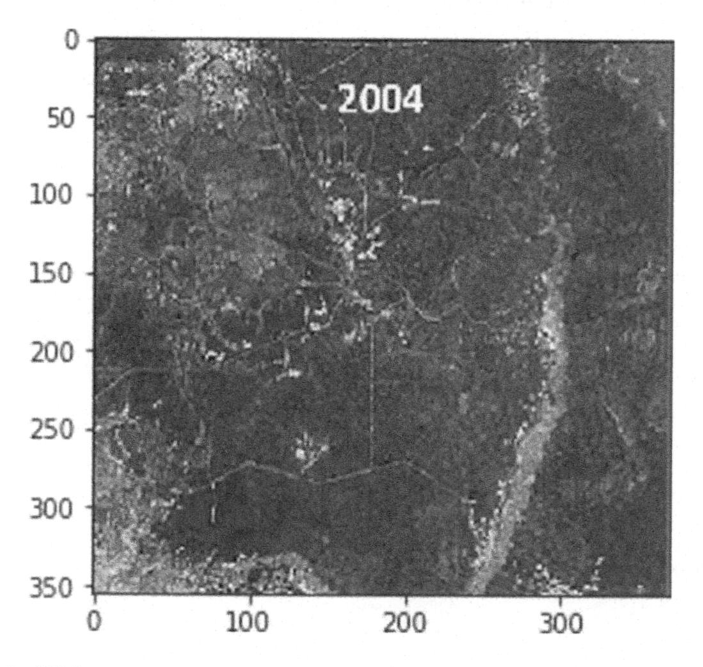

FIGURE 14.6 Density 2004.

Deforested by −0.09
Original image

FIGURE 14.7 Density 2007.

Afforested by 0.07084941445173426
Keep planting more trees for a better future!
Original image

FIGURE 14.8 Density 2010.

Deforested by −0.01
Original image

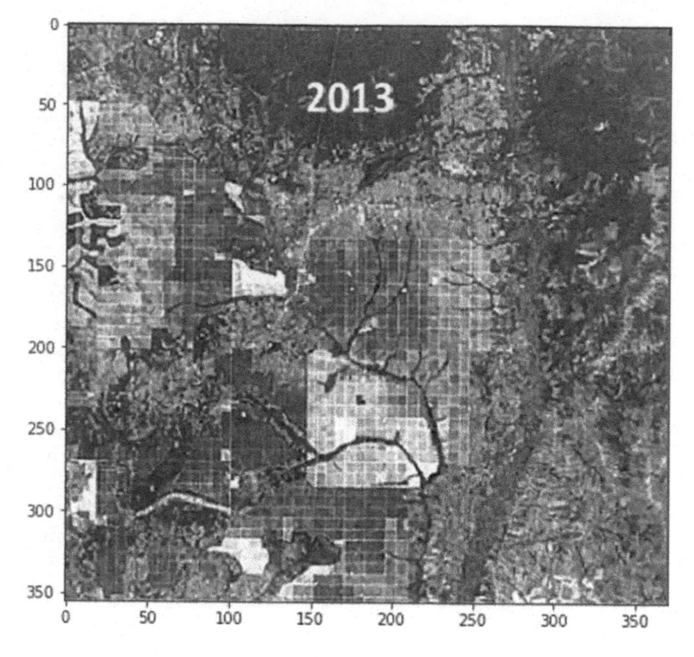

FIGURE 14.9 Density 2013.

Deforested by −0.16
ALERT: This is above the threshold value; an abnormal activity is detected!

Results:

TABLE 14.1

Comparison of various Automatic Alert System

Sr.no.	Year	Comment (Comparison Is Made from the Thresh Value)
1.	2001	Initializing variables
2.	2004	Deforested by –0.09
3.	2007	Afforested by 0.07084941445173426 Keep planting more trees for a better future!
4.	2010	Deforested by –0.01
5.	2013	Deforested by –0.16 Alert: This is above the threshold value; an abnormal activity is detected!

We opted for this approach of finding the median of the image pixel so as to be able to get a value of a particular image and then compare it with previous data and the threshold value. This comparison will be useful for providing an automatic alert system (Table 14.1).

Limitations of the Proposed Methodology: This methodology can provide an alert and measure the effect of deforestation for a static dataset. In order to deal with variance in images, and with a larger dataset, the algorithm needs to be modified for live data.

14.6 Conclusion

The chapter discussed the major environmental problems, their reasons, and causes. By using image processing of satellite images, deforestation density can be controlled. By analyzing and comparing the value of the particular image with the threshold value, one will come to know how much percent of deforestation took place and where and for forest fire detection by the above mentioned methods. Also, with the help of various machine learning algorithms, fire prediction can be done.

Future Scope: can be used for surveillance over a large area to monitor illegal movement and suspicious movement. Control deforestation density by live comparison with previous day's data to alarm about illegal activities similar to monitoring terror outbreak into a territory, farming land, desert areas, etc. This type of monitoring system can also be used by manufacturing units to monitor their open area production unit.

This product overcomes ancient monitoring systems by using live satellite data. It's an innovative solution as many of the monitoring devices are a constraint for a particular use or as per a specific area.

BIBLIOGRAPHY

Journal Article

1. Aditi Kansal, Yashwant Singh, Nagesh Kumar, Vandana Mohindru "Detection of forest fires using machine learning technique: A perspective" (2015).
2. Lan Downard "Predicting forest fires with spark Machine Learning" (2018).
3. Brigham Young University "New wildfire models to predict how wildfires will burn in next 20 minutes" (2019).
4. Kabir Uddin, Hammad Gilani "Forest conditioning monitoring using very high-resolution satellite imagery in a remote mountain watershed in Nepal" (2015).
5. Mubarak A. I. Mahmoud and Honge Ren "Forest fire detection using rule-based image processing algorithm and temporal variation" *Hindawi Journal* (2018). doi:10.1155/2018/7612487.
6. Ahmad A. A. Alkhatib "A review on forest fire detection techniques" *Sage Journal* (2014). doi:10.1155/2014/597368.
7. Mesgari Saadi and Ranjbar Abolfazl "Analysis and estimation of deforestation using satellite imagery and GIS" (January 2019).

News Article

1. Powered by Tensorflow "Fighting fire with machine learning: Two students use Tensor flow to predict wildfires" (2018).
2. Joey Hadden "Australian bush fires are ravaging the country. Here's how it all happened" (2020).

15

Application of Convolutional Neural Network in Feather Classifications

Milind Shah, Keval Nagda, Anirudh Mukherjee, and Pratik Kanani
Dwarkadas J. Sanghvi College of Engineering

CONTENTS

15.1 Introduction.. 223
15.2 Multilayer Neural Networks 224
15.3 Convolutional Neural Networks.. 224
15.4 Greedy Snake Algorithm 224
15.5 Dataset.. 225
15.6 Proposed Methodology 226
15.6.1 Base Model... 226
15.6.2 Data Augmentation.. 226
15.6.3 Implementation... 226
15.7 Experimentation and Measures 226
15.7.1 *K-Fold Validation*.. 226
15.7.2 Measures... 228
15.8 Results and Discussion.. 228
15.9 Conclusion...231
15.10 Future Scope 231
References.. 232

15.1 Introduction

Collisions with civilian airplanes are the cause of over 14,000 deaths of birds each year in the United States as shown in the data collected by the Federal Aviation Administration (FAA) [1]. Every time a bird strike occurs, the aircraft personnel voluntarily report the strike to the FAA. FAA has seen a steady growth in the number of reports that come in each year. In every report, they mention details such as date and time of the flight, aircraft model, which airport the plane came from, which airline it belongs to, engine type, damage done to the airplane, and the species of the bird. While all other information is readily available, identifying the species of the bird can be a tough task, and they are often not able to identify what species of bird was the victim of the collision. This information, however, is vital since researchers use this data to study the cause of these strikes, which can include changing migratory patterns of various bird species.

Bird species are identified by their remains and this identification is carried out by trained biologists working at the airports. Depending on the condition of the remains, these birds can be identified by physical characteristics, feather fragments, and in certain cases, DNA analysis. However, this process can get quite tedious and time-consuming since the remains might not be complete or in the best condition for analysis, in which case feathers might be the best way to identify the bird species. The task of identifying birds by feathers can be tedious and has the potential to be carried out by a well-performing model. It was decided to make a multilabel image classification CNN Model to identify the bird species using their feathers. The model was trained on a dataset consisting of feathers of over 300 species of birds.

15.2 Multilayer Neural Networks

If a supervised learning problem with access to labeled training examples was taken into consideration, a neural network would thus provide us with a way to define a hypothesis having a complex nonlinear form having parameters W and b that shall be able to fit the data.

This one singular unit of neuron takes a certain number of input parameters (x_1, x_2, x_3 along with an intercept) and gives the output as $hW, b(x) = f(\sum W_i x_i + b)$, where f is the activation function. The neuron corresponds to the input–output mapping which is defined by the said activation function.

A neural network is constructed by combining many such neurons together such that the output of one neuron can be the input of another. The neurons in the neural network have similar bias intercepts. The neurons are arranged in multiple layers wherein the first layer is called the input layer and the final layer is the output layer. The layers in between the input and output layers are called the hidden layers. A neural network can have many different architectures wherein they tend to work on different sets of problems. A feedforward neural network is one of the most basic kinds of neural networks in which the output of the member neurons is fed forward to the next layer and thereafter the output is evaluated. The error function makes a note of the difference in the actual value and the value computed by the neural network and adjusts the weights of the member neurons accordingly. This process of evaluating outputs by propagating the results of the neurons from the initial layers to the output layer and evaluating the output based on training data and thereafter adjusting the weights is known as forward propagation.

15.3 Convolutional Neural Networks

Convolutional Neural Networks are a class of deep neural networks, usually used for analyzing visual imagery. They are widely used in image and video recognition, classification for images, image analysis for medical purposes, NLP, and financial time series.

CNNs are derived by regularizing multilayer perceptions. Multilayer perceptrons are often fully connected networks, i.e. every neuron in a particular layer is connected to neurons in the following layer. A lot of methods of regularization consist of adding a form of magnitude measurement for weights of the loss function. CNNs use another method for this, hierarchical pattern in the data is used, and thus, more complex patterns are assembled using smaller and simpler patterns. Thus CNNs take inspiration from biological processes; this is seen in the connectivity patterns between neurons that are observed in the structure of the animal visual cortex. Single cortical neurons give a response to stimuli just in a restricted region of the visual field that is called the receptive field. The receptive fields of various neurons partially overlap such that they encapsulate the entire visual field.

CNNs require limited pre-processing when put up against other image classification techniques. This means that unlike in traditional algorithms where filters were hand-engineered, the network learns them. This independence from pre-existing knowledge and human effort is a major advantage.

CNNs consist of an input layer and an output layer, along with multiple hidden layers. The many hidden layers usually include a series of convolutional layers that convolve with a dot product or multiplication. RELU is usually the activation layer of choice, which is usually followed by additional convolutions such as fully connected layers or pooling layers and normalization layers. Even though the layers are referred to as convolutions, that is the case only by convention. Mathematically, it is actually a sliding dot product or cross-correlation. This has importance for the indices in the matrix; it affects how at each specific index point the weight is affected (Figure 15.1).

15.4 Greedy Snake Algorithm

Greedy snake algorithm, also known as the active contour model, is a framework in computer vision. It is primarily used for delineating an object outline from a possibly noisy 2D image in order to make it clearer and defining certain outlines. The greedy snake algorithm is popular in computer vision and is

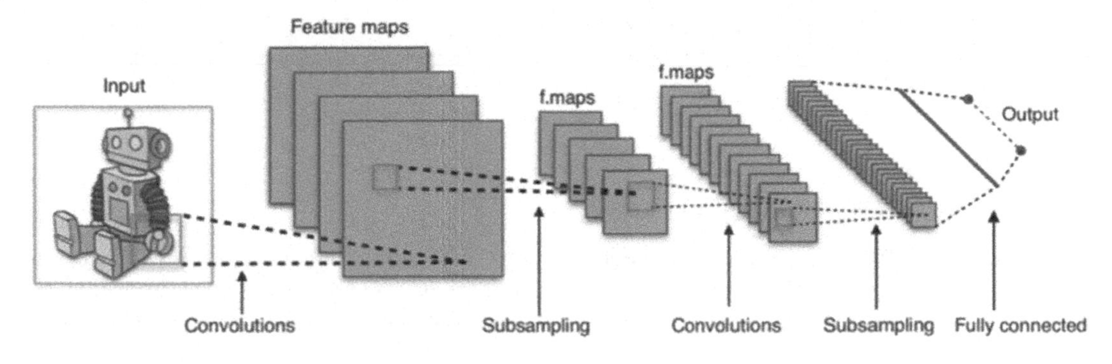

FIGURE 15.1 Example CNN architecture here.

widely used in applications like object tracking, shape recognition, segmentation, edge detection, and stereo matching.

A deformable spline influenced by constraint and image forces, a greedy snake is an energy minimizing algorithm that pulls the spline towards object contours and internal forces that try to resist deformation. Snakes may be understood as a special case of the general technique of matching a deformable model to an image by means of energy minimization. The active shape model represents a two-dimensional discrete version of this approach and takes advantage of the point distribution model to restrict the shape range to an explicit domain, which it learns from a training set.

As the method requires knowledge of the desired contour shape beforehand, the greedy snake does not solve the entire problem of finding contours in images. However, they depend on other mechanisms such as understanding higher-level image processes, interaction with users, or information from image data adjacent in time or space.

In computer vision, the boundaries of an image are described by contour models. Greedy snakes are particularly designed to solve a problem where the boundary shapes are known. Since it is a deformable model, greedy snakes are able to adapt to differences and noise in stereo matching and motion tracking. The method can also be used to find illusory contours by ignoring missing boundary information in the image.

On comparing with classical feature extraction techniques, the greedy snake has a number of advantages:

- They search for the minimum state autonomously and adaptively.
- In an intuitive manner the external image forces act upon the snake.
- The image energy function introduces scale sensitivity by incorporating Gaussian smoothing. They can be used to track dynamic objects.

Drawbacks of the traditional snake algorithm are

- Sensitive to local minima states. This can be counteracted by simulated annealing techniques.
- Minute features are often ignored along energy minimization over the entire contour.
- Their convergence policy largely determines the accuracy.

15.5 Dataset

For the dataset, the Feather Atlas was used, which was provided by the US Fish and Wildlife Service. They have a comprehensive set of images of over 390 species and includes feathers which include their primaries (outer wing feathers), secondaries (inner wing feathers), and rectrices (tail feathers) from the birds. The images were scraped from featheratlas.gov and then used a script (which employed OpenCV) to extract the individual feathers from the images that had been scraped [2,3]. The greedy snake algorithm was used for detecting contours along the feathers in the given set of images provided by Feather Atlas to create the dataset.

15.6 Proposed Methodology

This chapter provides the methods used to produce a strong baseline model in this section. This research work describes the adapted deep convolutional neural network (CNN) for multi-label classification, as well as the essential data augmentation techniques used for training a deep neural network model [4].

15.6.1 Base Model

Deep CNNs can be used to implement the model for the task of multilabel image classification with an image of the subject as the input to the CNN model and a C-dimensional score vector comprising of scores of multilabels or discrete classes as the output from the model [5]. In contrast to the traditional multilabel classification approaches, these deep neural network models integrate not only the feature extraction but also the classification task in a single framework, thus, enabling end-to-end machine learning [6]. More importantly, state-of-the-art deep CNN models are able to learn high-level visual representations of input training images and approximate very complex learning systems [7].

15.6.2 Data Augmentation

Data augmentation techniques affect multilabel image classification and, thus, are nontrivial since some commonly adopted data augmentation techniques such as random cropping will change the semantics in the original image. For example, random cropping of a multilabel image might result in image patches not containing all the objects in the original image [8]. Apart from the conventional data augmentation techniques, rotation, height, and width shift adaptations are made, shear shift, zoom in, and horizontal flip operations are done to further increase the data variability. The work implements the abovementioned data augmentation techniques due to the fact that it expands the target label space significantly which is quite different from other traditional data augmentation techniques [9].

15.6.3 Implementation

The deep CNN model used in this chapter's experiments is implemented in Keras 2 (TensorFlow 2). This chapter uses randomly initialized model weights and fine-tunes the weights of all layers as the model is trained on the dataset. The Adam optimizer is used for model training with an initial learning rate of 0.001 for the fully connected layers of the model. The learning rate decays to one-tenth after five epochs. The training was stopped after five epochs. The batch size is set as 32 in all the experiments. The proposed model includes five 2D-Convolutional layers, followed by max-pooling layers, and batch normalization at required positions. A dropout of 0.25 is also applied in order to avoid overfitting of the model on the training dataset. ReLU activation function is used for each layer and the Softmax function is used in the final layer for multilabel classification (Figures 15.2 and 15.3).

15.7 Experimentation and Measures

15.7.1 *K-Fold Validation*

K-Fold validation is a statistical method where a given data set is divided into *K* number of sections (also called folds), and each fold is made to be the testing set at a certain point. For example, if $K=7$ is taken, the dataset is divided into seven-folds, where the first fold is used as the test set and the rest as training

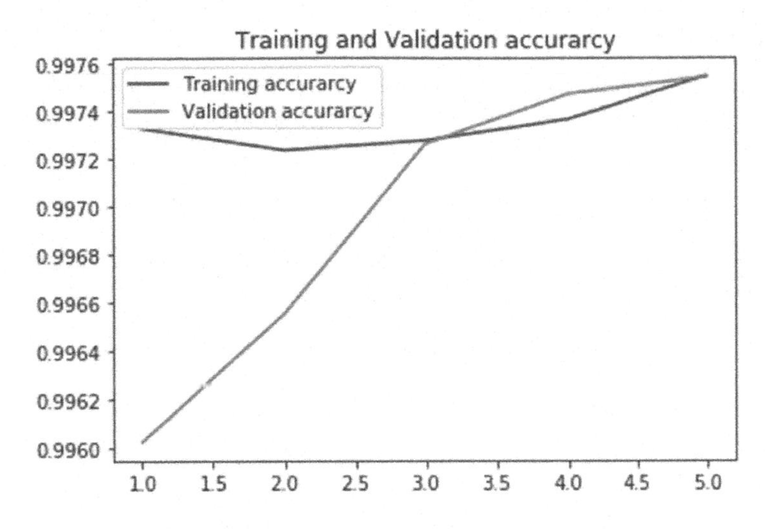

FIGURE 15.2 Training and validation accuracy here.

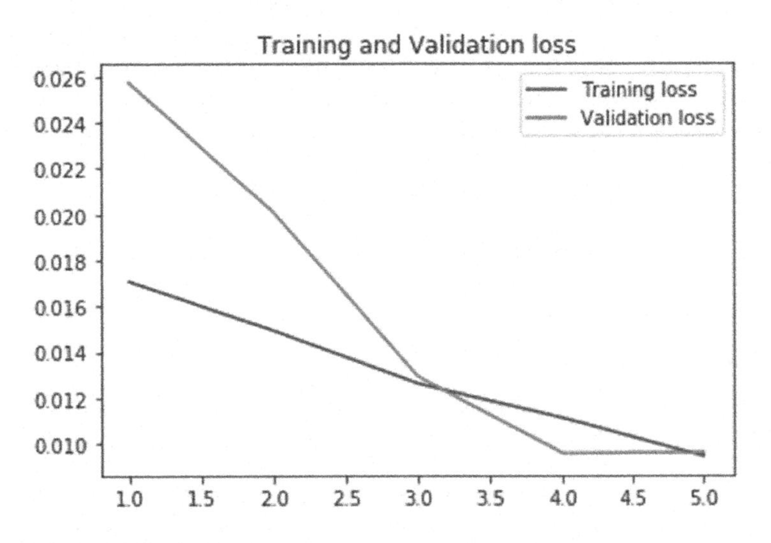

FIGURE 15.3 Training and validation loss here.

set in the first iteration, the second fold as test set in second iteration, and so on. This process is repeated till all folds are used as a test set once. It is a famous method as it is incredibly simple to grasp and results in a less-biased estimate of the model.

Generally, the following steps are taken:

1. Shuffle the dataset randomly.
2. Divide the dataset into K groups
    a. For every group:
    b. Reserve selected group as test set
    c. Use remaining groups as training set
    d. The model is fit on the training set and evaluated on the test set.
    e. Retain the scores (Model will be discarded)
3. Summarize the results of the model using the sample of the scores received above.

15.7.2 Measures

Accuracy: Accuracy is one of the most important parameters; it simply is a ratio of observations that are correct to the total observations. While high accuracy is a good indicator of performance, one must also look at other parameters too.

$$\text{Accuracy} = \text{True positive} + \text{true negative/true positive} + \text{false positive} + \text{false negative} + \text{true negative}$$

Precision: Precision is calculated as the ratio of predicted positive observations that are right to the total predicted positive observations. High values of precision equate to low false-positive rates (Figure 15.4).

$$\text{Precision} = \text{True positive/true positive} + \text{false positive}$$

A training precision value of 0.6 and a validation precision of 0.58 were seen.

Recall (Sensitivity): Recall is the ratio of positive observations that were predicted correctly to all observations in the class (Figure 15.5).

$$\text{Recall} = \text{True positive/true positive} + \text{false negative}$$

The training recall value was at 0.18 while the validation recall was 0.2.

F_1 Score: F_1 Score is the weighted average of Recall and Precision. F_1 takes into picture both false negatives and false positives. F_1 in most cases is more useful than pure accuracy. Accuracy works better if false negatives and false positives have similar costs. However, if the cost of false positives and false negatives are very varied, it's better to look at Recall and Precision (Figure 15.6).

$$F_1 \text{ score} = 2*(\text{recall}*\text{precision})/(\text{recall}+\text{precision})$$

The training F_1 score was 0.15 and validation F_1 was 0.3

15.8 Results and Discussion

Binary cross-entropy is used as a loss function for training the deep CNN model [10]. Figures 15.1 and 15.2 show experimental results on the dataset. The chapter first investigates the impact of input

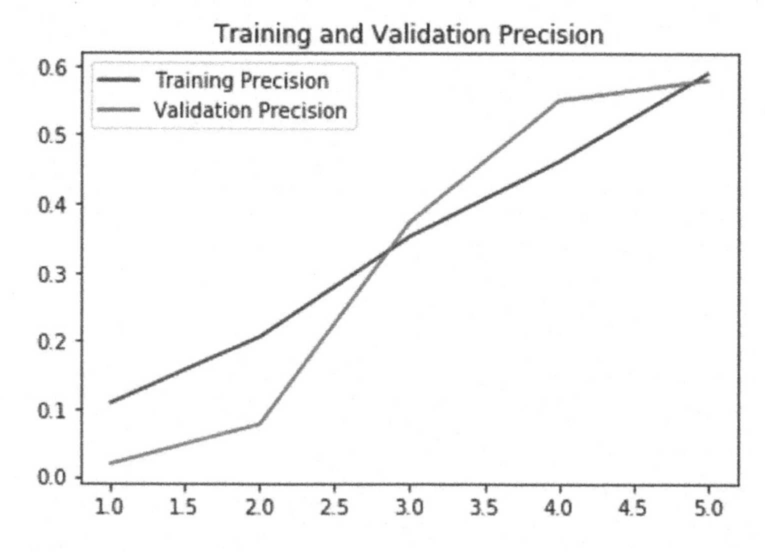

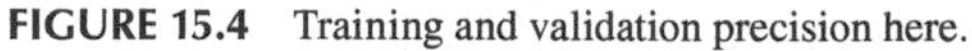

FIGURE 15.4 Training and validation precision here.

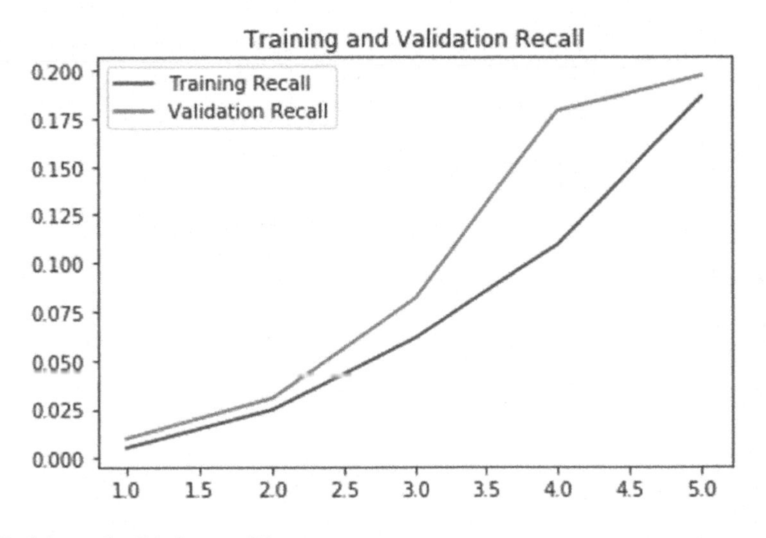

FIGURE 15.5 Training and validation recall here.

FIGURE 15.6 Training and validation F_1 score here.

image size and then appropriate size (i.e. 256) is employed for each experimental setting. The work also investigates the effectiveness of data augmentation by implementing a different set of operations, and experimental results show superior performance when using data augmentation strategies. The work also compares the proposed baseline performance against that of state-of-the-art approaches and without any tricks, only basic deep CNN models but achieve better performance (99.75% test accuracy and 0.0096 test loss) on the dataset [11]. As a result, this chapter's experimental results indicate that the basic deep models with proper training strategies have more capabilities than what have been explored for multilabel image classification, and a strong baseline is presented (Figures 15.7–15.9).

As you can see in the results above, our model is able to accurately predict the correct species by the bird's feather. We have obtained good scores in this experimental scenario, so there is a very good scope to prepare a robust model that will perform exceptionally well in real-life scenarios.

```
[ ]  image_path = '/content/Dataset/Feather_final/Acorn Woodpecker/ACWO_tail_adult_08.jpg'
     image_array = convert_image_to_array(image_path) #Acorn Woodpecker
     np_image = np.array(image_array, dtype=np.float16) / 225.0
     np_image = np.expand_dims(np_image,0)
     plt.imshow(plt.imread(image_path))
     result = model.predict_classes(np_image)
     print((image_labels.classes_[result]))
```

```
['Acorn Woodpecker']
```

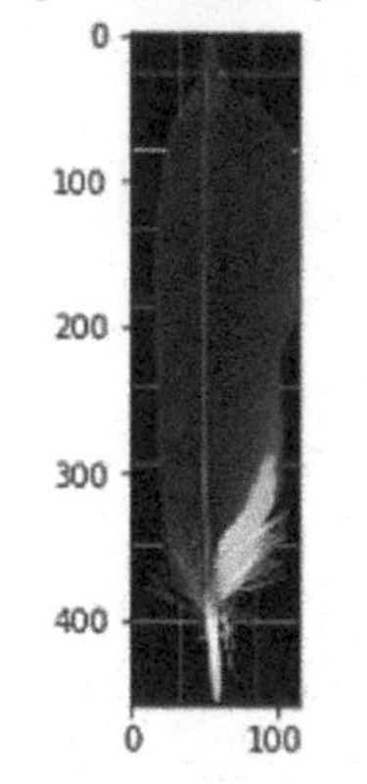

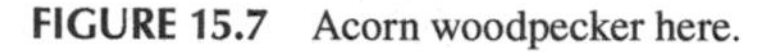
FIGURE 15.7 Acorn woodpecker here.

```
[ ]  image_path = '/content/Dataset/Feather_final/Brown Pelican/BRPE_primary_adult_14.jpg'
     image_array = convert_image_to_array(image_path) #Brown Pelican
     np_image = np.array(image_array, dtype=np.float16) / 225.0
     np_image = np.expand_dims(np_image,0)
     plt.imshow(plt.imread(image_path))
     result = model.predict_classes(np_image)
     print((image_labels.classes_[result]))
```

```
['Brown Pelican']
```

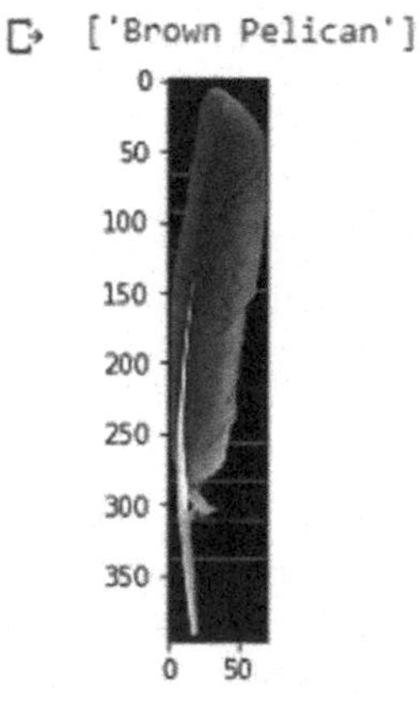

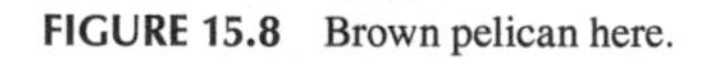
FIGURE 15.8 Brown pelican here.

```
[ ] image_path = '/content/Dataset/Feather_final/Elf Owl/83.png'
    image_array = convert_image_to_array(image_path) #Elf Owl
    np_image = np.array(image_array, dtype=np.float16) / 225.0
    np_image = np.expand_dims(np_image,0)
    plt.imshow(plt.imread(image_path))
    result = model.predict_classes(np_image)
    print((image_labels.classes_[result]))
```

```
['Elf Owl']
```

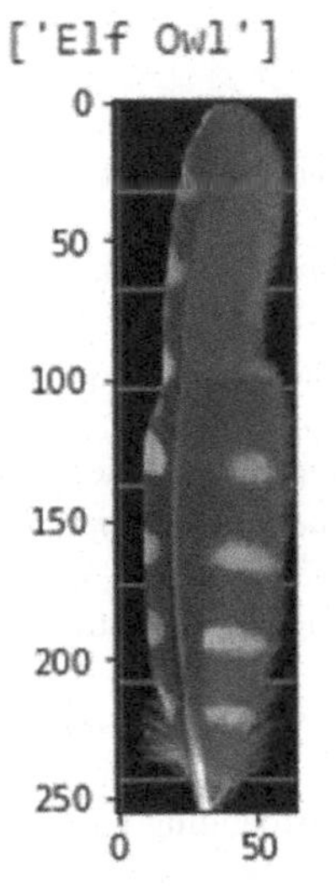

FIGURE 15.9 Elf Owl here.

15.9 Conclusion

The CNN model employed in this experiment gave solid, crisp, and clear results and is able to classify the feathers according to the species to which it belongs, with an accuracy of 99.75% averaged over all the species present in the dataset. The classifier can thus be used to accurately predict the bird species given any feather it has been trained on. This model currently considers ~58 million parameters and has been trained over five epochs to obtain the aforementioned results. The model does churn out correct results with the validation accuracy steeply rising to 99.75, thus being an indicator that the model avoids any and all forms of overfitting. The segmentation of feathers initially performed gave us a strong dataset with each image capturing the most out of what can be obtained from the segmented feathers. The result of the experiment overall has led us to come up with a well-built classifier to predict bird species with utmost accuracy.

15.10 Future Scope

The results obtained during the experiment are satisfactory, but they could still be improved if trained for a higher number of epochs and on a larger dataset. Employing this classifier on a large scale can be successful, and the model can be used to predict the bird species based on a single image of its feather. Further plans in the work are to train the model on a larger dataset to account for a wider variety of feathers. The classifier can be improved upon by using a greater number of data augmentation techniques in order to take into consideration the various distortions that can creep up in Figure 15.3. A little tweaking of the CNN architecture employed could also lead to better results, and once a bigger dataset is obtained, the model might have to be changed around, depending on the results of the experiment on a bigger dataset. Thus, this chapter's major plans involve obtaining a larger dataset and tuning the model according to the changes that the dataset might bring into the experiment along with certain data augmentation techniques that will count for the wide array of irregularities and distortion [9,12]. This system can, therefore, be used to predict the bird species and can find its applications in a large number of areas where such identification might be required, such as tracking migratory patterns, wildlife conservation, identifying bird hits on airplanes, which can be further used to track the height at which these birds venture out, evolutionary changes in these birds, etc. [13,14,15,16].

REFERENCES

1. Burger, J. (1985). Factors affecting bird strikes on aircraft at a coastal airport. *Biological Conservation*, 33(1), 1–13, 16–28.
2. Xie, G., & Lu, W. (2013). Image edge detection based on opencv. *International Journal of Electronics and Electrical Engineering*, 1(2), 104–106. doi: 10.12720/ijeee.1.2.104-106.
3. Brahmbhatt, S. (2013). Image segmentation and histograms. In *Practical OpenCV*, pp. 95–117. Apress: New York. doi: 10.1007/978-1-4302-6080-6_7.
4. Yim, J., Ju, J., Jung, H., & Kim, J. (2015). Image classification using convolutional neural networks with multi-stage feature. *Advances in Intelligent Systems and Computing Robot Intelligence Technology and Applications*, 3, 587–594. doi: 10.1007/978-3-319-16841-8_52.
5. Wang, Q., Jia, N., & Breckon, T. (2018). A baseline for multi-label image classification using ensemble deep CNN.
6. Tsoumakas, G., & Katakis, I. (2007). Multi-label classification: An overview. *International Journal of Data Warehousing and Mining (IJDWM)*, 3(3), 1–13.
7. Sharif Razavian, A., et al. (2014). "CNN features off-the-shelf: An astounding baseline for recognition." *Proceedings of the IEEE Conference on Computer Vision and Pattern Recognition Workshops*, Columbus, OH.
8. Zhong, Z., et al. (2017). Random erasing data augmentation. arXiv preprint arXiv:1708.04896.
9. Perez, L., & Wang, J. (2017). "The effectiveness of data augmentation in image classification using deep learning." arXiv preprint arXiv:1712.04621.
10. Brink, A. D., & Pendock, N. E. (1996). Minimum cross-entropy threshold selection. *Pattern Recognition*, 29(1), 179–188.
11. Wang, L., et al. (2015). Places205-vggnet models for scene recognition. arXiv preprint arXiv:1508.01667.
12. Sakalli, M., Lam, K.-M., & Yan, H. (2006). "A faster converging snake algorithm to locate object boundaries." *IEEE Transactions on Image Processing*, 15(5), 1182–1191.
13. Rudnick, J. A., et al. (2007). Species identification of birds through genetic analysis of naturally shed feathers." *Molecular Ecology Notes*, 7(5), 757–762.
14. Surani, M., & Mangrulkar, R. (2021). Online public shaming approach using deep learning techniques. *Journal of University of Shanghai for Science and Technology*. doi: 10.51201/jusst12675
15. Kotwal, A., Kotia, J., Bharti, R., & Mangrulkar, R. (2021). Training a feed-forward neural network using cuckoo search, In: Dey, N. (Ed.), *Applications of Cuckoo Search Algorithm and Its Variants*, pp. 101–122. Springer Nature: Singapore. doi: 10.1007/978-981-15-5163-5_5.
16. Kotwal, A., Bharti, R., Pandya, M., Jhaveri, H. & Mangrulkar, R. (2021). Application of BAT Algorithm for detecting malignant brain tumors. In: Dey N., Rajinikanth V. (eds) *Applications of Bat Algorithm and Its Variants*. Springer Tracts in Nature-Inspired Computing. Springer, Singapore. doi: 10.1007/978-981-15-5097-3_7.

16

Application of Deep Learning Coupled with Thermal Imaging in Detecting Water Stress in Plants

Saiqa Khan and Meera Narvekar
Dwarkadas J. Sanghvi College of Engineering

Anam Khan, Aqdus Charolia, and Mushrifah Hasan
M.H. Saboo Siddik College of Engineering

CONTENTS

16.1 Introduction 233
16.2 Related Work 235
16.3 Proposed Methodology 235
16.3.1 Data Set 237
16.3.2 Data Augmentation 237
16.3.3 Pre-Processing of Thermal Images 237
16.3.4 Feature Extraction 238
16.4 Results 241
16.5 Conclusion 243
References 243

16.1 Introduction

This chapter discusses how thermal imaging and deep learning can be used for stress detection in tomato plants and indirectly in the field of agriculture. Global warming has risen throughout the world, resulting in massive damage to the agricultural sector, especially in a country like India, which accounts for about 70% of Indian economy [1]. Thus, exploration of novel techniques is a must for combating global warming and simultaneously improving agricultural produce. One of the greatest impacts of global warming has been on water, a life supporting resource, resulting in droughts, water scarcity, and many more water-related problems. This has caused a great impact on agriculture, causing plants to undergo water stress and eventually wither. The reason for undergoing water stress is that water must be pumped to all the parts of the plant for it to carry out its functions normally. This is done by the process of transpiration where the plant releases water in the form of vapours to cool itself down from the warm climatic conditions and, secondly, create a suction force to pull the water and minerals to perform photosynthesis from the ground. It has been estimated that 98% of plants energy is used in the work of transpiration [2]. Rabi crops like tomato require optimum water and temperature conditions (21°C–23°C) to grow [3]. With global warming causing water scarcity and high temperatures, it prevents the leaves of the plant to transpire, causing a rise in temperature of the overall plant and water deficiency due to which the plants cannot carry out their activities at full potential and thus result in them getting stressed.

Water-stressed crops cannot be identified by the naked eyes of the farmers. Invasive strategies like temperature sensors, moisture sensors, and photo sensors can be and have been used to detect water stress, but they usually result in damaging the crops in the long run and are a costly approach. Another strategy is to employ non-invasive strategies to identify water stress in crops in early stages so that the

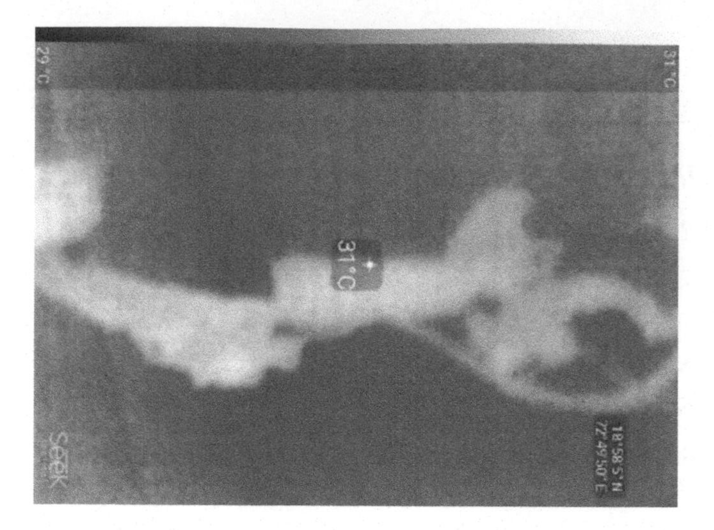

FIGURE 16.1 Image of stressed tomato leaf captured by thermal camera.

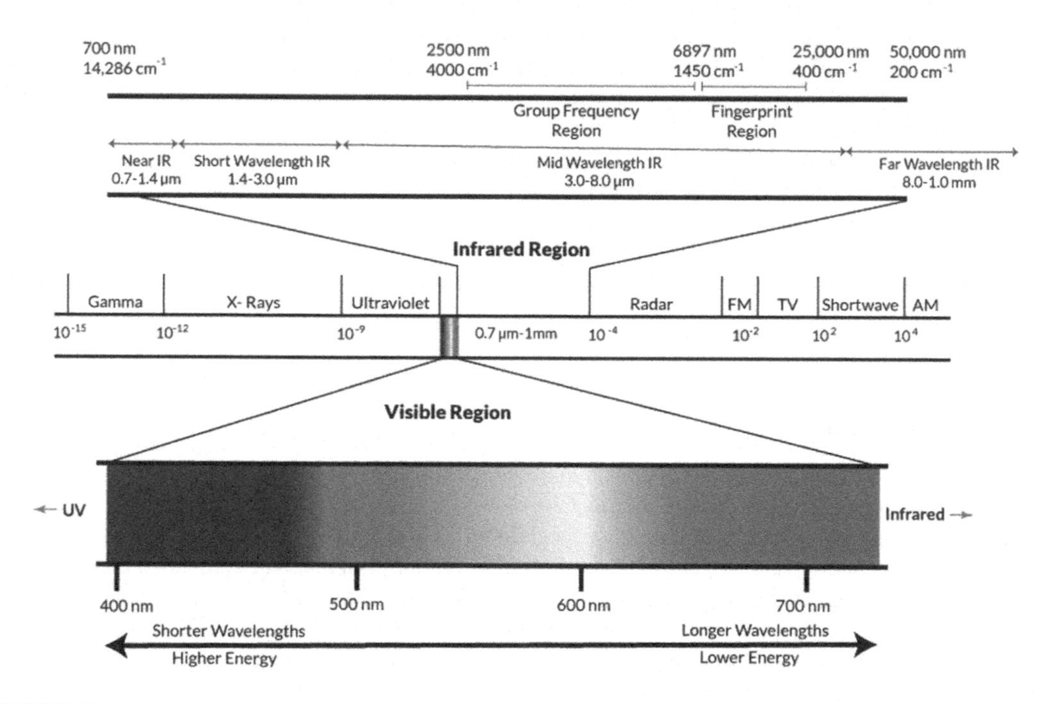

FIGURE 16.2 Electromagnetic spectrum showing the infrared region [5].

farmers can take the necessary actions to undo the damage caused by increased water stress levels in the crops. One such strategy is thermal imaging.

The approach behind thermal imaging is that every object emits an infrared radiation which is a function of its surface temperature. In lay man terms, the hotter the object, greater is the emission of infrared rays. Thermal cameras act as thermal sensors that collect rays of the infrared spectrum from objects and create images based on the difference in temperatures. Figure 16.1 depicts an image of a stressed tomato leaf captured by a thermal camera.

The rays of the infrared spectrum have the wavelength ranging between 700 nanometre (nm) and 1 millimetre (mm) in the electromagnetic spectrum [4]. This spectrum of radiation can be captured by thermal cameras which convert it into visible images. Figure 16.2 depicts the infrared spectrum.

The chronological order of the chapter is as follows: Section 16.2 presents the related work. Materials and methods are presented in Section 16.3. Experimental results and conclusions are explained in Sections 16.4 and 16.5, respectively.

16.2 Related Work

Thermal images have been used by researchers in the field of agriculture for stomatal conductance, stress detection, disease detection and many more [6]. Waseem Ahmed Bhat has proposed to differentiate between temperature variations in plants by using thermal imagery [7]. Camera that captures thermal images is used to focus on one leaf in the experimental setup. The soil is covered by a polythene material during capturing of images. Environment used for performing the experiment is dark and controlled and is supplied by Infrared light. It is observed that it is the central vein of the leaf along which the temperature varies. Results obtained show that the temperature captured at the centre of the leaf is greater than the temperature recorded on its edges.

Thermal imagery on soybean plants has been used by Alex Martynenko et al. as soybean plants are very sensitive to drought, which results in water stress [8,9]. Drought stress results in reduction of leaf size, stem extension and root proliferation, disturbs plant–water relations and reduces water-use efficiency. These factors result in an increase of leaf surface temperature. This results in reaching a threshold Crop Water Stress Index (CWSI) value which causes permanent leaf tissue damage. CWSI is mathematically represented as

$$\mathrm{CWSI} = \left(T_{\mathrm{canopy}} - T_{\mathrm{nws}}\right)/\left(T_{\mathrm{max}} - T_{\mathrm{nws}}\right) \tag{16.1}$$

Here, T_{max} is the temperature of a black body for capturing the maximum temperature of the environment; T_{nws} is temperature captured by watering the leaf to capture the minimum temperature of the surrounding environment; and T_{conopy} is the temperature of the leaf surface. The experiment uses a controlled environment to collect the data in the form of numerical values. The paper concludes that the CWSI calculated from the numerical data ($T_{max,}$ $T_{nws,}$ T_{conopy}) which is captured using thermal imaging can be used to indicate soil water deficit.

A.C. Hyperlink "https://www.sciencedirect.com/science/article/pii/S0048969717318612" W. Craparo et al. focus on the use of thermal imagery which is useful in the study of different plant phenomenon namely, abnormal stomatal closure, genotypic variations in stress tolerance and varied management strategies which impact crop water status [10]. They have used the CWSI index as an indicator for crop water stress. Thermal images of three random leaves are taken with sheet of aluminium foil placed behind the leaves to obtain constant environmental temperature.

Thus, various techniques have been proposed and implemented to determine water stress in plants. Factors like amount of sunlight plants are exposed to, water supply, climatic conditions and surrounding environmental areas help in determining the amount of stress in plants.

Finally, it can be concluded that stress detection in plants can be witnessed using thermal imagery which helps in capturing the difference in temperatures between the surface of plants and the surrounding environment, making it a non-invasive strategy [5,11]. Thus, coupled with deep learning, this chapter discusses the process to automate stress detection in tomato leaves, using the following proposed methodology mentioned in the next section of the chapter.

16.3 Proposed Methodology

The camera used for thermal imagery is Seek Thermal Compact Camera (Specifications: 206 × 156 Thermal Sensor, 36° Field of View, <9 Hz Frame Rate and −40°F to 626°F Detection). The experimental setup uses two tomato plants which are kept under normal environmental conditions as shown in Figure 16.3. The period of experimentation lasted for 3 months, from 1st November 2019 to 31st January 2020. To detect water stress, tomato plant labelled number 1 in the experiment was provided normal

FIGURE 16.3 Setup under normal environmental conditions.

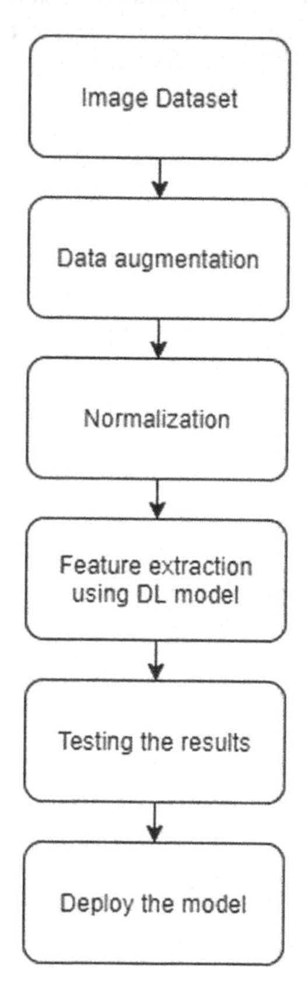

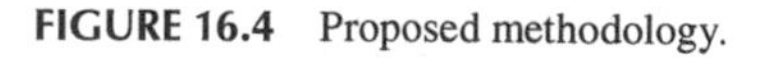

FIGURE 16.4 Proposed methodology.

amounts of water and sunlight for its growth, whereas plant labelled number 2 is only provided with sunlight; here the water acting as a controlling factor. Surface temperature readings of the leaves were taken twice daily, the first reading being taken between 9 am to 11 am and the second reading taken in between 4 pm to 5 pm along with thermal images of the leaves. As days passed, it was noticed that the temperature captured on the leaves surface of plant number 2 increased and indirectly changed the thermal images indicating the plant was under water stress. Figure 16.4 shows the flowchart of proposed methodology.

The steps of proposed methodology are as follows:

1. Collection of an image dataset which consists of stress and non-stress thermal images.
2. Perform data augmentation such as flip or rotate on the thermal images.
3. Perform normalization of images that is fed as input to the model.
4. Perform feature extraction using different deep learning models out of which choose the one which gives the best results.
5. Test the results on unseen data to check the performance of model.
6. Deploying the model for ease of use.

16.3.1 Data Set

The dataset of thermal images of tomato plants was collected over a period of 3 months with annotated dataset as stress and non-stress based on the CWSI calculated for each of the thermal image captured. It was empirically found that the cut-off for water stress in the tomato plant was 0.6 where the CWSI value ranges from 0 to 1. Table 16.1 shows few of the (CWSI) readings calculated using T_{canopy}, $T_{max,}$ and, T_{nws} respectively. Figure 16.5 depicts the thermal image dataset annotated as non-stressed and stressed. The dataset consists of a total of 3159 images, with 1603 as stress images and 1556 as non-stress images of plants. Table 16.2 breaks down the image dataset into training and testing data for analysis.

16.3.2 Data Augmentation

Data augmentation refers to randomly applying various kinds of transforms to the images. These transforms help to introduce varieties in the image dataset. Different techniques of data augmentation such as rotation of images at an angle, random cropping of images, brightness/contrast of images, etc. have been employed throughout the captured thermal image dataset. Data augmentation of captured thermal images is depicted in Figure 16.6.

16.3.3 Pre-Processing of Thermal Images

Pre-processing can be defined as the operations conducted on images at the lowest level of abstraction. Here both the input images and output images are intensity images. Pre-processing aims to transform the image intensities that suppress distortions which are obsolete or enhance certain important image features for further image processing. The following two approaches have been performed by us for

TABLE 16.1

Crop Water Stress Index to Classify Thermal Images as Stressed or Non-Stressed

T_{canopy}	T_{max}	T_{nws}	CWSI	Result
23	24	19	0.75	Stressed
24	25	21	0.75	Stressed
25	26	20	0.83	Stressed
26	31	25	0.17	Non-Stressed
.....	.....	.....	.....	.....

TABLE 16.2

Thermal Image Dataset

Target	Training	Testing	Total
Non-stressed	1244	312	1556
Stressed	1284	319	1603

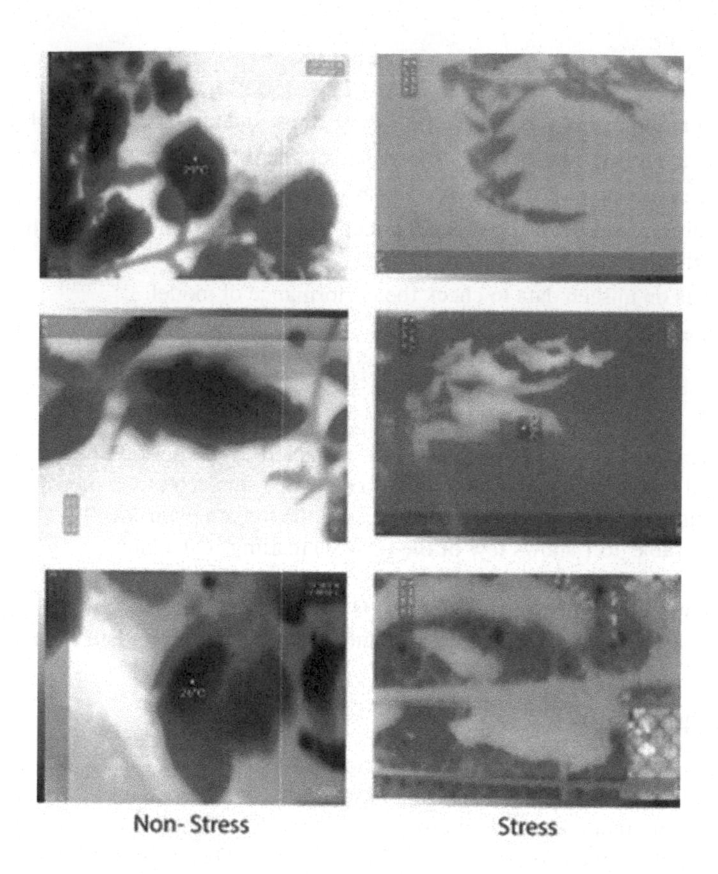

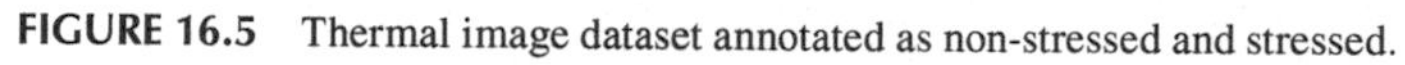

FIGURE 16.5 Thermal image dataset annotated as non-stressed and stressed.

executing the aforementioned step. In the first approach, basic pre-processing techniques are performed such as smoothening of images, removing noise before training the data, whereas in the next approach, add-on of image segmentation is performed by filtering out the background portion after which training of the image dataset is carried out. The results show that the second approach taken results in overfitting of data and is thus not suitable for implementation of the model. Figure 16.7 shows the image pre-processing steps for the second approach.

16.3.4 Feature Extraction

Feature extraction is helpful in describing the relevant shape information contained in a pattern so that the task of classifying the pattern is made easy by a formal procedure [12]. Feature extraction is a special form of dimensionality reduction in pattern recognition and image processing. Thus, feature extraction is a pivotal step in understanding the image. For feature extraction, ResNet-34 has been implemented as the pre-trained model since it generated the best results [13]. The dataset has been divided into training and testing with the ratio of 80:20. Next step involves the resizing of all the thermal images in the dataset to the size 224 by 224. These resized images are further passed along the different layers of the ResNet network layers. ResNet models consist of a combination of convolution layers along with batch normalization and dropout with ReLu as the activation function. Loss function has been used as cross entropy, and Adam optimizer has been selected [14]. Visual cues in an image can be detected using convolutional neural networks. Figure 16.8 shows the architecture and Figure 16.9 represents the output of the filters, highlighting the cues which the neural network determines to be significantly important.

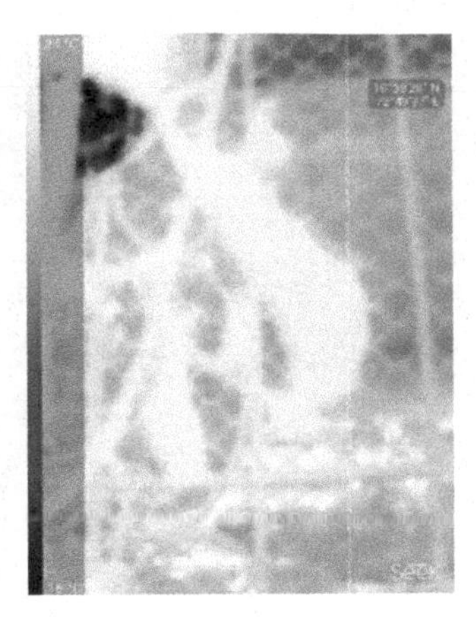

(a)

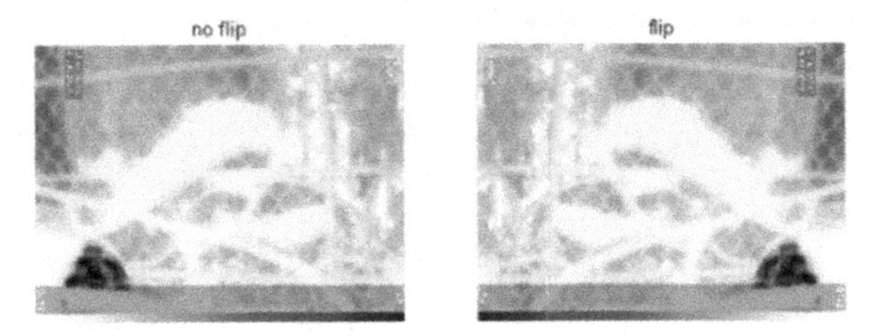

(b)

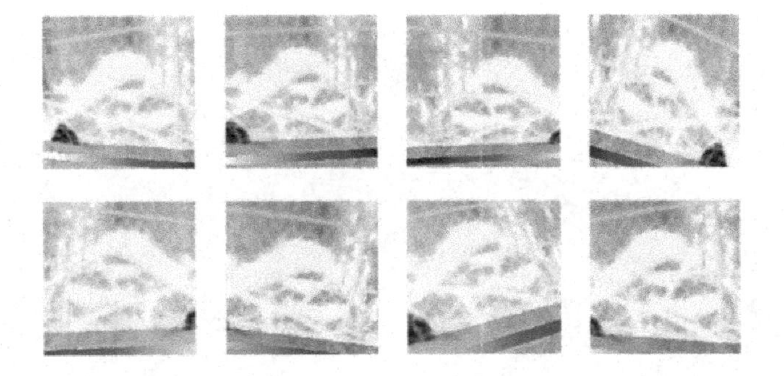

(c)

(d)

FIGURE 16.6 Augmented images. (a) Original image, (b) flipped image, (c) resized image, and (d) rotated images.

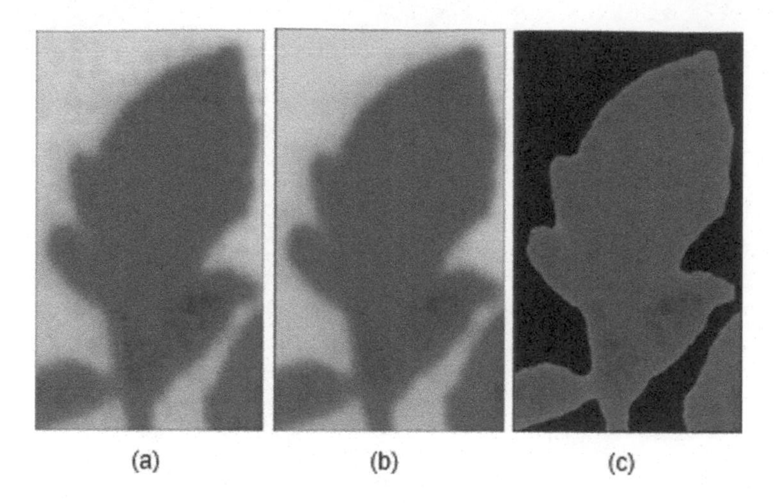

FIGURE 16.7 Image pre-processing. (a) Raw image, (b) noise removal, and (c) foreground extraction.

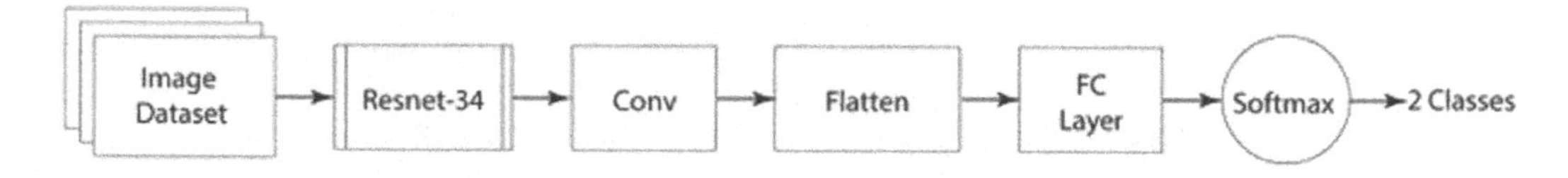

FIGURE 16.8 Architecture of proposed model layers.

FIGURE 16.9 Feature maps of ResNet-34 convolutional layers.

Researchers have observed that it makes sense to affirm that "the deeper the better", that is when it comes to convolutional neural networks, the deeper the network is the more capable it is, but at some particular stage, the problem of vanishing gradient was observed. This problem is solved using ResNets with the help of skip connections.

16.4 Results

The proposed methodology was tested on unseen images and on testing images obtained during the 80:20 ratio of the original dataset. The results are evaluated in the form of accuracy metrics namely precision, recall, accuracy and F1 score, respectively.

Precision is calculated as:

$$\text{TP} / \text{TP} + \text{FP} \tag{16.2}$$

where TP is the number of true positives and FP is the number of false positives.

Recall is calculated as:

$$\text{TP} / P \tag{16.3}$$

where P is the number of positive samples.

Accuracy is calculated as:

$$\text{TP} + \text{TN} / P + N \tag{16.4}$$

where TN stands for the number of true negatives and N stands for the number of negative samples. It is usually used to measure the overall performance of the model.

F1 score is calculated as:

$$2 \times (\text{Precision} \times \text{Recall} / \text{Precision} + \text{Recall}) \tag{16.5}$$

It is defined as the harmonic mean of precision and recall. Table 16.3 depicts the comparison between different models based on the accuracy metrics of the pre-processed images.

Figure 16.10 shows the confusion matrix of the above deep learning models used, that is VGG16 and ResNet-34.

Hence, from the accuracy metrics and the confusion matrixes depicted, it can be concluded that ResNet-34 is the best model to train the dataset. Stratified K-Fold was also performed using cross validation which gave an accuracy of about 90% for the given dataset.

Figure 16.11 depicts the UI of the application in which the user will upload a thermal image of the plant and get the results on whether it is stressed or not using the ResNet-34 model at the backend of the application.

TABLE 16.3

Comparative Analysis of Different Models

Model	Accuracy	Precision	Recall	F1 score
VGG16 [15]	0.893819	0.904025	0.89024	0.892966
VGG19 [15]	0.9223	0.9324	0.9225	0.9242
ResNet-34	0.932069	0.936508	0.924765	0.927090
ResNet-34 (foreground extracted dataset)	0.997	0.998	0.999	0.99
Proposed model	**0.9413**	**0.9462**	**0.9373**	**0.9417**

The local scope of our research will be confined to our experimental setup environment. The main scope of our project is early detection of stress in tomato plants. It also provides a user-friendly interface for users which would be simple to handle and easy to use, and we can also take the thermal images for analysis to predict the stressed and non-stressed plants. Lastly, we provide farmers or gardeners real-time assistance in farming or gardening of tomato plants.

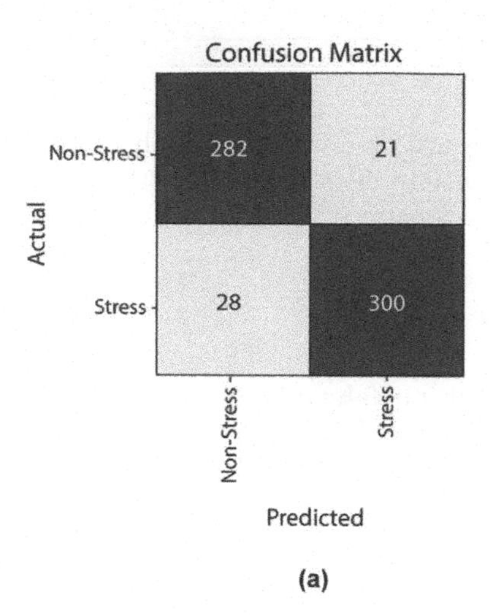

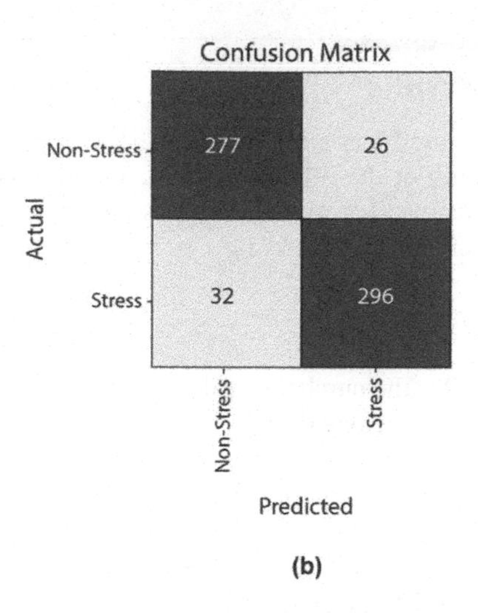

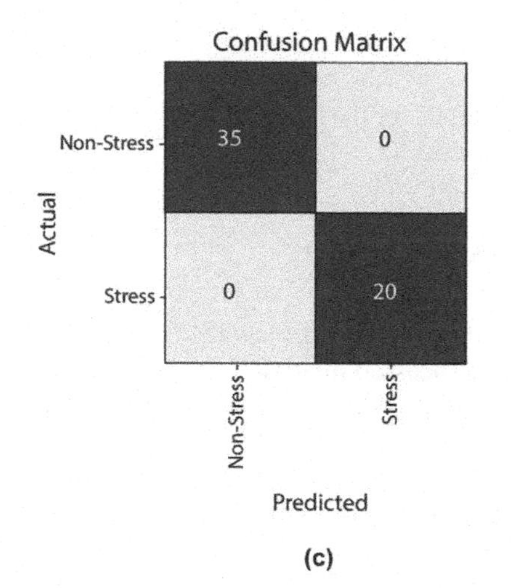

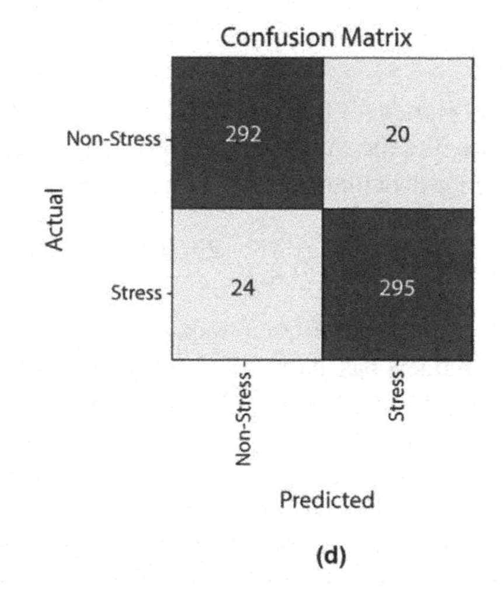

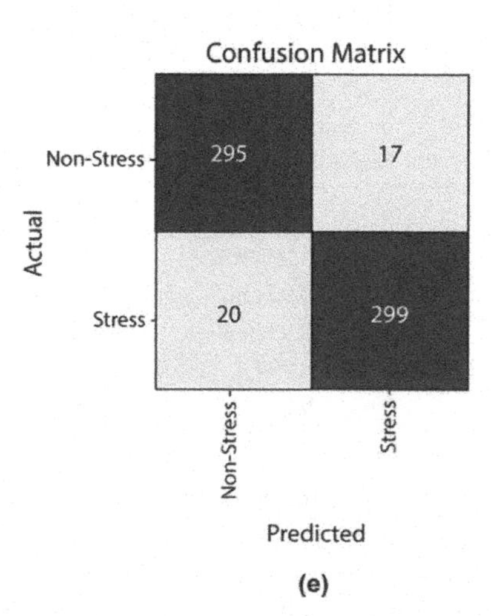

FIGURE 16.10 Results. (a) Vgg-16 confusion matrix, (b) Vgg-19 confusion matrix, (c) ResNet-34 foreground extracted dataset, and (d) Proposed model confusion matrix.

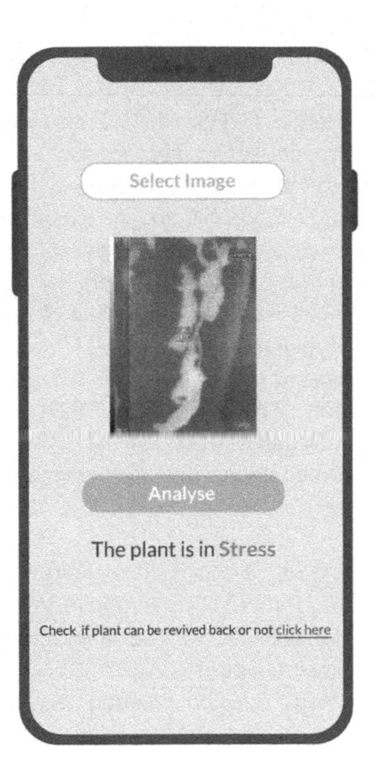

FIGURE 16.11 Application user interface.

16.5 Conclusion

This chapter presented that the deep learning model ResNet-34 gives the most accurate results and does not overfit the data. It was also inferred that raw thermal images should be used for more real-world specific accurate results rather than using images which have the background filtered out, since passing these images through the model caused overfitting of the image data. Accuracy of the results obtained can further be improved with the ever-evolving technological advances. Finally, the experimental analysis conducted and research methodologies implemented by us is a step forward in automating the process of water-stress detection in plants and indirectly in the field of agriculture, thus aiding farmers to increase the productivity of the crops.

To further the work presented in this chapter, an increase in the thermal image dataset should be done as it is one of the limiting factors faced while performing this experimental research.

REFERENCES

1. Madhusudhan, L. "Agriculture role on Indian economy." *Business and Economics Journal* 6, no. 04 (2015). doi: 10.4172/2151-6219.1000176.
2. Brawner, M. Transpiration and why it matters. https://harlequinsgardens.com/transpiration-and-how-it-matters (accessed January 11, 2020).
3. Dubey, K. Climatic and temperature requirements of tomato. January 19 (2012). http://agropedia.iitk.ac.in/content/climatic-and-temperature-requirement-tomato (accessed January 13, 2020),
4. TechTarget. What is Infrared Radiation (IR)? Definition from WhatIs.com. Search networking. May 02 (2017). https://searchnetworking.techtarget.com/definition/infrared-radiation (accessed December 05, 2020).
5. JAGER PRO Store. What is infrared? JAGER PRO Store (2020). https://jagerpro.com/help-center/what-is-infrared/ (accessed December 5, 2020).
6. Ishimwe, R., K. Abutaleb, and F. Ahmed. "Applications of thermal imaging in agriculture: A review." *Advances in Remote Sensing* 03, no. 03 (2014): 128–140. doi: 10.4236/ars.2014.33011.

7. Study of transient response of plant leaf using infrared. https://be.iisc.ac.in/BESTReports/WaseemBhat.pdf (accessed December 5, 2020).
8. Martynenko, A., K. Shotton, T. Astatkie, G. Petrash, C. Fowler, W. Neily, and A.T. Critchley. "Thermal imaging of soybean response to drought stress: The effect of ascophyllum nodosum seaweed extract." *SpringerPlus* 5, no. 1 (2016). doi: 10.1186/s40064-016-3019-2.
9. Khan, S., M. Narvekar, M. Hasan, A. Charolia, and A. Khan. Image processing based application of thermal imaging for monitoring stress detection in tomato plants. *2019 International Conference on Smart Systems and Inventive Technology (ICSSIT)*, (2019). doi : 10.1109/icssit46314.2019.8987900.
10. Craparo, A.C.W., K. Steppe, P.J.A. Van Asten, P. Läderach, L.T.P. Jassogne, and S.W. Grab. "Application of thermography for monitoring Stomatal conductance of Coffea Arabica under different shading systems". *Science of the Total Environment* 609 (2017), 755–763. doi: 10.1016/j.scitotenv.2017.07.158.
11. Chandler, N. How thermal imaging works. 21 May (2013). https://electronics.howstuffworks.com/thermal-imaging1.htm (accessed 9 October 2019).
12. Kumar, G., and P.K. Bhatia. "A detailed review of feature extraction in image processing systems." *2014 Fourth International Conference on Advanced Computing & Communication Technologies* (2014). doi: 10.1109/acct.2014.74.
13. He, K., X. Zhang, S. Ren, and J. Sun. "Deep residual learning for image recognition." *2016 IEEE Conference on Computer Vision and Pattern Recognition (CVPR)* (2016). doi: 10.1109/cvpr.2016.90.
14. Costa, J.M., O.M. Grant, and M. Manuela Chaves. Thermography to explore plant–environment interactions. *Journal of Experimental Botany* 64, no. 13 (2013), 3937–3949. doi: 10.1093/jxb/ert029.
15. Simonyan, K., and A. Zisserman. Very deep convolution networks for large-scale image recognition. arXiv:1409.1556.

17

Machine Learning Techniques to Classify Breast Cancer

Drashti Shah and Ramchandra S. Mangrulkar
Dwarkadas J. Sanghvi College of Engineering

CONTENTS

17.1 Introduction 245
17.2 Literature Survey 246
17.3 Proposed Methodology 247
17.3.1 Dataset 247
17.3.2 Machine Learning Algorithms 248
17.3.2.1 Logistic Regression 248
17.3.2.2 Random Forest 249
17.3.2.3 Decision Trees 250
17.4 Implementation 250
17.4.1 Confusion Matrix 251
17.4.2 Performance Metrics 251
17.4.3 Comparison among Classification Algorithms 252
17.5 Results and Discussions 252
17.6 Conclusion 255
References 255

17.1 Introduction

Cancer is one of the major leading causes of death globally, and according to the World Health Organization (WHO), in the year 2012, 8.2 million deaths globally were caused by cancer, and it is further assumed that the death ratio is likely to rise up to 27 million by the end of 2030 [1]. One of the most common types of cancer for women is breast cancer, which accounts for 17% of the deaths related to cancer in the United States. Hence, accurate detection and assessment of this cancer in its rudimentary stages is critical when it comes to reducing its mortality rate [2]. There are numerous risk factors such as sex, aging, estrogen, family history, gene mutations, and unhealthy lifestyle, which can increase the possibility of developing breast cancer. Most breast cancer occurs in women, and the number of cases is 100 times higher in women than that in men [3].

The onset of breast cancer takes place by an uncontrolled division of the cells from breast tissue, which results in the formation of a mass known as a tumor. This tumor can either be malignant or benign. Benign tumors are not a cause of worry, but malignant tumors can be fatal. Hence, to classify these tumors is of utmost importance to take the necessary precautions. Thus, the need for a system to accurately predict and classify these tumors is a cause of great concern for doctors. Machine learning, a subset of artificial intelligence, comes to the rescue to solve this dilemma. Various machine learning algorithms and different tools used for data mining and extraction have found their application in medical diagnosis since they perform classification accurately and effectively, given that classification tasks play a very crucial role in breast cancer detection.

There are many algorithms to predict and classify the outcomes of breast cancer diagnosis efficiently and accurately. The system proposed below makes use of three different algorithms of machine learning, namely, logistic regression, random forest, and decision trees, to classify the given cells as "malignant" or "benign" based on the dataset specifying its various properties such as radius, texture, smoothness, perimeter, etc. The dataset used contains 569 records of which 357 are benign breast cancer cases and the rest are those of malignant cancer. This dataset has been split such that 25% of the whole is our training set while the remaining is the test data. The models enable the system to classify based on the processed data. A detailed comparison between these three machine learning models and their results has been achieved.

This chapter is set out as follows:

In Section 17.1, related work and a literature review of papers encompassing similar topics are presented. In Section 17.2, all the relevant information about the methodology is presented, and an application of the three algorithms is explained, which would help in the diagnosis of breast cancer, followed by Section 17.3, which gives a detailed analysis of the results obtained along with a few graphical representations. This is followed by the conclusion of this topic.

17.2 Literature Survey

Breast cancer is an ubiquitous issue faced by people and doctors. Many systems and models have been prepared to classify these breast cells using various methods, the most common one being analyzing either a given dataset containing necessary information or analyzing mammographic images. Machine learning algorithms of support vector machines, K-Nearest Neighbors, and Naïve Bayes along with Artificial Neural Networks have been used the most for research purposes. Ilias Maglogiannis et al. presented a paper in which the use of an SVM-based classifier is proposed in comparison with Naive Bayes and artificial neural network for the diagnosis of breast cancer. The SVM algorithm used gave an efficiently high value of accuracy of up to 96.91%, specificity of 97.67%, and sensitivity of up to 97.84% [4].

Tolga Ensari et al. proposed the use of SVMs (Sequential Minimal Optimization and LibSVM) and artificial neural networks (Multi-Layer Perceptron and Voted Perceptron) to classify the breast cancer cells as either benign or malignant using WEKA tools. The accuracy of SVM was in the range of 95%–96%, and its precision was between 95% and 97%. Its artificial neural network model had its accuracy ranging from 88% to 95%, whereas precision varied between 89% and 91%. Based on its performance metrics, the sequential minimal optimization algorithm of SVM showed the best performance [5].

Lin et al. propose the research aimed at introducing a novel F1 score index by the construction of a Naïve Bayes classifier based on the F-score values. This classification model needs to have a low false-negative rate as well as low cost for which NBCOF methodology had the best evaluation results. However, the author contends that this method does not take the medical aspects of each feature into consideration, which dampens the practicality of the model [6].

Shubham Sharma et al. presented a paper that used the algorithms of K-nearest neighbors, random forest, and Naive Bayes to diagnose and classify breast cancer. K-fold cross-validation is used in which the data used is divided into K bits of equal size. In the testing phase, KNN gave an accuracy of 95.9% and a precision of 98.27%, the highest of all. K-fold method of cross-validation was used to examine the classifier. 10-fold technique was utilized out of which nine (398 observations) were used for the purpose of training while the last set (remaining 171 observations of 569) was used for testing and analysis purposes [7].

Anusha Bharat et al. presented a paper that employed the algorithms of SVM, decision tree, K-nearest neighbors, and Naive Bayes to classify breast cancer as benign or malignant. According to their algorithmic implementation, all of them except SVM gave a mean accuracy of above 92% whereas SVM performed surprisingly bad. Only after standardizing the dataset did its accuracy drastically improve [8]. Mandeep Rana et al. have also implemented the above algorithms in addition to logistic regression but decision trees in MATLAB® to study the accuracy of each for predictive analysis [9]. Naresh Khuriwal et al. have proposed an ensemble machine learning model which compares artificial neural networks and logistic regression to optimally diagnose and detect breast cancer [10]. Similarly, Naveen and Ramachandran Nair have comparatively studied decision trees, random forests, SVM, logistic regression, and K-nearest neighbors to predict the breast cancer result. Among all these ensemble methods, decision tress and K-NN algorithm surpassed all in precision, recall, and F1 score [11]. On carefully reviewing all these papers, the utmost importance of implementing machine learning and neural networks in early prediction of breast cancer is highlighted.

17.3 Proposed Methodology

The flowchart presented below in Figure 17.1 gives an idea of the methodology we have proposed via this chapter. The dataset used is cleaned and transformed before splitting it into test data and train data. Three different machine learning algorithms, namely, logistic regression, random forest, and decision trees are applied, and the accuracy and other parameters are subsequently calculated to see which algorithm gives the best output performance.

17.3.1 Dataset

The dataset used here is Wisconsin Diagnostic Breast Cancer (WDBC) dataset, which is obtained from the "UCI ML" repository. This dataset consists of a total of 569 breast cancer instances, 357 of which

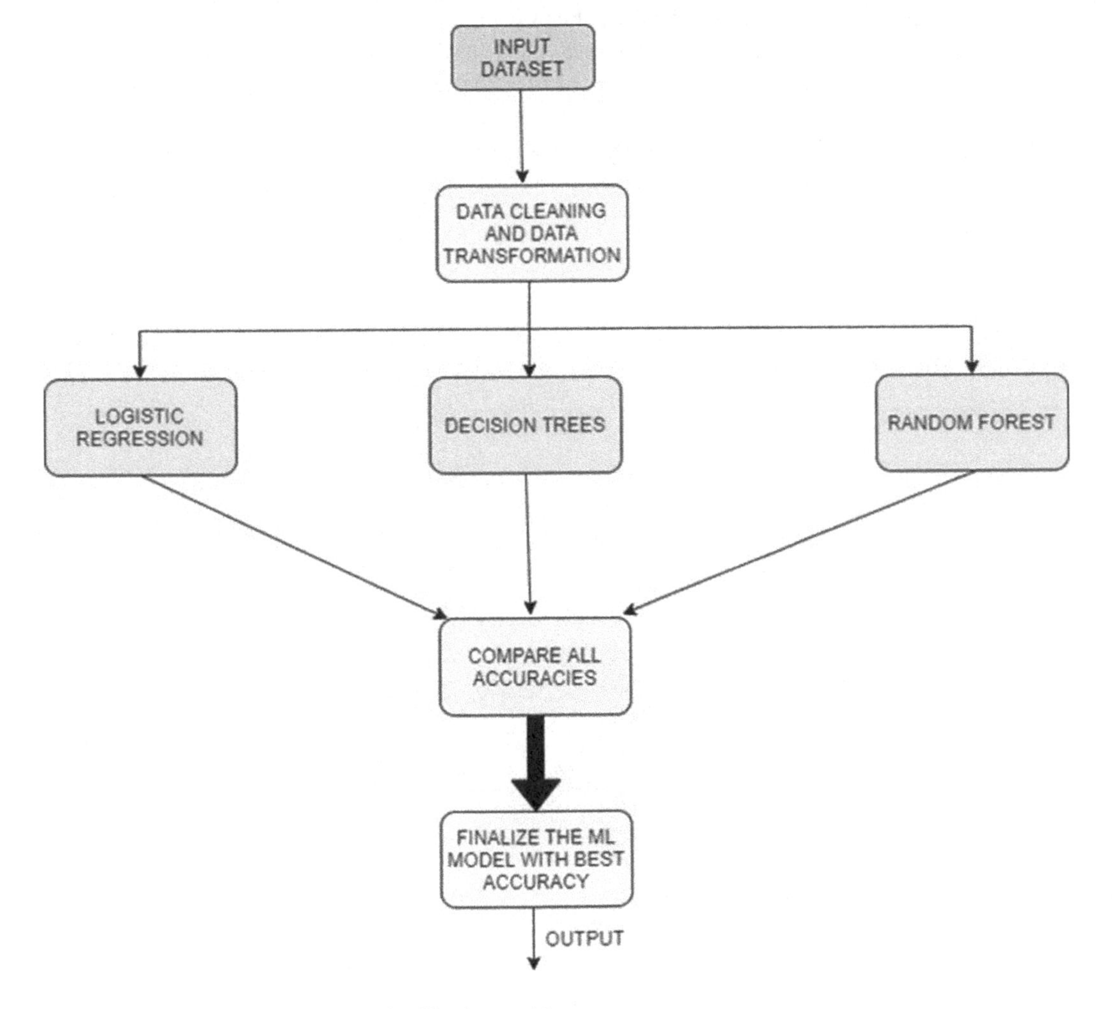

FIGURE 17.1 Proposed breast cancer classification model.

are benign and the remaining 212 are malignant. The variables that will enable the algorithm to classify these instances include radius_mean, area_mean, compactness_mean, and more, which help in accurately predicting the results as they explain in-depth the structure of the cells that are under the analysis.

This dataset is preprocessed and cleaned using various preprocessing techniques. Handling empty values is at the forefront of our data cleaning. An entire row with plethoric empty values had to be discarded as it did not add much value for the research purpose. Moreover, the categorical data needs to be encoded, and this transformed and clean dataset free of anomalies would be fed to the next phase of machine learning algorithms, which will be employed to classify the breast cancer cells into either benign or malignant classes.

17.3.2 Machine Learning Algorithms

Machine learning is an area of Computer Science and comes under the vast domain of artificial intelligence. It involves teaching the computer to turn data into information, i.e., the computer can learn from the data and make relevant decisions with least human intervention by making certain predictions. Machine learning is classified under the following three categories [12].

1) **Supervised Learning**: It generates a function based on input observations that predicts outputs. The training data is used to produce the function, and it guides the system to produce useful epiphanies for the implementation of new data sets.
2) **Unsupervised Learning**: In this technique, unlabeled dataset is deployed from which the machine is forced to train and then distinguished on the basis of certain characters, and sequentially, without any use of external guidance, the algorithm is acted upon it.
3) **Reinforcement Learning**: The learning process proceeds in iterative fashion from the setting. The machine gradually discovers all possible system states over an extended period of time (Figure 17.2).

17.3.2.1 Logistic Regression

One of the most extensively used algorithms in the field of machine learning is sigmoid function, also called logistic regression. It is a supervised classification algorithm. This algorithm is used for

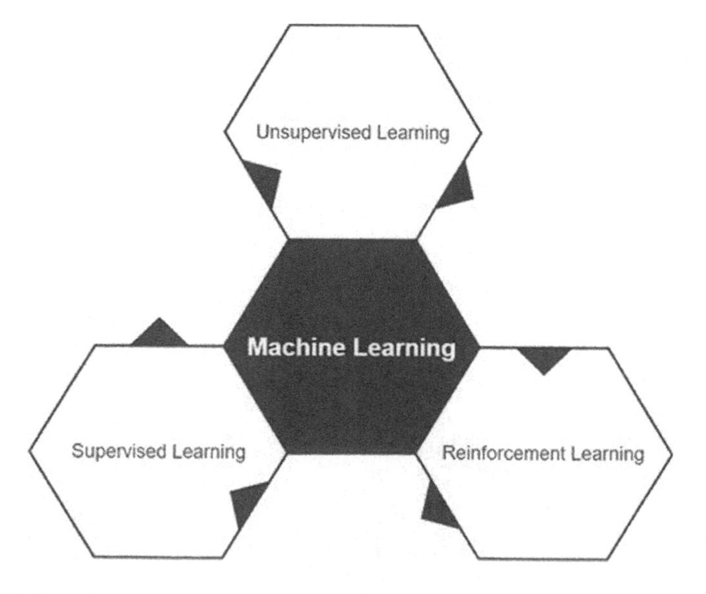

FIGURE 17.2 Machine learning types.

segregating the object to the class of instance where it might belong. While fitting the logistic regression model efficiently, the goodness of fit-statistics and its overall evaluation are few of the most important factors needed to be considered. Its values range from 0 to 1. The sigmoid function can be given as

$$\varnothing(z)=\frac{1}{1+e^{-z}} \tag{17.1}$$

We will find the b (or popularly β) parameters that reach the global maximum of the log likelihood function using maximum likelihood estimation [8].

Input value of x is combined linearly using either coefficient values or weights to predict an output value of y.

$$\ln\left(\frac{1}{1-p}\right)=b_0+b_1x_1+b_2x_2+\cdots \tag{17.2}$$

17.3.2.2 Random Forest

Random forest is a supervised learning algorithm that finds numerous applications in machine learning. It can be used for both regression as well as classification. Its algorithm working is as shown in the diagram given below in Figure 17.3.

To provide a more precise and stable prediction, it builds various trees and merges all of them together. This algorithm is used for segregating the object to the class of instance where it might belong. Instead of relying on just one tree, this algorithm takes a prediction from each tree and predicts the final output based on the majority.

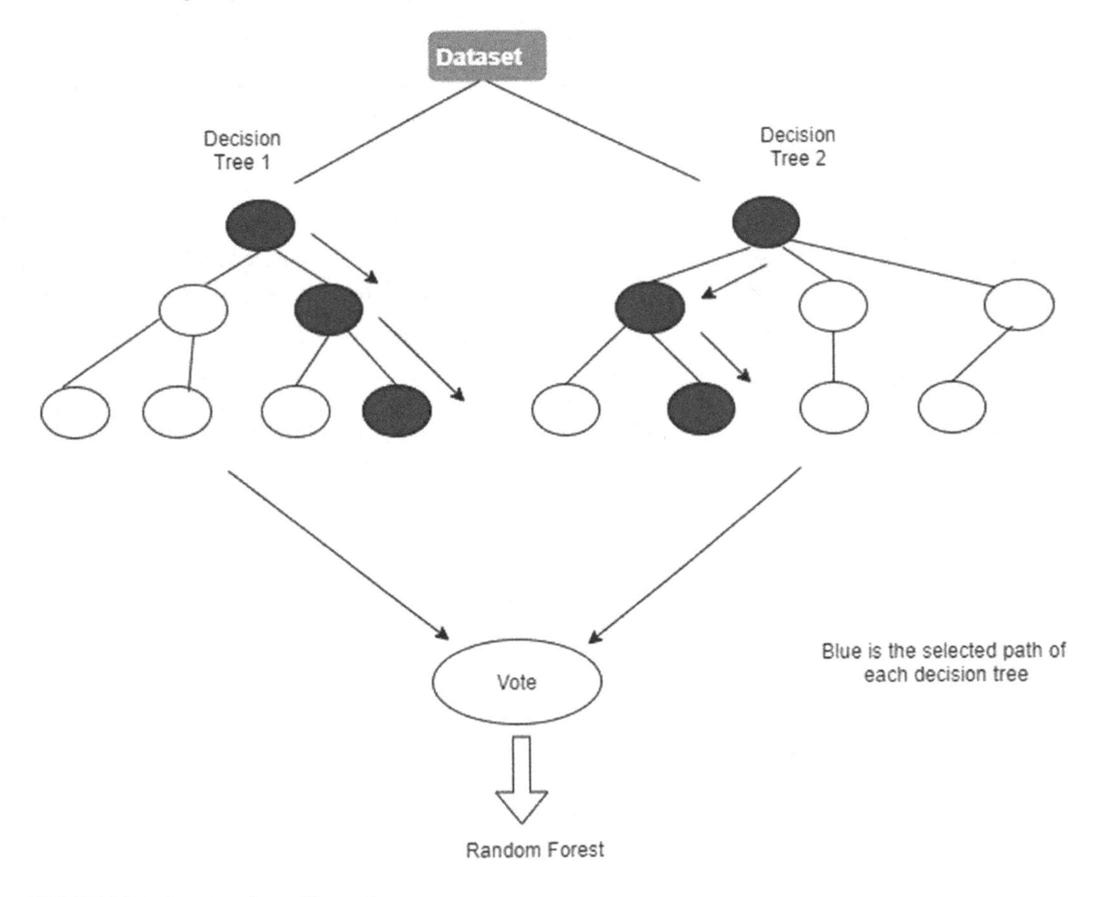

FIGURE 17.3 Random forest illustration.

This model provides additional randomness to a machine learning model. It searches for the best feature among a random subset of features instead of searching for the most important feature while splitting nodes, hence resulting in a wide diversity that leads to the generation of a better model [13].

17.3.2.3 Decision Trees

This algorithm again comes under supervised learning algorithm and can be known as building blocks of a random forest model. The structure of the decision trees is such that it consists of roots, internal nodes, and leaf nodes, which is used in classifying unknown data records [14]. It is a variation of the top-down greedy algorithm. It is a tree-structured classifier in which the leaf nodes contain the output category, and we can predict the class of the output based on the rules generated from this tree structure as is shown in Figure 17.4 [15].

We use the CART algorithm in order to build this tree, where CART stands for classification and regression tree algorithm. It simply asks a question whose answer can be either yes or no, and accordingly, the structure is further split. Gini index, information gain, and entropy are some of the factors which are used to determine the root node and thus the structure of the final decision tree.

$$\text{Gini} = 1 - \sum_{i=1}^{n} (p_i)^2 \tag{17.3}$$

Multiplying the above formula with the probability of every instance of a tuple gives Gini index. A feature that has the least Gini index is chosen for the split and becomes the root node. This procedure is carried out again after the remaining tuples are classified based on this new root node.

$$E(S) = \sum_{i=1}^{c} - p_i \log_2 p_i \tag{17.4}$$

Impurity and randomness of a dataset can be measured by a parameter known as entropy. A sample with an entropy of zero is completely homogeneous whereas the sample with an entropy of one is equally divided.

$$\text{Gain}(A, B) = \text{Entropy}(A) - \text{Entropy}(A, B) \tag{17.5}$$

The attribute that returns the highest value of information gain is selected as the root node of the decision tree. The weighted entropy of each branch is subtracted from the original entropy to calculate the gain.

17.4 Implementation

From this study, we have aimed to compare different performance metrics of three widely used algorithms in machine learning to classify the breast cancer types from the data available in the Wisconsin dataset.

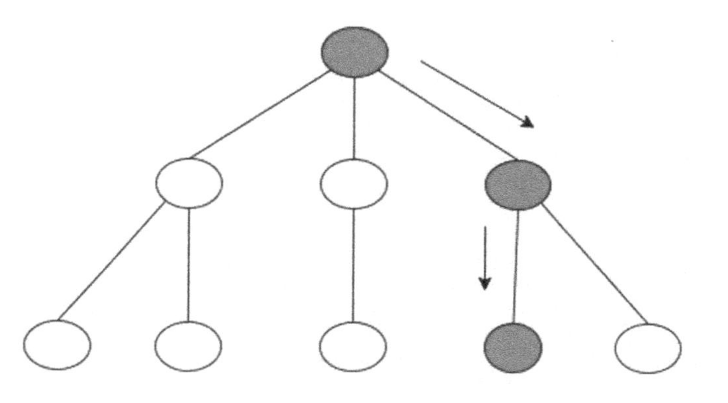

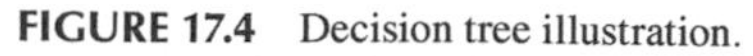

FIGURE 17.4 Decision tree illustration.

17.4.1 Confusion Matrix

As the name suggests, it is a matrix that accurately describes the performance of a classifier on the test data. It helps us to summarize the correct and incorrect predictions with the values broken down by each class (Figure 17.5).

Here,

True Positive (TP): Positive Observation, Positive Prediction
True Negative (TN): Negative Observation, Negative Prediction
False Positive (FP): Negative Observation, Positive Prediction
False Negative (FN): Positive Observation, Negative Prediction

17.4.2 Performance Metrics

a. **Accuracy**

Accuracy is one of the best predictors that gives us the degree of correctness and efficiency in the trained model. It is defined as the ratio of the total number of correct predictions to the total predictions made. Thus, the equation can be presented as follows:

$$\text{Accuracy} = \frac{\text{Total no. of correct predictions}}{\text{Total predictions made}} \tag{17.6}$$

$$\text{Accuracy} = \frac{TN + TP}{TP + FP + TN + FN} \tag{17.7}$$

b. **Precision**

Precision gives us the number of total positive predictions that truly belong in the positive class. It is given as the ratio of the true positives and the class of positives, i.e., true as well as false positives. However, it does not give valuable insight or information about the negative class.

$$\text{Precision} = \frac{TP}{TP + FP} \tag{17.8}$$

c. **Recall**

Recall, sometimes known as sensitivity, is the ratio of all the correct positive predictions to all the positive examples. Hence it gives us an estimation of all the relevant results returned in comparison to the pool of the relevant reasons that could have been returned.

$$\text{Recall} = \frac{TP}{TP + FN} \tag{17.9}$$

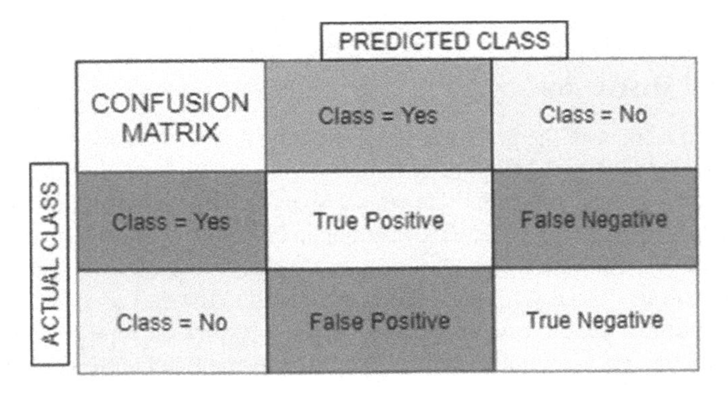

FIGURE 17.5 Confusion matrix.

d. **F1 Score**

Precision and recall when combined give an accurate performance evaluation of the model. The weighted average of precision and recall is known as the F1 score. Hence, this score takes both false positives and false negatives into consideration.

$$\text{F1 Score} = \frac{2*(\text{Recall} + \text{Precision})}{\text{Recall} + \text{Precision}} \tag{17.10}$$

Confusion matrix by applying the following algorithms to classify breast cancer is given as: (0 represents instances of benign cancer whereas 1 represents instances of malignant cancer).

a. Logistic Regression

	0	1
0	86	4
1	3	50

b. Random Forest

	0	1
0	87	3
1	2	51

c. Decision Trees

	0	1
0	83	7
1	2	51

The confusion matrix for each of the algorithms applied in this model gives an accurate number of the total positive class and the total negative class, thereby giving the accuracy, precision, and recall of each algorithm.

17.4.3 Comparison among Classification Algorithms

Parameters	**Logistic Regression**	**Decision Tree**	**Random Forest**
Library used	Scikit learn (python)	Scikit learn (python)	Scikit learn (python)
Accuracy	95.10	93.7	96.5
Precision	95.55	92.22	96.66
Recall	96.62	97.64	97.75

17.5 Results and Discussions

The following section deals with the graphical representation and visualization involved in the analysis of the results of this proposed model (Figure 17.6).

A heat map is a data-visualization tool that uses a specific color spectrum to show the correlation between different types of attributes. The above heatmap in Figure 17.7 gives a visual correlation between all the attributes of the Wisconsin Dataset. The individual values are present in a matrix and represented as colors in a definite range. Usually, higher values correspond to the darker shades in the chart.

This given graph summarizes the core of the entire dataset. It gives the total count of benign as well as malignant tumors (0 = benign, 1 = malignant). The *X*-axis in the graph in Figure 17.8 represents the cancer

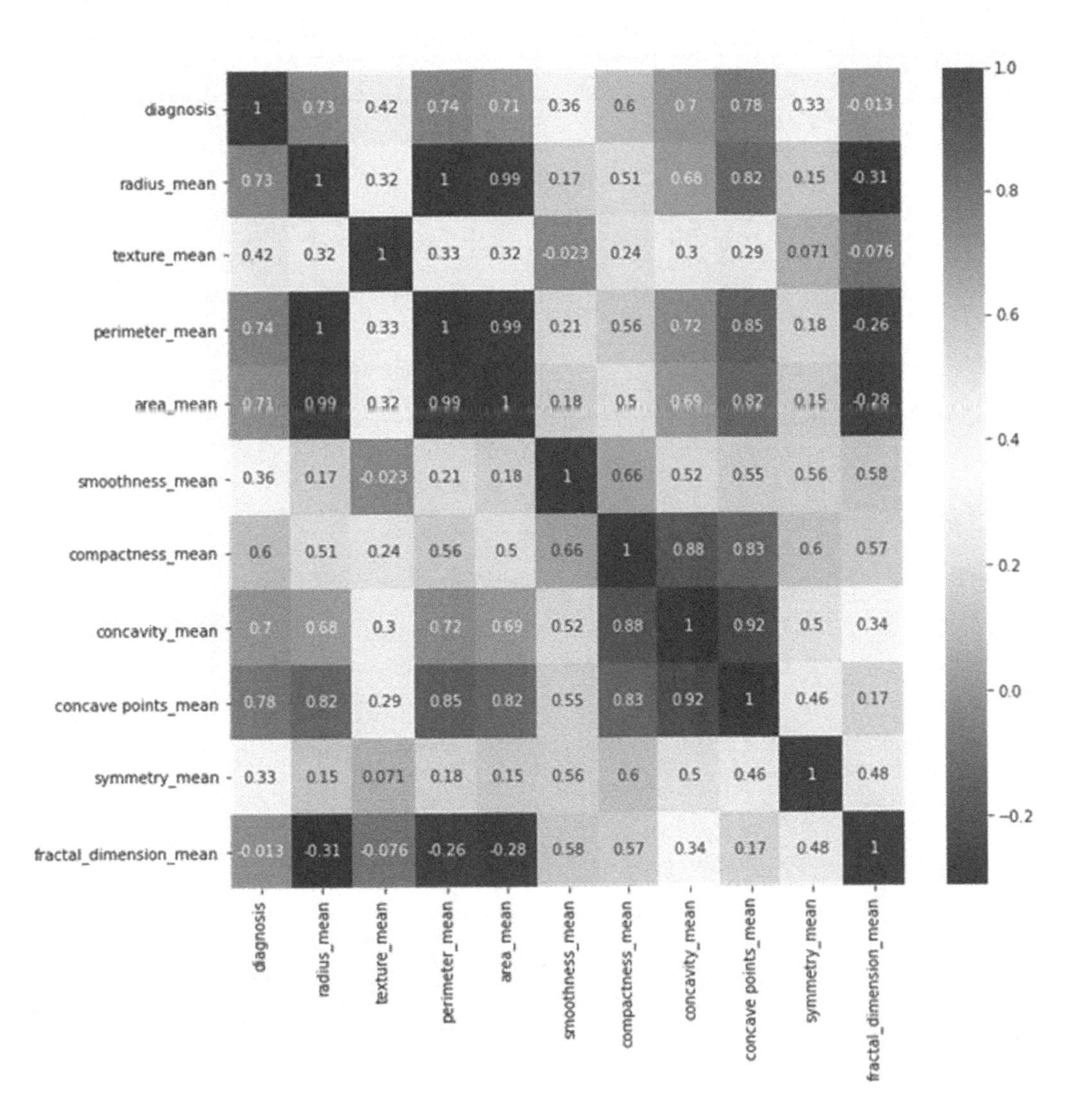

FIGURE 17.6 Heat map for the given dataset.

classification as its numerical value, and the *Y*-axis gives the total count for each. The aim of the applied machine learning algorithms was to classify these tumors as accurately as possible.

Figure 17.6 shows the correlation between four different attributes that help in classifying breast cancer, namely, texture_mean, radius_mean, area_mean, and perimeter_mean using scatter plots and graphs. In Figure 17.9, the light shade represent the numerical value of 0, and similarly, 1 represents the cases of malignant tumor cells.

To accurately depict this model's performance, in Figure 17.9, *X*-axis has been selected to represent the accuracy (*A*), Precision (*P*), and Recall (*R*) of each of the three models. The above graph sums up the results obtained by the implementation of the algorithms in this model. Here, LR stands for logistic regression, DT stands for decision tree, and RF stands for random forest. All the metrics give an above average performance, but it can be clearly seen from the above chart that the performance metrics for decision tree is highest in terms of its accuracy, precision as well as recall.

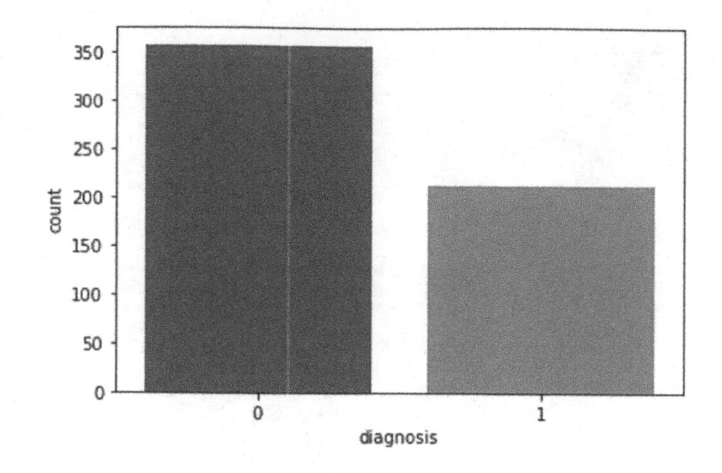

FIGURE 17.7 Count of benign/malignant tumors.

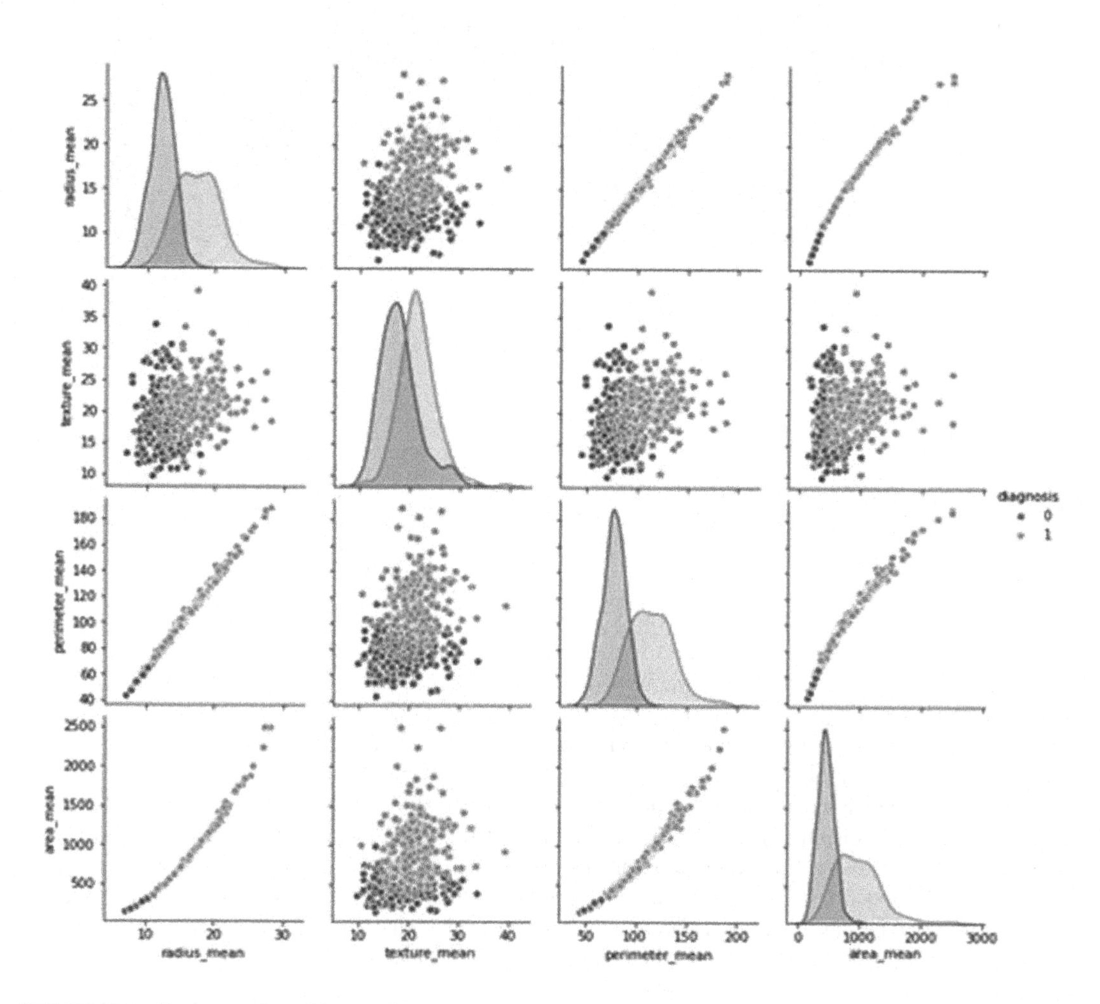

FIGURE 17.8 Representation of four attributes.

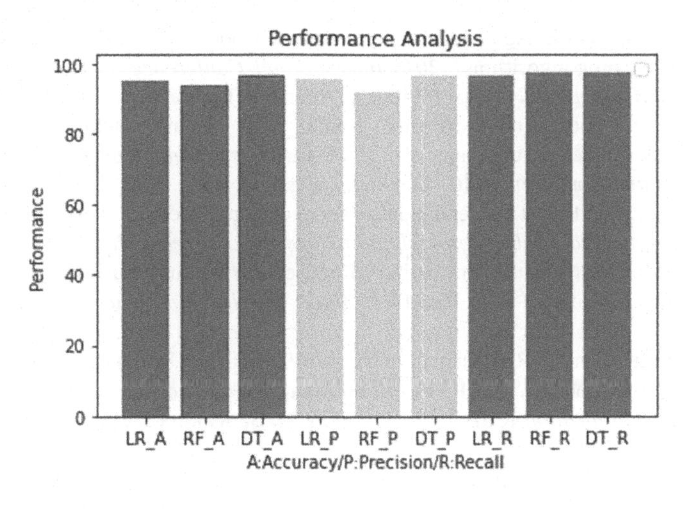

FIGURE 17.9 Comparing the performance of all the algorithms.

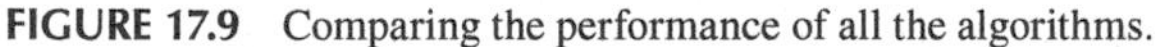

17.6 Conclusion

This chapter explores the working of three different machine learning algorithms to detect which one of them would give the best performance in one of the most crucial domains of health and medicine: breast cancer. The Wisconsin dataset has been utilized on which these algorithms are applied, and each of them have a different performance level based on the data trained and parameters selected. In this way, machine learning can be of utmost importance in healthcare, and when applied intuitively, it can give astounding results. According to the proposed model, decision tree has given the best results, but by further training the model or applying different algorithms, we might get a better performing model that classifies the cancer cells more accurately.

This methodology can be further hypothesized so that the model can be extended to reach higher levels of accuracy. The proposed model should help in accurately classifying the breast cancer cells in the upcoming years. The methodology can even be extrapolated to be implemented in the classification of other types of cancers through collaborative research work. The prospects of applying these methods for different medical classification decisions look promising and can be further probed and studied into.

REFERENCES

1. Hamed, Ghada, Mohammed Abd El-Rahman Marey, Safaa El-Sayed Amin, and Mohamed Fahmy Tolba. 2020. "Deep learning in breast cancer detection and classification." *In Proceedings of the International Conference on Artificial Intelligence and Computer Vision (AICV2020)*, Cham, pp. 322–333.
2. Tsochatzidis, Lazaros, Lena Costaridou, and Ioannis Pratikakis. 2019. "Deep learning for breast cancer diagnosis from mammograms: A comparative study." *Journal of Imaging* 5(3): 37.
3. Sun, Yi-Sheng, Zhao Zhao, Zhang-Nv Yang, Fang Xu, Hang-Jing Lu, Zhi-Yong Zhu, Wen Shi, Jianmin Jiang, Ping-Ping Yao, and Han-Ping Zhu. 2017. "Risk factors and preventions of breast cancer." *International Journal of Biological Sciences* 13(11): 1387–1397.
4. Maglogiannis, Ilias, Elias Zafiropoulos, and Ioannis Anagnostopoulos. 2007. "An intelligent system for automated breast cancer diagnosis and prognosis using SVM based classifiers." *Applied Intelligence* 30(1): 24–36.
5. Bayrak, Ebru Aydindag, Pinar Kirci, and Tolga Ensari. 2019. "Comparison of machine learning methods for breast cancer diagnosis." *2019 Scientific Meeting on Electrical-Electronics and Biomedical Engineering and Computer Science (EBBT)*, April, Istanbul, Turkey.
6. Lin, Xiaoli, Wei Huangfu, Fei Wang, Liyuan Liu, and Keping Long. 2016. "A breast cancer risk classification model based on the features selected by novel F-score index for the imbalanced multi-feature dataset." *2016 International Conference on Cyber-Enabled Distributed Computing and Knowledge Discovery (CyberC)*, October, Chengdu, China.

7. Sharma, Shubham, Archit Aggarwal, and Tanupriya Choudhury. 2018. "Breast cancer detection using machine learning algorithms." *2018 International Conference on Computational Techniques, Electronics and Mechanical Systems (CTEMS)*, December, India.
8. Bharat, Anusha, N. Pooja, and R. Anishka Reddy. 2018. "Using machine learning algorithms for breast cancer risk prediction and diagnosis." *2018 3rd International Conference on Circuits, Control, Communication and Computing (I4C)*, October, Bangalore, India.
9. Rana, Mandeep. 2015. "Breast cancer diagnosis and recurrence prediction using machine learning techniques." *International Journal of Research in Engineering and Technology* 4(4): 372–376.
10. Khuriwal, Naresh, and Nidhi Mishra. 2018. "Breast cancer diagnosis using adaptive voting ensemble machine learning algorithm." *2018 IEEMA Engineer Infinite Conference (ETechNxT)*, March, Noida, India.
11. Naveen, R. K. Sharma and A. Ramachandran Nair. 2019. "Efficient breast cancer prediction using ensemble machine learning models," *2019 4th International Conference on Recent Trends on Electronics, Information, Communication & Technology (RTEICT)*, Bangalore, India, pp. 100–104. doi: 10.1109/RTEICT46194.2019.9016968.
12. Srinivasaraghavan, Anuradha, and Vincy Joseph. 2020. *"Machine Learning"*, Wiley India Pvt. Ltd, New Delhi.
13. Mehta, Heeket, Shanay Shah, Neil Patel, and Pratik Kanani. 2020. "Classification of criminal recidivism using machine learning techniques", *IJAST*, 29(4): 5110.
14. Kharya, S., D. Dubey, and S. Soni. 2013. "Predictive machine learning techniques for breast cancer detection," *IJCSIT: International Journal of Computer Science and Information Technologies*, 4(6): 1023–1028.
15. Built In. 2019. "A complete guide to the random forest algorithm." https://builtin.com/data-science/random-forest-algorithm.

18

Application of Deep Learning in Cartography Using UNet and Generative Adversarial Network

Deep Gandhi, Govind Thakur, Pranit Bari, and Khushali Deulkar
Dwarkadas J Sanghvi College of Engineering

CONTENTS

18.1 Introduction ... 257
18.2 Dataset and Preprocessing ... 258
  18.2.1 *Dataset* ... 258
  18.2.2 Data Preprocessing ... 259
18.3 Generative Models ... 260
  18.3.1 UNet ... 260
  18.3.2 Generative Adversarial Networks ... 262
    18.3.2.1 BCELogItsLoss ... 263
  18.3.3 Residual Neural Networks ... 263
18.4 Experimentation ... 264
  18.4.1 Training of UNet ... 264
  18.4.2 Training of a GAN ... 265
    18.4.2.1 Training of Generator Module ... 266
    18.4.2.2 Training of Discriminator Module ... 267
    18.4.2.3 Combined Training ... 268
18.5 Results of the Experimentation ... 269
18.6 Future Applications ... 270
18.7 Conclusion ... 270
References ... 270

18.1 Introduction

The history of the term "machine learning" can be traced back to as far as the year 1959. During this period, an American named Arthur Lee Samuel, also regarded as the pioneer of artificial intelligence and computer gaming, came up and disseminated this term. All of this had been possible because of the Samuel checkers-playing program, one of the earliest self-learning algorithms in the field. However, the breakthroughs in the field of Deep Learning didn't come until the year 1967 when Alexey Ivakhnenko developed the first feedforward, multilayer perceptron network and revolutionized the field. The early days of computer vision started when Kunihiko Fukushima came up with the concept of Neocognitron in 1980. The field of cartography has been progressing from the first world map being introduced by Anaximander up to today when every direction can be accessed at a click of a button. With the invention and progress of modern technology, there has also been a significant boost in the creation of data. According to IBM, the population creates around 2.5 quintillion bytes of data every day and thus, it is necessary to find ways to save data in as many efficient ways as possible. This could be done with the fusion of two of the most prominent fields of the 21st century, cartography and computer vision.

Some of the reasons why Deep Learning plays an important role in cartography and the underestimated importance of computer vision for the same:

1. There is a pressing need for terrain identification in low network areas where professional equipment setup and testing are not possible and also in regions that stand as a dark spot for the satellites.
2. The UNet architecture was first invented by Ronneberger, Fischer, and Brox (Ref. [1]) for biomedical image segmentation to reduce human annotation efforts and for a long time, it was undervalued in other fields.
3. The determination of terrains using just hand-drawn images would be a big advantage to map areas that are considered inadvisable to explore because of the hostile groups settled there and thus, even a blur satellite image could upscale the image to a proper terrain map.
4. This could be further used to develop exoplanetary maps using low-resolution cameras and thus, saving a huge amount of money as well as the weight factor on various satellites.

However, a few limitations must be considered while developing such systems for terrain mapping. They are as follows:

1. The hardware accelerators requirement of the system can sometimes be monetarily and computationally expensive.
2. Lack of data or biased data can result in skewed inferences that would keep on degrading the systems developed.
3. These systems would still face a cold-start problem to the completely unknown kind of terrains.

The major problem that the authors are trying to solve here is the reconstruction of terrain images and underlying features of the said terrains from simple segmented images of the aforementioned regions and while doing so, present a juxtaposition of various generative image models such as the UNet and generative adversarial network (GAN). The dataset for the same can be found in Ref. [2].

18.2 Dataset and Preprocessing

18.2.1 Dataset

The dataset that the authors use to solve this problem has been used from Pappas' dataset (Ref. [2]), who created it by taking a random crop of the global map of the earth. Using these 512×512 images as the ground truth, he used MS Paint to create the seed images for the dataset. To create these seed images, the crops were quantized to five different colors, each corresponding to a feature of the terrain. These features were namely:

1. Blue: water bodies
2. Grey: mountains
3. Green: forests/jungles
4. Yellow: grasslands or desert
5. Brown: hills/badlands

However, successive mode filters were applied to these paintings to simplify the underlying dependencies. Along with that, various colors were used from these to mix and depict topology and other factors in the dataset. The dataset consists of 1360 images that cover the entire planet and thus, is used by the authors to generate a variety of landscapes in a single image itself due to the geographical diversity

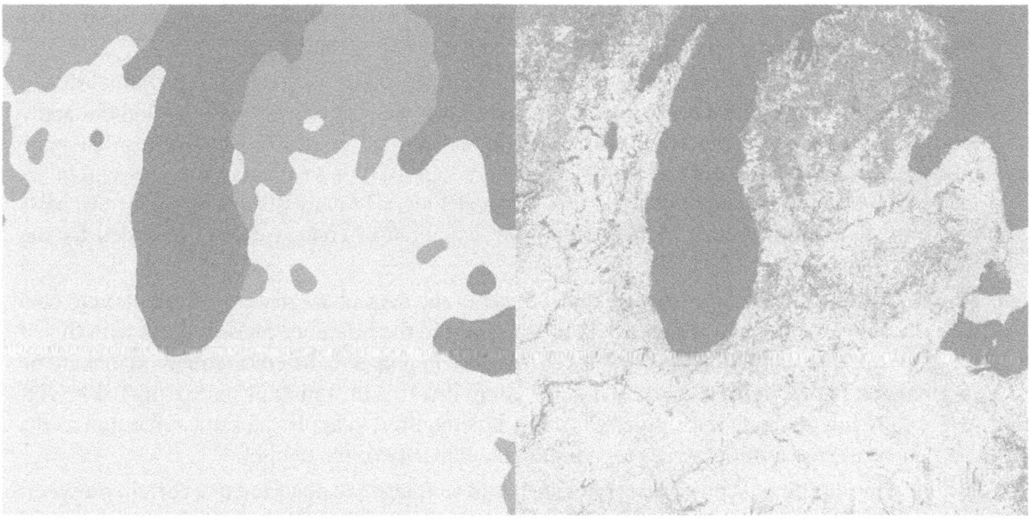

FIGURE 18.1 A sample image in the dataset chosen.

of Earth. This dataset can also be used to predict geographical topology after making a few tweaks. However, the authors would be discussing the reconstruction of terrain images from such paintings. An example of the image in the dataset is given in Figure 18.1.

18.2.2 Data Preprocessing

After the initial dataset setup, the dataset is divided into the training and validation sets since the 512×512 images are a collage of the training image and the ground truth, i.e., the map and the terrain. The images are first downsampled to a size of 256×256 and then converted to arrays to pull the map from the left side and the terrain from the right. Then, the dataset is randomly shuffled and then 10% of the total images are put into the validation set at any predetermined random seed so that the splitting remains deterministic at any given time. After this procedure is completed, there is a need to perform data augmentation to introduce some randomness in the dataset and also increase the data while doing so as the authors are dealing with a small dataset to solve this problem.

Data augmentation is a regularisation technique in Computer Vision as mentioned in Ref. [3], which is used to perform some random operations on the image before feeding it to the learner, which would then result in an increased variation of the data being used for learning and help the model to generalize better for inference. This technique involves various procedures such as random crop, zoom in, and padding the images.

For this dataset, the authors had to select the methods that they should not disturb the structural integrity of the maps. Thus, the techniques selected for the task include:

1. **Random Flip**: The images are randomly flipped in no particular order. This means that the mirror images of the given maps can be used that would then bring diversity to the same image. The authors use a probability of 0.5 to determine if a given image would undergo a flip or not. It is necessary to perform a flip on the terrain image in the prediction dataset if a flip is performed on the map as it makes sense to correlate the map to the terrain when the frame of visual is changed on it. It should be noted that these images are only flipped horizontally and not vertically.
2. **Rotation**: Even if a horizontal flip is not applied and the images are only rotated by a certain amount of degrees, it would not cause noise in the dataset and would be helpful as it remains in the frame of reference but alters the pixel values for a better generalization. The images in the

dataset are rotated between –10° and 10° and thus, along with the maps, the terrain images are also rotated at the same angle to keep the correlation between them intact.

3. **Zoom**: Along with rotation and identifying the specific terrains in the dataset, the model should also learn to recreate a random crop of the terrain on the planet and thus, the authors apply a random zoom on the dataset with a maximum zoom value set at 2.0. This means that some images are randomly selected and zoomed in at random parts up to a maximum 2× zoom in the map part as well as the terrain part so that the model identifies any given kind of terrain pairs in the image paired together. This is important when out-of-context data is provided for the inference part.
4. **Random Lighting**: The lighting and the contrast of the dataset are distorted at random to deal with the hues of the colors which would then be used in the inference phase. Along with this, it is made sure that this distortion doesn't go over 20% to preserve the color-coded regions in the original colors that were intended. This procedure involves dimming or increasing the brightness of any images at random. Again, for this, it is required to apply the same operation as the map images to the terrain images to maintain the integrity of the dataset.
5. **Image Warping:** In this process, certain shapes in an image are distorted to a certain prespecified extent. This may seem like an ambiguous choice for tasks such as dealing with terrestrial data; however, it would be good to introduce some randomness in the dataset. The former claim was made because the model needs to learn how to deal with the terrestrial changes on earth and also learn how to segment features based on shape and not just their color codings and thus, image warping is useful in dealing with that. Again, the authors apply the same transformation to the prediction image as the training image to keep it intact.
6. **Probability Factor:** The probability of a symmetric warp or the given affine transform being applied is determined by the probability factor. This is very necessary as it would lead to various combinations of transforms being applied which would themselves have random values to augment the amount of data that the authors deal with. In this case, a probability of 75% is considered to deal with a lot of variations in the transforms and thus, even if the amount of data supplied by the dataset is less, data augmentation succors to generalize the model to an acceptable extent.

A databunch is created after applying these transformations to the dataset according to the batch size specified by the authors. The batch size can be decided based on various factors related to the availability of computational resources such as GPU and RAM to avoid system crashes. After creating the databunch, ImageNet statistics are applied to these images. ImageNet statistics are the exact statistical functions applied to ImageNet when state-of-the-art ResNet models were being trained on them.

Generally, the ImageNet statistics used are [0.485, 0.456, 0.406] as mean and [0.229, 0.224, 0.225] as standard deviation, and the same was used while training the architectures. This would facilitate the use of transfer learning in the generative models during the downsampling path.

18.3 Generative Models

The authors discuss two generative models and compare their possible performance. They are a UNet and a GAN. Before training, it would be beneficial to understand how each of the models works.

18.3.1 UNet

Initially derived from a convolutional neural network (CNN), a UNet was used in the segmentation of biomedical images in the year 2015 by Ronneberger, Fischer, and Brox (Ref. [1]). As a convolutional network is known to perform tasks of classification, a UNet can perform these exact tasks on every pixel of the image and thus, it is highly used in the field of segmentation as the input and output of this network have the same size.

The architecture of a UNet, just like its name is a symmetric U-shaped one that consists of two different parts. The left part of the UNet is a general CNN and thus, it is the primary downsampling path of the architecture, whereas the right part of the UNet consists of a $2d$ convolutional network with half strides. This is often referred to as the upsampling part of the network.

As it can be observed in Figure 18.2, each stage of the UNet consists of two convolutional layers and, as the downward arrow indicates downsampling, the size of the image is halved due to the stride two configuration of the UNet. The right arrow indicates that a convolutional layer is applied to the image along with a rectified linear unit (ReLU). This process of convolution (Conv), ReLU, and max pooling is applied three times until the number of channels in the input vector becomes 1024. After this, a convolutional layer and ReLU are applied twice to the vector, but max pooling is not applied to the same (Figure 18.3).

Now, for the upsampling path, as can be observed in the image above, a transposed convolution is applied to the image to upsample it to its original size. A simple convolution operation means that a downsampled matrix is calculated using the addition of the product of the elements of the convolutional matrix and a kernel. Now for a transposed convolution, the model needs to perform a similar operation in a reverse manner, which means that it needs to establish a one-to-many relationship to reconstruct the image completely part by part. As it sounds, the transposed convolution network works in such a way that the transpose of the convolution matrix is taken and then the flattened image vector is used for the general operation of the sum of the product of the elements in both the matrices. This results in the formation of the bigger desired matrix and helps in the upsampling of the image. This step can also be referred to as the 0.5-stride step since it divides the original size by 0.5 to upsample it. After each upsampling operation, the resultant vector is concatenated with the corresponding image during the downsampling path to preserve the information and get more accurate predictions. The number of increased channels due to this concatenation is then balanced when it undergoes convolution operation and then ReLU is applied to it.

However, to increase the prediction accuracy of the model, the authors use a dynamic UNet for training. A dynamic UNet is an architecture in which a pretrained ResNet model is used for training the network on the required dataset. The authors use the pretrained architecture known as ResNet34. The ResNet is explained in detail herein. However, it can be safely assumed that the complete ResNet is used in the downsampling path of the UNet. However, for the upsampling part, the authors use a technique called Pixel Shuffle from Ref. [4]. This technique is responsible for rearranging the elements in an input vector of size $(N, C * r * r, H, W)$ to a vector of $(N, C, H * r, W * r)$. Here, r is the upscale factor (two in

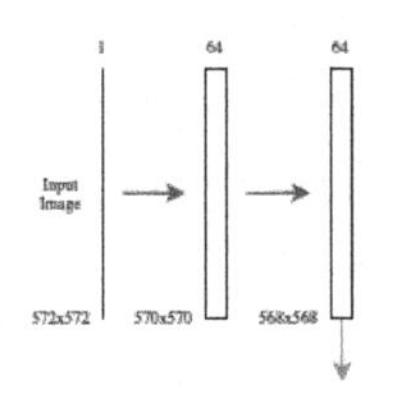

FIGURE 18.2 Architecture of the UNet.

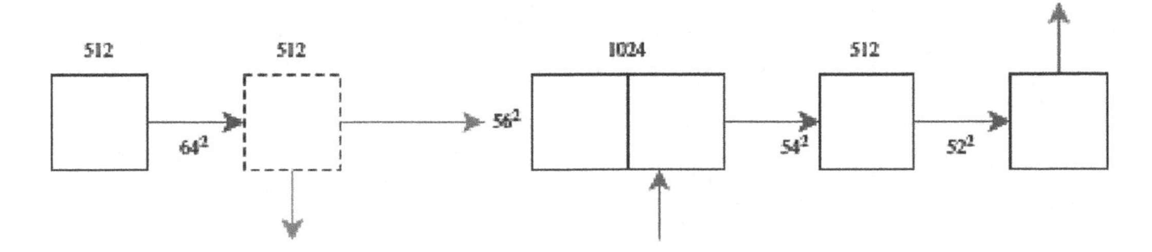

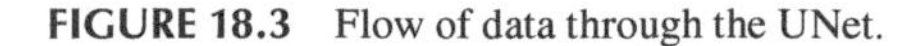
FIGURE 18.3 Flow of data through the UNet.

this case), C, H, W, and N are the dimensions of the original input layer. Thus, for upsampling stride of $1/r$, to increase the spatial resolution by a factor of r, the Pixel Shuffle technique is used to reduce the checkerboard artifacts that may form in the image due to the concatenation of the two layers.

This is how a UNet reconstructs images of the same size as the input. Since this model can make very good predictions even on small datasets with the excessive use of data augmentation techniques, as stated in Ref. [1], the authors have selected this as one of the models to be compared and used for the cartography task.

18.3.2 Generative Adversarial Networks

GANs are the other type of generative models that the authors would be used to reconstruct terrains from the color-coded map images. In principle, GANs also use the CNNs for a part of the whole generative system and thus can be said to be built upon the CNNs similar to the UNet.

As presented in Ref. [5], a GAN consists of two modules basically, a generator and a critic. The task of a generator, as it sounds, is to generate images just like any generative model would do such as a UNet. However, unlike a UNet, a GAN consists of a critic. A critic is a classifier that only makes sure that it can correctly identify and classify between the images generated by the generator and the real ground truth. A GAN was referred to as a novel approach in the field of Deep Learning as it uses two neural networks not as an ensemble but as in competition with each other to fool the other module and generate as real images as the ground truth, and the classifier tries to stop the generator by learning to identify these better images, which in turn helps the generator to generate even better images.

Figure 18.4 refers to a problem of super-resolution using GANs. However, the authors use it to explain the basic concept of a GAN. For super-resolution, a bad image is supplied to the model to get a high-resolution image as an output. Now as it shows, bad quality is supplied to the generator, and the generator tries to create a super-resolution image out of it. The prediction is given out by the generator and then the ground truth and the prediction are compared and a loss is calculated using pixel-wise mean squared error (MSE) to determine how different the generated images are from the original images, and this pixel MSE is supplied back to the adaptive switcher. An adaptive switcher is used in a GAN so that it can be efficiently determined when the generator needs to stop the generation of images and the critic needs to start classification on the present state or vice versa. The authors set a threshold of 0.65 in the switcher. This specifies that whenever the pixel MSE loss goes below 0.65, a switch needs to occur from the generator to the critic.

In the case of a critic, when the generator generates and sends an image to the critic for learning, along with the ground truth, the critic uses a basic convolutional network, generally with a convolutional layer,

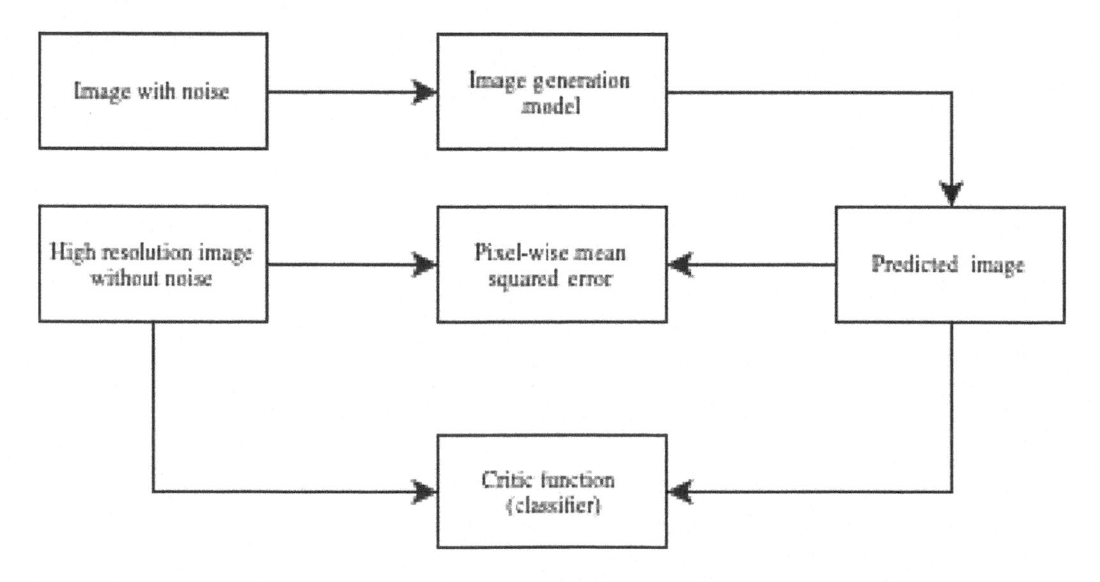

FIGURE 18.4 Super-resolution using GANs.

a dropout layer, and then multiple residual blocks (ResBlocks). A ResBlock is a general convolutional block with a convolutional layer, batch norm, and a ReLU applied to it as an activation function. The concept of a ResNet and Resblock is explained further in the ResNet section. However, the critic then uses a loss function to check the threshold and also to improve its classification. The loss function used by the critic is called BCELogItsLoss.

18.3.2.1 BCELogItsLoss

This loss function combines a sigmoid layer and the binary cross-entropy (BCE) loss in a single layer. The equation for BCELogItsLoss is given as [6]:

$$l(x, y) = \{l_1, \ldots, l_N\}^T, l_n = -w_n\left[y_n \log \sigma(x_n) + (1 - y_n)\log(1 - \sigma(x_n))\right]$$

The reason for using this over a general BCE loss is that numerical stability is achieved since the sigmoid is applied internally as observed in the loss function. However, a general BCE loss requires us to input the results of a sigmoid function to the BCE equation, which would result in skewed results in particular cases. This resulting stability is because the BCELogItsLoss allows the usage of the LogSumExp function.

Generally, LogSumExp is the logarithm of the sum of the exponentials of the arguments. It is proposed in Ref. [7]. The equation of the LogSumExp function and the trick that helps to achieve numerical stability is given below:

$$\log \sum_{k=1}^{n} e^{xk} = b + \log \sum_{k=1}^{n} e^{x\,k-b}$$

Thus, the authors determine BCELogItsLoss as a good loss function for the critic, and it is used for determining the switch when the loss goes below the threshold.

18.3.3 Residual Neural Networks

To understand how a residual neural network works, the authors start by explaining a basic convolution network. A basic convolution operation is as simple as can be imagined, calculating a product of the weight and the input and adding the bias to it to get the output. Thus, this can be referred to as a 2*d* convolutional layer and for a simple CNN, the model can be constructed by using a 2*d* convolutional layer, followed by batch normalization, and then applying a ReLU as an activation function. This is generally repeated for some layers as the stride is set to 2 for this task and then in the end layer, a Flatten layer is applied at the end, which flattens the continuous range of dimensions in a tensor. This is used to obtain a probability distribution for classification tasks.

This creates a decent network; however, to create a deeper network, the appropriate approach would be to always add a stride 1 convolutional block after the stride 2 convolutional block. Since this stride 1 block doesn't change the feature map size at all and thus, it was assumed that these stride 1 layers can be added in any quantity desired to make the models better. However, this was contradicted in Ref. [8]. The paper [8] compared a 56-layer model and a 20-layer model on the CIFAR-10 dataset. Since the 56-layer model has more parameters and a lot of the aforementioned stride 1 convolutional layers, it was expected to overfit and thus, drop the training loss to zero in a very short period. However, as observed in Ref. [8], it's not the case.

After this observation, as observed in Figure 18.5, the paper came up with a new technique to make the 56-layer network at least as good as the 20-layer network. According to this technique, for every two convolutions, the output would be a resultant of the input and the result of the two convolutional layers as observed in:

$$\text{Output} = x + \text{conv2}(\text{conv1}(x))$$

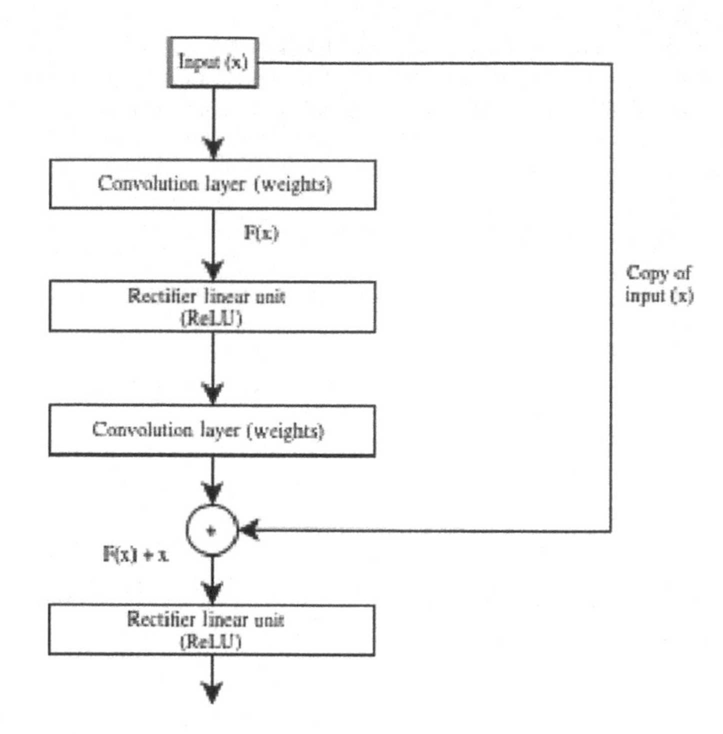

FIGURE 18.5 Residual block of ResNet.

Due to this, there is always going to be a guarantee that the model may set the input vector and the resultant of the convolutional layers to 0 weights except for the first 20 layers and thus, this would mean that after those 20 layers, the output would go right through them. These connections, also known as skip connections, would ensure that the created model would be better than or at least equivalent to the 20-layer model. Adding this technique resulted in the creation of the residual neural network, which produced state-of-the-art results.

The improvisation of the network due to the skip connections was also observed in Ref. [9], where the same thing was observed as the loss was smoothened due to the presence of skip connections.

18.4 Experimentation

In this section, the authors discuss the training of both the networks and also discuss various important hyperparameters for model tuning for making the network better. This module is divided into two parts, each discussing the minutiae of UNet and GANs, respectively.

18.4.1 Training of UNet

As discussed, while expounding the architecture and working of a UNet, the authors use a dynamic UNet for the training task and thus, the pretrained architecture used for the training of the UNet is ResNet34. A ResNet34 model has 34 layers with a 2-layer deep architecture, meaning the concatenation occurs at every 2 layers of the model.

For the hyperparameters to be used during the training and the importance of tuning each of these hyperparameters, the authors suggest the following:

1. **Blurring**: According to Ref. [10], it is an illustrious fact that a block type, chessboard effect is generated while using pretrained or even customized CNNs in generative models as they generally involve tasks related to super-resolution. While doing a PixelShuffle, the values are initialized

using an ICNR (Initialization to Convolution NN Resize) initialization and thus reduces a checkerboard effect (subpixel convolution) on the vector, as mentioned in Ref. [11]. However, to reduce this for further upscaling, the authors make use of the blur effect. Thus, to upscale an image using PixelShuffle as mentioned above, and then normalizing the borders of it, it is necessary to create a blur effect over the $H * W$ kernel as mentioned in the aforementioned paper and thus, to create this effect, initially, a replication padding is created on all four sides of the vector and then an average pooling function is applied to it. The equation of this function is given in Ref. [12]:

$$\text{out}\left(N_i, C_j, h, w\right) = \frac{1}{kH * kW} \sum_{n=0}^{kH-1} \sum_{n=0}^{kW-1} \text{input}\left(N_i, C_j, \text{stride}[0] \times h + m, \text{stride}[1] \times w + n\right)$$

2. **Weight Decay**: Weight decay is an effective technique to prevent the parameter from being set to 0 by the loss function as proposed in Ref. [13]. In the case of an increased number of parameters, the number of interactions increases and thus, the model can fit better on the data it requires. However, while calculating the loss, the loss function calculates the sum of the squares of these exact parameters to determine the loss of the models. There is always a possibility that this may result in a large number and thus, the parameters would be set to 0 due to this, as that would be the most optimum choice to reduce loss as understood by the model. To prevent this from happening, a constant number is multiplied to this sum to balance the model as well as the loss function. The number used to multiply here is called weight decay. Thus, the loss function can be represented as:

$$\text{Loss} = \text{MSE}\left(y_h, y\right) + wd * \text{sum}\left(w^2\right)$$

After this, the gradient descent step is carried out in which the weights are updated. This is written as:

$$w_t = w_{t-1} - \text{lr} * \frac{d(\text{Loss})}{dw}$$

Now, for the second term of the loss function, the derivative would be calculated as:

$$\frac{d\left(wd * w^2\right)}{dw} = 2 * wd * w$$

Thus, the authors would also be subtracting a portion of the present weight multiplied by a constant from the original weight. Therefore, the term is referred to as "weight decay." The authors find it best to take a weight decay of 0.01 and stop just before the model overfits for better generalization.

3. **Accuracy Metric**: The metric used for calculating the accuracy of this model is a simple mean squared error calculated using its pixel-wise values. The equation for the mean squared error is given as:

$$\text{MSE} = \frac{1}{n} \sum_{i=1}^{n} \left(y_i - y_h\right)^2$$

This is very useful as it informs how well the model generalizes and thus, in this case, it would be possible to determine if a particular model doesn't fit well over a certain kind of terrain. This can be possible with the use of hooks and extracting the tail layers of the model to find the same. Thus, the normal mean squared error metric is economically efficient and thus reliable for error calculation for the cartography UNet.

18.4.2 Training of a GAN

For the training of a GAN, as discussed above, there are two modules necessary, namely a generator and a critic. Now, for the generator module, the authors use a modified version of the dynamic UNet as discussed earlier.

18.4.2.1 Training of Generator Module

The generator module of the GAN consists of a dynamic UNet, which differs from the earlier UNet model used in the following hyperparameters:

1. **Normalization:** For the convolutional layers in the UNet being used for the generator layer, the authors use weight normalization instead of the regular batch normalization. As proposed in Ref. [14], weight normalization is a reparameterization approach used in generative models such as GANs. This helps in speeding up the faster occurrence and optimization of the stochastic gradients. This method is a modified version of batch normalization, but it is different in the way that the dependencies in a batch are eliminated by this procedure. Thus, this method of normalization is found more suitable for the current generator. However, it should be noted that even though the authors change the normalization, they still make use of the ResNet34 model for a comparative analysis between the dynamic UNet and the GAN.
2. **Self-Attention:** According to Ref. [15], a self-attention generative adversarial network has a discriminator that uses the attention layer, also highlighted in Ref. [16], to check the consistencies between the distant features in a highly detailed image. This means that the attention layer can figure out consistencies and thus, this same layer can be used by the generator to generate those consistencies in the image making it harder for the discriminator (critic) to classify the generated and ground truth images, thereby speeding up the training process (Figure 18.6).

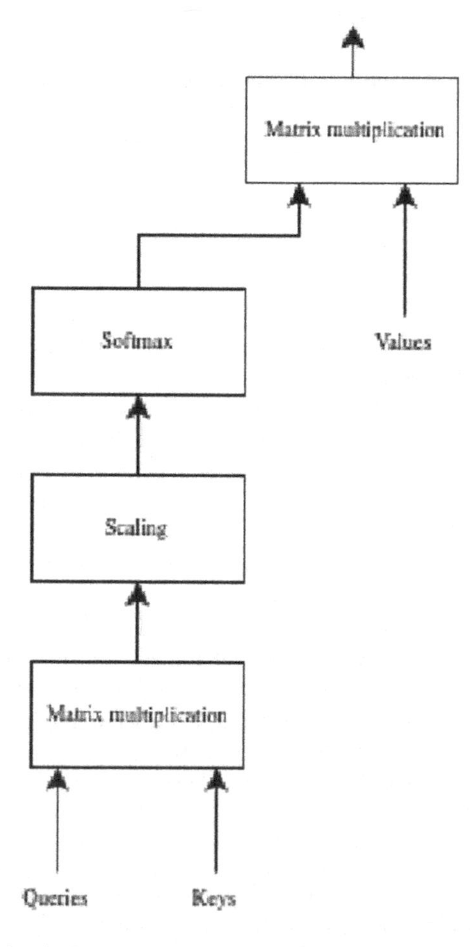

FIGURE 18.6 Attention mechanism in GANs.

As stated in Ref. [16], an attention block takes in queries, keys, and values are used to calculate the final probabilities. After this, a dot product of queries and keys is taken and the output is scaled by dividing the root of the key matrix dimension length. After this, a softmax layer is applied to this matrix and then the resultant is multiplied with the value matrix to get a probability of the resultant values as attention. This can be stated by the equation given below:

$$\text{Attention}(\text{Query, key, value}) = \text{softmax}\left(\frac{\text{Query} * \text{key}^T}{\sqrt{\text{dim}_k}}\right)\text{value}$$

3. **Ranged Sigmoid Function:** Unlike the older UNet, the authors also append a sigmoid layer at the end of the UNet to prevent discrepancies in the image. The sigmoid function may predict some values that may distort the image and thus, it is necessary to set a range in which the sigmoid function can predict the given values. The equation for the sigmoid can be given as:

$$\text{outp}_i = \frac{1}{1+e^{-\text{inp}}}$$

   This output is then multiplied by the range of the range specified and the base value is added to the output to keep it within the specified range.
4. **Loss Function:** The authors use the MSE loss for pixel-wise loss calculation in the generator itself. However, the only difference, in this case, is that the loss vector input and output are flattened for computational efficiency.

The other hyperparameters such as weight decay and blurring are kept the same as the earlier UNet. Thus, a generator is successfully built for the GAN.

Now, before the discriminator is set up, the authors follow an unconventional approach for the task. Instead of training and fine-tuning both the models simultaneously using the adaptive switcher, the generator (which contains a dynamic UNet) is first fine-tuned on the dataset without any discriminator declared. This is done because the GAN as an overall model during training takes a lot of time to generate images even as closely related to the ground truth. This extra time is equivalent to the extra computation resources required. Thus, with early fine-tuning, the generator reaches that level from early on and thus, it would be less expensive computationally and also fine-tune the weights in the correct direction. The images generated by this generator early on are saved for early fine-tuning of the discriminator.

18.4.2.2 Training of Discriminator Module

The discriminator (critic) module of the GAN is responsible for drawing distinctions between the real images and the generated images. Thus, it is made up using an architecture similar to a ResNet model. However, the authors find it useless to add a lot of layers to this model as a ResNet model since it would only be performing a basic classification task. For this very reason, it would be computationally efficient to create a custom neural network instead of using a pretrained model for classification.

The architecture of a critic consists of a convolutional layer of size 4-by-4 of stride 2 from the number of channels (3 in this case) to the number of features (128 in this case). This is followed by a dropout layer, as mentioned in Ref. [17], with a probability of 7.5% of dropout. This is followed by a dense block. A dense block is similar to a ResNet block, except that instead of adding the previous layer, it is concatenated and thus, it becomes dense. Due to this reason, the number of features is now doubled. This is followed by a series of blocks that consist of a convolutional layer and the dropout layer with the probability of dropout set at 15%. At the end, a convolutional layer of a size number of features by 1 is added and then the output is flattened. The hyperparameters used for the discriminator are as follows:

1. **Adaptive Loss:** As discussed earlier, the discriminator uses a loss function known as BCELogItsLoss for numerical stability. However, the output from a critic is a tensor of varying size and thus, it would be necessary to expand the target to the size of the output of the discriminator to calculate loss accurately using the given loss function. Thus, the loss function

is applied by expanding the target to the required dimensions and then calculating the loss is necessary. This adaptive loss function was proposed in Ref. [18].

2. **Weight Decay:** As discussed for the dynamic UNet, the weight decay is used in the same way for the convolutional network as it is used in the ResNet model of the UNet for loss calculation and weight update and thus, it doesn't require any further modifications for the discriminator.

Since the critic is a simple classifier, it does not require any additional special hyperparameter tuning other than the ones mentioned earlier. Just like the generator module, even the critic is fine-tuned before combined training to reach a similar accuracy as that of the fine-tuned generator. For this purpose, the output images from the fine-tuned generator are used to train the classifier for differentiating between the real image and the generated one.

18.4.2.3 Combined Training

After both the modules have undergone individual fine-tuning, they are loaded into an adaptive switcher. As discussed earlier, an adaptive switcher is responsible to determine the point to stop training either the discriminator or the generator and switch to the other module. This is determined by the accuracy threshold set in it and whenever the accuracy of generation or classification crosses this threshold, a switch is made with the former stopping training and the latter starting to train. Some of the special attributes used by the authors for the combined training are as follows:

1. **Modified Loss Function**: As discussed earlier, each of their loss functions for the specific tasks assigned to them. However, on an overall look, it can be safely said that the critic is the major loss function decider for the GAN. This would mean that if only the critic is considered as the loss function, the model would produce very real images, but they would be out of context with the dataset as required for the task. Thus, instead of just using the critic as a loss function; thus, it is necessary to consider the overall loss as a sum of the pixel loss as well as the critic loss. However, since both of these losses work on a different scale, it is necessary to multiply the critic loss with a constant of a certain range as it carries more weightage than the pixel loss. This helps factor meaningful as well as contextual images from the GAN.
2. **Optimizers**: The optimizer used for this model is Adam. As discussed in Ref. [19], Adam combines the approach of the exponentially weighted moving average [20] of a novel approach to momentum in stochastic gradient descent known as RMSProp, mentioned in Ref. [21], and the exponentially weighted moving average of the momentum divided by the exponentially weighted moving average of the square of the terms. An exponentially weighted moving average is given as:

$$S_t = \alpha g_t + (1-\alpha) S_{t-1}$$

   Due to the presence of Adam, it would be necessary to consider different learning rates during the fit stage of the model as the dynamic changes going on, the model needs to adapt to dynamic learning rates to determine which parameters need to be moved in which direction at what rate. This is determined as during the start of learning; the optimizer has a very low learning rate. However, it would not be damaging to increase the learning rate after gradually and keep it at the peak as the momentum gradually decreases, and it can be interpreted that the model is moving in the correct direction. However, it would be necessary to anneal this learning rate as optimum condition nears attainability to prevent learning process damage.
3. **Discriminative Learning Rates:** The authors use a method proposed in Ref. [22] called discriminative learning rates. According to this method, applying different learning rates to different layers of the model is very useful to control the loss of information. To train the models effectively, different learning rates can be applied to the generator and the critic, as the critic does basic classification work and could benefit computationally by applying a different learning rate to the critic. Thus, the authors multiply the learning rate of the generator by five times and supply it to the critic to speed up training and devote more time to the generator for accurate image generation.

Thus, after learning for a certain interval of time, the GAN would become capable of generating the required terrain images from the color-coded maps.

18.5 Results of the Experimentation

The training and validation results in several inferences to be drawn from the chapter. The first out of all them would be that for the task at hand, the GAN architecture [23] would be best suited for a complete end-to-end regeneration system. This can be observed in the learning rate vs loss graphs shown in Figures 18.7 and 18.8.

It can be inferred from these figures that upon using GAN, the learning rate and loss transition are very smooth rather than just the UNet architecture. This would mean that only the UNet architecture would not generalize very well and thus would result in weaker accuracy scores for the task.

Even though the scale of loss in both the figures is different, it can be extrapolated that upon further training, the models continue to achieve the same scale. This would mean that the UNet would start performing worse than the best scenario as observed in the graphs.

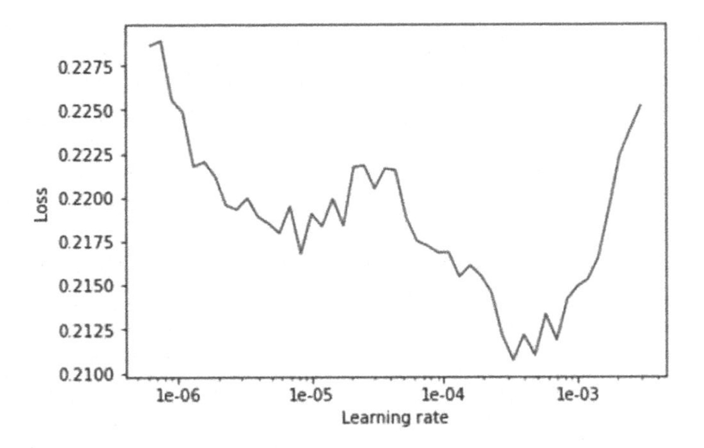

FIGURE 18.7 Learning rate vs loss for UNet.

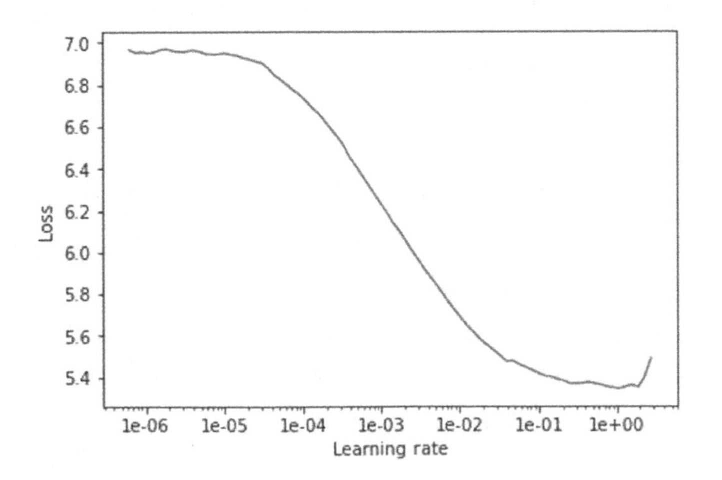

FIGURE 18.8 Learning rate vs loss for GANs.

18.6 Future Applications

The authors proposed the aforementioned methods to create the terrain images from color-coded maps because in regions of low connectivity when the required bandwidth is not available to transfer the whole terrain map in detail, the presence of such a system on the device can be used as an encoder–decoder, wherein the user encodes the map image they're trying to send in a color-coding predefined by the system, and the receiver gets a high-resolution terrain image due to the generative model present in the device. This would thus help in high-level compression in cartography. The only possible disadvantage of such a system would be that they wouldn't do lossless compression and thus, some data packets may be distorted upon transmission, resulting in a wrong prediction by the system. However, this can be avoided to a certain extent by a good transmission connection between the sender and receiver.

This is used to map not only the low-bandwidth regions but also unknown terrain surfaces of other planets with similar terrestrial features. This would be helpful to rovers or devices sending data over bleak network signals and thus would help them send data with low memory, resulting in the prevention of packet loss that pollutes the data quite often.

18.7 Conclusion

Cartography and Deep Learning can be used in conjunction to help the researchers and practitioners in the former field to a great extent. This can help put an end to certain obsolete practices in the field and also obliterates the margin of human error or machine error up to a great extent. Thus, the authors start with the prominent problem of converting color-coded maps to accurate terrain maps. In turn, that problem is then converted into a system for secure map data transmission and reduction of packet loss. This can be achieved by using either of the two models, a UNet or a GAN.

The authors demonstrate and compare the working of two prominent generative models in this chapter to make terrain maps from color-coded images. The working of a dynamic UNet and the optimal hyperparameters are also shown. This model is juxtaposed with the working of a GAN and the hyperparameters of the GAN are stated and discussed in detail. After studying the results of the experimentation, it can be believed to a certain extent that restrictions on the computational resources available during the training period of these models may be beneficial to use for the UNet based approach as it requires less training time and computational resources evident from the explanation. However, in case of no such restrictions, it proves advantageous to train both the models and also introspect various edge cases to determine which model should be chosen.

Although a lot of promising work has been done in cartography, the field demands further dedicated research to create more accurate models for solving case-specific problems that include dealing with unknown terrains, clustering of similar terrains and their differentiation, presence of different atmospheres on different planets if used for that purpose.

REFERENCES

1. Ronneberger, O., P. Fischer and T. Brox. "U-Net: Convolutional networks for biomedical image segmentation." MICCAI (2015).
2. Pappas, T. 'Terrain map image pairs'. Accessed 1 December 2020. https://kaggle.com/tpapp157/terrainimagepairs.
3. Mikołajczyk, A. and M. Grochowski. "Data augmentation for improving deep learning in image classification problem." *2018 International Interdisciplinary PhD Workshop (IIPhDW)*, Poland (2018), pp. 117–122.
4. Shi, W., J. Caballero, F. Huszár, J. Totz, A. Aitken, R. Bishop, D. Rueckert and Z. Wang. "Real-time single image and video super-resolution using an efficient sub-pixel convolutional neural , network." *2016 IEEE Conference on Computer Vision and Pattern Recognition (CVPR)*, Las Vegas, NV(2016), pp. 1874–1883.

5. Goodfellow, I.J., J. Pouget-Abadie, M. Mirza, B. Xu, D. Warde-Farley, S. Ozair, A.C. Courville and Y. Bengio. "Generative adversarial nets." NIPS (2014).
6. PyTorch. 'BCEWithLogItsLoss: PyTorch master documentation'. Accessed 1 December (2020). https://pytorch.org/docs/master/generated/torch.nn.BCEWithLogitsLoss.html.
7. Blanchard, P., D. Higham and N. Higham. "Accurate computation of the Log-Sum-Exp and Softmax functions." ArXiv abs/1909.03469 (2019).
8. He, K., X. Zhang, S. Ren and J. Sun. "Deep residual learning for image recognition." *2016 IEEE Conference on Computer Vision and Pattern Recognition (CVPR)*, Las Vegas, NV (2016), pp. 770–778.
9. Li, H., Z. Xu, G. Taylor and T. Goldstein. "Visualizing the loss landscape of neural nets." NeurIPS (2018).
10. Sugawara, Y., S. Shiota and H. Kiya. "Super-resolution using convolutional neural networks without any checkerboard artifacts." *2018 25th IEEE International Conference on Image Processing (ICIP)*, Athens, Greece (2018), pp. 66–70.
11. Aitken, A., C. Ledig, L. Theis, J. Caballero, Z. Wang and W. Shi. "Checkerboard artifact free sub-pixel convolution: A note on sub-pixel convolution, resize convolution and convolution resize." ArXiv abs/1707.02937 (2017)
12. PyTorch. 'AvgPool2d: PyTorch master documentation'. Accessed 1 December (2020). https://pytorch.org/docs/master/generated/torch.nn.AvgPool2d.html.
13. Krogh, A. and J. Hertz. "A simple weight decay can improve generalization." NIPS (1991).
14. Salimans, T. and D.P. Kingma. "Weight normalization: A simple reparameterization to accelerate training of deep neural networks." NIPS (2016).
15. Zhang, H., I.J. Goodfellow, D. Metaxas, and A. Odena. "Self-attention generative adversarial networks." ICML (2019).
16. Vaswani, A., N. Shazeer, N. Parmar, J. Uszkoreit, L. Jones, A.N. Gomez, L. Kaiser and I. Polosukhin. "Attention is all you need." NIPS (2017).
17. Hinton, G.E., N. Srivastava, A. Krizhevsky, I. Sutskever and R. Salakhutdinov. "Improving neural networks by preventing co-adaptation of feature detectors." ArXiv abs/1207.0580 (2012).
18. Ayyoubzadeh, S.M. and X. Wu. "Adaptive loss function for super resolution neural networks using convex optimization techniques." ArXiv abs/2001.07766 (2020).
19. Kingma, D.P. and J. Ba. "Adam: A method for stochastic optimization." CoRR abs/1412.6980 (2015).
20. Lucas, J. and M.S. Saccucci. "Exponentially weighted moving average control schemes: Properties and enhancements." *Quality Engineering* 36 (1990): 31–32.
21. Hinton, G. 2012. "Overview of mini-batch gradient descent". Cs.Toronto.Edu. https://www.cs.toronto.edu/~tijmen/csc321/slides/lecture_slides_lec6.pdf.
22. Howard, J. and S. Ruder. "Universal language model fine-tuning for text classification." ACL (2018).
23. Doshi, F., P. Doshi, J. Gandhi, K. Dwivedi, and R. Mangrulkar. (2020). "Image modification using text with GANs." *International Journal of Computer Applications Technology and Research* 9(2020): 287–294.

19

Evaluation of Intrusion Detection System with Rule-Based Technique to Detect Malicious Web Spiders Using Machine Learning

Nilambari G. Narkar and Narendra M. Shekokar
Dwarkadas. J. Sanghvi College of Engineering

CONTENTS

19.1 Introduction ... 273
19.2 Intrusion Detection System ... 274
19.2.1 Network Intrusion Detection System ... 274
19.2.2 Host-Based Intrusion Detection Systems ... 274
19.3 Web Spider ... 274
19.3.1 Well-Behaved Web Spider ... 274
19.3.2 Malicious Web Spider ... 274
19.4 Proposed System ... 275
19.4.1 Weka ... 277
19.4.2 C4.5 Algorithm ... 277
19.4.3 J48 Classifier ... 277
19.4.4 Multilayer Perceptron Classifier ... 278
19.5 Evaluation of Proposed System Using Machine Learning Algorithm ... 281
19.5.1 IBk Classifier ... 281
19.5.2 Random Tree Classifier ... 282
19.6 Conclusion ... 286
References ... 287

19.1 Introduction

Internet is requisite of the present era. It is an interlinked network of universal mobile devices or computer systems, which uses Internet protocol (IP) to link several billion systems globally. The idioms, Internet and World-Wide-Web (WWW or W3) are usually used interchangeably in routine conversation; it is common to hear of "surfing on the net" while opening a web browser to sight web sites. The advancement of the internet has revolutionized the functioning of conventional services into internet applications.

For searching information quicker and precisely on the internet, software or program named "web spider" is deployed [1]. A web spider is assigned with single or multiple core uniform resource locators (URLs). It fetches online pages linked with URLs and also obtains hyperlinks enclosed within it [2]. Therefore, internet crawling or spidering software is used by internet search engines and other online sites to update web information or web content indexes of online sites. Web spiders can perform denial-of-service active attacks or web scraping passive attacks on crawled web pages.

The chapter is structured as follows: Section 19.2 interprets IDS, Section 19.3 briefly explains the web spiders, Section 19.4 illustrates the proposed system, Section 19.5 infers the attainment of detection using machine learning algorithm, and Section 19.6 outlines the conclusion.

19.2 Intrusion Detection System

An IDS is designed to monitor network or system activities. It produces electronic reports to an administrator for malicious activities [3]. An IDS is classified into two main types: network intrusion detection systems (NIDS) and host-based intrusion detection systems (HIDS), based on where detection takes place (network or host) and the detection method that is employed on it [4].

19.2.1 Network Intrusion Detection System

NIDS are located to monitor traffic to and from all devices on the network at a planned point or points within the network.

19.2.2 Host-Based Intrusion Detection Systems

HIDS are located on individual hosts or devices on the network. Inward and outward packets from the device are scanned and, on the detection of malicious activity, the administrator is informed by sending alerts.

The IDS not only simply monitors and alerts but also performs an action or actions in response to a detected threat [5].

The features of the IDS are determined by its detection method. Behavior-based IDS utilizes data it views regarding the ordinary behavior of the system. Knowledge-based IDS utilizes information about the attacks.

The response to attacks determines the behavior of the IDS on detection. The active IDS responds to the attack actively by performing counteractive or preemptive actions. The passive IDS monitors activities and generates alerts on the detection of malicious activity.

19.3 Web Spider

A web spider systematically surfs the internet for online indexing. While surfing, a web spider facsimiles all surfed pages information for web search engine's upcoming processing. The web search engine uses it for indexing online pages so that users can perform search operations more rapidly. A web spider is also termed a web crawler or web robot [6,7].

Web spiders are categorized as:

19.3.1 Well-Behaved Web Spider

Google bot is a well-behaved web spider that is also recognized as a search bot. Google search engine constructs a searchable index from the information accumulated by Google bot from the web [8].

19.3.2 Malicious Web Spider

Malicious web spider performs malicious activities on internet-based applications [9]. It may either collect sensitive information from a web page and later collected information is misused, or it may perform attacks such as denial-of-service or distributed denial-of-service attacks on web pages.

Thus, malicious web spider needs to be precluded to preserve information confidentiality. Hence, HIDS is proposed to identify malicious and well-behaved web spiders. It utilizes the signatures and rules developed from distinctive surfing features of web spiders for identification.

19.4 Proposed System

The proposed novel system consists of two stages: study and categorize. The study stage consists of two phases: reviewing the behavior of web spiders and upgrading the IDS to improve its detection performance in the categorize stage by designing rules developed from the ranking of features. Figure 19.1 shows the study stage architecture.

In reviewing phase, web spiders sent a request to a web server that is hosted on the shared server. Then, the web server responds to each request of a web spider and registers all requests in its access log file to study its behavior.

Each requisition entry in the server log comprises data such as the date of the requisition, time of requisition, server site name, server computer name, server IP, client to server method of requisition, client to server-uri-stem specify file requested, client to server-uri-query specify query executed, server port number, client to server username, client IP, client to server version specify protocol version, client to server user agent specify client application used to access the website, client to server cookie, client to server referrer specify requested URL, client to server host specify host header name, server to client status specify HTTP status, server to client sub status specify protocol status, server to client win 32 status specify windows status, server to client byte specify byte sent by the server, client to server byte specify byte sent by client and time-taken specify the total time of the session in milliseconds.

A specimen requisition of server log is illustrated below:

2016-10-14 12:15:30 W3SVC614 MDIN-PP-WB2 103.21.58.28 POST /img/apps/container-box.gif msg=add 80 admin 60.243.30.127 HTTP/1.0 Mozilla/5.0+(Windows+NT+6.1)+AppleWebKit/537.36+(-KHTML, +like+Gecko)+Chrome/45.0.2454.101+Safari/537.36 ASP.NET_SessionId=xtba5wmqfcivuz5 23qdozkhw http://www.partdo.co.in / www.partdo.co.in 200 3 64 5677 344 499.

Detailing of various fields in specimen record of server access log file is given in Table 19.1.

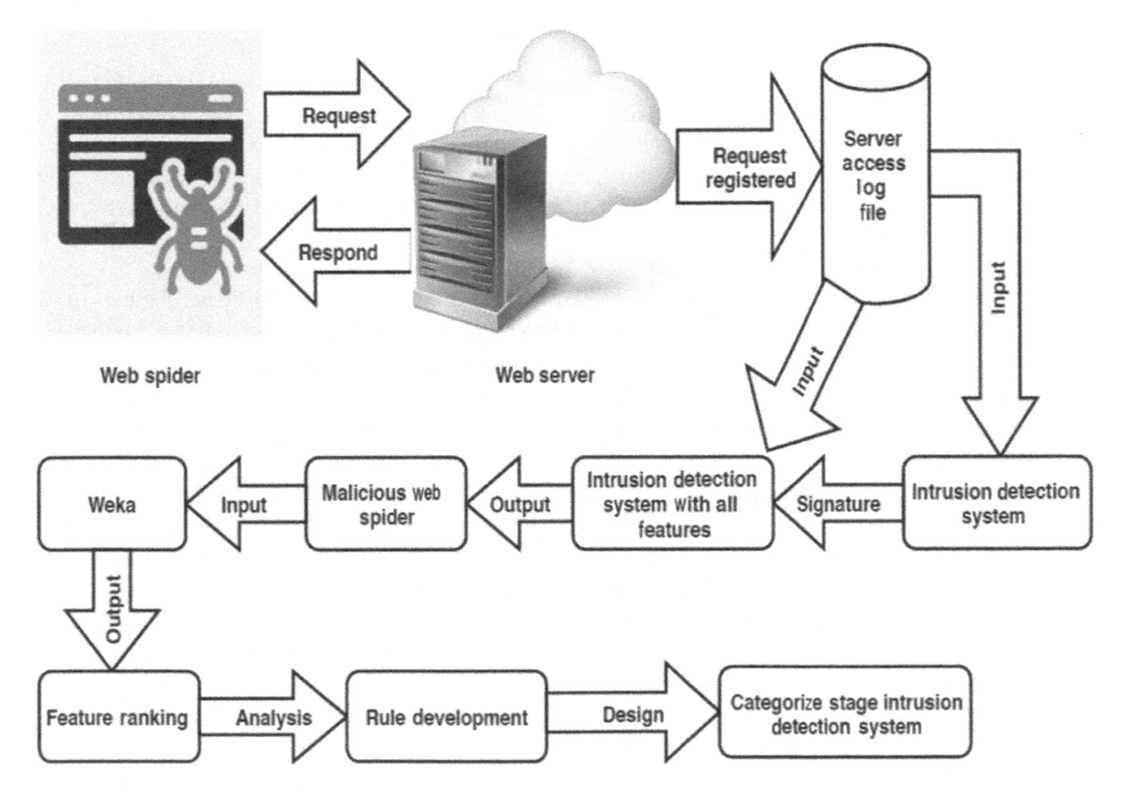

FIGURE 19.1 Study stage architecture.

TABLE 19.1

Server Log Record

Fields	Details
Date	2016-10-14
Time	12:15:30
Server Site Name	W3SVC614
Server Computer Name	MDIN-PP-WB2
Server IP	103.21.58.28
Client to Server Method	POST
Client to Server-uri-stem	/img/apps/container-box.gif
Client to Server-uri-query	msg=add
Server Port	80
Client to Server Username	Admin
Client IP	60.243.30.127
Client to Server Version	HTTP/1.0
Client to Server UserAgent	Mozilla/5.0+(Windows+NT+6.1)+AppleWebKit/537.36+(-KHTML,+like+Gecko)+Chrome/45.0.2454.101+Safari/537.36
Client to Server Cookie	ASP.NET_SessionId=xtba5wmqfcivuz523qdozkhw
Client to Server Referrer	http:// www.partdo.co.in /
Client to Server Host	www.partdo.co.in
Server to Client Status	200
Server to Client Substatus	3
Server to Client Win32status	64
Server to Client Byte	5677
Client to Server Byte	344
Time Taken	499

It is then inputted to an IDS for reviewing the behavior of a web spider. While reviewing the behavior of a web spider, an innovative feature of malicious web spider requests to a blank page is discovered. It has a decent likelihood of differentiating web spiders as well-behaved or malicious.

Then, each entry in the server log is evaluated to compute a median of previously identified eight [9–12] and one innovative features, which help in differentiating between surfing patterns of a web spider [13]. The features are as follows:

1. **Median of Error 4××:** A number is computed as the median of specious HTTP requisitions in a particular session. Mostly malicious web spiders request out-of-date or deleted pages, so they have a higher count of a specious request than well-behaved web spiders.
2. **Median of Successive Serial HTTP Requisition:** A number is computed as the median of succeeding serial HTTP requisitions in a particular session for web pages owned by the same website.
3. **Median of HTML-to-Image Ratio:** A number is computed as the median of HTML webpage requisitions over the image file requisition in a particular session. The HTML-to-Image ratio is mostly greater for malicious web spiders than well-behaved web spiders since well-behaved web spiders usually request HTML pages and overlook images on the website.
4. **Median of PDF File Requisition:** A number is computed as the median of PDF file requisitions in a particular session. Malicious web spiders mostly have higher PDF requisitions than well-behaved web spiders because well-behaved web spiders retrieve selectively, but malicious web spiders try to retrieve from every PDF files it encounters.
5. **Median of HTTP Requisitions of HEAD Type:** A number is computed as the median of HTTP HEAD type requisitions in a particular session. Mostly malicious web spiders' requisitions are of HEAD type and well-behaved web spiders' requisitions are of GET type.

6. **Median of Requisition Page's Depth Deviation:** A number is computed as the median of the page's requisition in a particular session. Malicious web spiders mostly have higher page requisition in a particular session than well-behaved web spiders.
7. **'Robots.txt' File Requisition:** Well-behaved web spiders never request robot.txt file as no linked pages are guiding to this file, but malicious web spiders mostly request for this file.
8. **Median of Unassigned Referrers Requisitions:** A number is computed as the median of unassigned referrer requisitions in a particular session. Malicious web spiders mostly pledge requisitions with unassigned referrer fields, whereas well-behaved web spiders pledge requisitions with assigned referrer fields.
9. **Median of Requisitions to Blank Page:** A number is computed as the median of requisitions to blank pages in a particular session. Malicious web spiders mostly have higher requisitions to blank pages as they request/creep blank pages to retrieve information if any.

Using the above features, signatures are developed for the detection of web spiders as well-behaved or malicious web spiders in the categorizing stage. Each feature signature is peak value estimated by the median count of the respective feature differentiating between surfing patterns of well-behaved and malicious web spiders.

Then, IDS is upgraded by designing signatures in it to detect a malicious web spider. The detection of a malicious web spider is evaluated using a machine learning algorithm: J48 classifier and multilayer perceptron (MLP) classifier on Waikato environment for knowledge analysis (Weka) to obtain web spider distinctive feature ranking [14].

19.4.1 Weka

Weka is a well-known set of machine learning software coded in Java. It is a worktop that contains a group of visualization tools and algorithms for data interpretation and prognostic modeling, jointly with graphical user interfaces for easy access to these functions. It supports various standard data mining functions, more precisely, data preprocessing, clustering, classification, regression, visualization, and feature selection [15]. In Weka, data needs to be transformed into a .arff (attribute-relation file format) file. It is an ASCII text file that describes a record of instances sharing a set of attributes that is inferred by Weka.

19.4.2 C4.5 Algorithm

C4.5 is an algorithm used to develop a decision tree. The decision tree developed can be utilized for categorization. C4.5 is frequently referred to as a statistical classifier [16]. It develops decision trees from a set of training data, using the theory of information entropy. The training data is a set $S=\{s_1, s_2, \ldots\}$ of previously classified samples. Each sample s_i comprises a p-dimensional vector $(x\{1,i\}, x\{2,i\},\ldots, x\{p, i\})$, where the x_j represents attribute values or features of the sample, as well as the class in which s_i cascades.

At each node of the tree, C4.5 selects the attribute of the data that most constructively break its set of samples into subsets improved in one class or the other. The breaking criterion is the normalized information gain. The attribute with the maximum normalized information gain is selected to make the decision. The C4.5 algorithm then returns on the smaller sublists.

19.4.3 J48 Classifier

J48 is an open-source Java implementation of the C4.5 algorithm in the Weka data mining tool. For the proposed system machine learning algorithm: J48 classifier in Weka,

A. Binary split value is set "false" so that binary split attributes are not used over nominal attributes while building a tree.
B. Confidence factor value is set to 0.25 that is used for pruning.

C. MinNumObj is the minimum number of instances per leaf and its value is set to 2.
D. NumFolds determines the amount of data used for reduced-error pruning and its value is set to 3.

A total of 285 instances are used for evaluation. J48 classifier evaluation for feature ranking is shown in Figure 19.2.

19.4.4 Multilayer Perceptron Classifier

An MLP classifier uses a back propagation technique for classification. It is a feed-forward machine learning model. It consists of a completely connected directed graph with input nodes, hidden layers, and output nodes.

The grey color square nodes represent input nodes. It is labeled with inputted distinctive features of a web spider. The hidden layers of MLP are represented by dark grey color nodes, and the output nodes are represented by light grey color nodes. They are labeled with classification output as malicious or well-behaved [17].

A. **Activation Function:** An MLP classifier has a linear activation function in all neurons, to be precise, a linear function that maps the weighted inputs to the output of each neuron.
B. **Layers:** The MLP consists of eight layers (one input layer and one output layer with six hidden layers) of nonlinearly activating nodes and is thus considered a deep neural network. Since an MLP is a completely connected network, each node in one layer connects with a certain weight to every node in the following layer.

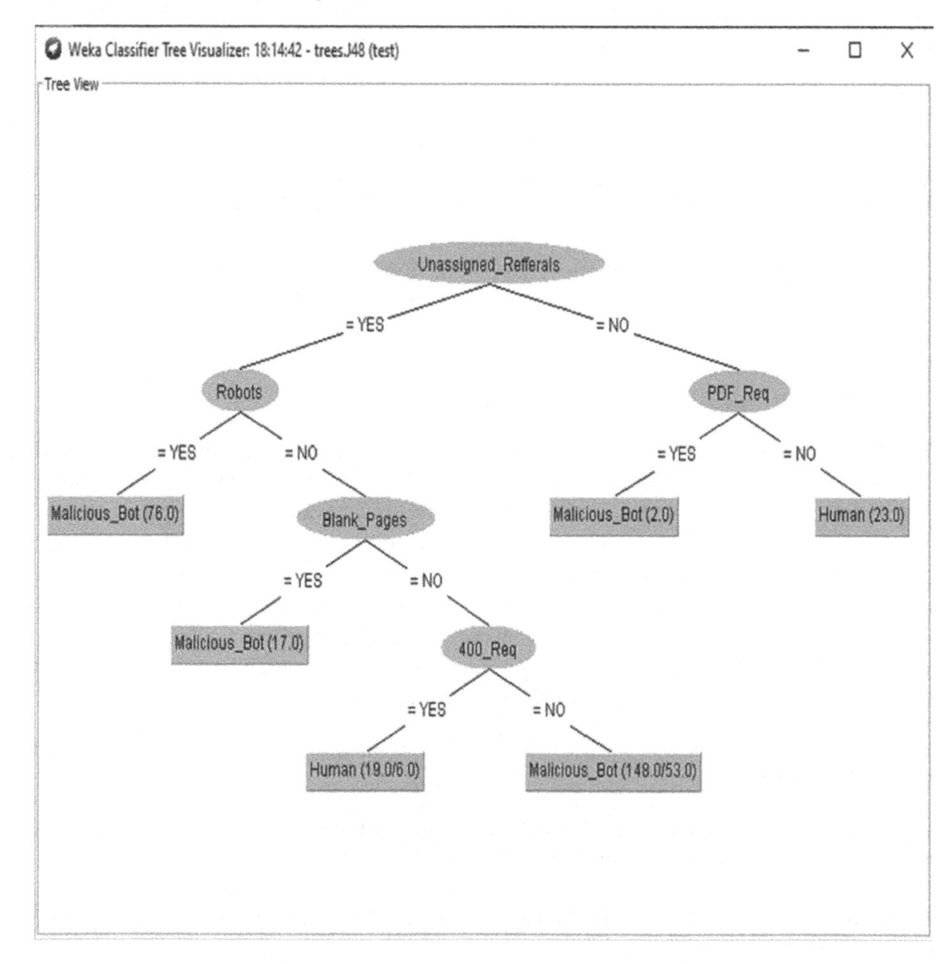

FIGURE 19.2 J48 Classifier evaluation for feature ranking.

C. **Learning through Back Propagation:** Error is computed by comparing real output with projected output. It is then backpropagated from the output layer node to the hidden layer node to change its connection weight.

For the proposed system machine learning algorithm: MLP in Weka,

A. Auto-build value is set "true," so that an MLP classifier network is added and connected to hidden layers automatically.
B. Hidden layer value is set to "a," where a=(attributes+classes)/2.
C. Learning rate is the number of weights updated, which is set to value 0.3 as it should be among 0 to 1.
D. Momentum value is set to 0.2 as it should be among 0 to 1. While updating, it is multiplied with weights.
E. Training time is the number of epochs to train, which is set to value 500.
F. Seed value is set to 0 as it should be ≥0. It makes a random number generator ready.
G. Validation threshold value is set to 20. It ends perceptron's learning process.

MLP classifier evaluation for feature ranking is shown in Figure 19.3.

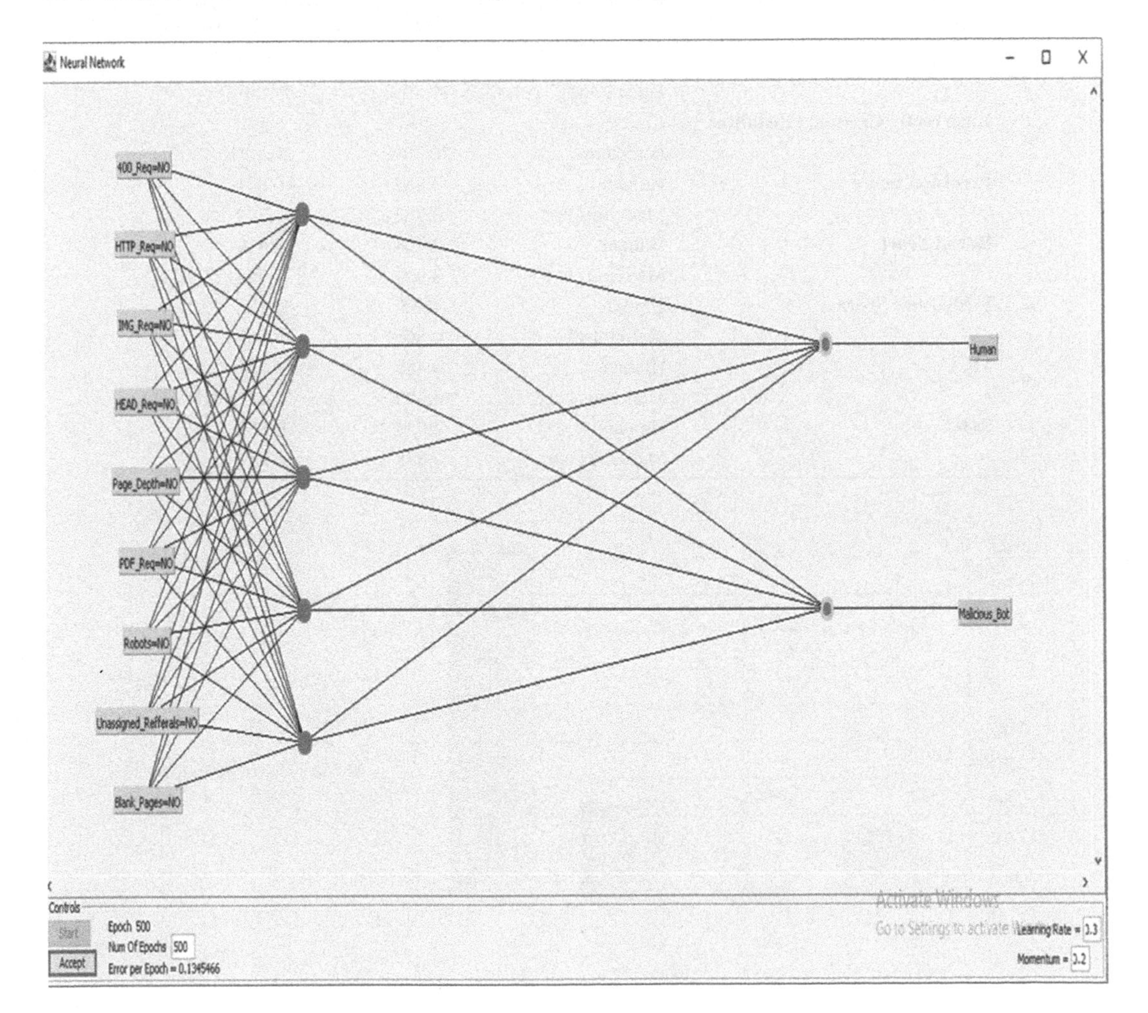

FIGURE 19.3 MLP Classifier evaluation for feature ranking.

Machine Learning Algorithms: J48 classifier and MLP classifier execution generate tally and percentile for the classification of instances as precise and imprecise. In-depth classification accuracy is generated in terms of precision, recall, F-measure, Matthews's correlation coefficient (MCC), and receiver operating characteristic (ROC) area score for well-behaved and malicious web spiders [18].

Table 19.2 exemplifies execution data of machine learning algorithms: J48 classifier and MLP classifier.

Machine Learning Algorithms: J48 classifier and MLP classifier execution generate confusion matrix [19]. It is a summary of the precise and imprecise classification of web spiders as a well-behaved and malicious using tally.

Figure 19.4 exemplifies machine learning algorithms: J48 classifier and MLP classifier confusion matrix. After experimentation, it is observed that out of 285 instances, 226 instances are correctly classified and 59 instances are incorrectly classified.

Output is evaluated on Weka to obtain distinctive feature ranking, which is shown in Figure 19.5. All nine distinctive features are ranked in descending order. The first three top-ranked features are robot.txt, PDF, and blank page.

From the feature ranking outcome, the first three features are selected and three innovative rules are derived, which has a decent likelihood for differentiating well-behaved and malicious web spiders.

TABLE 19.2

Execution Data of J48 and Multilayer Perceptron Classifier

Terms		J48	Multilayer Perceptron
Correctly Classified Instances	Count	226	226
	Percentage	79.29%	79.29%
Incorrectly Classified Instances	Count	59	59
	Percentage	20.70%	20.70%
Precision Score	Human	0.857	0.841
	Malicious bot	0.782	0.784
Recall Score	Human	0.404	0.416
	Malicious bot	0.969	0.964
F-Measure Score	Human	0.55	0.556
	Malicious bot	0.866	0.865
MCC	Human	0.489	0.487
	Malicious bot	0.489	0.487
ROC	Human	0.835	0.859
	Malicious bot	0.835	0.859

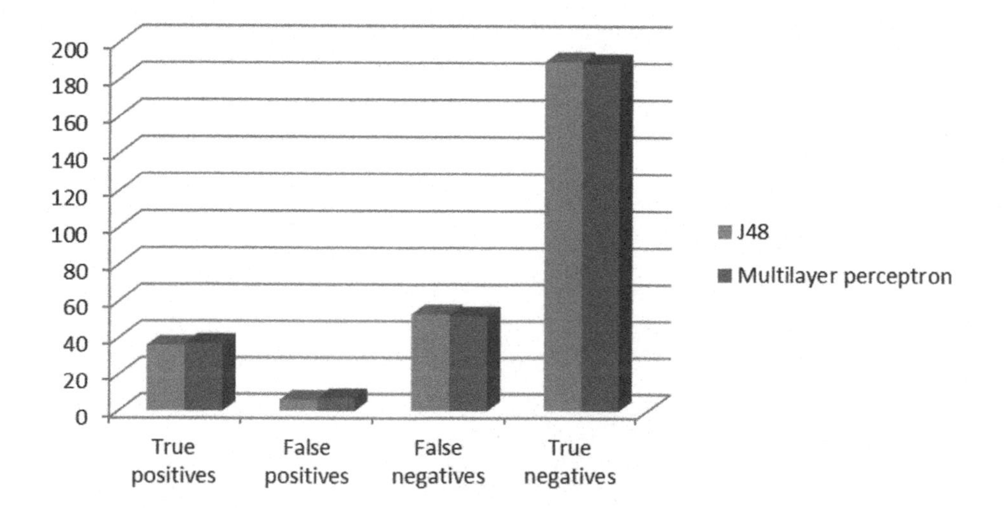

FIGURE 19.4 J48 and MLP Classifier confusion matrix tally.

```
Search Method:
        Attribute ranking.

Attribute Evaluator (supervised, Class (nominal): 10 class):
        Information Gain Ranking Filter

Ranked attributes:
 0.17979    7 Robots
 0.159696   6 PDF_Req
 0.149558   9 Blank_Pages
 0.114887   8 Unassigned_Refferals
 0.03528    3 IMG_Req
 0.005678   1 400_Req
 0.001899   2 HTTP_Req
 0.000193   5 Page_Depth
 0          4 HEAD_Req

Selected attributes: 7,6,9,8,3,1,2,5,4 : 9
```

FIGURE 19.5 Feature ranking.

Three innovative rules are illustrated below:

1. If a requisition is a robot.txt file and a PDF file, the requisition is classified as malicious.
2. If a requisition is a PDF file and a blank page, the requisition is classified as malicious.
3. If a requisition is a robot.txt file and a blank page, the requisition is classified as malicious.

Then, the IDS is again upgraded by designing rules in it to detect malicious web spiders.

Then, in the categorizing stage, a malicious web spider is detected using rule-based IDS. Figure 19.6 shows categorize stage architecture.

Web spider requisitions to a web server are registered in the webserver log. An IDS evaluates the requisitions on basis of nine features signature and three innovative rules that differentiate well-behaved and malicious web spiders. The requisition is classified as a malicious web spider by the IDS if it surpasses the peak tally of any one signature or fulfills any one rule.

19.5 Evaluation of Proposed System Using Machine Learning Algorithm

Attainment of proposed IDS in the detection of well-behaved and malicious web spiders is evaluated using machine learning algorithms: J48 classifier, IBk (instance-based learner) classifier, K-nearest neighbor (KNN) classifier, random tree classifier, and MLP on Weka.

Weka and J48 classifier are already discussed in Sections 19.4.1 and 19.4.3, respectively. Also, MLP is discussed in Section 19.4.4. Two hundred and forty-one instances are used for the attainment of evaluation.

19.5.1 IBk Classifier

KNN algorithm is implemented in the Weka data mining tool as an IBk classifier. In this, a distance measure is used to find k "close" instances for every test instance in the training data set; then, it uses

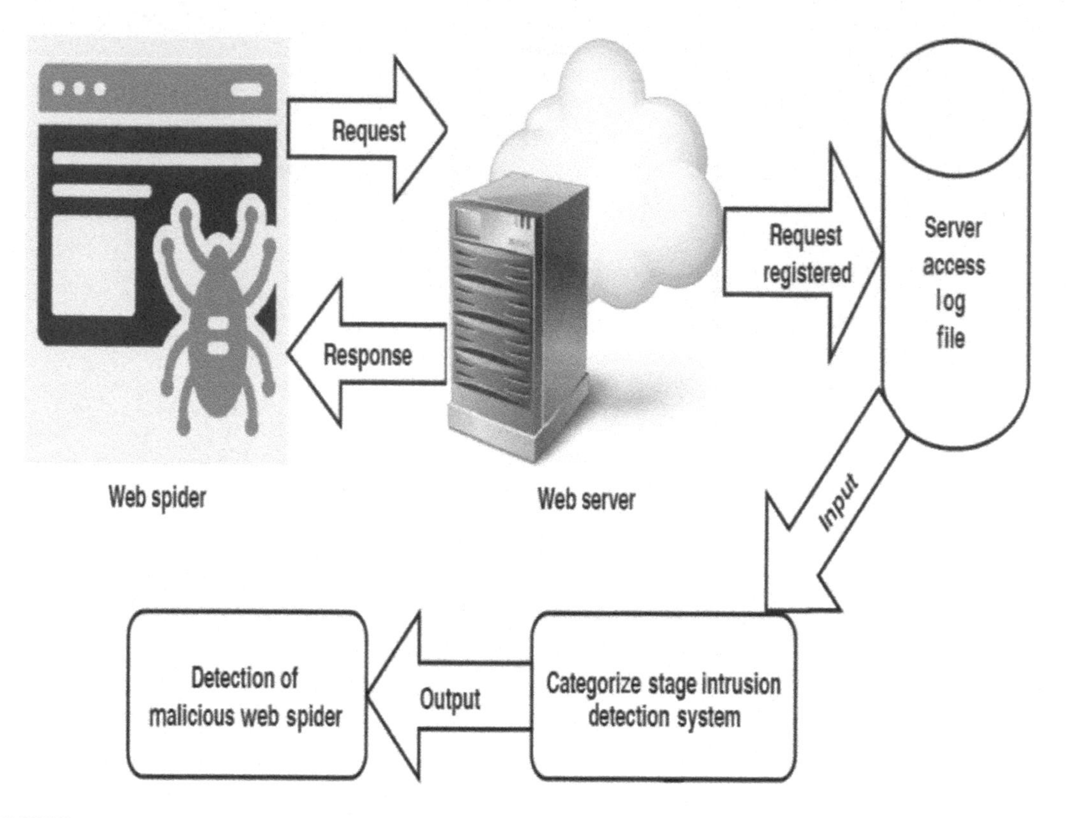

FIGURE 19.6 Categorize stage architecture.

those particular instances to make a classification. For the proposed system machine learning algorithm: IBk classifier in Weka,

- **A.** KNN specifies the number of neighbors to be used, and its value is set to 1.
- **B.** Batch-Size is the favored count of instances to process if a batch prediction is to be performed, and its value is set to 100.
- **C.** Cross-Validate value is set to false. The finest *K* value can be selected using it.
- **D.** If debug value is set to true, the classifier may output additional information to the console, so its value is set to false.
- **E.** Distance-Weighting acquires the distance weighting method used, and its value is set to no distance weighting.
- **F.** If the do-Not-Check-Capabilities value is set, the classifier capabilities are not checked before the classifier is built, and its value is set to false.
- **G.** Mean-Squared value is set to false. During cross-validation, it determines which error is utilized either mean squared or absolute error.
- **H.** Nearest-Neighbor-Search-Algorithm is set to LinearNNSearch. It searches the nearest neighbor.
- **I.** Num-Decimal-Places value is set to 2. It determines the digit of decimal places for the output.
- **J.** Window-Size value is set to 0. It indicates an unlimited count of instances for training.

19.5.2 Random Tree Classifier

It is a collaborative learning algorithm that generates numerous individual learners. It employs bagging knowledge to produce a random set of data for building a decision tree. For the proposed system machine learning algorithm: random tree classifier in Weka,

A. K-value sets the number of randomly chosen attributes. If 0, int(log_2(#predictors) + 1) is used.
B. Allow-Unclassified-Instances decide whether to allow unclassified instances, and its value is set to false.
C. Batch-Size is the favored number of instances to process if a batch prediction is being performed, and its value is set to 100.
D. Break-Ties-Randomly is used to break ties arbitrarily when several attributes appear similarly good, and its value is set to false.
E. If debug value is set to true, the classifier may output additional information to the console, so its value is set to false.
F. If the do-Not-Check-Capabilities value is set, the classifier capabilities are not checked before the classifier is built, and its value is set to false.
G. Max-Depth determines the maximum depth of the tree, and its value is set to 0 for unlimited.
H. Min-Num value is set to 1.0. It determines the least count of the instances on every leaf.
I. minVarianceProp value is set to 0.001. It is the minimum proportion of variance needed to perform splitting.
J. Num-Decimal-Places value is set to 2. It determines the digit of decimal places for the output.
K. Num-Folds value is set to 0, which indicates no back-fitting.
L. Seed value is set to 0 as it should be ≥0. It makes a random number generator ready.

Attainment of proposed IDS in the detection of well-behaved and malicious web spiders is evaluated using machine learning algorithms: J48 classifier, IBk - KNN classifier, random tree classifier, and MLP classifier on three terms: 1–8 (previously identified features differentiating between well-behaved and malicious web spiders), 1–9 (previously identified features and innovative feature differentiating between well-behaved and malicious web spiders) and Fully (previously identified features, innovative feature, and rules developed differentiating between well-behaved and malicious web spiders).

Classifier execution generates tally and percentile for the classification of instances as precise and imprecise. In this classification, accuracy is generated in terms of precision, recall, F-measure, MCC, and ROC area score for well-behaved and malicious web spiders.

Table 19.3 exemplifies execution data of machine learning algorithms: J48 classifier.

Table 19.4 exemplifies execution data of machine learning algorithms: IBk - KNN classifier.

Tables 19.5 exemplifies execution data of machine learning algorithms: random tree classifier.

Tables 19.6 exemplifies execution data of machine learning algorithms: MLP classifier.

TABLE 19.3
Execution Data of J48 Classifier

Terms		1–8	1–9	Fully
Correctly Classified Instances	Count	192	194	194
	Percentage	79.67%	80.50%	80.50%
Incorrectly Classified Instances	Count	49	47	47
	Percentage	20.33%	19.50%	19.50%
Precision Score	Human	0.829	0.879	0.879
	Malicious bot	0.791	0.793	0.793
Recall Score	Human	0.403	0.403	0.403
	Malicious bot	0.964	0.976	0.976
F-Measure Score	Human	0.542	0.552	0.552
	Malicious bot	0.869	0.875	0.875
MCC	Human	0.477	0.505	0.505
	Malicious bot	0.477	0.505	0.505
ROC	Human	0.827	0.859	0.859
	Malicious bot	0.827	0.859	0.859

TABLE 19.4

Execution Data of IBk–K-Nearest Neighbor Classifier

Terms		1–8	1–9	Fully
Correctly Classified Instances	Count	194	195	195
	Percentage	80.50%	80.91%	80.91%
Incorrectly Classified Instances	Count	47	46	46
	Percentage	19.50%	19.09%	19.09%
Precision Score	Human	0.838	0.641	0.641
	Malicious bot	0.799	0.913	0.913
Recall Score	Human	0.431	0.819	0.819
	Malicious bot	0.964	0.805	0.805
F-Measure Score	Human	0.569	0.720	0.720
	Malicious bot	0.874	0.855	0.855
MCC	Human	0.502	0.588	0.588
	Malicious bot	0.502	0.588	0.588
ROC	Human	0.884	0.898	0.898
	Malicious bot	0.884	0.898	0.898

TABLE 19.5

Execution Data of Random Tree Classifier

Terms		1–8	1–9	FULLY
Correctly Classified Instances	Count	194	195	195
	Percentage	80.50%	80.91%	80.91%
Incorrectly Classified Instances	Count	47	46	46
	Percentage	19.50%	19.09%	19.09%
Precision Score	Human	0.838	0.641	0.641
	Malicious bot	0.799	0.913	0.913
Recall Score	Human	0.431	0.819	0.819
	Malicious bot	0.964	0.805	0.805
F-Measure Score	Human	0.569	0.720	0.720
	Malicious bot	0.874	0.855	0.855
MCC	Human	0.502	0.588	0.588
	Malicious bot	0.502	0.588	0.588
ROC	Human	0.884	0.898	0.898
	Malicious bot	0.884	0.898	0.898

TABLE 19.6

Execution Data of Multilayer Perceptron Classifier

Terms		1–8	1–9	Fully
Correctly Classified Instances	Count	194	195	195
	Percentage	79.67%	80.91%	80.91 %
Incorrectly Classified Instances	Count	49	46	46
	Percentage	20.33%	19.09%	19.09 %
Precision Score	Human	0.621	0.641	0.641
	Malicious bot	0.911	0.913	0.913
Recall Score	Human	0.819	0.819	0.819
	Malicious bot	0.787	0.805	0.805
F-Measure Score	Human	0.707	0.72	0.72
	Malicious bot	0.844	0.885	0.885
MCC	Human	0.568	0.588	0.588
	Malicious bot	0.568	0.588	0.588
ROC	Human	0.881	0.895	0.895
	Malicious bot	0.881	0.895	0.895

It is interesting to note that evaluating the attainment of proposed IDS in the detection of well-behaved and malicious web spiders using machine learning algorithms presents improvement in the percentile of accurately classified instances ~81%.

Machine Learning Algorithms: J48 classifier, IBk - KNN classifier, random tree classifier, and MLP classifier execution generates confusion matrix. It is a summary of the precise and imprecise classification of web spiders as a well-behaved and malicious using tally. It encompasses a tally of false positives, false negatives, true positives, and true negatives instances of classification.

Figure 19.7 exemplifies machine learning algorithms: J48 classifier confusion matrix. After experimentation, it is observed that out of 241 instances, 194 instances are correctly classified and 47 instances are incorrectly classified on term Fully (previously identified features, innovative feature, and rules developed differentiating between well-behaved and malicious web spiders). It concludes that there is a reduction in misclassification of web spiders.

Figure 19.8 exemplifies machine learning algorithms: IBk classifier, KNN classifier confusion matrix. After experimentation, it is observed that out of 241 instances, 195 instances are correctly classified and 46 instances are incorrectly classified on term Fully (previously identified features, innovative feature, and rules developed differentiating between well-behaved and malicious web spiders). It concludes that there is a reduction in misclassification of web spiders.

Figure 19.9 exemplifies machine learning algorithms: random tree classifier confusion matrix. After experimentation, it is observed that out of 241 instances, 195 instances are correctly classified and 46 instances are incorrectly classified on term Fully (previously identified features, innovative feature, and

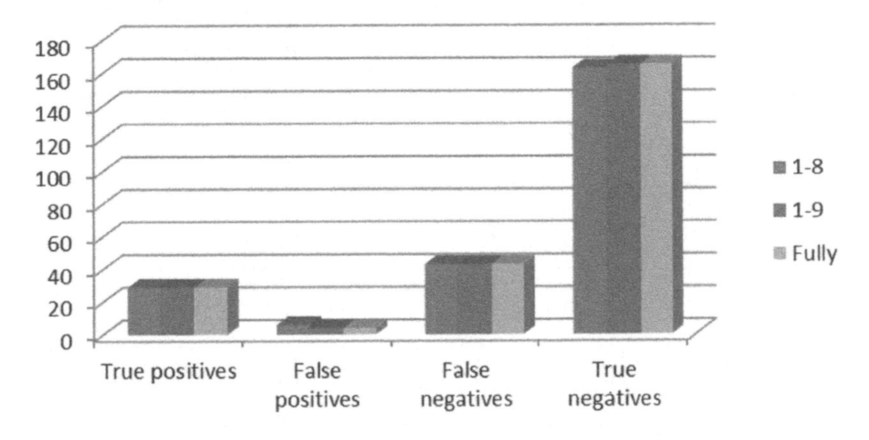

FIGURE 19.7 J48 Classifier confusion matrix tally.

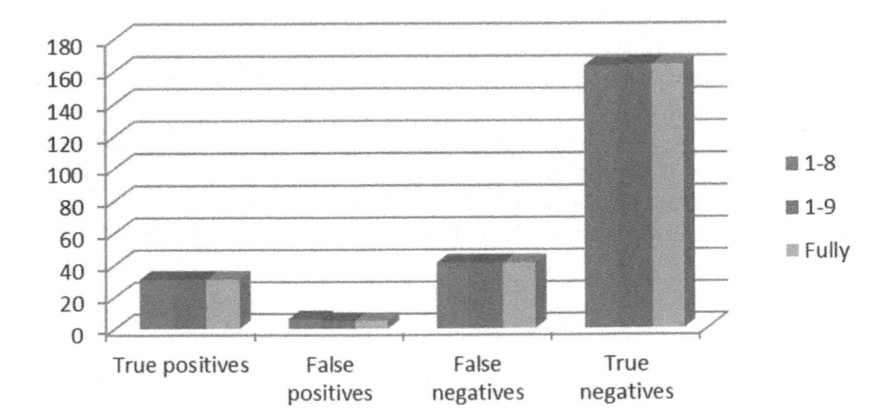

FIGURE 19.8 IBk - K-nearest neighbor's classifier confusion matrix tally.

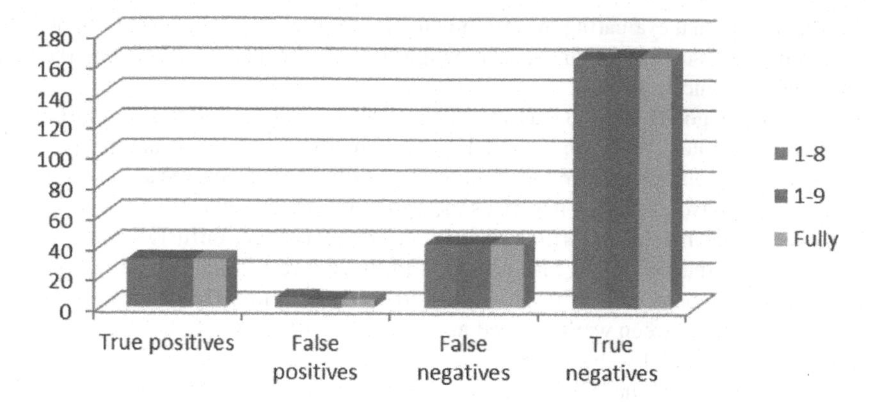

FIGURE 19.9 Random tree classifier confusion matrix tally.

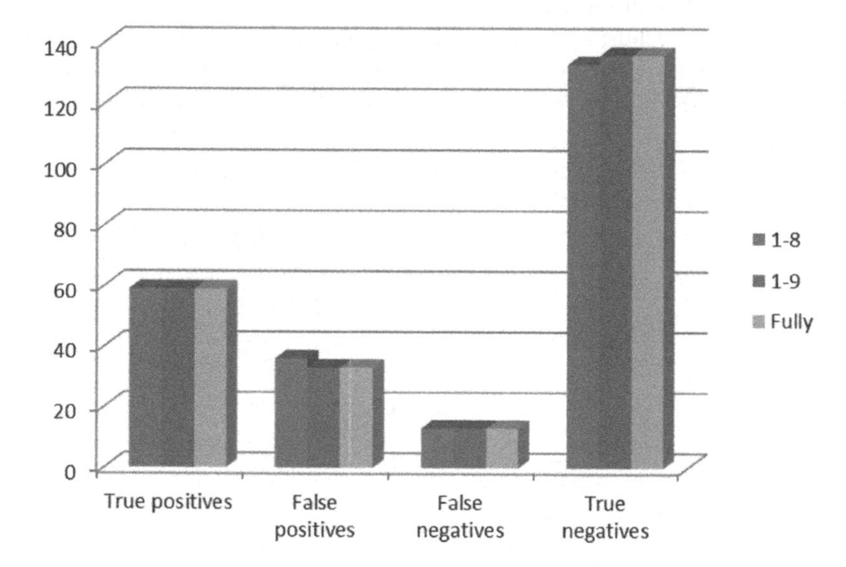

FIGURE 19.10 MLP classifier confusion matrix tally.

rules developed differentiating between well-behaved and malicious web spiders). It concludes that there is a reduction in misclassification of web spiders.

Figure 19.10 exemplifies machine learning algorithms: MLP classifier confusion matrix. After experimentation, it is observed that out of 241 instances, 195 instances are correctly classified and 46 instances are incorrectly classified on term Fully (previously identified features, innovative feature, and rules developed differentiating between well-behaved and malicious web spiders). It concludes that there is a reduction in misclassification of web spiders.

It is remarkable to note that the evaluation of IDS with rule-based technique to detect malicious web spiders using machine learning algorithm reports that there is a reduction in misclassification of web spiders.

19.6 Conclusion

Today, most traditional facilities are delivered as mobile applications, so that they can be accessed at anytime from anywhere via the internet. Also, most services and data are accessible online. Hence, data security is a prerequisite of the online world. A program named web spider is designed to procure data from online pages. It is initialized with a seed uniform resource locator of a web page. Then, it skims a web page and navigates all pages linked to obtaining data. Therefore, a system is proposed to detect malicious web spiders.

In the beginning, an IDS is interpreted along with its classification. Then, web spider definition, its working, and categorization are briefly explained. Features differentiating between surfing patterns of web spiders are summarized. Origin of novel unique feature blank page requisition of malicious web spiders that enhances detection precision is studied.

Using feature ranking, the discovery of three outstanding features differentiating between surfing patterns of web spiders is outlined. Deriving three rules from feature ranking outcome and its designing in the proposed system are illustrated.

Then, the detection of malicious web spider skimming websites using the proposed IDS with rule-based technique is illustrated. Attainment of IDS with rule-based technique to detect malicious web spiders is evaluated using machine learning algorithms: J48 classifier, IBk classifier, KNN classifier, random tree classifier, and MLP on Weka. Nearly 81% of requisitions are precisely detected. The confusion matrix reports a reduction in misclassification of web spiders.

REFERENCES

1. DeXiang Zhang, DiFan Zhang and Xun Liu, "A novel malicious web crawler detector: Performance and evaluation," *IJCSI International Journal of Computer Science*, 10(1), 1–6, 2013.
2. Andoena Balla and Marios D. Dikaiakos, "Real-time web crawler detection," *IEEE 18th International Conference on Telecommunications (ICT)*, Ayia Napa, Cyprus, 2011.
3. Giovanni Vigna, William Robertson, Vishal Kher and Richard A. Kemmerer, "Stateful intrusion detection system for world-wide web servers," *IEEE Computer Security Applications Conference*, Las Vegas, NV, December 2003.
4. wikipedia.org. "Intrusion detection system," accessed February 26, 2020. https://en.wikipedia.org/wiki/Intrusion_detection_system.
5. Herv_e Debar, "An introduction to intrusion-detection systems," *Proceedings of Connect*, Citeseer, 2000.
6. wikipedia.org, "Web_crawler," accessed February 1, 2020. https://en.wikipedia.org/wiki/Web_crawler.
7. Raja Iswary and Keshab Nath, "Web crawler," *International Journal of Advanced Research in Computer and Communication Engineering*, 2, 4009–4012, 2013.
8. Marios D. Dikaiakos , Athena Stassopoulou and Loizos Papageorgiou, "Characterizing crawler behavior from web server access logs," *4th International Conference on E-Commerce and Web Technology, EC-Web*, Prague, Czech Republic, 2003.
9. Dusan Stevanovic, Aijun An and Natalija Vlajic, "Feature evaluation for web crawler detection with data mining techniques," *Elsevier Expert System with Application*, 39(10), 8707–8717, 2012.
10. Aviral Nigam, "Web crawling algorithms," *International Journal of Computer Science and Artificial Intelligence*, 4, 63–67, September 2014.
11. Rahul kumar, Anurag Jain and Chetan Agrawal. "Survey of web crawling algorithms," *International Journal of Advances in Vision Computing*, 1, 1–8, 2014.
12. Junsup Lee, Sungdeok Cha, Dongkun Lee, and Hyungkyu Lee, "Classification of web robots: An empirical study based on over one billion requests," *Computers and Security*, 28(8), 795–802, 2009.
13. Nilambari Narkar, and Narendra Shekokar, "A rule based intrusion detection system to identify vindictive web spider," *IEEE CAST*, Pune, India, 2016.
14. Arief Rakhman, Goeij Yong Sun and Rama Catur APP, "Building artificial neural network using Weka software," *Information System Department, Sepuluh Nopember Institute of Technology*, Surabaya, Indonesia.
15. wikipedia.org, "Weka," accessed February 25, 2020. https://en.wikipedia.org/wiki/Weka_(machine_learning).
16. wikipedia.org, "C4.5Algorithm," accessed February 25, 2020. https://en.wikipedia.org/wiki/C4.5_algorithm.

17. wikipedia.org. "MultilayerPerceptron," accessed February 27, 2020. https://en.wikipedia.org/wiki/Multilayer_perceptron.
18. Madhavi Dhingra , S C Jain and Rakesh Singh Jadon, "Malicious intrusion detection using machine learning schemes," *IJEAT*, 8(6), 1–5, 2019.
19. Mohammad Almseidin and Maen Alzubi, "Evaluation of machine learning algorithms for intrusion detection system," *International Symposium on Intelligent Systems and Informatics (SISY),* Subotica, Serbia, 2017.

20

Application of Machine Learning to Improve Tourism Industry

Krutibash Nayak and Saroj Kumar Panigrahy
VIT-AP University

CONTENTS

20.1 Introduction 290
20.1.1 Types of Tourism 290
20.2 Industries Related to Tourism 291
20.2.1 Hotels 291
20.2.2 Restaurants 291
20.2.3 Retail and Shopping 292
20.2.4 Transportation 292
20.2.5 Travel Agencies 292
20.2.6 Tour Operator 292
20.2.7 Cultural Industries 292
20.3 Challenges of Tourism Industry 292
20.3.1 Taxation 292
20.3.2 Globalization 292
20.3.3 Marketing 292
20.3.4 Security 293
20.3.5 Terrorism 293
20.3.6 Infrastructure 293
20.3.7 Economy 293
20.3.8 Culture 293
20.3.9 Environment 293
20.3.10 Competition 294
20.3.11 Social Media 294
20.3.12 Pandemic 294
20.3.13 Price 294
20.4 Machine Learning and Deep Learning in Tourism 294
20.4.1 Machine Learning 294
20.4.1.1 Supervised Learning 295
20.4.1.2 Unsupervised Learning 295
20.4.1.3 Reinforcement Learning 295
20.4.2 Deep Learning 295
20.5 Sentiment Analysis 296
20.5.1 Block Diagram of Sentiment Analysis 296
20.5.2 Different Types of Sentiment Analysis 297
20.5.2.1 Fine-Grained 297
20.5.2.2 Emotion Detection 297
20.5.2.3 Aspect-Based 297
20.5.2.4 Intent Analysis 297

20.5.3 Sentiment Analysis Datasets ... 297
20.5.4 Sentiment Analysis Methods ... 298
20.5.4.1 Rule-Based Sentiment Analysis ... 298
20.5.4.2 Automated Sentiment Analysis ... 298
20.5.5 Sentiment Classification Techniques ... 299
20.5.6 Models for Sentiment Analysis ... 300
20.5.6.1 Logistic Regression ... 300
20.5.6.2 Random Forest ... 301
20.5.6.3 Support Vector Machine (SVM) ... 301
20.5.6.4 Naive Bayes Classifier ... 301
20.6 Sentiment Analysis-Based Research on Tourism ... 301
20.6.1 Sentiment Analysis Based on Environment ... 303
20.6.2 Sentiment Analysis Based on Locations and Hotels ... 303
20.6.3 Sentiment Analysis Based on Language ... 304
20.6.4 Sentiment Analysis Based on Trip Length ... 305
20.7 Summary ... 305
References ... 306

20.1 Introduction

Tourism is defined as "a collection of activities, services, and industries which deliver a travel experience comprising transportation, accommodation, eating and drinking establishments, retail shops, entertainment businesses, and other hospitality services provided for individuals or groups traveling away from home" [1]. According to the United Nations World Tourism Organization, "Tourism comprises the activities of persons travelling to and staying in places outside their usual environment for not more than one consecutive year for leisure, business, and other purposes" [1].

Tourist is an individual who travels the places for his/her interest and pleasure. So the behavior of tourists is an important indicator for future planning of the tourism sector. Tourism is the growing industry, and it heavily affects the GDP of a country. It is considered as an important sector in a country as it provides good opportunity for employment directly and indirectly. As the social media has evolved rapidly in recent years, tourism becomes the largest industry in India and globally. The industry contributed around 2300 billion USD to the economy globally [2]. The tourism industry provides 8.78% of job opportunities in India and also contributes toward 6.23% of GDP. According to a report by the World Travel and Tourism Council, India is going to be a hot spot of tourism, with a 10-year highest growth potential [3]. According to the report, India's tourism and travel industry was the second largest employer in the world in 2019.

20.1.1 Types of Tourism

Tourism is classified as many types under different circumstances and using different criteria. There are six important types of tourism defined by Tureac and Turtureanu [4] and are listed as follows:

i. Visiting tourism
ii. Professional tourism
iii. Relaxing (leisure) tourism
iv. Reduced distance tourism
v. Healthcare tourism
vi. Transit tourism

Some tourism is related to enjoying holidays, and some fall under productivity and personal or official activity. Leisure tourism is the major economic center for many countries, which solely depend on the tourism industry for its economic activity [5,6]. According to the Federation of Indian Chambers of Commerce and Industry, more than 11 million people visit foreign countries for medical treatment [7]. The United States and Europe are the major destinations for medical tourism there due to world standard facilities available. India also generates revenue around 3 billion USD for the year 2015 for medical travel of the tourists. The plan for 2020 is to increase 9 billion USD [7].

Tourism has different forms on the basis of the purpose of visit and nature of visit [8]. These are as follows:

- Ecotourism
- Geotourism
- Space tourism
- Atomic tourism
- Virtual tourism
- Beach tourism
- Rural tourism
- Sports tourism
- Medical tourism
- Religious tourism
- Industrial tourism
- Adventure tourism
- Sex tourism
- Cultural tourism
- Bicycle tours
- War tourism
- Wildlife tourism
- Nature-based tourism
- Sustainable tourism
- Meetings, Incentives, Conventions, and Exhibitions (MICE) tourism

20.2 Industries Related to Tourism

This section describes different industries that are closely associated with tourism. These industries are the core strength of the tourism industries. They generate huge revenues and employment along with their due business. They target the tourism industry for the customers and apply the best business model for it.

Depending upon the characteristics and objectives of the travel, tourist demand for facilities, and services, there are several wide ranges of commercial activities that have been associated with the tourism industry [8]. So, the following industries are associated with the tourism sector for its closeness and proximity of business models.

20.2.1 Hotels

Hotels are the mercantile institutions that provide accommodations, other amenities, and helps. Hotel industry takes a significant role in the tourism industry. All the tourists need a hotel to stay at their destinations and require more amenities, guest services, and facilities for their needs.

20.2.2 Restaurants

The restaurants are merchandise business establishments that serve food and beverages to consumers. Restaurants are helpful for the tourists to have food and beverages and to experiment with local cuisines.

20.2.3 Retail and Shopping

It is very important for tourists to shop and fulfill their needs for necessary items and souvenirs. It also offers world-class experience for the tourist during shopping. Cities like London, Paris, Dubai, and Milan are famous for shopping outlets and fashion.

20.2.4 Transportation

Transportation is the movement of goods and people from a place to different places. A well-developed transportation system is the integral part of tourist destinations. It also builds tourist attraction, and good infrastructure helps to attract tourists to the destination with a proper planning and blissful journey.

20.2.5 Travel Agencies

It is also a retailing business that deals with products and services related to travel by customers on top of different other services like airlines, car rental, cruise, hotels, and sightseeing. They plan the travel itinerary for the customers, tourists, and clients for the necessary arrangements of their travel, stay, and sightseeing with other facilities as well.

20.2.6 Tour Operator

Tour operator is the person who assembles various elements of travel. It consists of tour and travel elements for a holiday. His role signifies in such a way that most of the tourists prefer a tour operator during the travel.

20.2.7 Cultural Industries

This industry is accountable for the creation, production, and distribution of products and services, which come under intellectual property rights.

20.3 Challenges of Tourism Industry

This section explains different challenges that are encountered by the tourism sector. These challenges are the external and internal factors that must be regulated properly to make the industry profitable. The chapter mainly describes those parameters that influence the tourist and stakeholders.

The tourism sector also faces some challenges that hinder its growth. The various challenges are taxation, globalization, marketing, security, terrorism, infrastructure, economy, culture, environment, competition, social media, pandemic, and price.

20.3.1 Taxation

The tourism sector is heavily taxed as airline tickets and hotels are paid huge tax. These taxes increase the cost of the travel for the tourist. In the competitive market like this industry, it is too difficult to continue with sustained growth.

20.3.2 Globalization

Globalization is also another challenge for the industry. Due to global standardization, it is difficult to maintain uniqueness in luxury. It decentralizes the local market for the tourists. It creates a less distinct locale.

20.3.3 Marketing

Marketing is the key role in this industry. The Internet usage and social media usage increased heavily in recent years, which also invite fraudulent agents to grow. The people take advantage of the technology to show the false information as well as inadequate data, and the tourist falls in the trap of such activities. It also increases monetary fraud, which decreases the tourist entry of those places. It is also too difficult to create a perception in the tourist attitude by using new technology. It requires high creativity and endurance.

20.3.4 Security

Security is the important factor in tourist footfall. In recent surveys conducted by different travel organizations, it is found that tourists are not interested to visit those places, which are volatile in terms of security. Security includes terrorism, civil war, political instability, racism, human trafficking, and conflict with neighbors. Tourists are susceptible to security threats related to crime, violence, and health risk [9]. The footfall of the tourist decreases rapidly with security issues.

20.3.5 Terrorism

Terrorist attacks in different countries affect the tourism of the country as the visits by tourists to those countries decrease drastically. The threats for tourists from terrorism are high as visitors are soft targets and pronounced symbols of foe. Tourist attractions are crucial targets to attack by the adversaries [10]. After the Paris terrorist attack in 2015, France has witnessed reduced foreign as well as domestic tourists to Paris [11]. The industry has to conduct recovery marketing or marketing integrated with full crisis management activities. Tourists were dislocated from Portugal to other regions due to terrorist attacks in Spain and Russia [12].

20.3.6 Infrastructure

Infrastructure of some of the tourist destinations is ill maintained. Some tourist site infrastructure is underdeveloped or outdated. In some tourist places, basic amenities are not provided, so tourists avoid visiting these places. The infrastructure challenge includes better public transportation, immigration at airports, state-of-the art facilities in guesthouses or hotels nearby.

20.3.7 Economy

Another challenge that the tourism industry faces is the economic situation of the country. If the economy of the country is growing, then it spends a lot on infrastructure of the places and it provides facilities for the industry to develop. It shows that that country that is developed and developing category attracts large tourists compared to underdeveloped countries [9,10].

20.3.8 Culture

Different cultures also add stress to the industry growth. Some cultures that are not accepted by the locals put obstacles for the inflow of tourists in that area.

20.3.9 Environment

The environment factor also adds challenges to the tourism industry. This factor, as unpredictable sometimes, impacts the huge economic crisis in this sector. Recently, the oil spilling crisis near Mauritius has had a huge impact in the tourism industry of the country. The environment and the ecosystem are affected by this incident. The coral reef, mangrove forest, and special species, which are only found in the region, are the main tourist attraction. Now due to the degradation of the environment, mainly marine life will reduce the tourist footfall in the country, which heavily depends on the tourism sector for its economy. The natural disasters have a huge impact on tourism due to fear and anxiety among tourists for the visit of the tourist places. Australian bush fire that happened in the year 2019–2020 is the factor of climate change that attracts the attention of the globe and the visitors have dropped the plan to visit Australia and it costs around 4.5 billion AUD [13].

20.3.10 Competition

To provide the facilities to attract the tourists is also a challenging task for the industry players. In the competitive market, different stakeholders compete among themselves for the marginal profit to sustain.

20.3.11 Social Media

Social media with increase in Internet and easy accessibility also provide the challenge among the stakeholders to formulate a proper business plan. The rising role of social media in influencing travel habits and destinations also makes it a key powerhouse of travel-related insights. Now, instant availability of the details about the places and services makes the challenge for the players to think and add creativity in the business. Social media provides a huge amount of data daily, and it is difficult to analyze the data and identify the tourist behavior. Twitter trends show that around 500 million tweets a day are tweeted [14]. It also shows a pattern of sentiments for the particular service or places. During the pandemic situation, Twitter provides the actual sentiments regarding the prediction of industry growth.

20.3.12 Pandemic

Pandemic is also another measure now to be considered in the development of the industry. Recently, corona affects and almost destroys the tourism and hospitality industry. It affects almost all the countries in terms of hotel booking and airline reservation [15,16]. So, the tourist flow is reduced miserably and it will take a lot of time to show the normalcy. The Ebola outbreak in West Africa had impacted heavily in the tourist spots in Africa during 2014 [17]. The travel risk perception by the tourist due to use of social media networks has resulted in the downfall of tourist visits to the nations affected by Ebola [18].

20.3.13 Price

Price challenge is the factor that is considered seriously in the industry. Due to high competition in the market, the hotel industry that is the part of the tourism industry faces the challenges for sustainability. Hoteliers or proprietors have implemented revenue management for their assets inventory optimization and maximizing revenues and profits using price offers. The same method is also applied in the transport industry, which is involved in this sector and grows with respect to tourism sectors [19,20].

20.4 Machine Learning and Deep Learning in Tourism

This section describes the types of machine learning (ML) and deep learning (DL) methods, which are used in the tourism industry. The sector applies different algorithms to improve the service for the customer. Different methods are applied to find the opportunity for the improvement and to take the competitive advantages. This section covers different types of ML methods and DL applications in tourism sectors.

20.4.1 Machine Learning

ML can be stated as the procedures of emulating the learning process of the human brain by machines. Machines can recognize patterns, develop the skills and knowledge to improve by themselves, and do most of the work on their own without being explicitly programmed. It predicts based on its previous knowledge and experience. It enables machines to make data-driven decisions. A model is created by ML algorithms, which is trained using sample training data, and test input is given to predict the output. The predicted output is examined against the acceptable accuracy. If the prediction is satisfactory, then the model is implemented for use.

There are three types of ML models—supervised, unsupervised, and reinforcement learning, which are used. These are described as follows.

20.4.1.1 Supervised Learning

It is the one where learning is guided by previous knowledge and with trained dataset. We have the training dataset that is used to train the model, and once the model is ready, it can start making predictions and decisions when test data or new data is given to it.

20.4.1.2 Unsupervised Learning

The model learns through observation and identifies the structures in the given dataset. It automatically finds the pattern and relationships in the dataset. It only separates the data and clusters them in different groups according to their similarity, but it cannot assign any label to the dataset. It is used where the structure of data changes dynamically.

20.4.1.3 Reinforcement Learning

It interacts with the environment and explores the best outcome. It follows the hit and trial concepts. An agent is rewarded or penalized based upon the correct and wrong answers. The model trains itself if it gets a positive reward point. If the prediction is not favorable, the algorithm is forced to reiterate until it finds the result better than the previous result. This algorithm is used in the shortest route between two places on a map.

ML algorithms are used by town planners to take steps to identify the places where tourists have shown interest. It is used to identify proper tourism recommendations [21]. By using recommendation systems, hotels, guides, and tour operators are generating revenues.

20.4.2 Deep Learning

DL is the part of ML. With huge amounts of data produced daily in terms of text, audio, and video news articles, it is a cumbersome task to find the actual data. So ML algorithms also fail sometimes to analyze such huge data and predict a business model. DL is related to the learning that allows a system to rediscover itself by the means to organize raw data. Different ML algorithms are used and improved to implement the DL methods.

DL is the statistical learning that extracts features or attributes from raw datasets. It uses multilayer neural networks with many hidden layers stacked one after another. It uses sophisticated algorithms and requires high computation power resources. DL is used in many complicated and sophisticated data analysis like stock market prediction, gambling, weather forecasting, economic activities, sales, customer review analysis, health care, education, transport, construction, and insurance. DL provides the facility to identify the sentiment of the tourist by analyzing the tweets from the tourist.

Applications of DL in travel sectors are as follows:

- Fraud detection
- Customer support
- Recommender systems
- Customer in-stay experience
- Intelligent travel assistance
- Optimized disruption management
- Flight and hotel fare forecasting
- Sentiment analysis in social media
- Personalized offers for royal customers
- Dynamic pricing in the hospital industry
- Prediction of seasonal demands for services

20.5 Sentiment Analysis

This section covers sentiment analysis in detail. It includes sentiment analysis applications in the tourism sector. Further, it focuses on different types of sentiment analysis, different sentiment analysis techniques, and models of sentiment analysis. Sentiment analysis is the standard method to predict the mood of the customer, which helps the industry player to make decisions and find viable solutions to the existing problems.

Sentiment analysis is the procedure of identification and analysis of customer sentiments by using natural language processing (NLP), text mining, and statistics [22]. The best business learns the sentiments of the customers—what customers are conveying, how they are representing it, and what is the meaning of a message. The sentiment can be found in tweets, blogs, comments reviews, and other media platforms where the customer mentions the brand names. With advancements in technology and evolution of analysis techniques, sentiment analysis is now the key factor for the success of an organization. With use of the ML and DL techniques, we can extend and classify the sentiment of words into positive, negative, and neutral classes. Sentiment analysis deals with the perception of the product and understanding of the market by using sentiment data. It gives the decision-makers an opportunity to dapple the strong and weak points of the product from the tourist's point of view [23]. Sentiment analysis is the most influential factor in case of market research, which will help to find a niche and establish the product in the market.

Sentiment analysis is used for the following applications [24]:

- Brand monitoring
- Customer service, support, and reviews
- Market research and analysis
- Reputation management
- Precision targeting
- Product reviews
- Determining the polarity
- Public relations
- Product feedback
- Net promoter scoring
- Product analytic
- Define the general and specifically matter
- Identify the correlation with an existing customer segment

Sentiment analysis algorithms can be used for the following scopes:

i. **Document-Level** sentiment analysis classifies an opinion document as a negative or positive sentiment by considering the whole document as a basic source of information.
ii. **Sentence-Level** sentiment analysis classifies sentiment defined in every sentence. The initial step is to determine if the sentence is objective or subjective. If it finds the sentence subjective, then it identifies the sentence indicating negative or positive opinion.
iii. **Aspect-Level** sentiment analysis classifies the sentiment based on specific aspects of entities. The user can give different opinions for different perspectives for the same entity.

20.5.1 Block Diagram of Sentiment Analysis

The block diagram in Figure 20.1 describes the workflow of the sentiment analysis for the given input data. The input is the text data of reviews and tweets given by the tourists. The reviews are assigned with specific labeled sentiments. The input is then fed for the feature extraction process. After the feature extraction process, the ML model will take the input data with several parameters. The model then

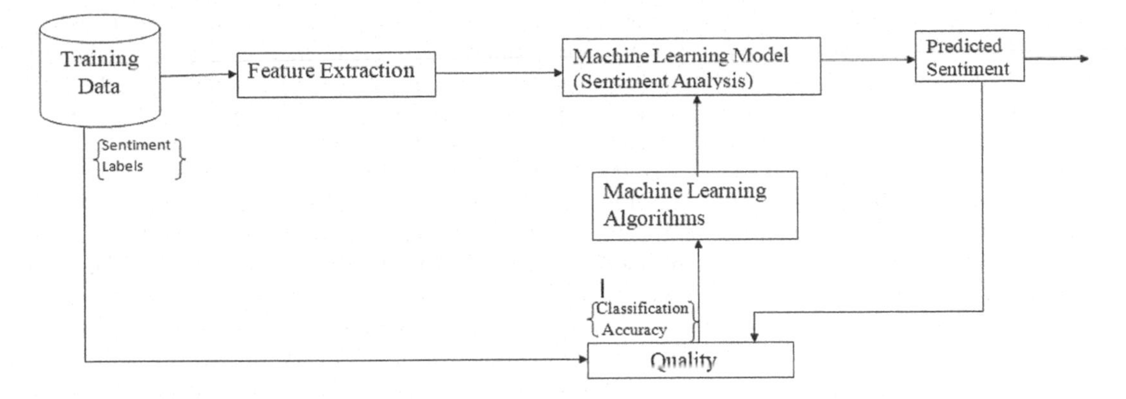

FIGURE 20.1 Block diagram of sentiment analysis.

predicts the output, which is the sentiment for that given input. The output will be evaluated, and it is given to the algorithm for improvement of the parameters. The main improvements are happened to be weights. The predicted sentiment is compared with the true labels for the sentiments. To get more accurate classification of sentiment, the process continues by updating the parameters at each cycle.

20.5.2 Different Types of Sentiment Analysis

The four different types of sentiment analysis are briefly outlined as follows.

20.5.2.1 Fine-Grained

This type of analysis involves determining the polarity of the opinion. This may be a binary sentiment category like positive and negative. There can be higher specifications that totally depend on the use case.

20.5.2.2 Emotion Detection

It is used to identify signs of specific emotional states embedded in the text. It is the combination of lexicons and ML algorithms, which resolve what and why queries.

20.5.2.3 Aspect-Based

The purpose of this analysis is to identify an opinion of a specific element. This analysis is specifically used for product analytic to take care on how the product is perceived and what is the strong and weak point from the customer point of view.

20.5.2.4 Intent Analysis

Its purpose is to identify what kind of intention is expressed in the message. It is highly used for customer support systems.

20.5.3 Sentiment Analysis Datasets

The primary step in making any model is collecting an appropriate source of data for training. Some standard datasets are frequently used to benchmark models and compare accuracy. There are an enormous number of articles published every year that use different datasets. But there are also some standard datasets that are used to verify the models. One of these datasets is Stanford Sentiment Treebank

[25]. It contains 11,000 sentences that are extracted from the reviews of different movies. These are properly parsed into a labeled parse tree. This helps recursive models to be trained on each level in the tree and allows them to predict the sentiment for sub-phrases in the sentence first and then for the sentences as a whole. Another dataset is Amazon Product Reviews Dataset [26], which consists of more than 142 million reviews regarding Amazon products, including metadata. It allows the researchers to train sentiment models using product ratings. The IMDB Movie Reviews Dataset [27] records 50,000 movie reviews, which are highly polarized and can be used for training and testing data. The next standard dataset is Sentiment140 Dataset [28], which provides important data for training the models of sentiment analysis to work with posts from social media and other unorganized texts. The dataset has 1.6 million training samples that are classified as negative, positive, or neutral. Google News Article dataset [29] contains millions of news article sentences, which can be used for training data in sentiment analysis.

The dataset scope is general from different Web forums, movie reviews, news articles, hotel reviews, product reviews, mobile customer reviews, Chinese training data, book reviews, restaurant reviews, social media polls, social media reviews, GPS dataset, tweets, stock market, car reviews, blog posts, stock news, smartphone reviews, and e-book reviews [30].

20.5.4 Sentiment Analysis Methods

There are two important sentiment analysis methods—*rule-based approach* and *automatic sentiment analysis*.

20.5.4.1 Rule-Based Sentiment Analysis

In rule-based sentiment analysis algorithms, the description of opinion is clearly defined. It includes identification of subject, polarity, or the context of opinion. It is the general-purpose algorithm to determine the tone of the message [30,31]. It involves the following operations with the text corpus:

- **Stemming** is the procedure of producing morphological form of a root/base word. It is generally referred to as stemming algorithms or stemmers. In this procedure, the words are chopped off from the end till the stem is found. It is used for searching keywords. For instance, travel, travels, and travelling have the same meaning and root travel.
- **Lemmatization** considers the morphological analysis of the words. It is mandatory to have complete dictionaries so that the algorithm can check back to its lemma. Lemmatization is more informative than stemming. For instance, it finds the words "be", "is", "was", and "were", which are the same with different morphological meanings.
- **Tokenization** is the conversion of text into tokens before transforming it into vectors. It is used to segregate unnecessary tokens.
- **Part of Speech (PoS) Tagging** is the process of identifying the structural elements of a text document like nouns, adjectives, verbs, and adverbs. It is the mechanism of breaking a document down into its components involving PoS tagging.
- **Syntactic Parsing** is about obtaining the internal structure of sentences in NLP. It is a critical job for AI applications that finds the meaning of the natural language speech or text [20].
- **Lexicon Analysis (Context Dependent):** It requires calculating the sentiment from the semantic orientations. By assigning different labels like positive and negative sentiment to the words from a text message to predict the overall sentiments of the message.

20.5.4.2 Automated Sentiment Analysis

It is the real-time application of sentiment analysis. This type of sentiment analysis uses ML methods to figure out the summary of the messages. The accuracy for this type of analysis is more than the rule-based analysis. In this type, the processing of information depends on numerous criteria. It involves two approaches, namely supervised and unsupervised ML approaches. Supervised approach is used to classify the data to maximum effect, whereas an unsupervised approach is used to explore data.

Sentiment analysis generally involves the following algorithms:

i. Linear regression
ii. Naive Bayes
iii. Support vector machines (SVMs)
iv. Recurrent neural networks derivatives—long short-term memory (LSTM) and gated recurrent unit (GRU)

20.5.5 Sentiment Classification Techniques

Sentiment classification methods are divided into ML approach and lexicon-based approach. Figure 20.2 depicts the classification of sentiment analysis methods. The ML approach uses ML algorithms and linguistic features. The model predicts a class label for an input instance of unknown class. The sentiment analysis, which applies mainly classification methods for texts using ML approach, is classified into unsupervised and supervised learning methods. The unsupervised methods are applied for unlabeled training data. The supervised methods use large training data, which are labeled.

The supervised learning methods rely on already defined labeled training data. The most popularly used supervised classifiers are described in brief.

1. **Probabilistic Classifier***:* It uses a hybrid model based on the assumption that each class label is an element of the hybrid and each hybrid component is a generative model that provides the sampling probability of a particular term for that component.
2. **Naive Bayes Classifier***:* This model uses a collection of words for feature extraction that ignores the location of words in the document. Given the distribution of words in the document, this model calculates the posterior probability of a class as per the following equation.

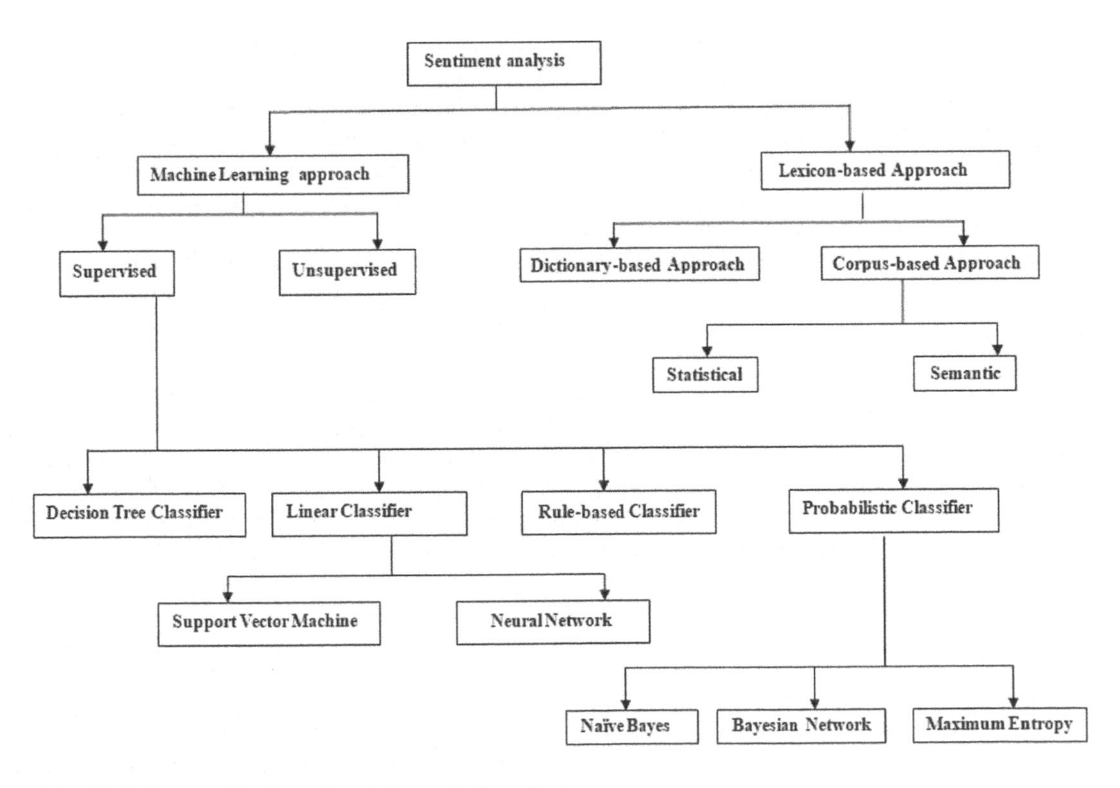

FIGURE 20.2 Classifications of sentiment analysis methods.

$$P(\text{label}|\text{features}) = \frac{P(\text{label})*P(\text{features}|\text{label})}{P(\text{features})} \tag{20.1}$$

where P(label) is the prior probability of the likelihood that a random feature set the label. P(features|label) is the prior probability that a given feature set is being classified with the label. P(features) is the probability for a given feature set present. Hermanto et al. [23] used Naive Bayes model to consider sentiment analysis of tourist destinations using Twitter data.

3. **Bayesian Network***:* The independence of features is the main assumption of this classifier. The features are independent from each other, which lead to the model that is a directed acyclic graph whose nodes are the random variables and edges are conditional dependencies [23,30]. This model is widely used for real-world problems in text classification.
4. **Maximum Entropy Classifier***:* This is also known as conditional exponential classifier. It converts the labeled feature set to a vector using the method of encoding. This vector is used for calculating weights for each of the features that can be combined to identify the most possible label for the feature set. The probability of each label is computed using the following equation:

$$P(f_s|\text{label}) = \frac{\text{Dotprod}(\text{weights,encode}(f_s,\text{label}))}{\text{Sum}(\text{dotprod}(\text{weights,encode}(f_s,\text{label}))\text{forlinlabels})} \tag{20.2}$$

where f_s is the feature set, dotprod is the dot product of weights and encoding, and forlinlabels is for linear labels.

5. **Support Vector Machine (SVM):** Its main principle is to determine separator lines in the search space, which subsequently separate multiple classes. The best-suited data for SVM models is text data due to its sparse nature, where some features are correlated with each other and separable even if they are irrelevant [30].

The lexicon-based approach relies on a sentiment lexicon and a set of sentiment words [32]. The lexicon-based approach finds the opinion lexicon for analyzing the text document. It is further divided into a dictionary-based and corpus-based approach. The dictionary-based approach finds opinion root words and then searches the dictionary for their synonyms and antonyms. The corpus-based approach starts with an initial list of opinion words and then finds other opinion words in the corpus related to specific context of orientation. This approach is carried out by two methods—*statistical approach* and *semantic approach*—to find sentiment inclination.

The hybrid approach is the combination of both approaches that is used by the majority of users.

20.5.6 Models for Sentiment Analysis

The main idea of sentiment analysis models is to select the fact that a word correlated with text having a particular sentiment and use this information to predict further unlabeled texts.

Logistic regression is a model that trains rapidly on large datasets and gives better results. Other models that are used include random forest, SVM, and naive Bayes. These models are further improved by using bi-gram, tri-gram, or n-grams training. This allows the classifier to find short phrases, which may carry sentiment details that individual words do not pick. Although it provides accuracy, it increases the complexity of the model.

20.5.6.1 Logistic Regression

Logistic regression model is used for probability of classes. It is used in many fields like ML, medical and social science. Logistic regression can be used as binomial or multinomial. It is a generalized linear model. In binary logistic regression, the result is given as "0" or "1" format, which is the most direct interpretation [33].

20.5.6.2 Random Forest

Random forest was created by Tin Kam Ho [34], and it was later extended by Leo Breiman and Adele Cutler. It is a statistical algorithm that is mostly used for clustering data points. Random forest performs efficiently on large datasets. Random forest overcomes the main limitation of complexity of tree classifier. Random forests do not overfit the training data. It is the stochastic modeling that increases accuracy over decision tree classifiers.

20.5.6.3 Support Vector Machine (SVM)

SVM, which is also known as a support vector network, was developed by Vapnik with other colleagues in AT&T Bell Laboratories [35]. SVM performs linear classification as well as nonlinear classification using kernel-SVM. SVM is efficient in high-dimensional data. As it uses a subset of training points, known as support vectors, it is memory efficient.

SVM can further be extended to support vector clustering (SVC) that is an unsupervised learning method.

20.5.6.4 Naive Bayes Classifier

It is a simple probabilistic classifier that applies Bayesian theorem with high independence assumptions among features [36]. It provides higher accuracy than other classifiers like random forest and boosted trees [37].

To achieve more accuracy, DL techniques are used that provide a new way of measuring sentiment analysis models and also address common model architectures. These models are prototyped and adapted quickly to a particular dataset. These models transform the input text into an embedded representation. The models attain more accuracy by using pretrained embeddings like FastText [38], Word2Vec [39], Bidirectional Encoder Representations from Transformers (BERT) [40], or Glove [41].

20.6 Sentiment Analysis-Based Research on Tourism

This section describes different studies and research conducted by researchers and industry to get the insight of the tourism industry. The researchers identified several parameters to find solutions for the problems. In this section, we have summarized different research papers and put our view on the recent trends in the tourism industry. The subsections describe the parameters that affect the sentiment of tourists.

A new data source and process of tourism research can be found by analyzing sentiments on social media data to take out opinions provided by the public [42]. Sentiment analysis is the study of public perceptions, attitudes, and evaluation of emotions concerning problems, events, and individuals. The objective of a sentiment analysis is to pull the emotional orientation of text, picture, and videos, which are generally unstructured. Nowadays, user-generated information or content has allowed the researchers to understand tourist's perception of the overall factors of tourist destinations [43]. Using the sentiment analysis of different attributes is also another factor for examining.

Sentiment analysis has attained more interest among researchers recently. The context of sentiment swings from user to user, so it increases difficulty in finding the sentiment. The sentiment analysis methods can be divided into two categories—lexicon matching method based on sentiment dictionary- and corpus-based ML methods. ML methods can be classified into three categories: (i) the SVM method that searches for the most relevant linear separation between the data expressing positive emotions and negative emotions; (ii) the Naive Bayesian method that estimates the probability of certain emotions based on different attributes of text messages, and (iii) the artificial neural network (ANN) method, using self-organized networks [23,31,42]. Sentiment analysis differs with the phases of processed information such as words, sentences, and documents [44]. The ML methods and dictionary matching methods are being used heavily in tourism studies. The dictionaries used include English dictionaries like WordNet and SentiWordNet, and some dictionaries like Valence Aware Dictionary for Sentiment Reasoning (VADER) used in studies that include tourism vocabulary [31].

Tao et al. [44] describe the effect of air quality on the tourism sector of China. They have collected the text for the corpus from Sina Weibo, Facebook, and Twitter. They have studied the data from January 1, 2010, to December 31, 2017. They have used ROST CM, BosonNLP, and Goseeker to analyze the dataset. The result from the research is analyzed with respect to different attributes. They have analyzed the result to identify the top five provinces in terms of air quality and their location.

Customer feedback that is the sentiment in terms of text and visuals can help hotels and restaurants to increase recommendations along with frequent visitors. Sentiment is the core of the tourism sector as tourists anticipate to enjoy the visit or holiday. In the tourism sector, customer experience is the critical factor that leads to the industry throughput. The tourists and customers generally give informal feedback, which requires big data solutions. Thelwall [45] identifies different tasks and considerations for analysis. They are i) the impact of topic domain on accuracy of algorithm, ii) language, iii) image sentiment analysis, and iv) accuracy and bias. These factors influence the outlook toward the sentiment as a different context gives different analyses. The author discussed different core sentiment analysis techniques like ML-based sentiment analysis and lexical-based sentiment analysis. All end users of sentiment analysis for big data like social media should be conscious of the self-selection bias limitations innate in the medium.

Chaniotakis and Antntoniou [46] identify two factors that influence in extracting information about transportation systems from different social media platforms, which are the focus of the social media and data availability. The building blocks of the context of transportation are presence, sharing, relationships, identity, interactions, groups, and reputations [47]. The authors collected data from public API, and sometimes, they have built apps and asked the user to join for giving feedback for the data collections. The authors used Strengths, Weaknesses, Opportunities, and Threats analysis (SWOT) technique of using social media data on transportation studies. They have also identified the threats for privacy issue, social media usage, and data inaccessibility of social media of users. Finally, the authors identified from the empirical evidence that social media cannot displace the existing survey methods used.

Vallikannu and Meyyappan [14] used ConceptNet technique and created a domain-specific ontology for Oman tourism. They have identified four factors, which are domain-specific ontology, entity-specific opinion extraction, combined lexicon-based approach, and conceptual semantic sentiment analysis, to learn the sentiment analysis of tweets regarding Oman tourism.

Jain and Tiwari [19] identified socioeconomic well-being score (SEWS), which is used to acknowledge the travel behavior variation of people belonging to divergent income groups in an Indian city named Visakhapatnam. Chi-square test was conducted to examine the robustness of the score. In the case study, they have found that both trip length and mode of choice vary significantly with respect to the score. Low-income travelers travel shorter distances and use low-carbon modes of transportation as compared to others.

Lopez and Wong [48] identified that living standards of the country are improved by implementing sustainable mobility of promoting usage of a large amount of public transport options and active modes of transports like cycling and walking. The factors drive travel behavior due to change in user's moves from conventional modes of transport to active modes of transport.

Schmude et al. [11] conducted research where the economic results of the terrorist attacks on the tourism industry for the periods from 2016 to 2018 were carried out. It is found that due to terrorist attacks, the footfall of tourists decreases, which means it lowers the income in the areas of shopping, catering, accommodation, transport, and leisure.

Dinis et al. [49] found the Google trends in data usage in the tourism and hospitality sector. They collected data from the year 2012 to 2017 of tourism forecasting. They were keen to know the user's search interest for tourist attraction, tourist destinations. They have identified a relation between tourism statistics and the search volume index of Google trends. They have identified that the forecasting model for casino revenues is significantly improved.

Kesorn et al. [50] used a personalized tourism information service (PTIS) model to analyze user interests and execute personalized attraction recommendations using check-in information expressed in Facebook. As the method directly recommends the places based on the personalized data, it will be helpful to the tourism industry. The system used close friend's detailed information to detect attractions in which the target user is interested.

Martin et al. [51] identified how consumer's individual differences are associated with their attitude for trip advice. They have used individual factors such as anthropomorphic tendency, need for cognition, and life satisfaction for research. They have found that anthropomorphic tendency has a positive association with trip advice compared to other factors.

Pikkemaat et al. [52] identified many opportunities in small enterprises. It is also found that to give tourism a boost, microenterprise should have been given more attention. Innovation in behavior of microenterprises must be a factor that will provide sustainability in eco-innovations and tourist-driven innovation in the tourism industry.

20.6.1 Sentiment Analysis Based on Environment

Tao et al. [44] in their research found that the air quality is the reason for sentiment in different provinces across China. Locations are the major factors to identify tourist footfall in the provinces. The air quality near the industrial area is poor. So it is difficult to attract tourists to such places.

20.6.2 Sentiment Analysis Based on Locations and Hotels

The location of the tourist places is also an important factor while considering the business model. The destination nearer to a major transport facility increases the tourist footfall. Hotels nearer to the destination give an opportunity to the stakeholders to formulate a business model easily. The evaluation criteria are represented as sentiment values for hotel and travel reviews in Figures 20.3 and 20.4, respectively.

In Figures 20.3 and 20.4, the Recall values are better compared to other factors of calculation. The authors also identified that the air quality is a factor because the sentiment values across the reviews

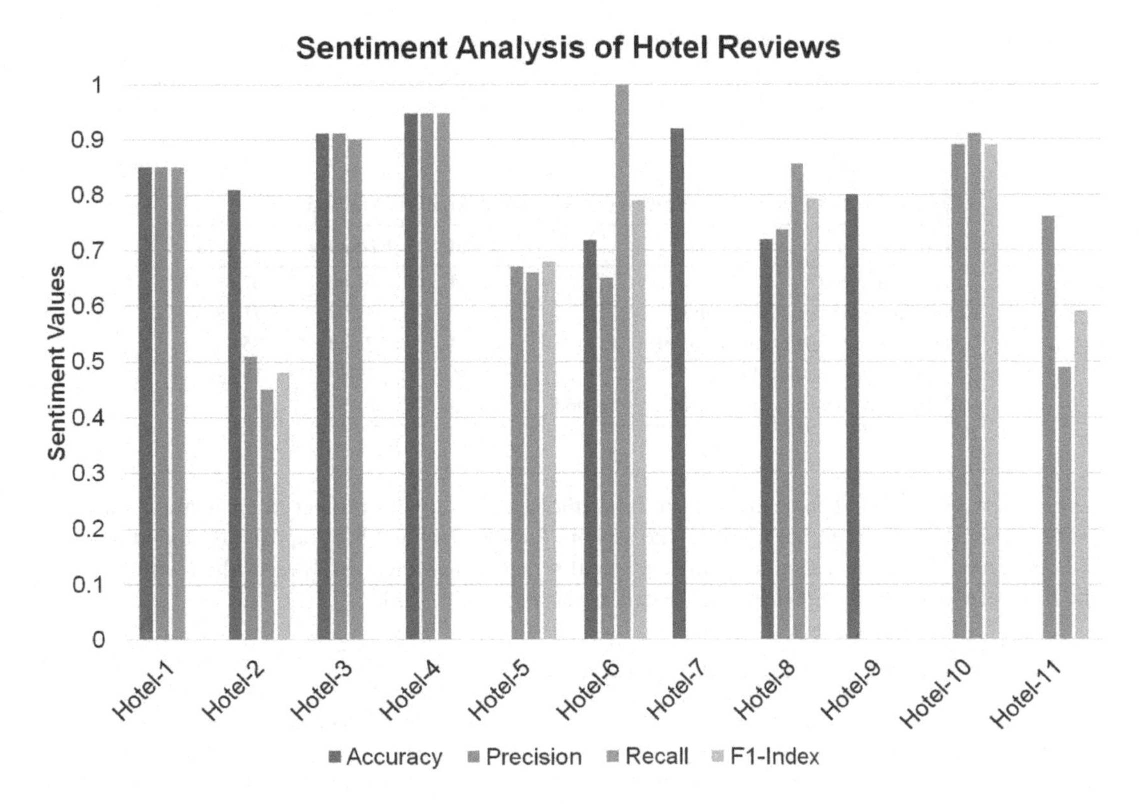

FIGURE 20.3 Sentiment values in reviews of different hotels in China.

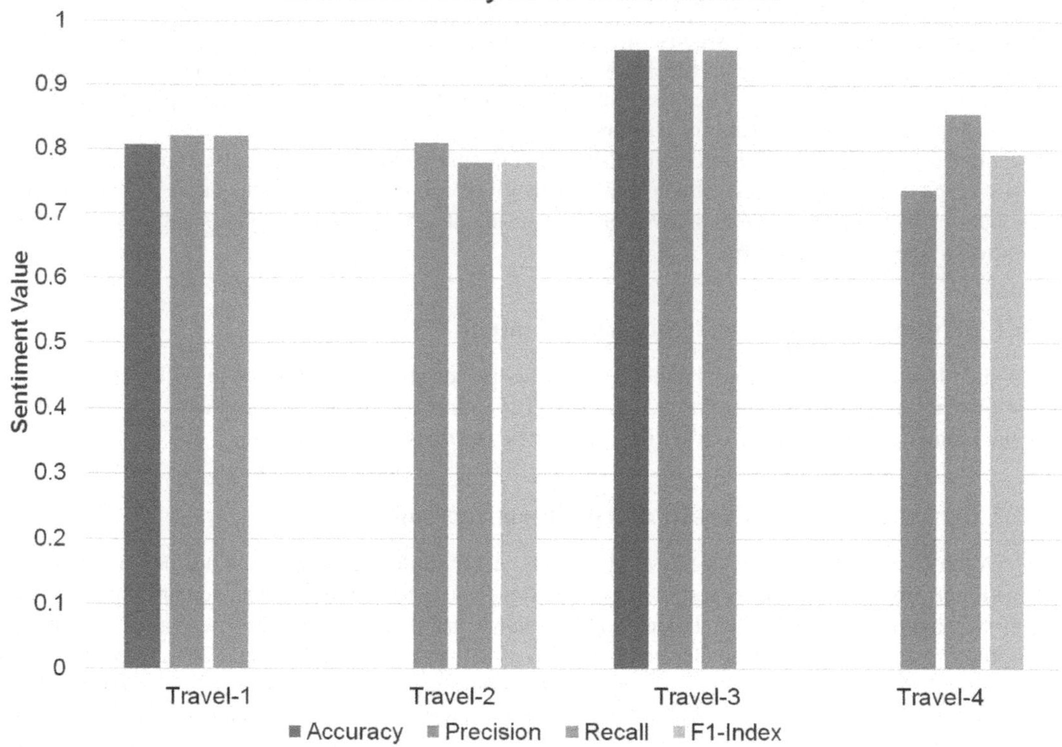

FIGURE 20.4 Sentiment values in travel reviews of different provinces across China.

TABLE 20.1

Comparison of Parameters in Different Sentiment Analysis Methods

Method	Recall	F1-Score	Precision
Baseline method	65.34	66.62	67.83
Domain-specific ontology method	73.42	72.77	71.38
Entity-specific opinion extraction method	74.23	75.44	75.45
Combined lexicon-based method	76.56	79.43	79.23
Conceptual semantic sentiment analysis method	84.23	85.54	83.34

are precisely consistent. This leads to identifying the sentiments with respect to air quality, which is used to make policies to attract tourists. The authors Ramanathan and Meyyappan [14] found that the F1-Score value is the most significant in conceptual semantic sentiment analysis method with 85.54%. The F1-Score outperforms Precision and Recall as shown in Table 20.1.

20.6.3 Sentiment Analysis Based on Language

Language is also a factor that influences and deviates sentiment results. The sentiment algorithms are language specific since they learn from codes written by humans from a specific language. The limitation of social media analysis is that customer bias is based on their age-group [4]. It is also difficult to identify the fake reviews that swing the sentiment toward a particular class.

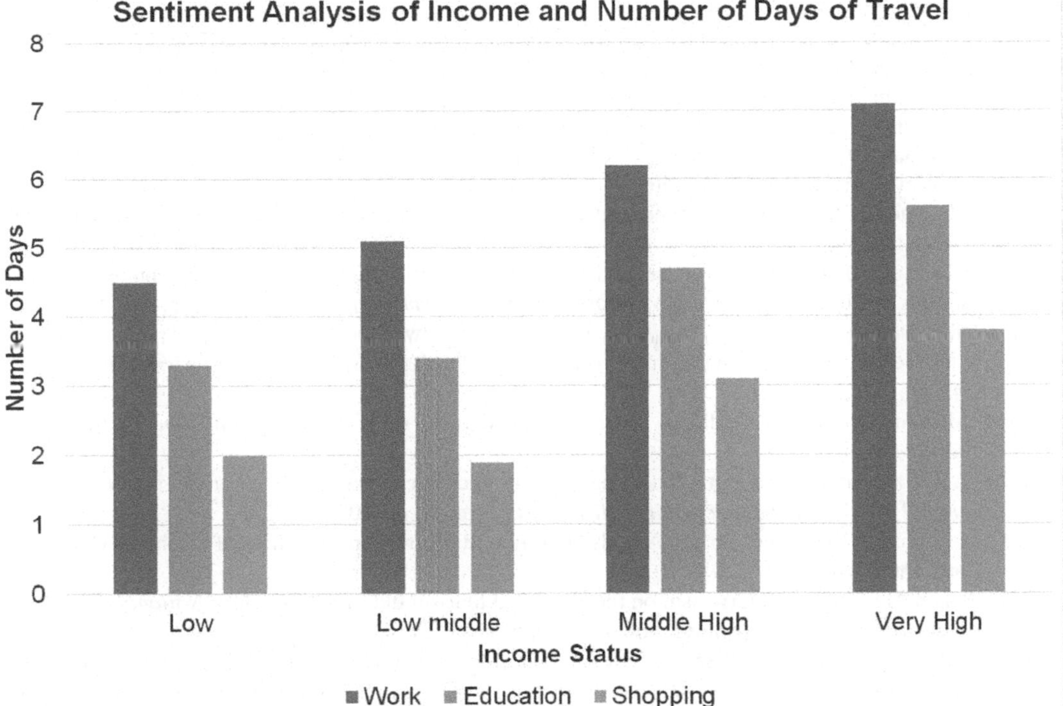

FIGURE 20.5 Sentiment analysis of income and number of days of travel.

20.6.4 Sentiment Analysis Based on Trip Length

Figure 20.5 shows the variation in trip lengths with respect to SEWS groups, the study carried out by Jain and Tiwari [19]. From the research, the authors found that people move from one place to another for the work purpose and daily routine livelihood. For shopping purposes, the movement is restricted in the lower-income category, whereas the very high-income category moves more distance. The trip length also differs significantly from lower-income category to very high-income category.

20.7 Summary

This chapter presented an introduction to tourism industries and outlined the contemporary research in sentiment analysis and ML for the tourism industry. The researchers in recently published papers have found that most sentiment analysis methods are effective in classifying positive sentiment compared to negative and neutral ones in real-world problems. It is also identified that most of the classifications are feedback based where the feedback was taken from social media and Web scraping. For the research work, all the data collected are homogeneous and unilingual, that is, the feedback given in only one language. The authors have segregated the sentiments based on bilingual or multilingual as the feedback given may contain more than one language and different structures. The feedback may contain other languages with Latin script, which produces different meanings with respect to the context. So, the improvement will provide different aspects of the sentiment while working with multiple languages and formats. The study can further be extended to include other industries like restaurant, hotel, and transport industries to identify the correlation of each sector, which are part of the tourism industry. The other industries that are indirectly linked to the tourism industry should be explored and analyzed for further improvements and options of opportunities.

REFERENCES

1. C. R. Goeldner and J. B. Ritchie, *Tourism Principles, Practices, Philosophies*. John Wiley & Sons: Hoboken, NJ, 2007.
2. M. Al Shehhi and A. Karathanasopoulos, "Forecasting hotel room prices in selected GCC cities using deep learning," *Journal of Hospitality and Tourism Management*, vol. 42, pp. 40–50, 2020.
3. D. Amutha, "Development and impact of tourism industry in India." https://papers.ssrn.com/sol3/papers.cfm?abstractid=2825213, 2016.
4. C. E. Tureac and A. Turtureanu, "Types and forms of tourism," *Acta Universitatis Danubius: Œconomica*, vol. 4, no. 1, pp. 99–103, 2010.
5. N. F. Dale and B. W. Ritchie, "Understanding travel behavior: A study of school excursion motivations, constraints and behavior," *Journal of Hospitality and Tourism Management*, vol. 43, pp. 11–22, 2020.
6. E. Ulker-Demirel and G. Ciftci, "A systematic literature review of the theory of planned behavior in tourism, leisure and hospitality management research," *Journal of Hospitality and Tourism Management*, vol. 43, pp. 209–219, 2020.
7. M. Memon, A. Bajaj, S. Dadhich, and S. Patel, "Medical value travel in India," *in FICCI Health Conference*, 2014.
8. M. T. K. Rana and M. S. Singh, "Emerging trends of mice in context of Indian tourism industry," *CLIO An Annual Interdisciplinary Journal of History*, vol. 6, no. 4, pp. 400–416, 2020.
9. C. M. DePuma, Terrorized tourists: A study of the impact of terrorism on tourism. PhD thesis, University of Southern Mississippi, 2015.
10. D. Baker, "The effects of terrorism on the travel and tourism industry," *The Inter National Journal of Religious Tourism and Pilgrimage*, vol. 2, no. 1, pp. 58–67, 2014.
11. J. Schmude, M. Karl, and F. Weber, "Tourism and terrorism: Economic impact of terrorist attacks on the tourism industry: The example of the destination of paris," *Zeitschrift f¨ur Wirtschaftsgeographie*, vol. 64, no. 2, pp. 88–102, 2020.
12. C. Seabra, P. Reis, and J. L. Abrantes, "The influence of terrorism in tourism arrivals: A longitudinal approach in a mediterranean country," *Annals of Tourism Research*, vol. 80, p. 102811, 2020.
13. T. Thiessen, "Australia bushfires burn tourism industry: $4.5 billion as holiday ERS cancel." https://www.forbes.com/sites/tamarathiessen/2020/01/20/australiabushfires-hit-tourism-industry-as-holidayers-cancel, 2020.
14. V. Ramanathan and T. Meyyappan, "Twitter text mining for sentiment analysis on people's feedback about Oman tourism," *in* 4th *MEC International Conference on Big Data and Smart City (ICBDSC)*, Muscat, Oman, pp. 1–5, IEEE, 2019.
15. E. Becker, "How hard will the coronavirus hit the travel industry?" https://www.nationalgeographic.com/travel/2020/04/how-coronavirus-isimpacting-the-travel-industry, 2020.
16. R. Dean, "The corona effect on the tourism industry." https://www.outlookindia.com/outlooktraveller/explore/story/70362/expertsshare-insights-on-the-effects-of-coronavirus-on-the-tourism-industry, 2020.
17. Centers for Disease Control and Prevention. "2014–2016 ebola outbreak in west Africa." National Center for Emerging and Zoonotic Infectious Diseases (NCEZID), Division of High-Consequence Pathogens and Pathology (DHCPP), Viral Special Pathogens Branch (VSPB). https://www.cdc.gov/vhf/ebola/history/2014-2016-outbreak/index.html.
18. I. Mizrachi and G. Fuchs, "Should we cancel? An examination of risk handling in travel social media before visiting ebola-free destinations," *Journal of Hospitality and Tourism Management*, vol. 28, pp. 59–65, 2016.
19. D. Jain and G. Tiwari, "Explaining travel behaviour with limited socio-economic data: Case study of Vishakhapatnam, India," *Travel Behaviour and Society*, vol. 15, pp. 44–53, 2019.
20. T. H. Tran and V. Filimonau, "The (de) motivation factors in choosing airbnb amongst vietnamese consumers," *Journal of Hospitality and Tourism Management*, vol. 42, pp. 130–140, 2020.
21. A. Dewangan and R. Chatterjee, "Tourism recommendation using machine learning approach," *in Progress in Advanced Computing and Intelligent Engineering*, pp. 447–458, Springer, Singapore, 2018.
22. P. Aurchana, R. Iyyappan, and P. Periyasamy, "Sentiment analysis in tourism," *International Journal of Innovative Science, Engineering & Technology*, vol. 1, no. 9, pp. 1–8, 2014.

23. D. Hermanto, M. Ziaurrahman, M. Bianto, and A. Setyanto, "Twitter social media sentiment analysis in tourist destinations using algorithms naıve Bayes classifier," *Journal of Physics: Conference Series*, vol. 1140, no. 1, pp. 1–6, 2018.
24. V. Bilyk, "What is sentiment analysis: Definition, key types and algorithms." https://theappsolutions.com/blog/development/sentiment-analysis.
25. "Stanford dataset treebank." https://nlp.stanford.edu/sentiment/treebank.html.
26. "Amazon dataset review 1996–2016 updated on 2018." http://jmcauley.ucsd.edu/data/amazon/, 2018.
27. S. Ashirwad, "IMDB dataset version 6." https://www.kaggle.com/ashirwadsangwan/imdbdataset, 2019.
28. Kazanova, "Sentiment140 dataset version 2." https://www.kaggle.com/kazanova/sentiment140, 2017.
29. "Google news article dataset." https://research.google/tools/datasets/, 2016.
30. W. Medhat, A. Hassan, and H. Korashy, "Sentiment analysis algorithms and applications: A survey," *Ain Shams Engineering Journal*, vol. 5, no. 4, pp. 1093–1113, 2014.
31. C. Gomez-Rodrıguez, I. Alonso-Alonso, and D. Vilares, "How important is syntac tic parsing accuracy? An empirical evaluation on rule-based sentiment analysis," *Artificial Intelligence Review*, vol. 52, no. 3, pp. 2081–2097, 2019.
32. A. Garcıa, S. Gaines, M. T. Linaza, et al., "A lexicon based sentiment analysis retrieval system for tourism domain," *Expert Systems with Applications of International Journal*, vol. 39, no. 10, pp. 9166–9180, 2012.
33. D. W. Hosmer and S. Lemeshow, Applied *Logistic R*egression. Wiley: New York, 2000.
34. T. K. Ho, "Random decision forests," *in Proceedings of 3rd International C*onference on *Document Analysis and Recognition*, Montreal, QC, Canada, vol. 1, pp. 278–282, IEEE, 1995.
35. C. Cortes and V. Vapnik, "Support-vector networks," *Machine Learning*, vol. 20, no. 3, pp. 273–297, 1995.
36. S. J. Russell and P. Norvig, *Artificial Intelligence: A Modern* Approach, 3 edn, Pearson: London 2009.
37. R. Caruana and A. Niculescu-Mizil, "An empirical comparison of supervised learning algorithms," *in Proceedings of the 23rd International C*onference on Mac*hine L*earning, New York, pp. 161–168, 2006.
38. S. Reddy, "FastText method.", 2019. https://cai.tools.sap/blog/glove-and-fasttext-twopopular-word-vector-models-in-nlp/.
39. B. A. Jayesh, "Word2vec algorithm." https://medium.com/@jayeshbahire/introductionto-word-vectors-ea1d4e4b84bf.
40. M. Chris, "Word2vec algorithm." 2019. https://mccormickml.com/2019/05/14/BERTword-embeddings-tutorial/.
41. R. Sanjana, "Glove algorithm." 2019. https://cai.tools.sap/blog/glove-and-fasttext-twopopular-word-vector-models-in-nlp/.
42. A. R. Alaei, S. Becken, and B. Stantic, "Sentiment analysis in tourism: Capitalizing on big data," *Journal of Travel Research*, vol. 58, no. 2, pp. 175–191, 2019.
43. C. Van Vuuren and E. Slabbert, "Travel motivations and behaviour of tourists to a south African resort," *Tourism & Management Studies*, vol. 1, pp. 295–304, 2012.
44. Y. Tao, F. Zhang, C. Shi, and Y. Chen, "Social media data-based sentiment analysis of tourists' air quality perceptions," *Sustainability*, vol. 11, no. 18, p. 5070, 2019.
45. M. Thelwall, "Sentiment analysis for tourism," In M. Sigala, M. A. Thelwal, R. Rahimi (eds), Big Data and Innovation in Tourism, Travel, and Hospitality, pp. 87–104, Springer: Berlin, Heidelberg, 2019.
46. E. Chaniotakis, C. Antoniou, and F. Pereira, "Mapping social media for transportation studies," *IEEE Intelligent Systems*, vol. 31, no. 6, pp. 64–70, 2016.
47. J. H. Kietzmann, K. Hermkens, I. P. McCarthy, and B. S. Silvestre, "Social media? Get serious! understanding the functional building blocks of social media," *Business Horizons*, vol. 54, no. 3, pp. 241–251, 2011.
48. M. C. R. Lopez and Y. D. Wong, "Process and determinants of mobility decisions– a holistic and dynamic travel behaviour framework," *Travel Behaviour and Society*, vol. 17, pp. 120–129, 2019.
49. G. Dinis, Z. Breda, C. Costa, and O. Pacheco, "Google trends in tourism and hospitality research: A systematic literature review," Journal of Hospitality and Tourism Technology, vol. 10, pp. 747–763, 2019.
50. K. Kesorn, W. Juraphanthong, and A. Salaiwarakul, "Personalized attraction recommendation system for tourists through check-in data," *IEEE Access*, vol. 5, pp. 26703–26721, 2017.

51. B. A. Martin, H. S. Jin, D. Wang, H. Nguyen, K. Zhan, and Y. X. Wang, "The influence of consumer anthropomorphism on attitudes towards artificial intelligence trip advisors," *Journal of Hospitality and Tourism Management*, vol. 44, pp. 108–111, 2020.
52. B. Pikkemaat, M. Peters, and B. F. Bichler, "Innovation research in tourism: Research streams and actions for the future," *Journal of Hospitality and Tourism Management*, vol. 41, pp. 184–196, 2019.

21

Training Agents to Play 2D Games Using Reinforcement Learning

Harshil Jhaveri, Nishay Madhani, and Narendra M. Shekokar
Dwarkadas J. Sanghvi College of Engineering

CONTENTS

21.1 Introduction 309
21.1.1 Description 309
21.1.2 Problem Formulation 310
21.2 Review of Literature 311
21.3 Proposed Method/Design 314
21.3.1 Experimental Setup 315
21.3.2 Results and Discussion 316
21.3.3 Conclusion and Future Scope 318
References 318

21.1 Introduction

21.1.1 Description

Effective emulation, simulation and agent-based application of learning is a very sophisticated criterion of establishing the supremacy and applicability of visual and spatial parameters. The advent of artificial intelligence can be effectively used to create advanced systems that are capable of decision-making, solve complex issues and emulate human processes of learning and decision-making virtually and comparably accurate. Reinforcement learning is a highly promising field of artificial intelligence, which is now rampantly used in resources management in computer clusters, robotics, web system configuration, bidding and advertising.

A special type of algorithm called reinforcement learning enables an agent to learn within an environment and earn the maximum reward available. Other types of algorithms such as deep learning require a training dataset, on which the model trains and tries to accomplish a maximum accuracy. It does so by tweaking a combination of parameters and hyperparameters to process the given input dataset and obtain results, as close to the actual results as possible. Now however with the major advancements in the fields of computer vision (CV), reinforcement learning and deep learning research is driven towards multi-modal learning (Figure 21.1).

The gym incorporated here on is powerful enough to emulate a real game situation in its environment, capture frames at short intervals of a few milliseconds as input, preprocess each collection of frames for a particular time frame, and then give it as an input to a 2-dimensional Convolutional Neural Network which gives predictions regarding the detection of the object in the image that will be treated as the agent playing for the CPU. Now, all movements defined as the points between the starting and ending points are divided into n states and each m movement leading from one state to another is called a set of actions and each action here is assigned a cost. Each movement from the first to final state along with the list of states entered is documented, and the final cost is calculated.

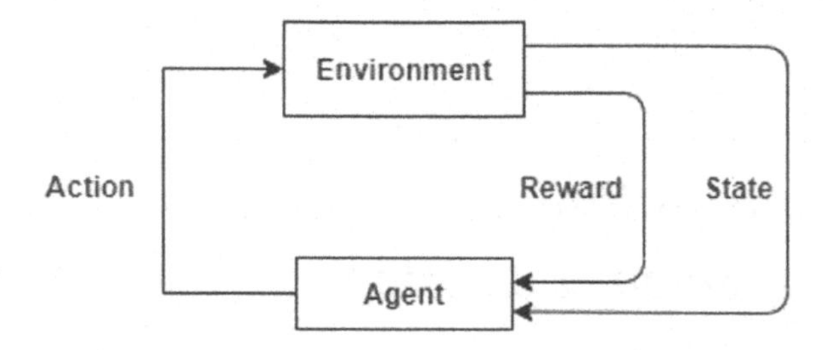

FIGURE 21.1 Reinforcement learning flowchart.

The two steps involved with Q-learning here are exploration and exploitation.

Exploitation involves using the state-action table for reference and doing a Depth First Search of all possible actions for that particular state. From that set of actions, the selection is done on the basis of the maximum reward awarded for an action and that action is chosen.

Exploration involves taking actions randomly instead of selecting actions based on the max future reward. Acting randomly removes the path of taking only actions with maximum rewards and allows the agent to further explore and comprehend new states that are otherwise bypassed during exploitation. Balance is struck between these two processes using a parameter epsilon and using a combination of ε and $(1-\varepsilon)$ for exploring and exploiting accordingly.

The formula for updating the table for each combination of state and action is denoted by

$$Q[s,a] = Q[s,a] + \text{learning\_rate}*\left(\text{reward} + \text{gamma} * \text{maximum}\left(Q\left[n_s,(\text{all\_actions})\right]\right) - Q[s,a]\right),$$

where *s means State, a means Action, and n_s means New State.*

Rate of Learning denoted by α is the rate at which the newer values are being updated in the table as compared to the older ones. The difference between the new and old states in product with the old q-value propagates the sum in line with the updated value.

Reward is the value received after completing a certain action at a given state. Applied at all time steps. **Gamma**, the discount factor is used to strike a balance between short- and long-term rewards ideally ranging between 0.8 and 0.99. **Maximum** is the maximum of all later rewards and adding it to the current reward (Figure 21.2).

21.1.2 Problem Formulation

AI-based training of a CPU agent to simulate and play games in diverse environments has been thought of for quite some time but has been relatively paused due to the unavailability of GPUs and fast processors owing to its high computation requirement. Hence, the reinforcement learning methods that were thought of as early as the 1980s were never practically executed due to a lack of computational capacity.

With the advent of GPUs, developed to handle lots of parallel computations using thousands of cores, they have a large memory bandwidth to deal with the data for these computations. With this, they are now the perfect processing unit used by the deep learning and computer vision communities. Another important breakthrough in these communities that has well supported the training and execution of the model is by Google. Google released a research tool Google Colab used for machine learning applications and research. It provides the environment of the Jupyter Notebook with no setup required. Google Colaboratory provides the Tesla K80 GPU's computing services for perusal.

OpenAI offers a list of environments and simulations that this chapter uses to train the agent. The environment is then rendered, and all observations are captured as frames from this rendered environment. Hence, the frames from these are provided as the **input** for the neural network to process. The Convolutional Neural Network processes the input, makes its predictions after activation and presents the most probable action to be performed. To calculate the reward, the chapter makes use of the Q-Learning table, and this reward is in turn used as output for the current batch. This reward along with

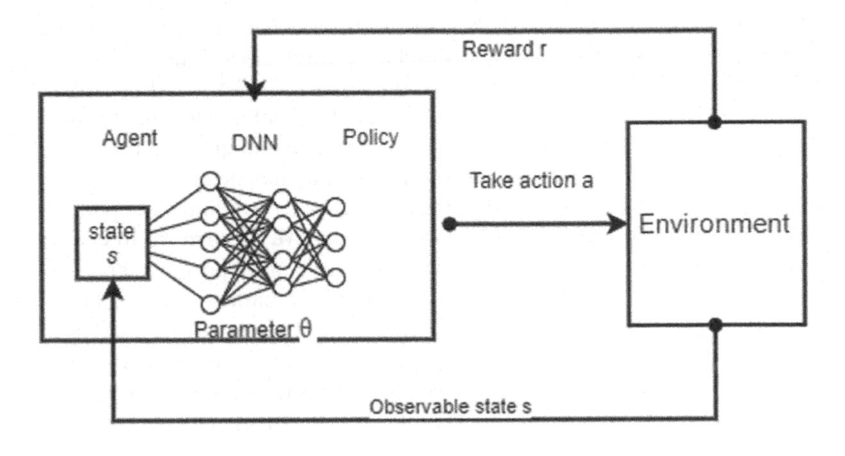

FIGURE 21.2 Deep Q-network architecture.

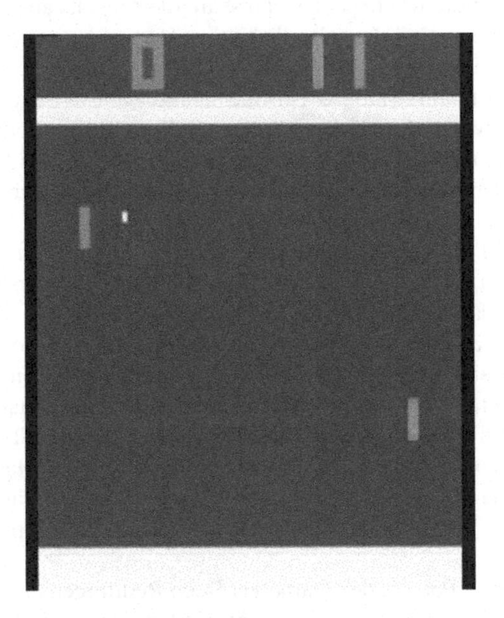

FIGURE 21.3 Frame generated after preprocessing.

the action is further fed to the ConvNet again, for example, given a game selection "Pong". The system generates frames of the following type (Figure 21.3):

In order to preprocess these generated frames, four of them are stacked together, each of dimension $210 \times 160 \times 3$ RGB arrays which need to be significantly downsized. Hence, each array is downsized to $84 \times 84 \times 3$ whilst still maintaining the RGB channels. Since the four frames are now stacked together, the input shape is now $84 \times 84 \times 3 \times 4$.

21.2 Review of Literature

The paper "An Introduction to Deep Reinforcement Learning" by Vincent Francois-Lavet et al. [1] hopes to provide the general reinforcement learning framework along with the case of Markov decision process capabilities to make computationally fast decisions at every node.

In this paper by having a smaller action set, the agent learns more quickly since there are fewer Q-table entries to update. They demonstrate the ability to combine reinforcement learning and deep learning for exploration and exploitation to learn from experience through trial and error. However, the disadvantage of this approach is that the smaller action set might not be sufficient for achieving the desired behaviour.

The second paper "Agent57: Outperforming the Atari Human Benchmark" is by Adrià Puigdomènech Badia et al. [2]. This Deep RL agent outperforms the standard human benchmark on all Atari games. To achieve this result, they train a neural network which parameterizes a family of policies ranging from very exploratory to purely exploitative. They propose an adaptive mechanism to choose which policy to prioritize throughout the training process. This paper marks an important step in the field of reinforcement learning, and their agent outperforms all human benchmarks on Atari games. They do so by training an agent for prioritization as well as exploration and exploitation. The SARSA (and on-policy learning in general) has lower per-sample variance than Q-learning and does not suffer from problems converging as a result. SARSA is more conservative – if there is risk of a large negative reward close to the optimal path, whilst SARSA will tend to avoid a dangerous optimal path and only slowly learn to use it when the exploration parameters are reduced.

The third paper, "OpenAI Gym" by Greg Brockman et al. [3] describes the OpenAI Gym as a toolkit for reinforcement learning research. It includes a growing collection of benchmark problems that expose a common interface and a website where people can share their results and compare the performance of algorithms. OpenAI Gym focuses on the episodic setting of reinforcement learning, where the agent's experience is broken down into a series of episodes. The goal in episodic reinforcement learning is to maximize the expectation of total reward per episode, and to achieve a high level of performance in as few episodes as possible. However, the tasks are meant to be solved from scratch, so transfer learning for the algorithm to train on one task after the other is not possible. Real-world operation i.e. integration of the Gym API with robotic hardware, testing reinforcement learning algorithms in the real world is not configured yet. Inability and insufficient support to multi-tier agents.

The question on how to address more complex environments, in which the reward is sparse and the state space is huge, is answered in "Deep Reinforcement Learning for Doom using Unsupervised Auxiliary Tasks" by Georgios Papoudakis et al. [4]. In this paper, they propose a divide-and-conquer-based DeepRL solution and test an agent in the first person shooter (FPS) game of Doom. The method clearly stakes advantage since it requires only the reward signal to be used in the environment to train the agent. The unsupervised auxiliary tasks are an interesting method, both for enhancing the learning capabilities of the agent and for learning features of the trained system. As a result, there is sufficient reason to believe that the divide and conquer method combined with the auxiliary tasks could solve a variety of problems. Conversely, the biggest drawbacks of Policy Gradients, i.e. the high variance in estimating the gradient, are complemented by the DQN which usually shows a better sample efficiency and more stable performance.

The paper "Learning to Play Pong Video Game via Deep Reinforcement Learning" by Ilya Makarov et al. [5] considers deep reinforcement learning algorithms for playing a game based on the video input. Careful choice of proper hyperparameters like rewards and penalty in the deep Q-network model and comparison with the model-free episodic control focused on reusing successful strategies are discussed here. The evaluation was made based on the Pong video game implemented in Unreal Engine 4. Noticeably, this Deep RL-based controller outperforms the standard rule-based AI, whilst using raw pixels as an input. Applicability of episodic control in similar tasks conducted with Variational AutoEncoder (VAE) in the future work for more complex 3D games is possible. The only training takes over 6 hours on GPU. It takes more than 10 million iterations to beat the built-in scripted AI.

"Model-Based Reinforcement Learning for Atari" by Kaiser et al. [6] entails a system enforcing a state-of-the-art deep reinforcement model that successfully emulates a wide range of 2D games in the Arcade Learning Environment benchmark. A wide variety of prediction techniques have been used for verification of results including stochastic techniques for prediction in videos based on a custom model including distinct hidden variables. A novel technique called SimPLe from Simulated Policy Learning is presented uses methods for video prediction in order to perfect the policy for training the agent to play a game. After several rounds for assembling data in order to train the policy, it is able to emulate several games and play them in the emulated environment. For a majority of similar game environments, it was noticed during the evaluation that the SimPLe algorithm works much better on a defined sample size compared to the Rainbow algorithm's highly tuned version.

The paper "Deep Reinforcement Learning in Strategic Board Game Environments" by Xenou et al. [7] generates Q-values based on the action as generated as a result of the state features and thus approximates its value. A subsequent deep reinforcement learning algorithm is suggested that uses LSTM (Long Short-Term Memory) units parallelly, each unit used for approximation of the Q-value based on a state-dependent action value. Thus, all calculations of the sort occur concurrently doing away with the need to replay experience as generally seen in deep reinforcement networks. The network, furthermore, does not require a pretrained model and can train itself during the game play itself. The aforementioned approach was used in the strategic domain multiplayer game Settlers of Catan verifying the efficacy subsequently outperforming related algorithms including the Settler heuristic algorithm.

OpenSpiel, by DeepMind, as introduced in "OpenSpiel: A Framework for Reinforcement Learning in Games" by Lanctot et al. [8], is a general framework and set of environments for emulation of games and general research in the fields of deep reinforcement learning. The varied settings supported by the framework include single and multiplayer agents, interdependent and individualistic, single move and sequential, regular grid-based settings and social situational settings. Several assessment metrics to judge performance as well as learning dynamics have been included in the paper as well. The paper serves the dual purpose of introducing the features of the OpenSpiel framework as well as making certain jargons and terminologies clear. It also introduces some core algorithms commonly used in deep reinforcement learning and computer gaming concepts.

"Deep Reinforcement Learning Agent for Playing 2D Shooting Games" by Lee and McNair [9] primarily aims at discussing the varied methods in deep reinforcement learning and analysing their performance on certain 2D shooting games. The efficacy of various agents on different fronts. A systematic and individualized approach is used in the paper, where the agents, although trained independently, share modules. For assessment, various features and metrics are used for the agents, within each game. It has been finally suggested that the combination of Long Short-Term Memory units and Asynchronous Advantage Actor Critic algorithm was much quicker in learning compared to the others.

In "2D Racing game using reinforcement learning and supervised learning" by Henry Teigar et al. [10], the central aim is to make an environment, where the car which is dictated by algorithms would be able to drive around perfectly, and this was achieved. The paper presents two reinforcement algorithms, which proved to be challenging but the results were satisfactory. One supervised learning algorithm was also used; however, its results were undermined due to the data acquisition process as mentioned. The model was then trained on CarRacing-v0, and supervised learning was successful, but reinforcement learning was not. The paper in the end makes a suggestion to change the reward system to improve performance but does not provide information on how to.

"Playing FPS Games with Deep Reinforcement Learning" by Lample and Chaplot [11] presents a novel approach to 3D environments for FPS games which include partially observable states. They contrast their approach with conventional methods by presenting that during training they exploit game information such as enemies or items to build up on conventional models. During the training phase, the model is not only maximizing the reward obtained by the agent but also learning the game features, and this has shown to increase training speed and performance of the architecture. They conclude by showing that their agents outperform built-in AI agents as well as average humans.

OpenAI Gym is a well-known and preferred way of interfacing reinforcement learning algorithms with environments; however, there are limitations to this, and this paper describes a way to interface General Video Game AI to OpenAI. The paper "Deep Reinforcement Learning for General Video Game AI" by Ruben Torrado et al. [12] analysed the performance of several implementation of Deep RL on QVGA Games. These results show and compare the challenges faced by QVGA games against each other as well as Arcade Learning Environment. The paper test algorithms like A2C, DQN, planning and learning agents and concluded that planning agents have an advantage, though there exist large differences between games.

"Reinforcement Learning and Video Games" by Yue Zheng [13] focuses on the effects and influence of batch normalization on reinforcement learning algorithms. T-rex Runner is the game of choice for this paper. The paper presents four types of algorithms to play it, and they then go on to compare the difference among these agents and investigate the effects of their primary objective. The paper accomplishes their goal; however, it was not able to investigate on experience replay due to game environment issues. The paper concluded by great success they achieved with DQN and duelling DQN, both having better results than human experts.

The central aim of the paper "A Survey of Deep Reinforcement Learning in Video Games" by Kun Shao et al. [14] is to review deep reinforcement learning methods and their successful applications. This paper used 2D as well 3D environments on which single as multi-agents are trained on, and they achieve performances comparable to humans if not better in various games. This paper reviews the fundamental aspects of reinforcement learning such as sample efficiency, and exploration and exploitation.

Hence based on the papers reviewed above, this chapter identifies that some methods use SARSA algorithms whereas some use Q-learning methods. Hence, there is no clear decision as to whether on-policy methods are better suited or off-policy. Some methods use the greedy approach whereas the other uses the current policy to determine the updated value of Q. Similarly, the emulation of the games on OpenAI and using custom optimizer functions and custom loss functions is also a challenge. Further, selection of appropriate rewards and penalties to the agents is also a constantly evolving method, not boilerplate for all sorts of games. Hence, these gaps need to be overcome with specifically designed methods.

21.3 Proposed Method/Design

The chapter presents a Deep RL-based solution that allows emulation of games into frames. CNN and DQN in combination both provide expansive power, and a linear collection of frames can be used to form a story, which is then contextualized into a sequence of images (Figure 21.4).

The first layer in the architecture is the input layer X formed of multiple elements X_1, X_2, X_3, etc. Each of these elements of the input layer is a raw RGB pixels image with a total of three channels initially. Each of these inputs X is now processed with a convolution filter of certain dimensions (m, m) and c number of channels, thereby forming outputs of dimensions $(n-m+1, n-m+1)$. These are now further processed in the next hidden layer with filters of dimensions (k, k) and $c_i$number of channels where i is the number of that hidden layer.

At the end of each of these convolutions using an m-dimensional filter, the corresponding $(n-m+1)$ shaped pixels are activated using an activation function. The activation used for non-terminal layers is ReLU activation and Softmax for the terminal hidden layer. The chapter uses the max-pooling method for pooling the output of each processing cell in a hidden layer.

Furthermore, at the end of the three hidden layers used, there are two fully connected layers which is essentially a feed forward neural network. These two layers form the last few layers in the network. The input to the first fully connected layer is the output of the max pooling from the last hidden layer. This

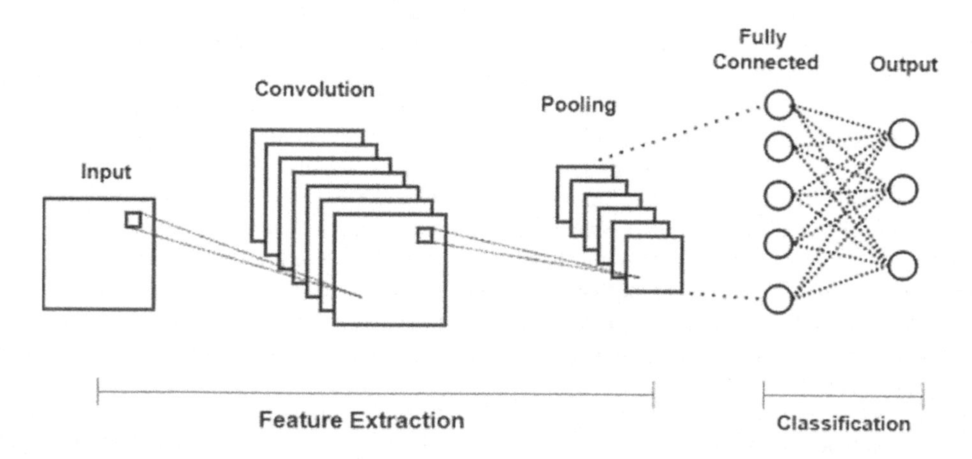

FIGURE 21.4 Overall architecture of the proposed system.

input is flattened and then submitted to the fully connected layer. The output of the last fully connected layer forms a set of outputs Y_1, Y_2, Y_3, etc. Based on these prediction outputs, an action is taken whether to move right, move left, etc., and based on the action, a reward is awarded to the DQN and this is sent back as a feedback to the state predicting it.

The process of Reward Generation is simple for a game like Pong, and the agent is simply awarded a unit positive point for every time it scores a point and a negative unit point for every time it concedes a point. Hence, the agent always tends to repeat the kind of policy where reward is maximized and avoid moves where reward is depleted. This chosen policy guides the agent in selecting plausible actions, like moving up, down or staying put. This feedback goes on as long as the reward is maximized for each state in each different epoch, and the AI reaches the final state in the emulation.

21.3.1 Experimental Setup

Of late, a lot of research has gone into building efficient reinforcement learning-based genetic algorithms with deep learning. A Convolutional Neural Network is a deep learning variant which takes in a raw RGB image input, assigns parameters in form of learnable weights and biases to various regions/quadrants in the image and is able to detect and recognize these different objects.

The following aspects are being tackled in the problem statement:

Emulation of the 2D Game in the OpenAI Gym: OpenAI Gym has integrated the Arcade Learning Environment within itself for simulation of all games available in the original Atari set. OpenAI Gym is then responsible for serving as the interface between the code and the ALE.

Generating Rewards for every time the agent makes a profitable move, either long term or short term. Different games have different metrics of measuring an agent's performance, and some require a life-based approach where the agent loses a point for wrong moves or sequences of such type. In this approach, reward is calculated based on the final score achieved by the agent when playing the game.

Computing and Minimizing Loss with every epoch and maximizing rewards, until the agent finally reaches terminal state, that is the point of victory in the game. By using multiple loss functions, one can tune the ConvNet and its weights to make better decisions based on the input available to it. ConvNet may recognize features such as ball direction, speed, and paddle position to optimize its chances of a reward.

The working is divided into two scripts:

Frame Processing: At the heart of the model's architecture lies a complex custom-made ConvNet architecture. It comprises three hidden layers of a variable number of neurons, followed by two fully connected layers of 80 neurons each. The learning rate is fixed at 0.001 with the choice of an Adam Optimizer. The cross-entropy methods are used for computing losses as opposed to the standard mean square error method. Several hyperparameters cause faster convergence to an optimal maximum, and plausible high scores. The batch size is fixed at a hundred and twenty eight for each epoch. Non-terminal and terminal layers have ReLU and Softmax as their activation functions respectively. On processing the images, the model returns a set of probabilities for each command from the set of actions.

Agent Performance Optimization: The initial Q-value is generated randomly at the start of the game. As the processor loops through the set of states and reaches the present states, it executes either an exploratory action, decided randomly, or an exploitative one, chosen on basis of the highest reward assigned to it. In the first training phase, exploration is followed by the system thereby choosing random actions from the set to maximize reward and reach implausible high scores, not possible when the highest reward action is chosen. Following this, the system starts using the exploitation segment. The agent is awarded a reward every time it chooses and performs an action, by the environment.

The Bellman equation helps compute the new Q-values as the agent reaches a new state based on the aforementioned Q-value. There is a proper documentation of the initial state, corresponding action performed, and the state reached by effect of the action, alongside the reward awarded and a Boolean of whether the terminal state has been reached. This documented cache is fed into the ConvNet architecture. This procedure is popularly known as Replay Memory (Figure 21.5).

21.3.2 Results and Discussion

The rewards will be generated in the following format for each epoch of a fixed batch size (Figure 21.6):
Epochs 1–9:

After every certain number of epochs, it will record the performance of the agent for those particular epochs (Figure 21.7).

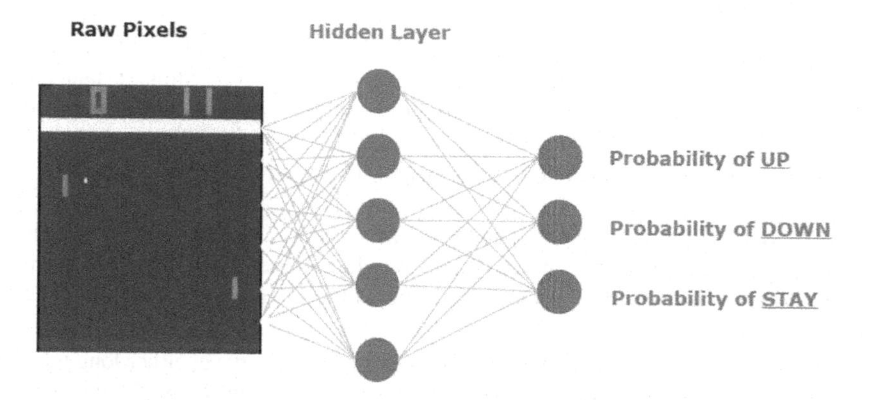

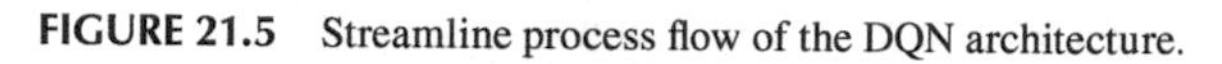

FIGURE 21.5 Streamline process flow of the DQN architecture.

```
(2020-02-02 11:28:22,255] Making new env: ppaquette/DoomCorridor-v0
[2020-02-02 11:28:22,738] Starting new video recorder writing to /
Users/Nisha/Desktop/Doom/videos/openaigym.video.0.1773.video000000.mp4
[2020-02-02 11:28:24,000) Starting new video recorder writing to /
Users/Nisha/Desktop/Doom/videos/openaigym.video.0.1773.video000001.mp4
Epoch: 1, Average Reward: -41.9083557129
Epoch: 2, Average Reward: 1118.30119324
Epoch: 3, Average Reward: 697.771152496
Epoch: 4, Average Reward: 415.120799473
(2020-02-02 11:29:03,694) Starting new video recorder writing to /
Users/Nisha/Desktop/Doom/videos/openaigym.video.0.1773.video000007.mp4
Epoch: 5, Average Reward: 287.70327301
Epoch: 6, Average Reward: 220.423992157
Epoch: 7, Average Reward: 172.363989694
Epoch: 8, Average Reward: 136.319223404
Epoch: 9, Average Reward: 121.477974387
Epoch: 10, Average Reward: 98.4545199746
Epoch: 11, Average Reward: 525.821778564
```

FIGURE 21.6 Result documentation of the proposed system (Epochs 1–9).

Epochs 19–26:

The rewards keep increasing as the agent get trains, and finally, thc output stage is reached when the agent reaches the final state in minimum cost (Figure 21.8):

The balancing factor between short- and long-term rewards gradually comes down as the agent reaches closer to its goal (Figure 21.9).

Performance Metrics: To evaluate the model, it is compared to a series of evaluation criteria that define the certain aspects needed to be covered.

Reward Maximization: Each RL agent has only one task that is to increase its rewards over each episode. The model is playing a game which is limited in its reward. Pong ends when one player has received the score of 20, and as seen from the results presented, the agent can get an average score of greater than 19 in pong and 1500 in Doom.

Epoch: 19, Average Reward: 1144.46448166
Epoch: 20, Average Reward: 1187.87156412
Epoch: 21, Average Reward: 1258.35586466
Epoch: 22, Average Reward: 1303.79177838
Epoch: 23, Average Reward: 1360.59863423
Epoch: 24, Average Reward: 1407.97487823
Epoch: 25, Average Reward: 1436.69232084
Epoch: 26, Average Reward: 1502.09423706
Congratulations, your AI wins

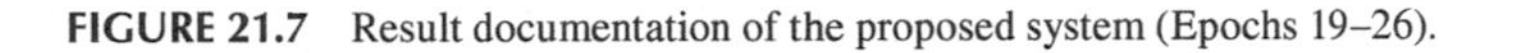

FIGURE 21.7 Result documentation of the proposed system (Epochs 19–26).

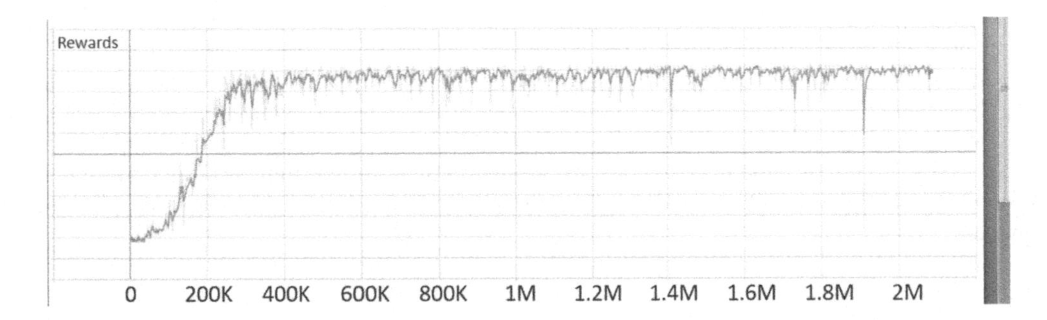

FIGURE 21.8 Rewards trend across various epochs for game Pong.

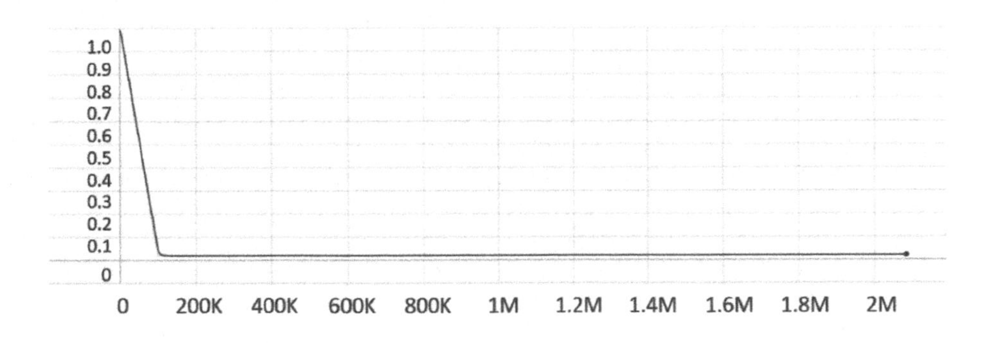

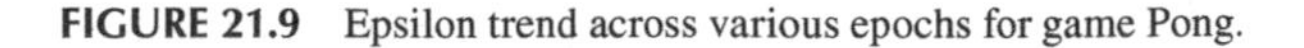

FIGURE 21.9 Epsilon trend across various epochs for game Pong.

Randomization: RL agents should be able to show exploration within them, that is to say agents should perform actions where chances of reward might decrease or increase based on this. By performing new actions and exploring new states, the agent can either optimize its rewards or achieve them quickly. Epsilon is used as a factor to denote the random actions performed. The agent gradually reduces its epsilon over time ensuring a degree of randomness.

21.3.3 Conclusion and Future Scope

The game has been successfully emulated and rendered, and the agent has been made to learn and reach the winning point. The results have been successfully documented, and it can be concluded that the CPU agent with its new high score rewards has successfully completed the game and reached the final state in an almost optimized route. The combination of ReLU for all non-terminal states and Softmax function for the terminal state as activation have worked well.

OpenAI Gym is the most efficient environment for rendering a two-dimensional game environment. The custom-made three hidden layered and two fully connected layered ConvNet for processing the frames extracted has worked very well. The system of value-based models rather than policy-based models for rewards function has given very promising results. Apart from that, a Q-state learning in the form of a DQN has proven most useful for ensuring higher rewards for better performance as well as a good balance between long-term and short-term rewards. Finally, the system of cross-entropy loss in place of mean square error loss computation has also given very good results and in all, both the games, Pong and Doom, have been efficiently and successfully completed by the RL agent.

The future scope of the project lies not in continuing to play games but in its massive ability to refocus. The agent makes clever use of a single agent system where an agent's performance only depends on the fixed hyperparameters and is independent of other scenarios. However when the performance of agents affects each other, the system is relatively tougher to design. Multi-agent RL can potentially redress several problems from simultaneous control, telecommunications, robotics and finance. These multiple agents should be capable of driving a solution from their mistakes without a dataset. Several multi-agent reinforcement learning algorithms can be applied to put in practical use concerning the synchronized movement of an object by two interdependent agents.

Predicting the 2D structure of an RNA protein using a hydrophobic polarity model is a classic non-polynomial indeterministic problem. Optimization of these models is currently done using traditional brute force inefficient methods and Greedy methods. Using Markov Decision Making Algorithm, RL can maximize the overall cumulative return, which usually does not converge to its lowest conformations. It is specifically suited at building overall optimization solutions of these biological genetic sequences.

REFERENCES

1. V. François-Lavet, P. Henderson, R. Islam, M. Bellemare and J. Pineau. 2018. An introduction to deep reinforcement learning. *Foundations and Trends in Machine Learning.* arXiv:1811.12560.
2. A.P. Badia, B. Piot, S. Kapturowski, P. Sprechmann, et al. 2020. Agent57: Outperforming the Atari human benchmark. arXiv:2003.13350 [cs.LG].
3. G. Brockman, V. Cheung, L. Pettersson, et al. 2018. OpenAI Gym. *OpenAI Conference.* Bay Area.
4. G. Papoudakis, K. Chatzidimitriou and P. Mitkas. 2018. Deep reinforcement learning for doom using unsupervised auxiliary tasks. arXiv:1807.01960 [cs.LG].
5. I. Makarov, A. Kashin and A. Korinevskaya. 2017. Learning to play pong video game via deep reinforcement learning. Supplementary Proceedings of the Sixth *International Conference on Analysis of Images, Social Networks and Texts.* AIST 2017, Moscow.
6. L. Kaiser, M. Babaeizadeh, P. Milos, et al. 2019. Model-based reinforcement learning for Atari. *International Conference on Learning Representations*, Addis Ababa.
7. K. Xenou, G. Chalkiadakis and S. Afantenos. 2019. Deep reinforcement learning in strategic board game environments. In: Slavkovik M. (eds), *Multi-Agent Systems: EUMAS 2018.* Lecture Notes in Computer Science, vol. 11450, 3–4. Springer, Cham.

8. M. Lanctot, E. Lockhart, J. Lespiau, et al. 2019. OpenSpiel: A framework for reinforcement learning in games. arXiv:1908.09543 [cs.LG].
9. D. Lee and J. McNair. 2018. Deep reinforcement learning agent for playing 2D shooting games. *International Journal of Control and Automation*, 13, 193–200.
10. H. Teigar, M. Storožev and J. Saks. 2017. 2D Racing game using reinforcement learning and supervised learning. Institute of Computer Science, University of Tartu. Neural Networks.
11. G. Lample and D. Chaplot. 2017. Playing FPS games with deep reinforcement learning. In *Proceedings of the Thirty-First AAAI Conference on Artificial Intelligence. AAAI'17*, San Francisco, CA, AAAI Press. pp. 2140–2146.
12. R. Torrado, P. Bontrager, J. Togelius, J. Liu and D. Perez Liebana. 2018. Deep Reinforcement Learning for General Video Game AI. *2018 IEEE Conference on Computational Intelligence and Games (CIG)*, Maastricht, pp. 1–8, doi: 10.1109/CIG.2018.8490422
13. Y. Zheng. 2019. Reinforcement learning and video games. arXiv:1909.04751 [cs.LG].
14. K. Shao, Z. Tang, Y. Zhu, N. Li and D. Zhao. 2019. A survey of deep reinforcement learning in video games. arXiv:1912.10944 [cs.MA].

22

Analysis of the Effectiveness of the Non-Vaccine Countermeasures Taken by the Indian Government against COVID-19 and Forecasting Using Machine Learning and Deep Learning

Akash Shah
JP Morgan Chase & Co

Romil Shah, Manan Gandhi, Rashmil Panchani, Govind Thakur, and Kriti Srivastava
Dwarkadas J. Sanghvi College of Engineering

CONTENTS

22.1 Introduction....321
22.2 Related Work....322
22.3 Data Analysis....322
22.3.1 Top Ten States with Most Fatalities....322
22.3.2 Highest Fatality Rates....322
22.3.3 Population to Confirmed Cases Ratio....323
22.3.4 Analysis of Five-Phased Lockdown in India....323
22.4 ARIMA (Autoregressive Integrated Moving Average)....324
22.4.1 Mathematical Explanation of the ARIMA Model....325
22.4.2 Daily Death Count Forecasting....327
22.4.3 Daily Active Count Forecasting....333
22.4.4 Daily Confirmed Cases Count Forecasting....340
22.5 Recurrent Neural Networks and Long Short-Term Memory Networks....342
22.6 Conclusion....351
Bibliography....351

22.1 Introduction

Most of the infected people from COVID-19 suffer from respiratory illness and recover without the need for any special treatment. So far (as of July 2020), India seems to have been successful in flattening the curve and tackling the situation effectively. The strategies implemented by India are acknowledged by the leaders of various nations and WHO itself. This chapter embarks on a journey of exploring insights predicting the future of the COVID-19 situation in India.

The key objective of this chapter is to demonstrate the power of statistical analysis and deep learning models in predicting the future cases and understanding the situation. This helps with performing a risk assessment and mitigating the imminent threat to the economy and lives of the people. There exist several statistical and deep learning models that play a huge role in future trend prediction based on the past and current data. This chapter explores models such as ARIMA (autoregressive integrated moving average) and LSTM (long short-term memory). The first step to any prediction task is to study the data and note the inferences observed. To do this, the chapter first dives into the Data Analysis phase. Using certain data processing techniques, the data is first cleansed by removing irregularities and noise and then composed in some readable format. Then, the chapter dives into forecasting approaches via ARIMA and LSTM models and assessing their output performances with different parameters.

22.2 Related Work

The papers [3–5] focus on predicting the trend analysis of the number of cases in India. The study in [3] has based their findings by using ARIMA modeling and Exponential Smoothing methods which are the most widely used methods for quick forecasting and trend analysis. The studies were done in [4] use the SEIR model and Regression modeling to achieve future predictions. The study [5] is an incremental version of the previous studies where they implement network modeling and pattern mining to predict their results.

The paper [10] uses models such as exponential, logistical, SIS model, and the daily infection rate (DIR). These models have already been applied and tested in past pandemic situations. The study [11] uses machine learning techniques instead of the standard SIR and SEIR models. The techniques used comprise Multi-Layer Perceptrons (MLP) and Adaptive Network-based Fuzzy Inference System (ANFIS). In the study [12], exponential models without the logistic control support are implemented. This paper only predicts the number of infections in the upcoming days. This chapter focuses on the latest data available until 23rd July 2020.

22.3 Data Analysis

Data were gathered from Kaggle, COVID-19 API, and John Hopkins CSSE repository and later cleaned as per the requirements. Data cleaning is the most essential step before analyzing the data at hand as it becomes more convenient and gives more efficient outcomes. Data analysis helps in building the foundation for implementing various algorithms on the data.

Let us start the analysis by first understanding the situation in the country concerning certain parameters. They are as follows.

22.3.1 Top Ten States with Most Fatalities

The effect of an epidemic or a pandemic is primarily judged by the number of deaths that it causes. Due to the massive population of India, it was feared that once the infection starts spreading locally, the number of deaths was going to increase. As of 23rd July, 28,732 fatalities have been recorded due to the coronavirus in India. The visualization below shows the top ten states in India concerning the reported deaths. Maharashtra has the highest fatalities. Uttar Pradesh, being the most populated state in India, has fewer deaths in comparison. Although Bihar is one of the most populated states, it has the least deaths among them (Figure 22.1).

22.3.2 Highest Fatality Rates

Although the number of fatalities gives us an overview of how under control the situation is, the actual scenario can be understood by considering the fatality rate of the region. Around the world, India has been regarded as one of the most successful countries in limiting the spread of the virus. This is because of the fatality rate in India. The left visualization shows the top ten states with the highest fatality rate. Maharashtra has the highest fatality rate among all the states. Although Delhi has the second most fatalities in the top ten states, it is ninth when it comes to the fatality rate. The right visualization shows the fatality rates of the seven most populated states of India. It can be inferred that over these several months of lockdown, the rate in all these states has gone down despite there being a spike in April (Figure 22.2).

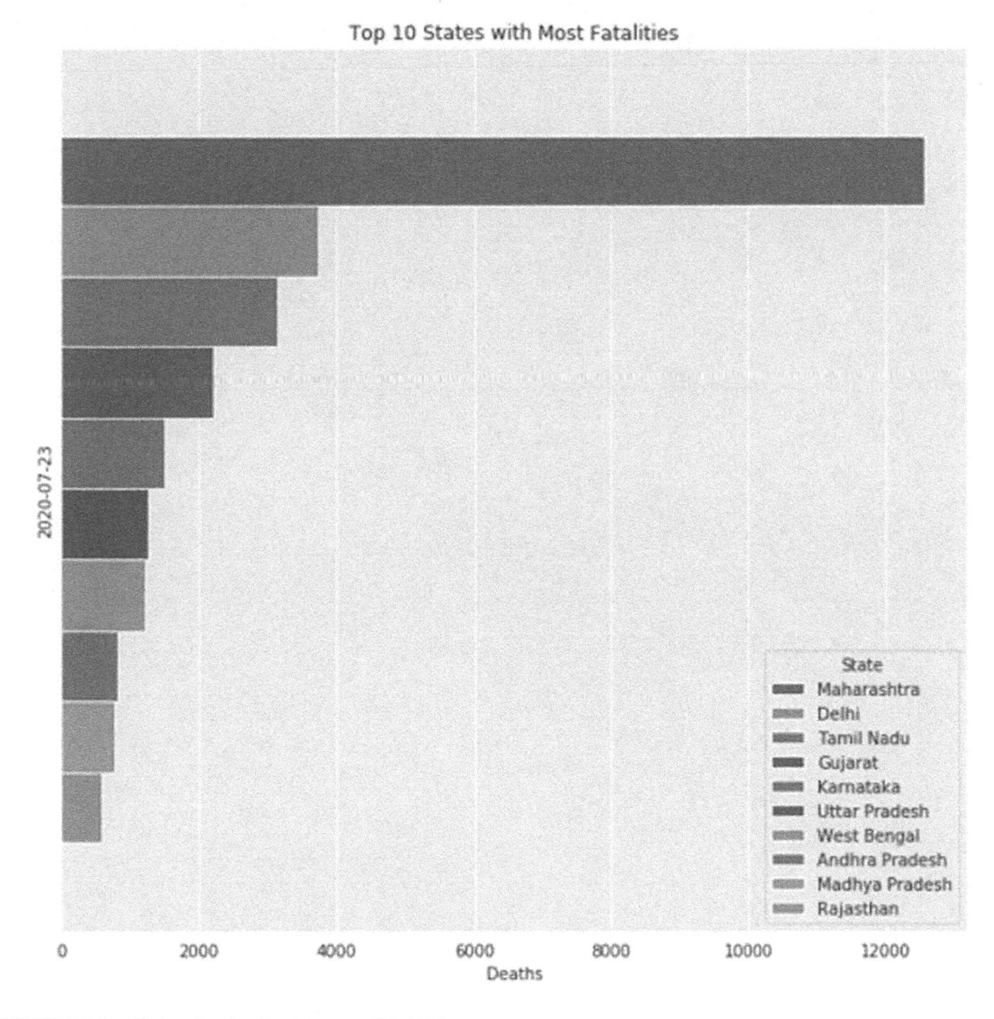

FIGURE 22.1 State-wise death plot as per 23rd July.

22.3.3 Population to Confirmed Cases Ratio

The first and the foremost parameter that the layman notices while analyzing the situation of an epidemic or a pandemic is the number of confirmed cases but the actual estimation of the severity of the situation can be found by the ratio of the population to the confirmed number of cases. The above visualization shows this ratio concerning different states in India (Figure 22.3).

22.3.4 Analysis of Five-Phased Lockdown in India

The Government of India on 4th March enforced a nationwide 21-day lockdown as a preventive measure against the spread of the coronavirus. The lockdown was placed when the number of confirmed positive coronavirus cases in India was ~500. After this 1st lockdown, more such lockdowns have been enforced with conditional relaxations. It was stated that India has emerged as one of the most stringent countries in terms of limitations enforced on its citizens (Figures 22.4 and 22.5).

As the lockdown in the country progressed, the number of tests conducted increased, and as a result, more confirmed cases were recorded. Although this was alarming, the rate at which tests are being conducted is a big positive. It can be seen in the visualization that the percent change in the number of tests conducted has increased in every lockdown phase.

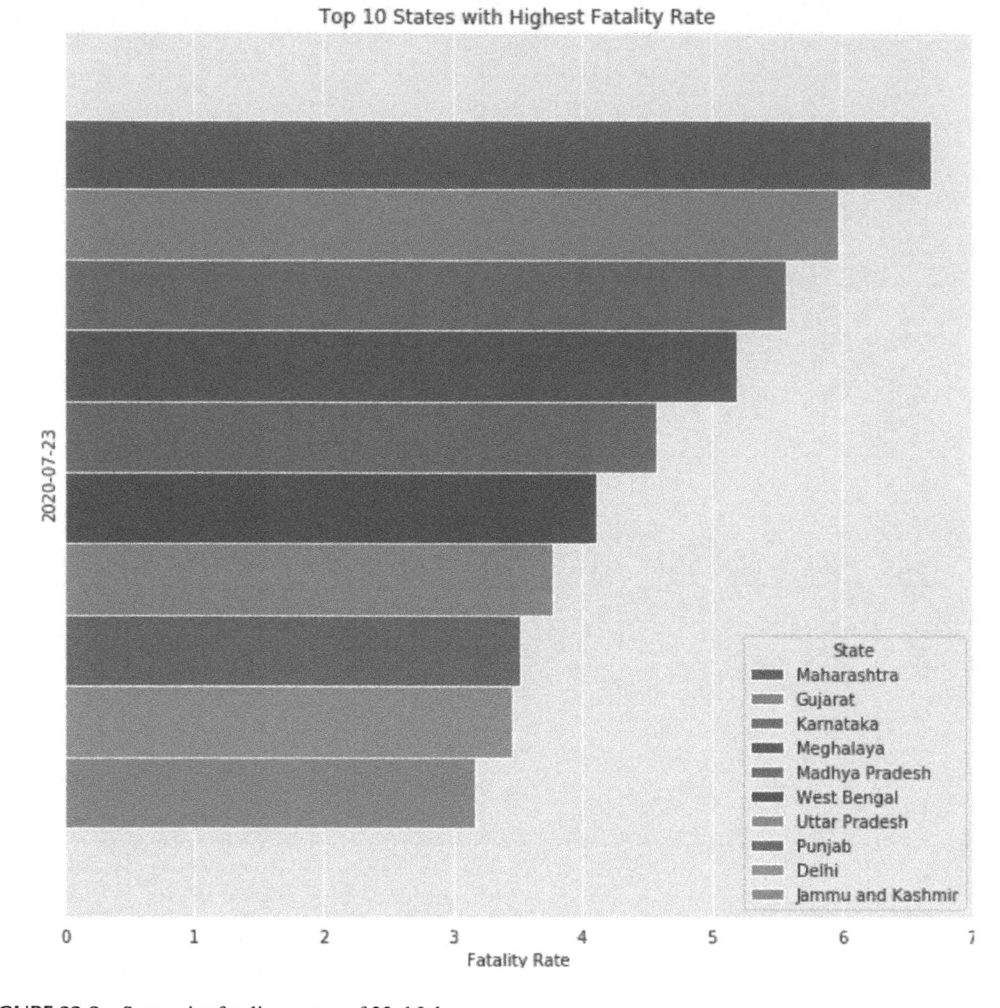

FIGURE 22.2 State-wise fatality rate as of 23rd July.

Due to this, the percent change in the number of confirmed cases has been almost zero through all the lockdown phases. The same has happened with the percent change in the number of recovered cases.

22.4 ARIMA (Autoregressive Integrated Moving Average)

India is at a crucial stage where one miscalculation in estimating the resource allocation for the appeasement of the current situation may cause adverse effects on the current grasp we have over the situation. It becomes increasingly necessary in forecasting the number of cases and thus devises a strategy to handle the influx of new cases. This chapter tries to project the number of cases by applying models such as ARIMA and LSTM, which has proven to be very effective in forecasting time series data.

ARIMA, a machine learning technique, is a class of models that helps us understand the time series based on its past values, i.e., its lags and its own lagged forecast errors so that the equation can be used to forecast future values.

ARIMA model is characterized by the following three parameters: p, d, and q where "p" is the order of AR (Auto Regression) term which governs the number of lags of Y to be used as predictors, "q" is the order of MA (Moving Average) term which governs the number of lagged forecast errors, and "d" is the number of differences required to achieve stationarity. The term autoregressive in ARIMA basically

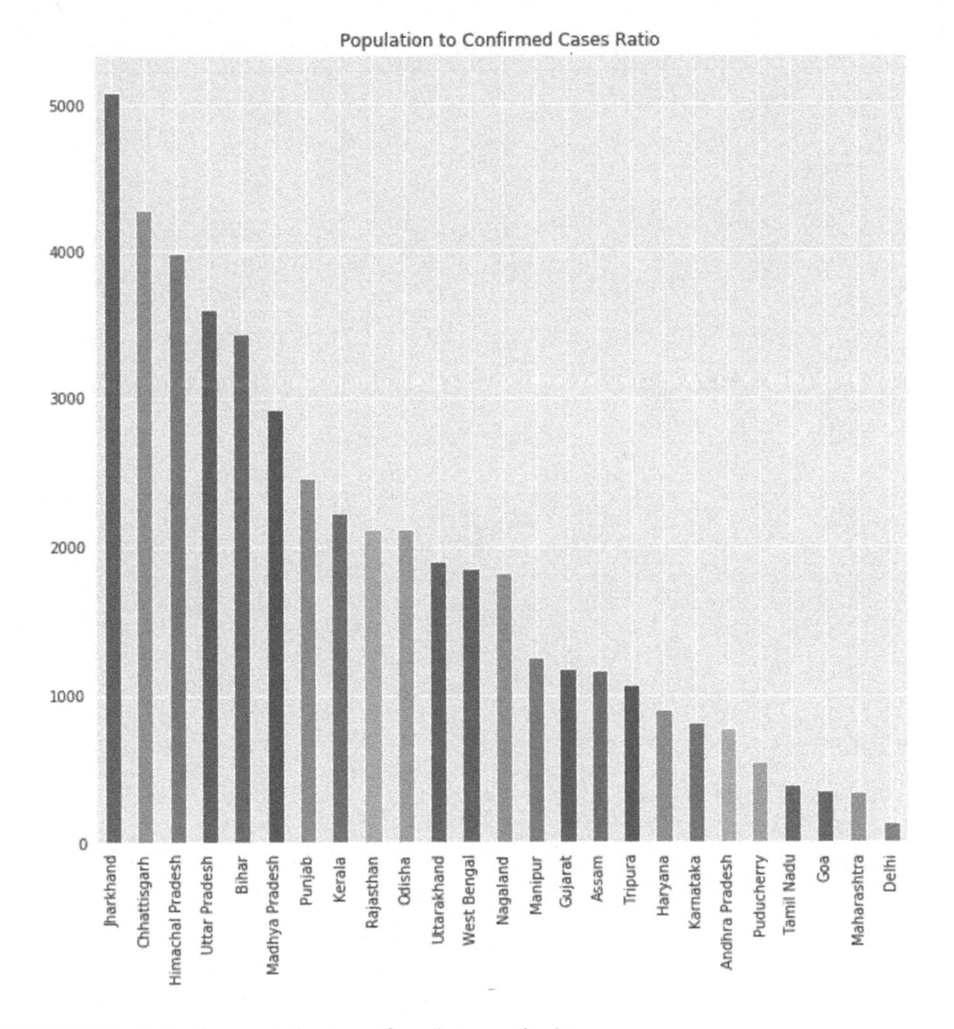

FIGURE 22.3 State-wise population to confirmed cases ratio plot.

means it is a linear regression model that uses its lags as predictors and they work best when the predictors are not correlated, i.e., they are independent of each other. Hence, it is necessary to ensure that the series has been transformed to achieve stationarity. The dataset we have considered is non-stationary, and thus, we have used "differencing" methods to achieve stationarity.

22.4.1 Mathematical Explanation of the ARIMA Model

A pure autoregressive model is one in which Y_t is only dependent on its lags.

$$Y_t = \alpha + \beta_1 Y_{t-1} + \beta_2 Y_{t-2} + \cdots + \beta_p Y_{t-p} + \varepsilon_1$$

where

Y_{t-1} is the lag,

β_1 is the coefficient of the corresponding lag, and

α is the intercept term

Similarly, a pure Moving Average model is the one in which Y_t depends only on the lagged forecast errors.

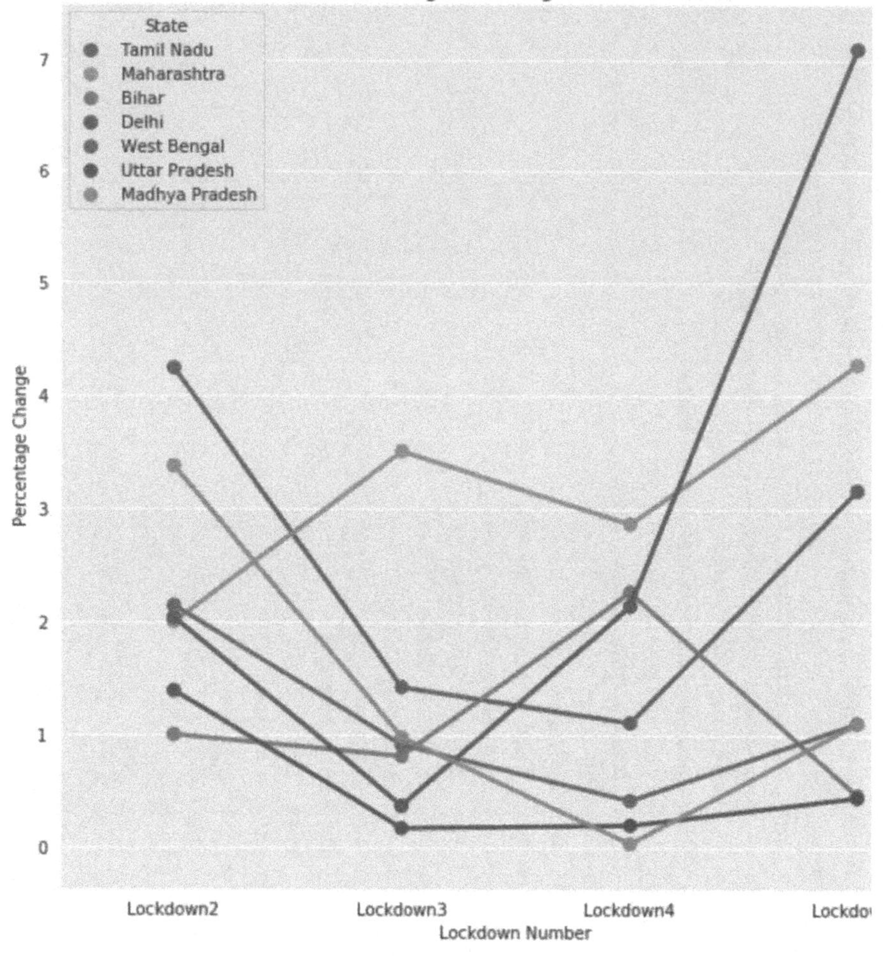

FIGURE 22.4 Change in testing as the lockdown.

$$Y_t = \alpha + \varepsilon_t + \phi_1\varepsilon_{t-1} + \phi_2\varepsilon_{t-2} + \cdots + \phi_q\varepsilon_{t-q}$$

where the error terms are the errors of the autoregressive models of the respective lags. The errors ε_t and ε_{t-2} are the errors from the following equations:

$$Y_t = \beta_1 Y_{t-1} + \beta_2 Y_{t-2} + \cdots + \beta_0 Y_0 + \varepsilon_t$$

$$Y_{t-1} = \beta_2 Y_{t-2} + \beta_3 Y_{t-3} + \cdots + \beta_0 Y_0 + \varepsilon_{t-1}$$

Hence, the final ARIMA model equations are a combination of both the AR and the MA equations.

$$Y_t = \alpha + \beta_1 Y_{t-1} + \beta_2 Y_{t-2} + \cdots + \beta_p Y_{t-p}\varepsilon_t + \phi_1\varepsilon_{t-1} + \phi_2\varepsilon_{t-2} + \cdots + \phi_q\varepsilon_{t-q}$$

ARIMA model in words:

Predicted Y_t=Constant+Linear combination Lags of Y (up to p lags)+Linear Combination of Lagged forecast errors (up to q lags)

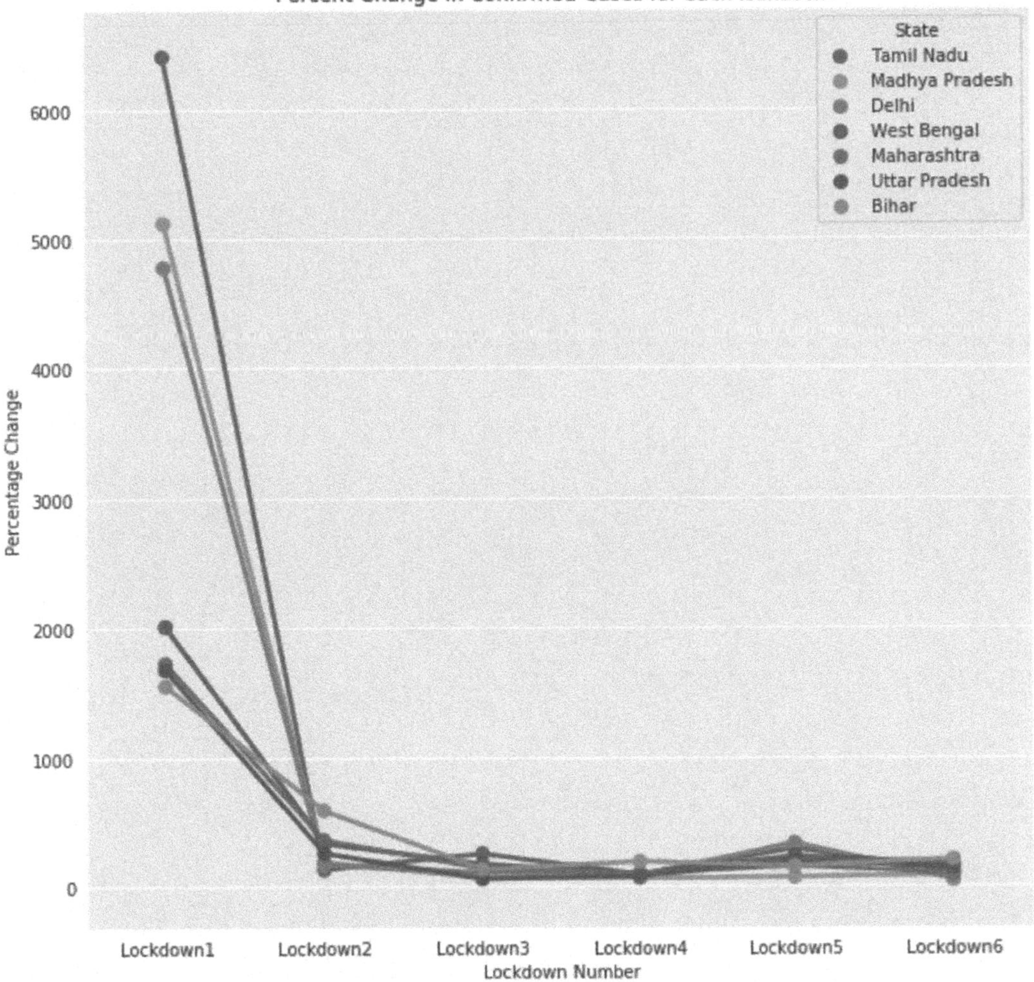

FIGURE 22.5 Change in confirmed cases during the lockdown.

This chapter forecasts data for several aspects such as daily death count, confirmed cases, active cases, and recovered cases.

22.4.2 Daily Death Count Forecasting

Below is the plot of the data for deaths occurring in India from 22 January 2020 till 13 June 2020 which clearly shows that the series under observation is non-stationary.

On observing this plot, it was clear that there is a need to perform the difference between the series to make the series stationary. Below are the plots that depict the series after two successive differences (Figure 22.6).

The reason behind choosing the order of difference "*d*" as 2 is because the autocorrelation plot for $d=1$ swings way too quickly and far into the negative zone, whereas for $d=2$ the autocorrelation value does not go far into the negative zone, which indicated that the series has fairly achieved stationarity. The reason for some irregularities in the stationarity of the data can mainly be accounted for the noise present in the data due to various reasons (Figures 22.7–22.9).

To determine the Autoregressive order (p), partial autocorrelation (PACF) plot is used (Figure 22.10).

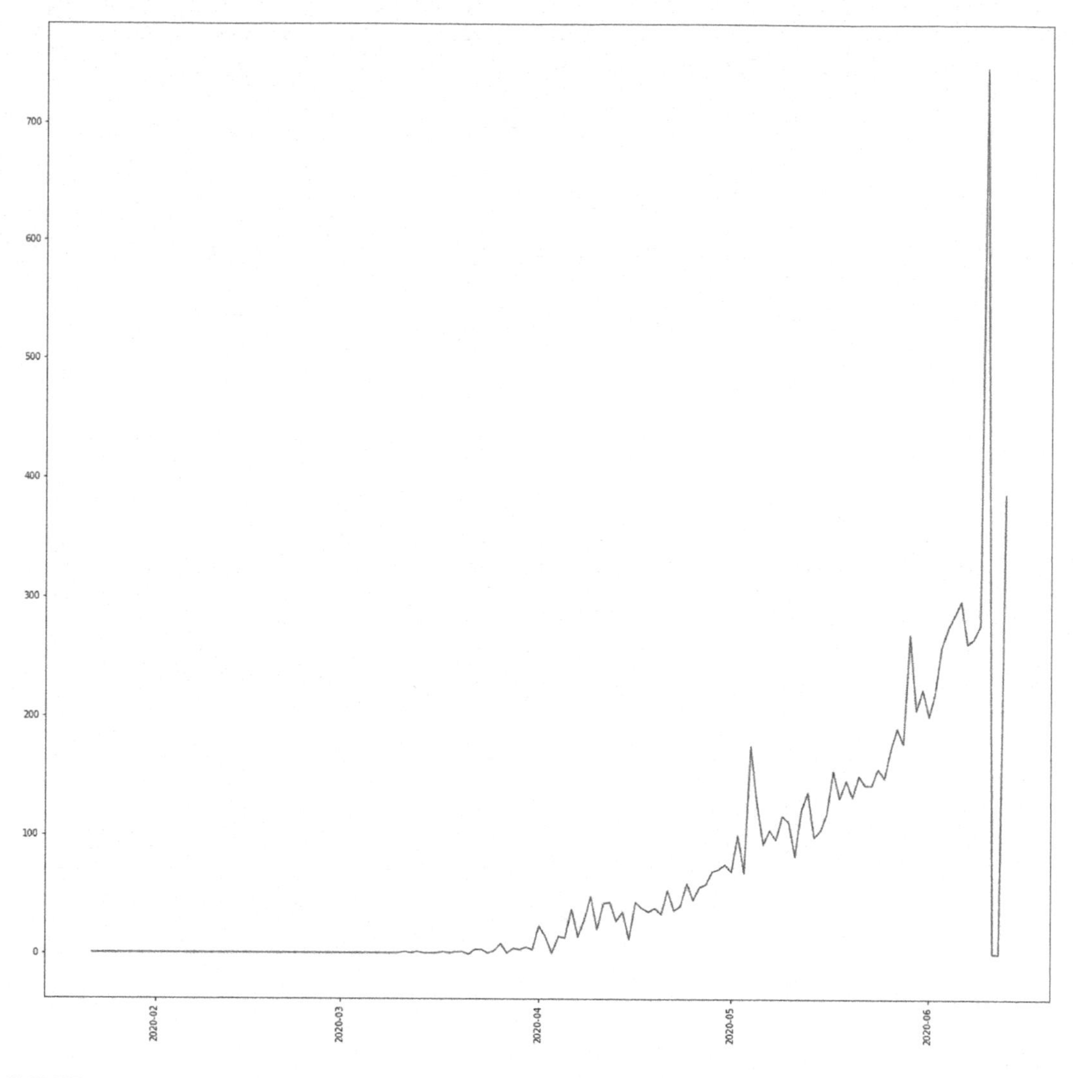

FIGURE 22.6 Number of deaths vs days.

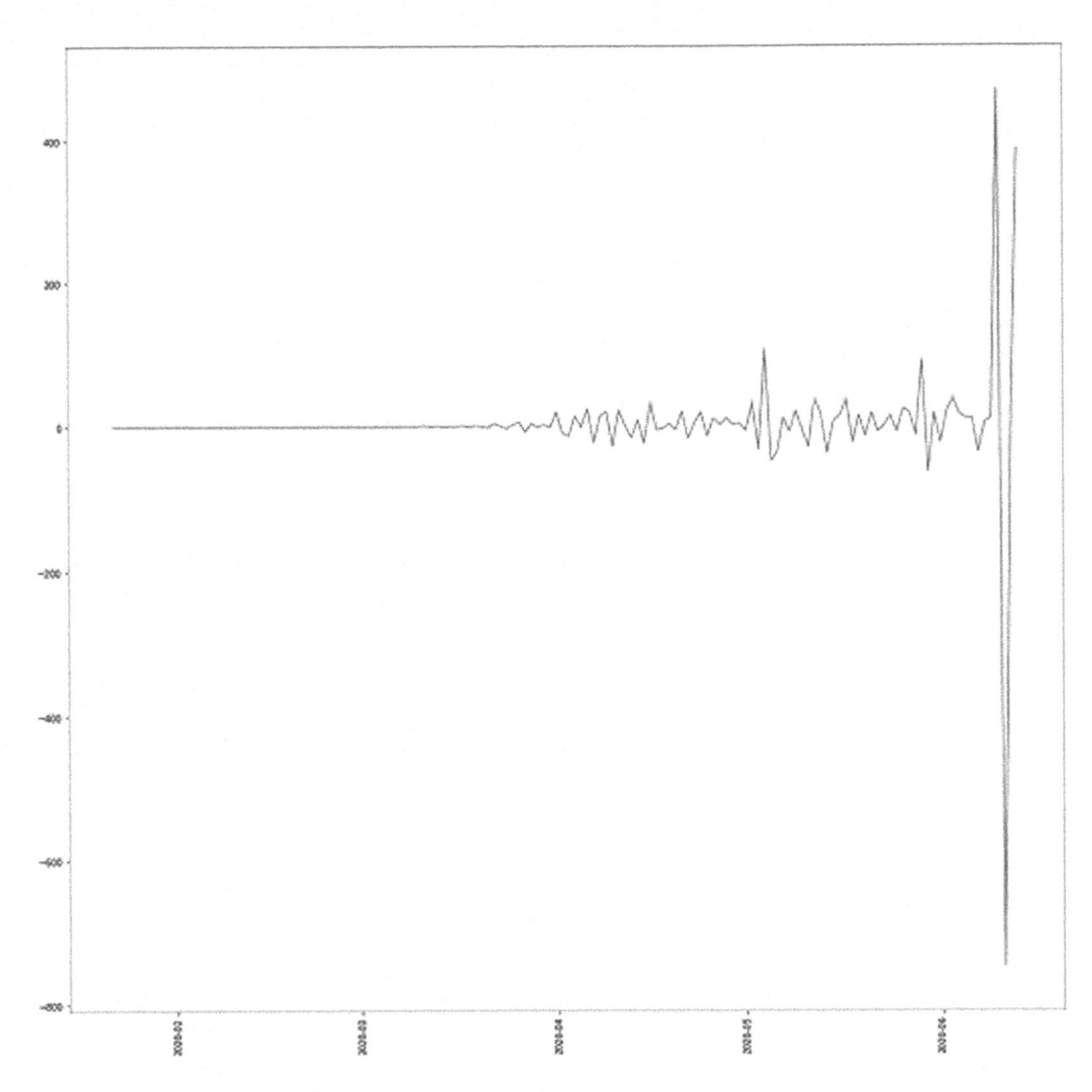

FIGURE 22.7 After the first difference.

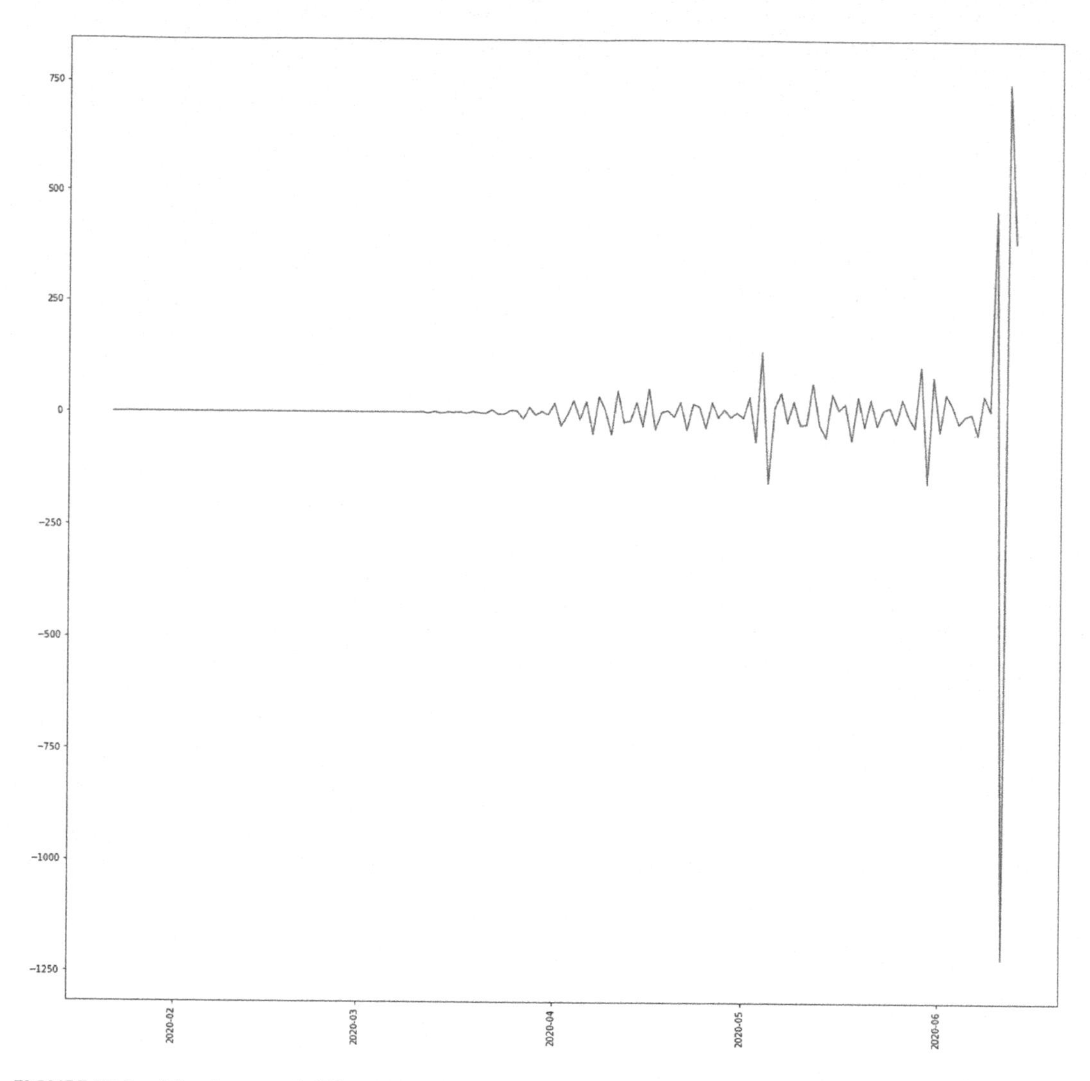

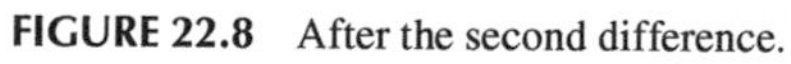
FIGURE 22.8 After the second difference.

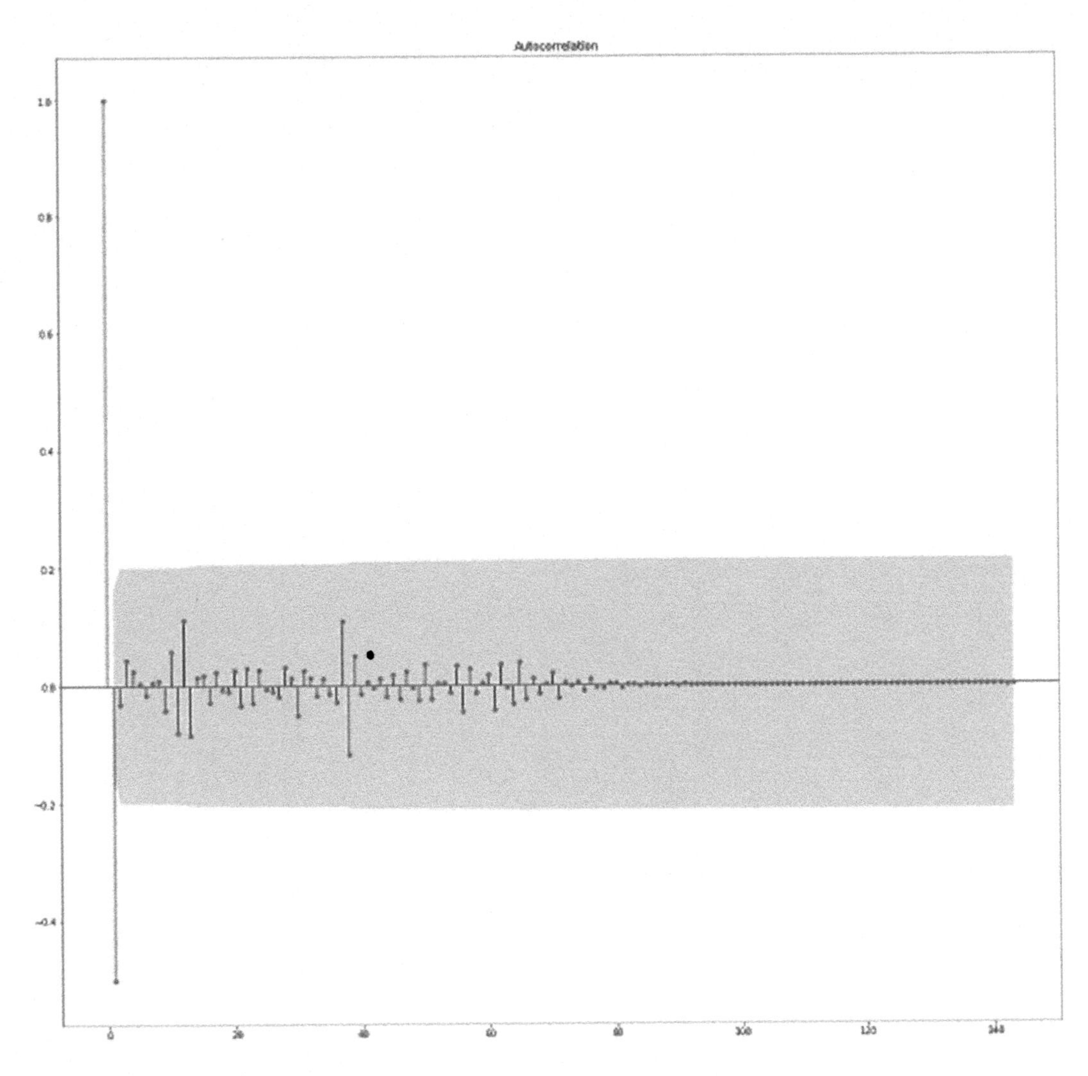

FIGURE 22.9 Autocorrelation plot.

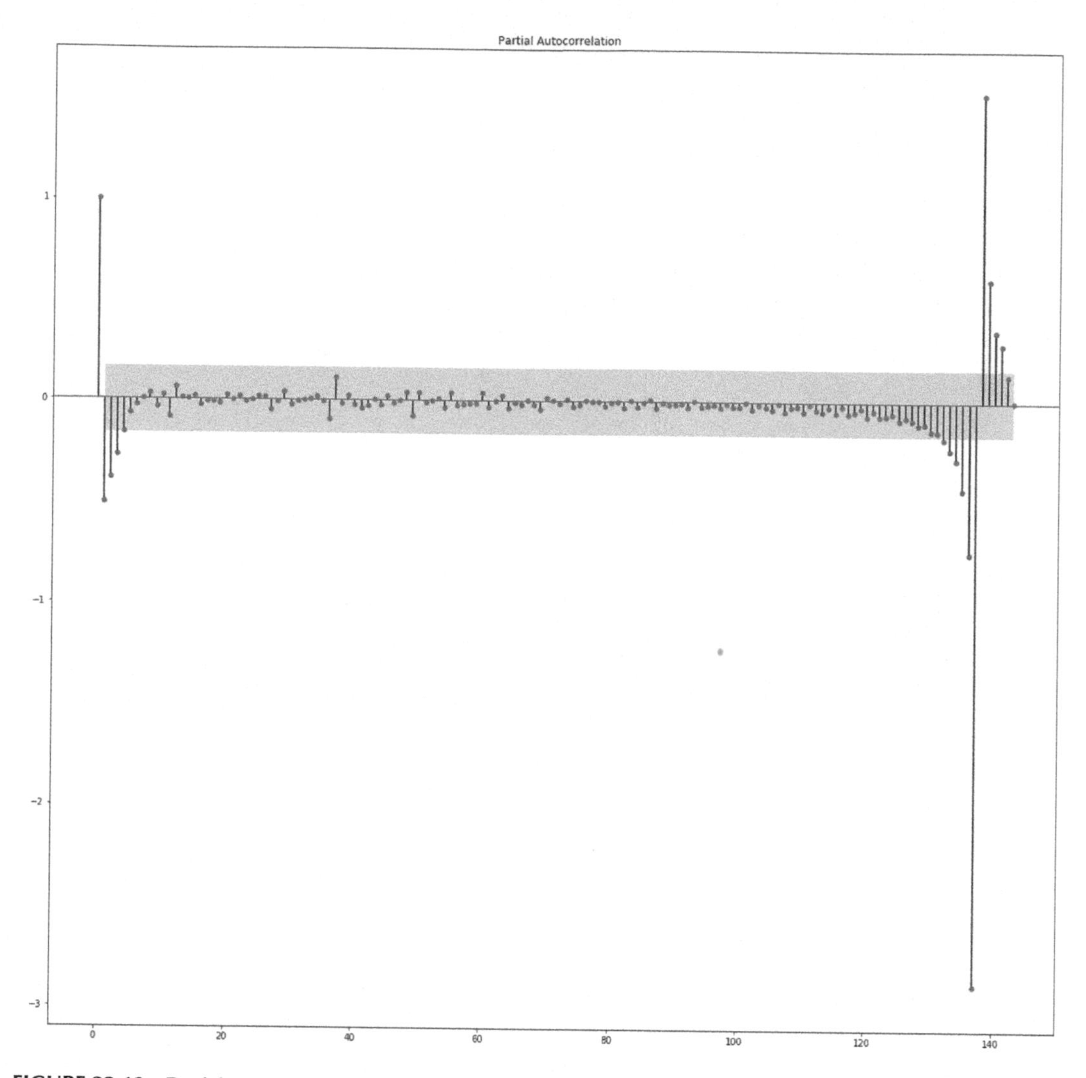

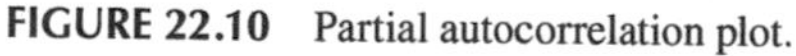
FIGURE 22.10 Partial autocorrelation plot.

As it is evident from the PACF plot, the majority of the points lie in the grey region which means that the autocorrelation is significant. Thus, the parameter p can be set to the value 2. Further to determine the value of q (Moving Average order), one can see the ACF plot to determine its value. This chapter tries out different values of p and q which gave optimal AIC (Akaike information criteria) and BIC (Bayesian information criteria) values. After some trials and errors, $q=4$ provides the least value for AIC and BIC.

At this stage, we have determined the values of the variables: p, d and q, let us build the ARIMA model and feed the data. The orders fed were $p=2$, $d=2$, and $q=4$ and set the "transparams" parameter to false.

The following is the summary of the ARIMA model (Figure 22.11).

As seen in the model summary, the AIC and the BIC values are comparatively less than any other model parameters tested. Also, the coefficients of all the variables are highly significant as the p-value is 0 for all the variables.

Using this model to predict the values of the time frame till 25th July 2020 from the beginning keeping in mind that the model was only fed till 13th July 2020 (Figure 22.12).

On evaluating the performance of the model using the root mean square error, the error comes out to be around 46.51369 which is high but still, this model will give a rough estimate of the provisions required to cater to the patients so that this quantity may not be reached and in turn saving as many lives as possible.

ARIMA Model Results

Dep. Variable:	D2.India	**No. Observations:**	142
Model:	ARIMA(2, 2, 4)	**Log Likelihood**	-755.660
Method:	css-mle	**S.D. of innovations**	46.306
Date:	Sun, 14 Jun 2020	**AIC**	1527.320
Time:	20:11:06	**BIC**	1550.967
Sample:	01-24-2020 - 06-13-2020	**HQIC**	1536.929

	coef	**std err**	**z**	**P>|z|**	**[0.025**	**0.975]**
const	0.0543	0.010	5.287	0.000	0.034	0.074
ar.L1.D2.India	0.0105	0.001	7.617	0.000	0.456	0.771
ar.L2.D2.India	-0.9639	0.033	-29.446	0.000	-1.028	-0.900
ma.L1.D2.India	-2.9381	0.083	-35.523	0.000	-3.100	-2.776
ma.L2.D2.India	3.8107	0.217	17.568	0.000	3.386	4.236
ma.L3.D2.India	-2.6789	0.244	-10.990	0.000	-3.157	-2.201
ma.L4.D2.India	0.8062	0.105	7.651	0.000	0.600	1.013

Roots

	Real	**Imaginary**	**Modulus**	**Frequency**
AR.1	0.3183	-0.9676j	1.0186	-0.1994
AR.2	0.3183	+0.9676j	1.0186	0.1994
MA.1	0.5737	-0.8521j	1.0273	-0.1557
MA.2	0.5737	+0.8521j	1.0273	0.1557
MA.3	0.9999	-0.0000j	0.9999	-0.0000
MA.4	1.1756	-0.0000j	1.1756	-0.0000

FIGURE 22.11 Summary of ARIMA model (2,2,4).

Testing for different models and parameters yields variegated results which have been summarized in the Table 22.1.

22.4.3 Daily Active Count Forecasting

Taking inferences from the daily death predictions, it would be equally important to have an idea of the active number of cases in the future. For this, let us start by visualizing the active cases count across time (Figure 22.13).

This plot is till the 18th of July, and the count on that day was around 400 thousand. It is evident in the next plot which depicts the plot of the natural values of the active count cases (Figure 22.14).

Observing this, the next step is to achieve stationarity in the series. Following the same steps as before, we get the following plots (Figure 22.15).

After the second differencing, it can be observed that the series has finally achieved stationarity to some extent, and further differencing might disrupt the stationarity. Also, from the below plot of autocorrelation, the values do not swing quickly into the negative zone which is a clear indication that the series has achieved near stationarity. Hence, it is admissible to set the value of $d=2$ (Figure 22.16).

To determine the Autoregressive order (p), the partial autocorrelation (PACF) plot is checked for its significance (Figure 22.17).

As it is clear from the partial autocorrelation plot that the series is not quite significant, setting the order of $p=0$, the computation of the q value can be done by observing the ACF plot and testing the model for the least values of AIC and BIC. By experimenting with different values, it comes out to be around $q=9$.

Hence, the order of the ARIMA model is $p=0$, $d=2$, and $q=9$ (Figure 22.18).

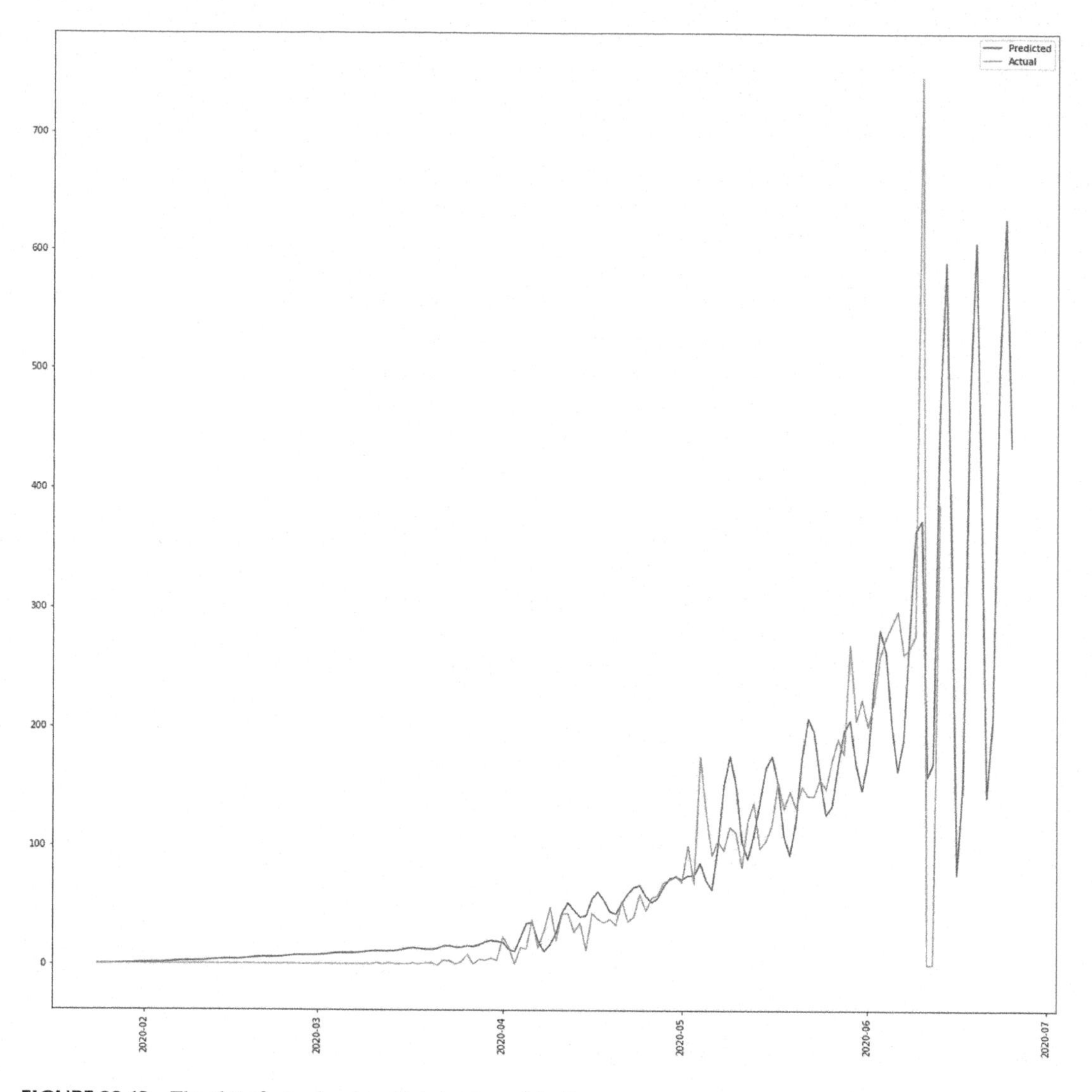

FIGURE 22.12 The plot of actual and predicted values of deaths.

TABLE 22.1

Evaluation Summary of ARIMA Model for Death Count Cases Forecasting

Sr no.	AR (p)	MA (q)	*d*	AIC	BIC	RMS-Error
1	1	1	2	2342	2355	150
2	0	1	2	2345	2360	150
3	4	0	2	2354	2373	154
4	3	0	2	2366	2382	160
5	1	0	2	2395	2405	175

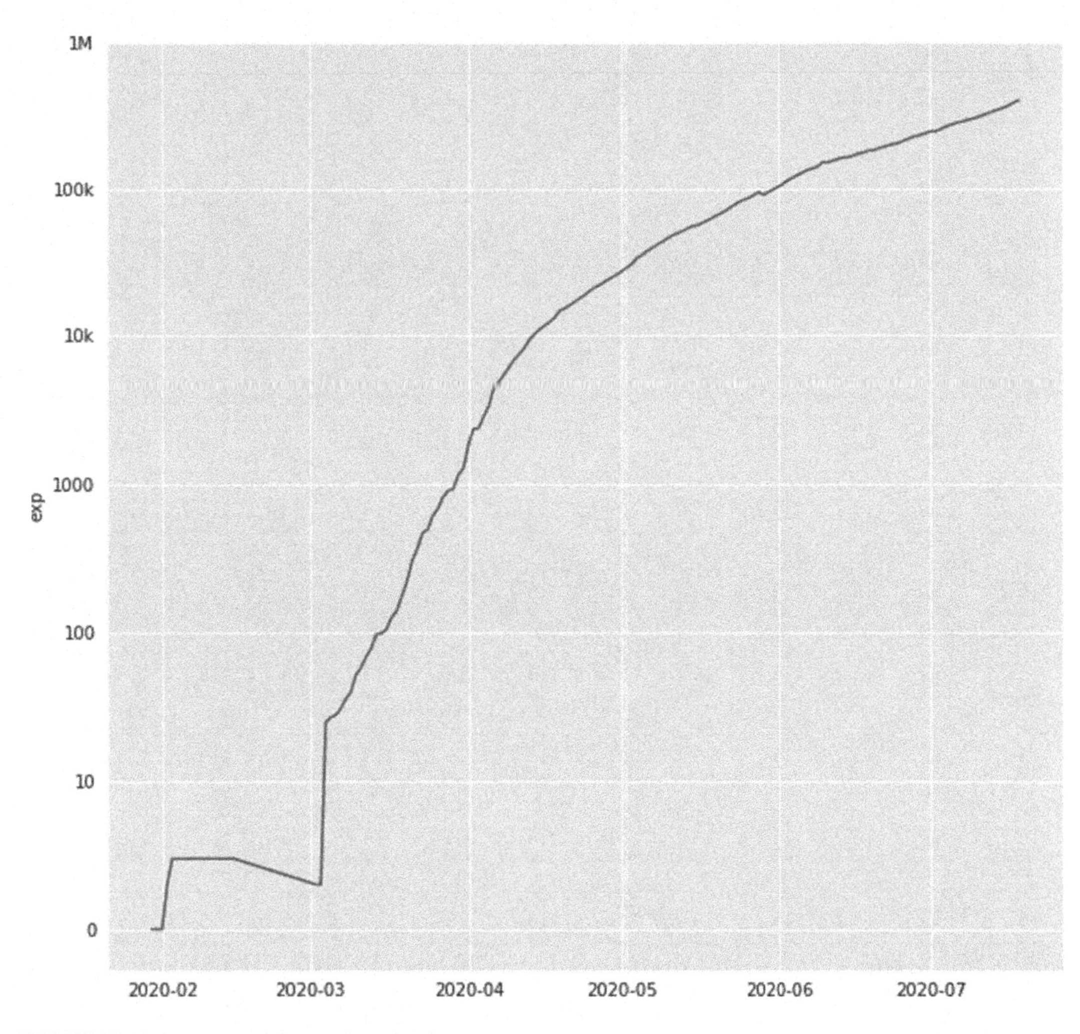

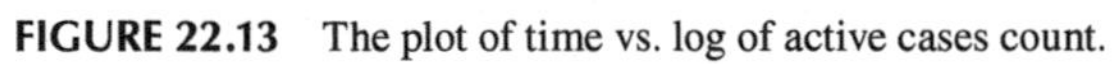
FIGURE 22.13 The plot of time vs. log of active cases count.

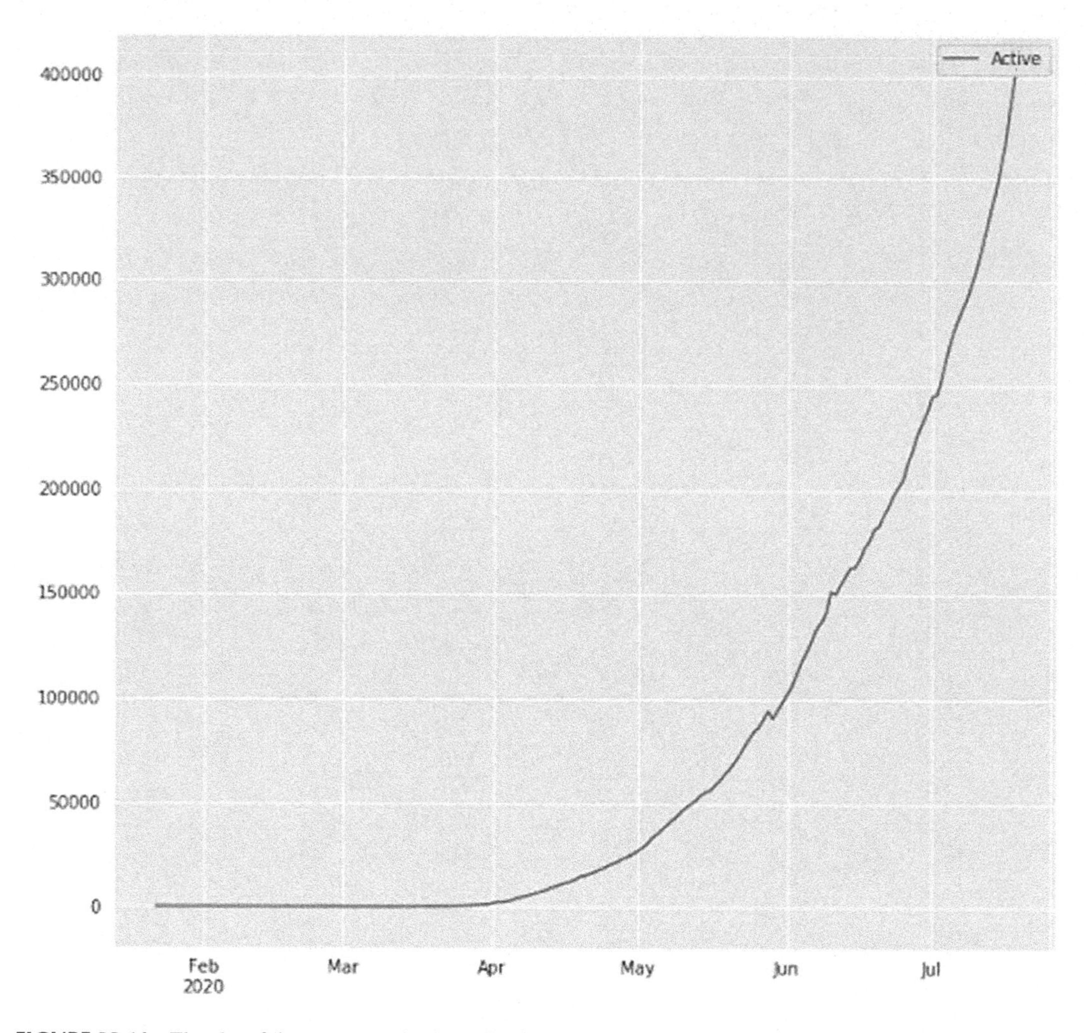

FIGURE 22.14 The plot of time vs. natural values of active cases.

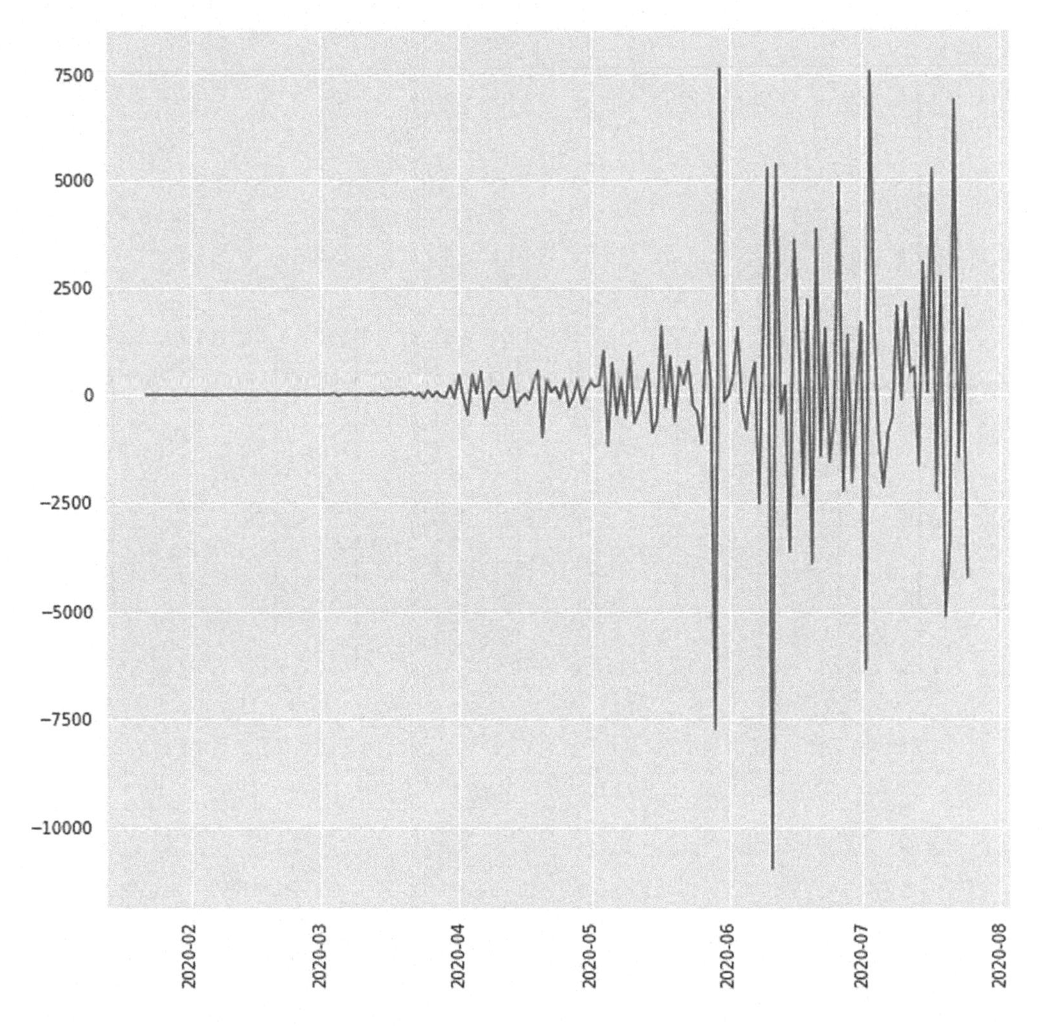

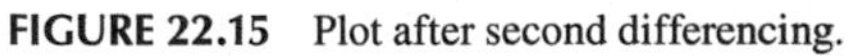

FIGURE 22.15 Plot after second differencing.

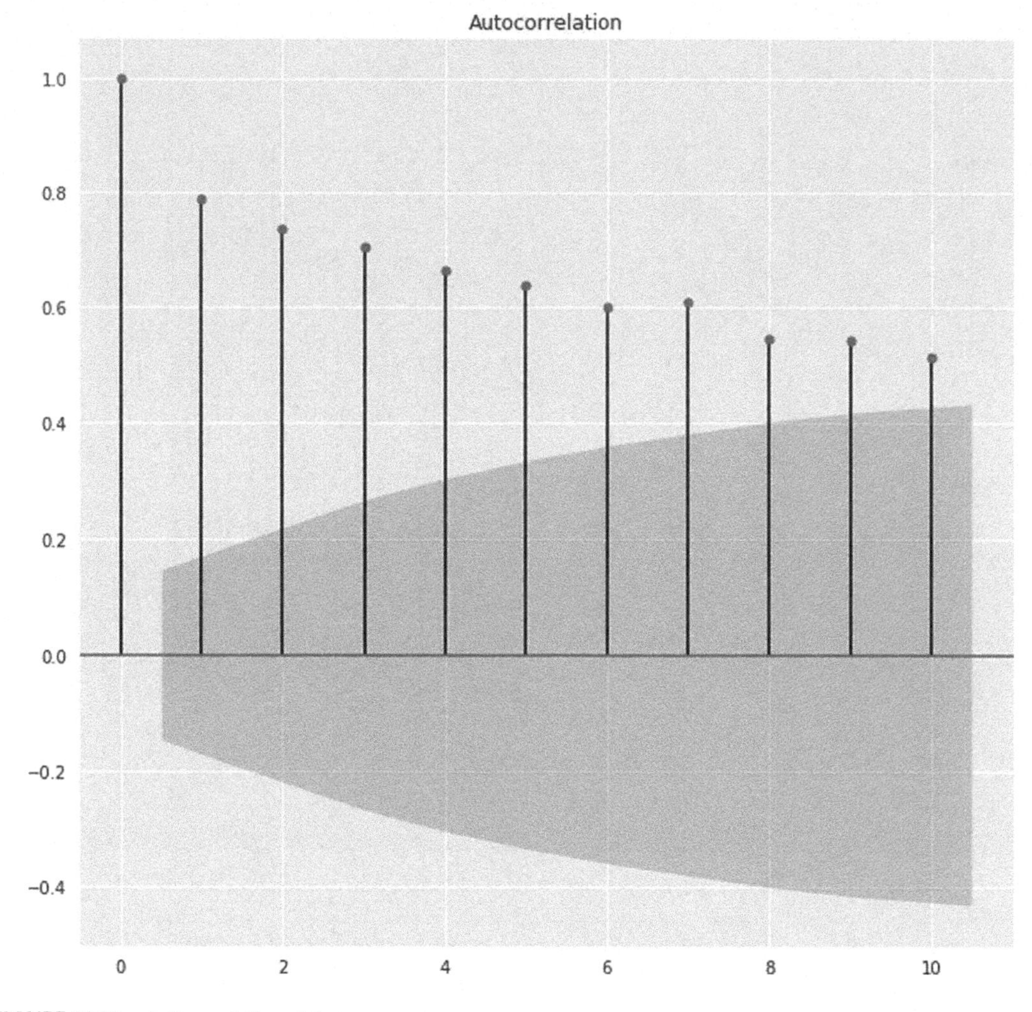

FIGURE 22.16 Autocorrelation plot.

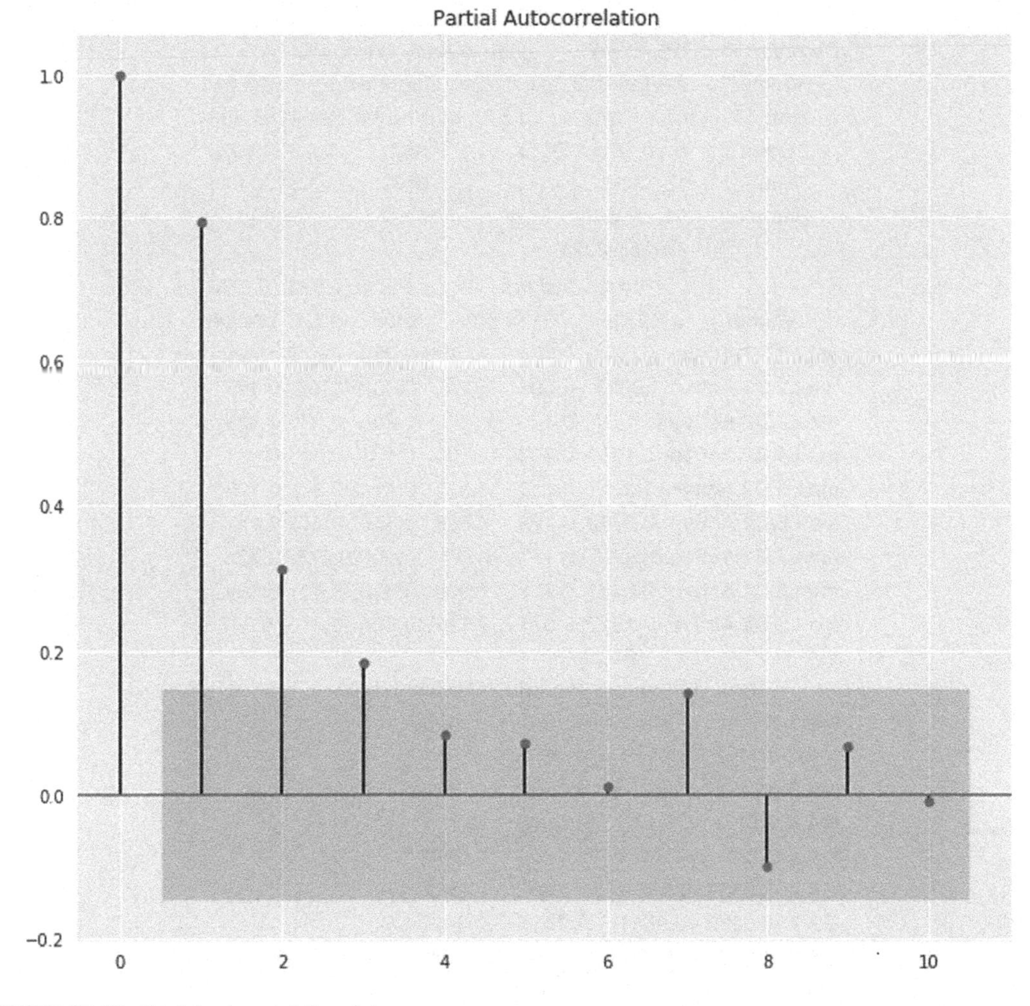

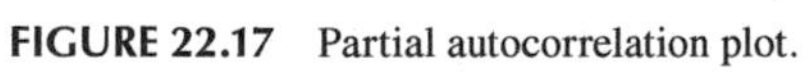

FIGURE 22.17 Partial autocorrelation plot.

ARIMA Model Results

Dep. Variable:	D2.Active	**No. Observations:**	177
Model:	ARIMA(0, 2, 9)	**Log Likelihood**	-1522.018
Method:	css-mle	**S.D. of innovations**	1284.588
Date:	Sun, 19 Jul 2020	**AIC**	3066.036
Time:	16:13:32	**BIC**	3100.973
Sample:	01-24-2020	**HQIC**	3080.205
	- 07-18-2020		

	coef	**std err**	**z**	**P>|z|**	**[0.025**	**0.975]**
const	82.0859	42.174	1.946	0.052	-0.573	164.745
ma.L1.D2.Active	-0.8276	0.074	-11.199	0.000	-0.972	-0.683
ma.L2.D2.Active	0.3015	0.099	3.057	0.002	0.108	0.495
ma.L3.D2.Active	0.1267	0.100	1.265	0.206	-0.070	0.323
ma.L4.D2.Active	-0.1421	0.109	-1.308	0.191	-0.355	0.071
ma.L5.D2.Active	-0.0917	0.112	-0.817	0.414	-0.312	0.128
ma.L6.D2.Active	-0.0733	0.122	-0.598	0.550	-0.313	0.167
ma.L7.D2.Active	0.5057	0.117	4.317	0.000	0.276	0.735
ma.L8.D2.Active	-0.2131	0.111	-1.920	0.055	-0.431	0.004
ma.L9.D2.Active	-0.1627	0.077	-2.115	0.034	-0.313	-0.012

Roots

	Real	**Imaginary**	**Modulus**	**Frequency**
MA.1	1.1942	-0.0000j	1.1942	-0.0000
MA.2	0.9496	-0.5873j	1.1166	-0.0882
MA.3	0.9496	+0.5873j	1.1166	0.0882
MA.4	0.3201	-0.9474j	1.0000	-0.1981
MA.5	0.3201	+0.9474j	1.0000	0.1981
MA.6	-0.6369	-0.9651j	1.1563	-0.3428
MA.7	-0.6369	+0.9651j	1.1563	0.3428
MA.8	-1.2032	-0.0000j	1.2032	-0.5000
MA.9	-2.5663	-0.0000j	2.5663	-0.5000

FIGURE 22.18 Summary of ARIMA (0,2,9) model.

Some of the variables have crossed the threshold of the *p*-value of being <0.05, which indicates the existence of some non-stationarity and more scope of improvement but adjusting various data points that may be present as noise and affecting the overall capacity of the model.

Using this model to further predict the number of active cases in India gives us an estimation of 350 thousand cases by the 15th of July, and when tested with the actual data, the figures are very close (Figure 22.19).

The mean squared error for this prediction comes out to be 1539.5157, which is very close to the accurate predicted value.

Further testing for various other parameters, the results for the models are summarized in Table 22.2.

22.4.4 Daily Confirmed Cases Count Forecasting

Using similar approaches from Sections 22.3.2 and 22.3.3, the problem of forecasting daily confirmed cases can be done similarly.

Visualizing the confirmed cases counts as a function of time (Figure 22.20).

Performing series differencing because of the presence of non-stationarity (Figure 22.21).

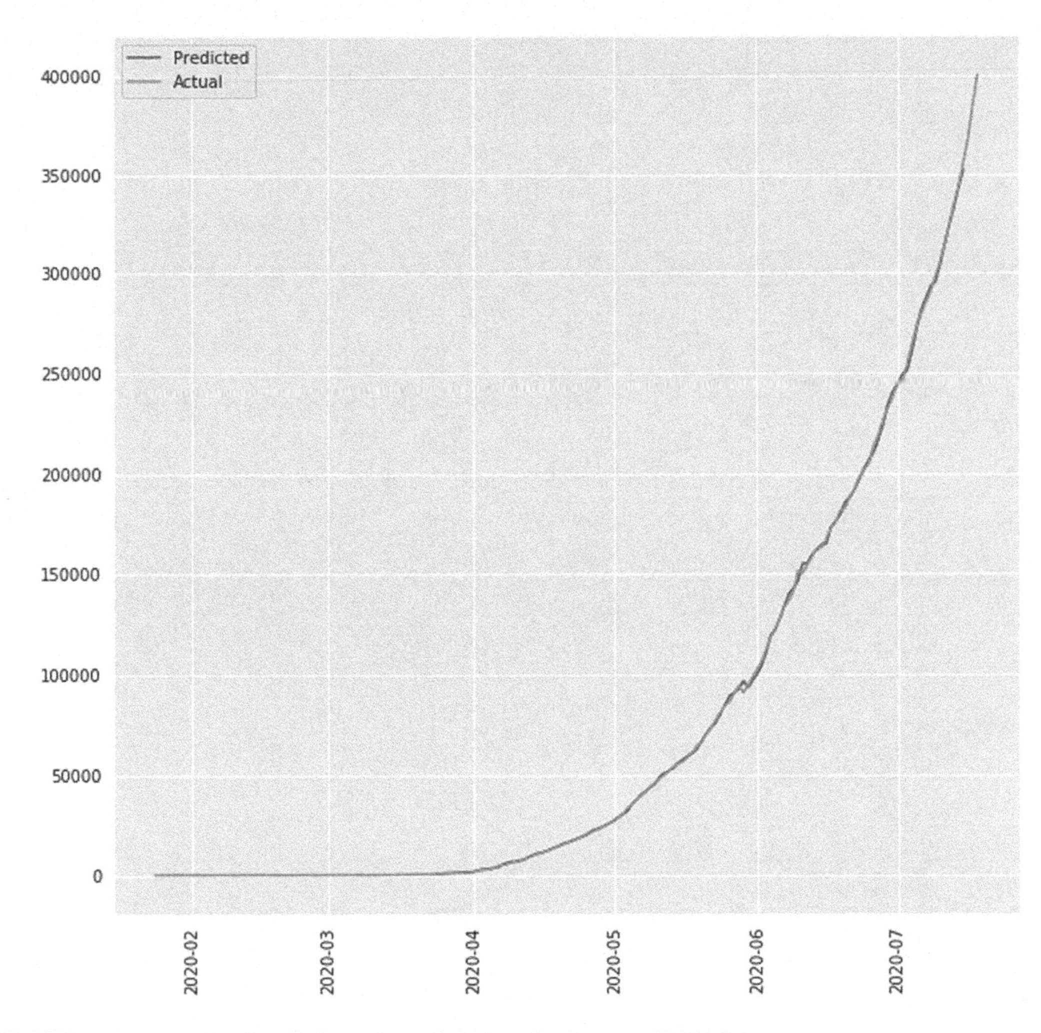

FIGURE 22.19 The plot of prediction and actual number of active cases till 15th July.

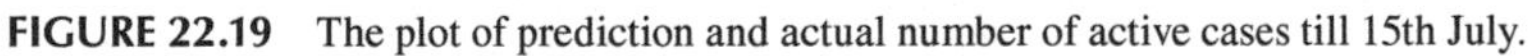

TABLE 22.2

Evaluation Summary of ARIMA Model for Active Cases Count Forecasting

Sr no.	AR (p)	MA (q)	*d*	AIC	BIC	RMS-Error
1	6	2	2	3221	3253	1645
2	7	1	2	3221	3254	1646
3	6	1	2	3225	3265	1652
4	4	0	2	3288	3308	1707
5	5	0	2	3289	3310	1709

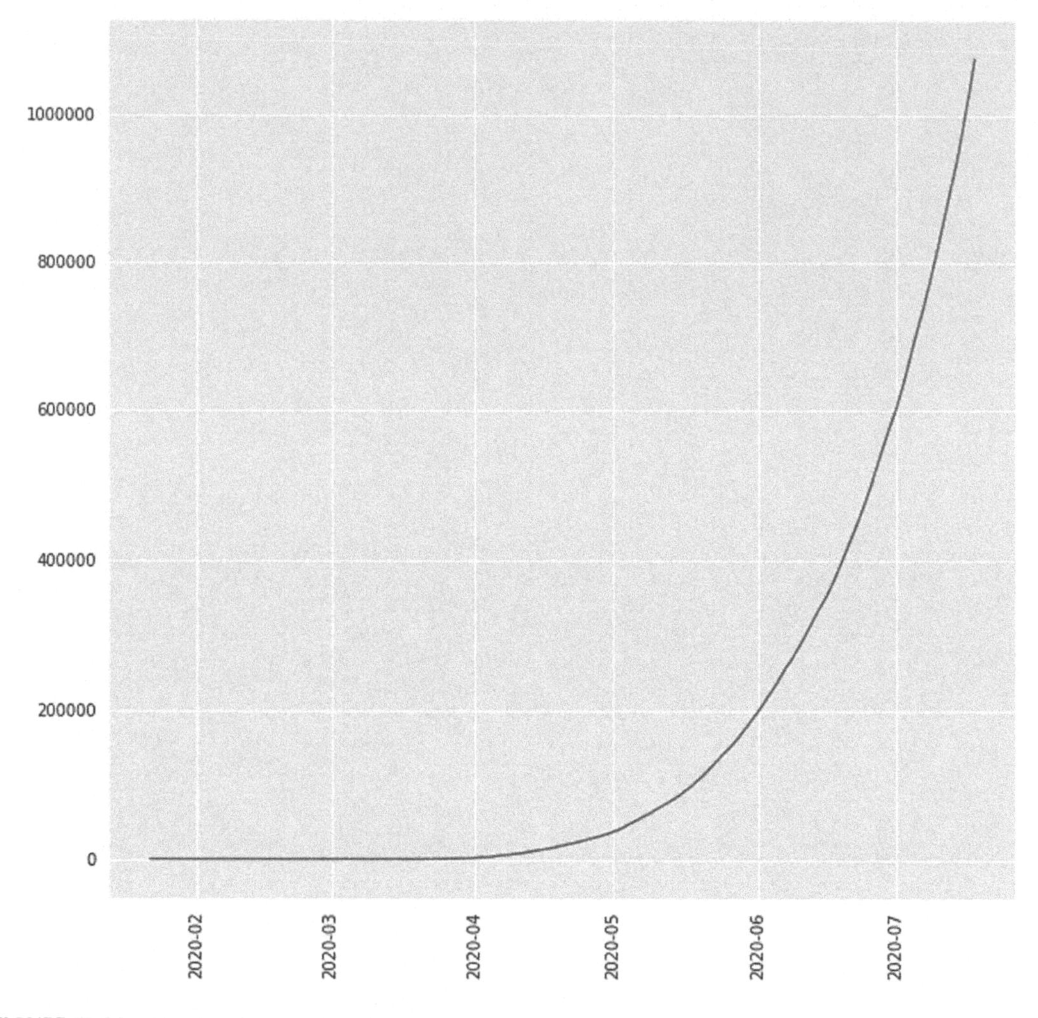

FIGURE 22.20 The plot of the number of confirmed cases vs. time.

It is clear from the plots that after differencing for two times, the series achieves stationarity to some extent. Hence setting the order of differencing, $d=2$. To decide the order of p and q, one must observe the corresponding autocorrelation and partial correlation plots for their significance (Figures 22.22 and 22.23).

By observing the autocorrelation plot and partial autocorrelation plot since the values are looking quite significant, setting the order of $p=5$ and $q=2$ leads to achieving the best possible AIC and BIC values.

Therefore, the resulting ARIMA model parameters are going to be $p=5$, $d=2$, and $q=2$, which gets us the following summary of the ARIMA model (Figure 22.24).

It can be noted here that the p-values of each coefficient are well below the threshold level of 0.05 which indicates that the resulting model is good. Below is the plot for the predictions by the model which indicates the predictions till 15th July (Figure 22.25).

Further, testing the model with various parameters, the results come out to be as follows (Table 22.3).

22.5 Recurrent Neural Networks and Long Short-Term Memory Networks

Feedforward neural networks are the most basic type of artificial neural networks. In feedforward neural networks, the data and information move only in one direction, i.e., the forward direction. RNN is a special case of a feedforward neural network. RNNs use their internal state (memory component) to process the sequence of inputs. This helps make the input related to each other.

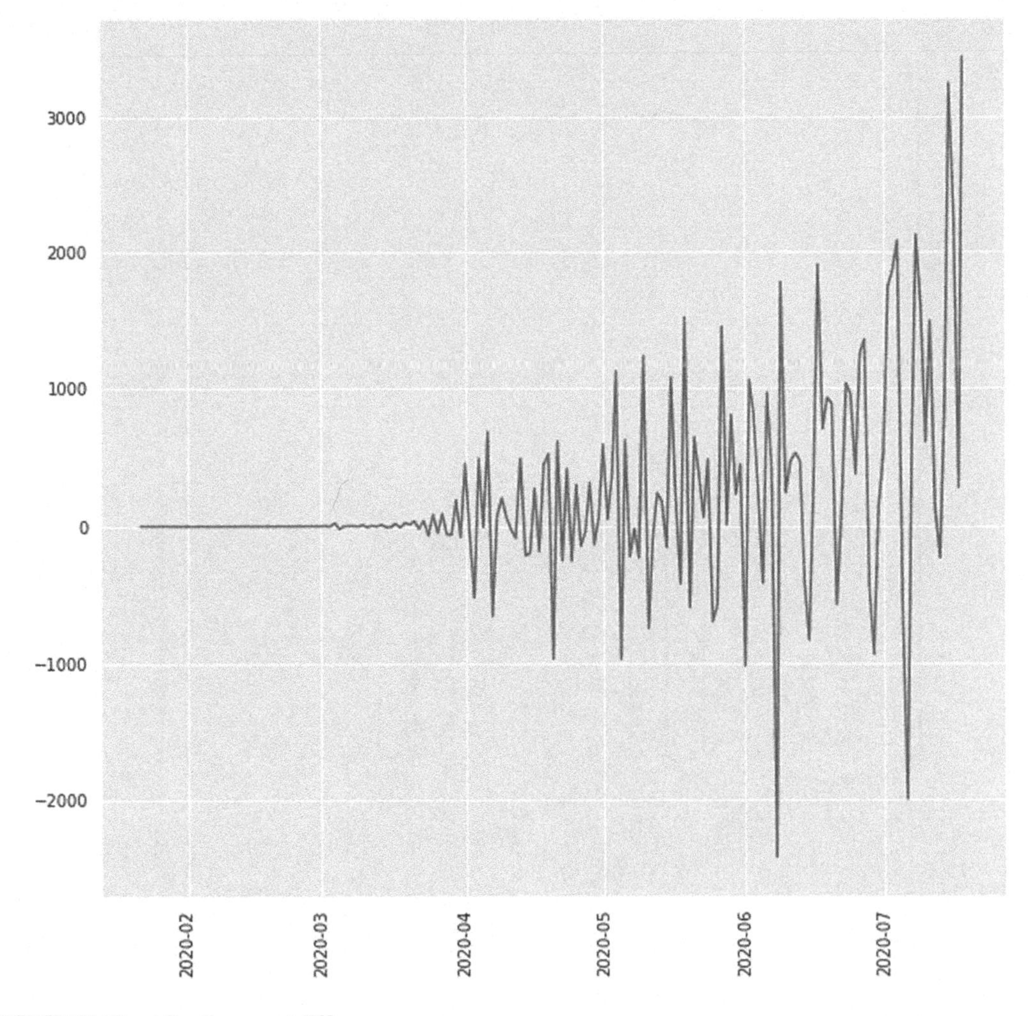

FIGURE 22.21 After the second difference.

Forecasting is the process of making future predictions based on past data, by analyzing the trends. For forecasting problems, time series data are used. Time series data are a sequence of data points indexed in time order. So, a technique that can process data sequentially gives the most efficient result, and thus, RNNs are one of the most preferred options to solve forecasting problems.

But, processing long sequences of data is not possible using RNN. For that, LSTM networks are used. LSTM networks are a variant of RNNs, which makes it easier to remember past data in memory. They are well-suited to process and predict time series given time lags. Their ability to infer over longer sequences of input enables the retention of information across the longer input sequences.

The most important parameters for the LSTM class are as follows:

- **Hidden Layer Size**: Defines the number of hidden units in an LSTM model.
- **Input Size**: Defines the number of inputs to the LSTM model.
- **Output Size**: Defines the number of outputs to be generated by the model.

Along with these parameters, forward pass behavior for the LSTM model needs to be defined. This is defined in the forward function. Implementation of the same in Python using PyTorch library is shown in Figure 22.26.

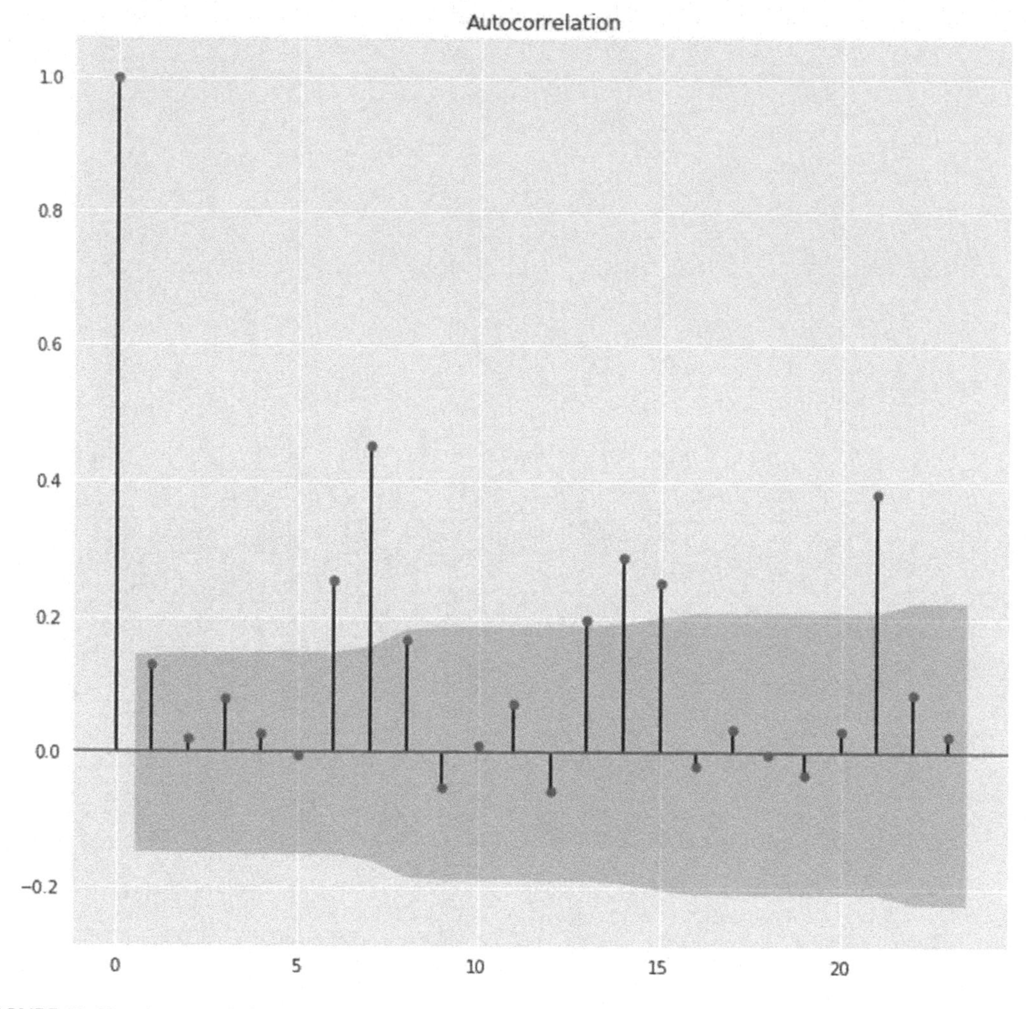

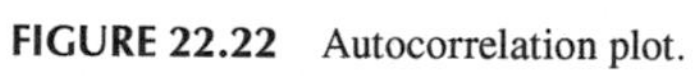
FIGURE 22.22 Autocorrelation plot.

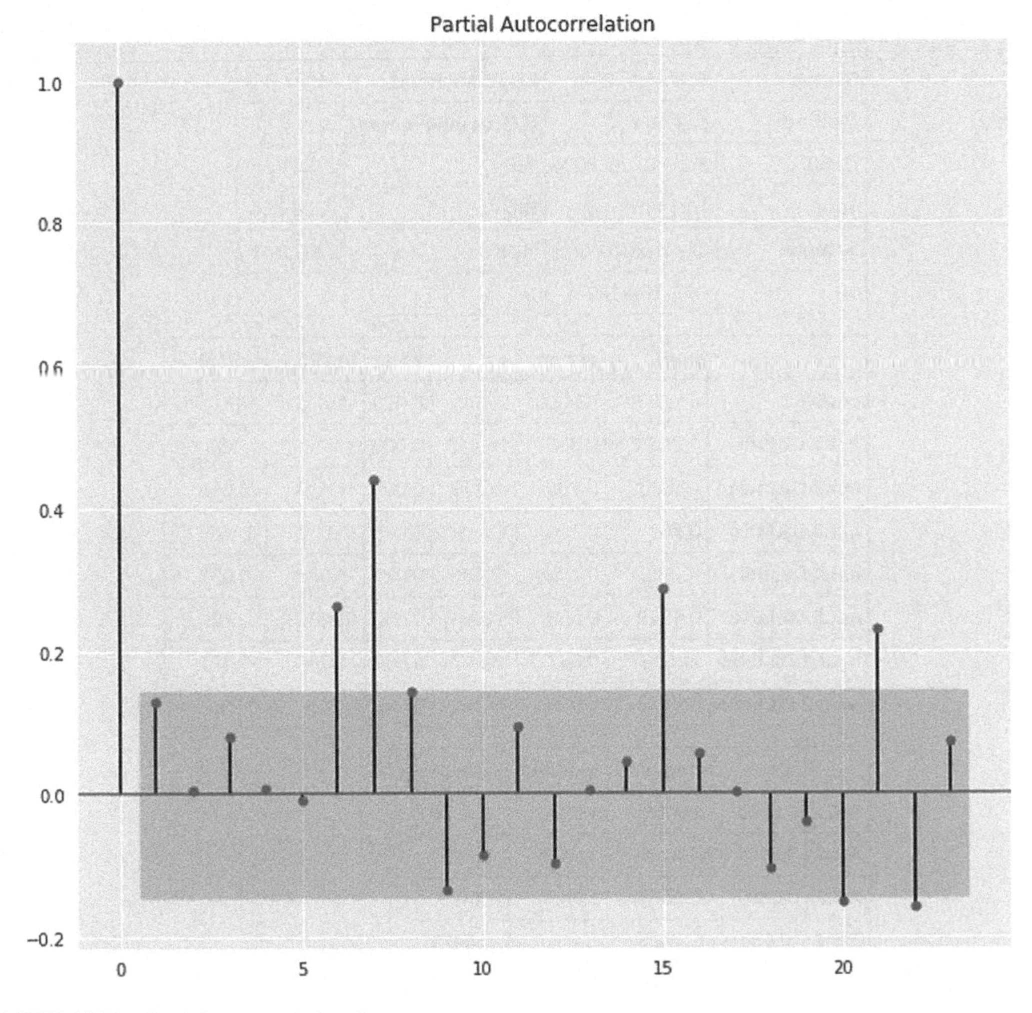

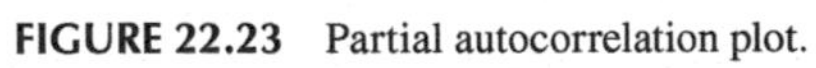

FIGURE 22.23 Partial autocorrelation plot.

Dep. Variable:	D2.India	No. Observations:	177
Model:	ARIMA(5, 2, 2)	Log Likelihood	-1381.084
Method:	css-mle	S.D. of innovations	nan
Date:	Sun, 19 Jul 2020	AIC	2780.168
Time:	16:14:07	BIC	2808.754
Sample:	01-24-2020	HQIC	2791.761
	- 07-18-2020		

	coef	std err	z	P>|z|	[0.025	0.975]
const	559.9361	694.631	0.806	0.420	-801.515	1921.387
ar.L1.D2.India	1.3938	0.085	16.412	0.000	1.227	1.560
ar.L2.D2.India	-0.8332	0.136	-6.119	0.000	-1.100	-0.566
ar.L3.D2.India	0.2677	0.150	1.781	0.075	-0.027	0.562
ar.L4.D2.India	-0.2981	0.135	-2.205	0.027	-0.563	-0.033
ar.L5.D2.India	0.4248	0.083	5.091	0.000	0.261	0.588
ma.L1.D2.India	-1.5127	0.047	-32.477	0.000	-1.604	-1.421
ma.L2.D2.India	0.9119	0.042	21.595	0.000	0.829	0.995

	Real	Imaginary	Modulus	Frequency
AR.1	1.0293	-0.0000j	1.0293	-0.0000
AR.2	0.6669	-0.8065j	1.0465	-0.1400
AR.3	0.6669	+0.8065j	1.0465	0.1400
AR.4	-0.8307	-1.1825j	1.4451	-0.3475
AR.5	-0.8307	+1.1825j	1.4451	0.3475
MA.1	0.8294	-0.6393j	1.0472	-0.1045
MA.2	0.8294	+0.6393j	1.0472	0.1045

FIGURE 22.24 Summary of ARIMA (5,2,2) model.

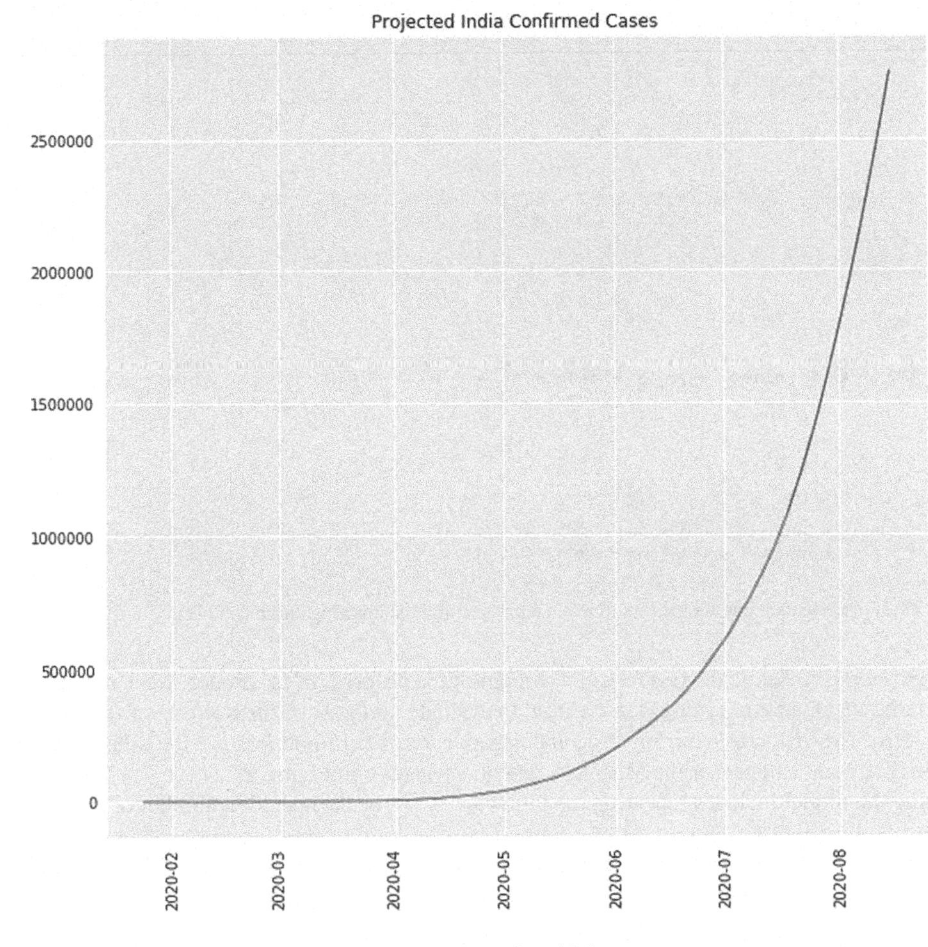

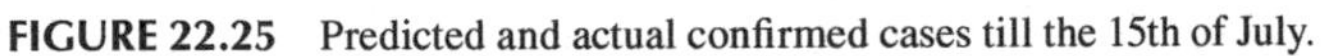

FIGURE 22.25 Predicted and actual confirmed cases till the 15th of July.

TABLE 22.3

Evaluation Summary of ARIMA Model for Confirmed Cases Count Forecasting

Sr no.	AR (p)	MA (q)	*d*	AIC	BIC	RMS-Error
1	7	0	2	2979	3008	793
2	3	5	2	2986	3019	822
3	2	4	2	2999	3025	870
4	2	1	2	3041	3057	995
5	1	0	2	3057	3066	1053

```
class LSTM(nn.Module):
    def __init__(self, input_size=1, hidden_layer_size=100, output_size=1):
        super().__init__()
        self.hidden_layer_size = hidden_layer_size

        self.lstm = nn.LSTM(input_size, hidden_layer_size)

        self.linear = nn.Linear(hidden_layer_size, output_size)

        self.hidden_cell = (torch.zeros(1,1,self.hidden_layer_size),
                            torch.zeros(1,1,self.hidden_layer_size))

    def forward(self, input_seq):
        lstm_out, self.hidden_cell = self.lstm(input_seq.view(len(input_seq) ,1, -1), self.hidden_cell)
        predictions = self.linear(lstm_out.view(len(input_seq), -1))
        return predictions[-1]
```

FIGURE 22.26 Code example of the LSTM defined.

```
model = LSTM()
loss_function = nn.MSELoss()
optimizer = torch.optim.Adam(model.parameters(), lr=0.001)
```

FIGURE 22.27 Code example for the loss function employed and the learning parameters set.

The next step is to define the loss function to monitor the progress of the model. Since numeric data are involved, the mean square error (MSE) is used to evaluate and govern the training of the model. Mean absolute error (MAE), cross-entropy loss, and negative loss likelihood loss are the other loss functions available. The code snippet for the MSE loss function is shown in Figure 22.27.

Training the model includes repeating the following steps for the specified number of epochs:

i) Clearing the PyTorch accumulated gradients using the "zero_grad()" method,
ii) Running the predefined forward pass using the "forward()" method,
iii) Calculating loss and gradients using "loss_function()" method,
iv) Updating the parameters using "optimizer.step()" method.

The code snippet to train the model for 250 epochs and the corresponding output is shown in Figure 22.28.

Depending upon the device configuration and GPU availability, the training time differs. Once the training is completed, predictions are made for the next 15 days. Results are shown in Figure 22.29.

As it is seen from the plot, the predictions are not so accurate. The errors are a clear amplification of the noisy dataset and the imbalance of the model parameters, thereby indicating a scope of improvement.

Due to the less amount of data available, it is difficult for the LSTM to recognize the underlying patterns and trends. Nevertheless, a quite close trend can be observed; hence, a general idea of the further steps required to mitigate the situation can be inferred.

For the prediction of active cases, a new model using the previous approach is defined. Results on testing the model to predict the number of active cases are shown in Figure 22.30.

It is evident that this model is very close to the actual data and therefore can be employed to determine further strategies to counter this exasperating situation.

There might be multifarious such models that can be determined by adjusting the hyperparameters to attain higher accuracy models. To achieve the best fit model, models with different parameters are implemented and trained. This process is usually called "hyperparameter tuning". The model summary for the same is as below.

```
epochs = 250

for i in range(epochs):
    for seq, labels in train_inout_seq:
        optimizer.zero_grad()
        model.hidden_cell = (torch.zeros(1, 1, model.hidden_layer_size),
                        torch.zeros(1, 1, model.hidden_layer_size))

        y_pred = model(seq)

        single_loss = loss_function(y_pred, labels)
        single_loss.backward()
        optimizer.step()

    if i%25 == 1:
        print(f'epoch: {i:3} loss: {single_loss.item():10.8f}')

print(f'epoch: {i:3} loss: {single_loss.item():10.10f}')
```

```
epoch:   1 loss: 0.01623761
epoch:  26 loss: 0.00000030
epoch:  51 loss: 0.00063443
epoch:  76 loss: 0.00040369
epoch: 101 loss: 0.00000111
epoch: 126 loss: 0.00000152
epoch: 151 loss: 0.00000409
epoch: 176 loss: 0.00000067
epoch: 201 loss: 0.00045728
epoch: 226 loss: 0.00000457
epoch: 249 loss: 0.0000007302
```

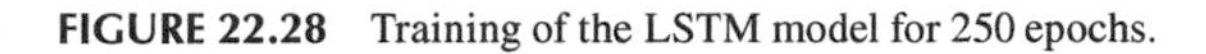

FIGURE 22.28 Training of the LSTM model for 250 epochs.

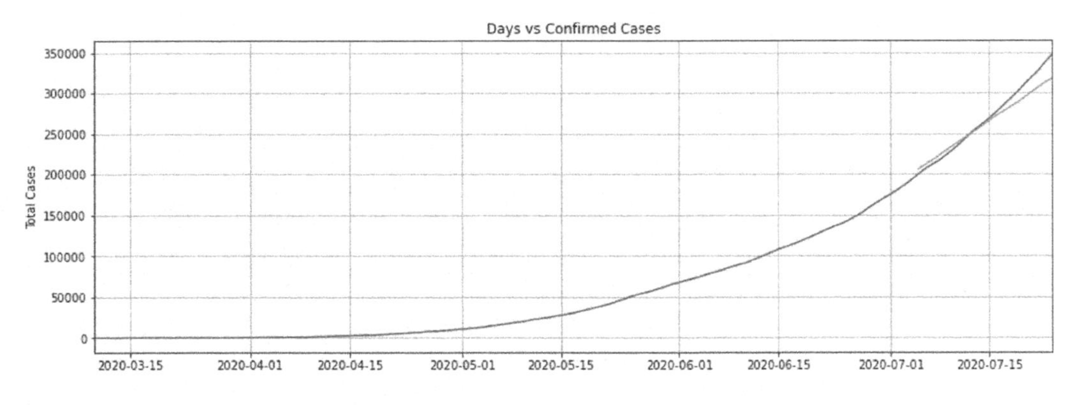

FIGURE 22.29 Prediction plot of confirmed cases using LSTMs.

1. Confirmed Cases:

Hidden Layers	Epochs	Learning Rate	R2 Loss	RMSE
100	250	0.0001	0.982785575	5872.474912
100	250	0.01	0.98054854	6242.391547
75	250	0.001	0.946377258	10364.52526
75	250	0005	0.892763629	14657.02938
50	250	0.0001	0.900639204	14108.55172

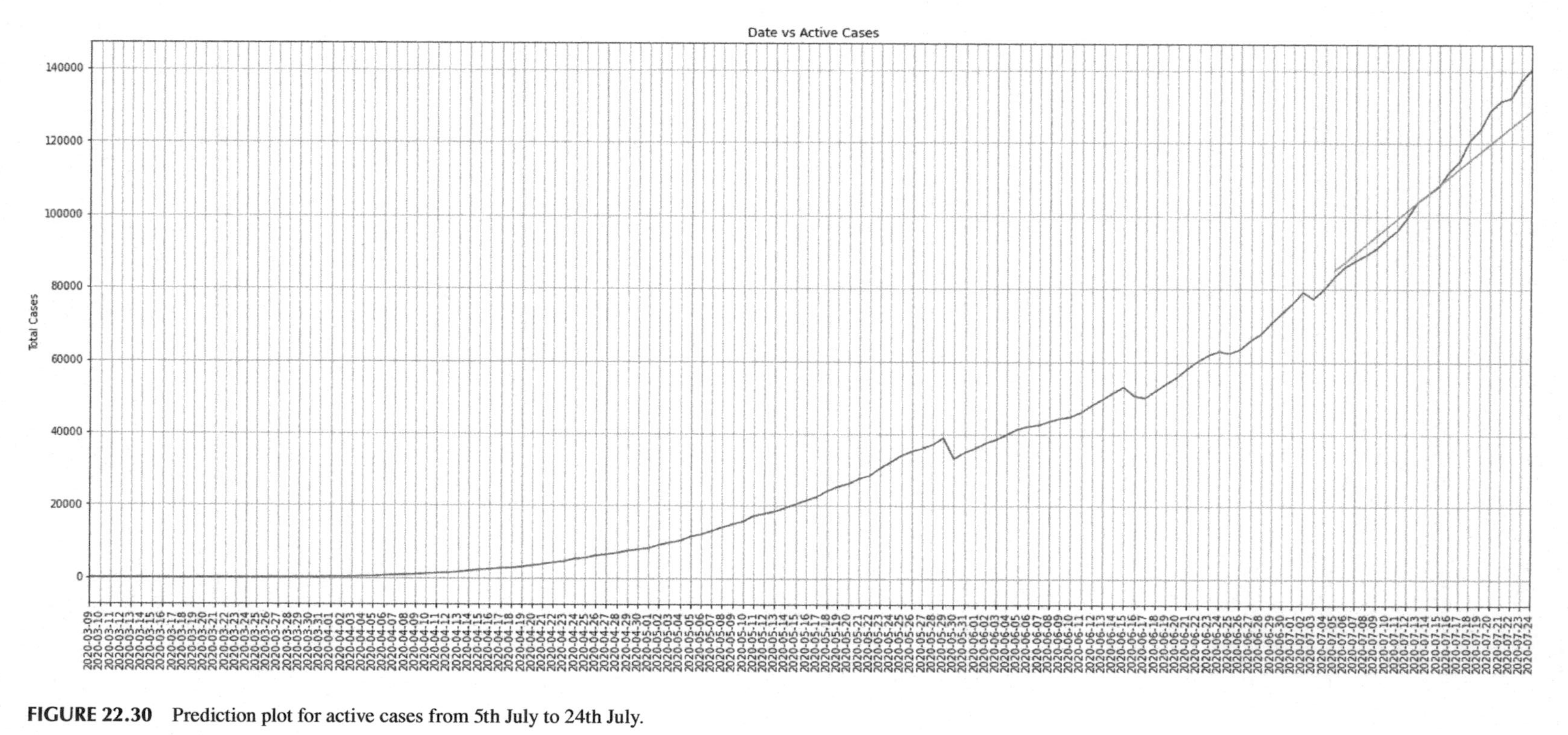

FIGURE 22.30 Prediction plot for active cases from 5th July to 24th July.

2. Active Cases:

Hidden Layers	Epochs	Learning Rate	R2 Loss	RMSE
200	250	0.01	0.970257566	3116.667524
50	250	0.1	0.950217115	4032.201708
200	250	0.0005	0.290753547	15219.51477
200	250	0.0001	-69.89192001	152160.0315
100	250	0.0001	-2.540764143	34005.62173

3. Total Deaths:

Hidden Layers	Epochs	Learning Rate	R2 Loss	RMSE
50	250	0.0005	0.983454	161.900843
200	250	0.001	0.981872	169.465195
100	250	0.005	0.972688	208.009759
75	250	0.0001	0.917135	362.320556
75	250	0.0005	0.900409	397.207256

22.6 Conclusion

Given the limited available data, the ARIMA and LSTM model performances are quite appreciable. If more data were available, then LSTM networks would have been able to perform much better. Nevertheless, better results have been achieved using the hyperparameter tuning. It can be inferred that India is still on a steep upward slope on the growth curve, indicating India is yet to curb the pandemic. Albeit India has been successful in flattening the curve and controlling the deaths, it is far from overcoming the pandemic. This highlights how the advancements in the field of data science and data analytics can prove useful for government and other decision-making bodies to understand the situation better and strategize their action plans effectively.

BIBLIOGRAPHY

1. Wikipedia, 2020 Tablighi Jamaat coronavirus hotspot in Delhi, last updated: 15th June 2020. Available at https://en.wikipedia.org/wiki/2020_Tablighi_Jamaat_coronavirus_hotspot_in_Delhi.
2. Anand Mishra, Migrant workers in India on move as COVID-19 stalls world, 28th March 2020. Available at https://www.deccanherald.com/national/migrant-workers-in-india-on-move-as-covid-19-stalls-world-818723.html.
3. Gupta, Rajan, and Saibal Kumar Pal. "Trend analysis and forecasting of COVID-19 outbreak in India." medRxiv (2020).
4. Pandey, Gaurav, Poonam Chaudhary, Rajan Gupta, and Saibal Pal. "SEIR and regression model-based COVID-19 outbreak predictions in India." arXiv preprint arXiv:2004.00958 (2020).
5. Gupta, Rajan, Saibal Kumar Pal, and Gaurav Pandey. "A comprehensive analysis of COVID-19 outbreak situation in India." medRxiv (2020).
6. Singh, Rajesh, and Ronojoy Adhikari. "Age-structured impact of social distancing on the COVID-19 epidemic in India." arXiv preprint arXiv:2003.12055 (2020).
7. For population details, https://www.populationpyramid.net/.
8. Prem, Kiesha, Alex R. Cook, and Mark Jit. "Projecting social contact matrices in 152 countries using contact surveys and demographic data." *PLoS Computational Biology* 13, no. 9 (2017): e1005697.
9. Ghosh, Palash, Rik Ghosh, and Bibhas Chakraborty. "COVID-19 in India: State-wise analysis and prediction." medRxiv (2020).
10. Ardabili, Sina F., Amir Mosavi, Pedram Ghamisi, Filip Ferdinand, Annamaria R. Varkonyi-Koczy, Uwe Reuter, Timon Rabczuk, and Peter M. Atkinson. "COVID-19 outbreak prediction with machine learning." Available at SSRN 3580188 (2020).

11. Shekhar, Himanshu. "Prediction of spreads of COVID-19 in India from current trend." medRxiv (2020).
12. Schüttler, Janik, Reinhard Schlickeiser, Frank Schlickeiser, and Martin Kröger. "COVID-19 predictions using a Gauss model, based on data from April 2." *Physics* 2, no. 2 (2020): 197–212.
13. McCall, Becky. "COVID-19 and artificial intelligence: Protecting health-care workers and curbing the spread." *The Lancet Digital Health* 2, no. 4 (2020): e166–e167.
14. Lin, Leesa, and Zhiyuan Hou. "Combat COVID-19 with artificial intelligence and big data." *Journal of Travel Medicine* 27, no. 5 (2020): taaa080.
15. Hu, Zixin, Qiyang Ge, Li Jin, and Momiao Xiong. "Artificial intelligence forecasting of COVID-19 in China." arXiv preprint arXiv:2002.07112 (2020).
16. Kavadi, Durga Prasad, Rizwan Patan, Manikandan Ramachandran, and Amir H. Gandomi. "Partial derivative nonlinear global pandemic machine learning prediction of COVID 19." *Chaos, Solitons and Fractals* 139 (2020): 110056.
17. Lalmuanawma, Samuel, Jamal Hussain, and Lalrinfela Chhakchhuak. "Applications of machine learning and artificial intelligence for COVID-19 (SARS-CoV-2) pandemic: A review." *Chaos, Solitons and Fractals* 139 (2020): 110059.
18. Sufian, Abu, Anirudha Ghosh, Ali Safaa Sadiq, and Florentin Smarandache. "A survey on deep transfer learning to edge computing for mitigating the COVID-19 pandemic." *Journal of Systems Architecture* 108 (2020): 101830.
19. Vaishya, Raju, Mohd Javaid, Ibrahim Haleem Khan, and Abid Haleem. "Artificial Intelligence (AI) applications for COVID-19 pandemic." *Diabetes and Metabolic Syndrome: Clinical Research and Reviews* 14, no. 4 (2020): 337–339.
20. Pirouz, Behrouz, Sina Shaffiee Haghshenas, Sami Shaffiee Haghshenas, and Patrizia Piro. "Investigating a serious challenge in the sustainable development process: Analysis of confirmed cases of COVID-19 (a new type of coronavirus) through a binary classification using artificial intelligence and regression analysis." *Sustainability* 12, no. 6 (2020): 2427.
21. https://www.maa.org/press/periodicals/loci/joma/the-sir-model-for-spread-of-disease-the-differential-equation-model.
22. https://sites.me.ucsb.edu/~moehlis/APC514/tutorials/tutorial_seasonal/node4.html.
23. Grant, Alastair. "Dynamics of COVID-19 epidemics: SEIR models underestimate peak infection rates and overestimate epidemic duration." medRxiv (2020).
24. He, Shaobo, Yuexi Peng, and Kehui Sun. "SEIR modeling of the COVID-19 and its dynamics." *Nonlinear Dynamics* 101, no. 3 (2020): 1667–1680.
25. PyTorch RNN & LSTM Documentation, https://pytorch.org/docs/stable/generated/torch.nn.RNN.html and https://pytorch.org/docs/stable/generated/torch.nn.LSTM.html.
26. Li, Qiang, Wei Feng, and Ying-Hui Quan. "Trend and forecasting of the COVID-19 outbreak in China." *Journal of Infection* 80, no. 4 (2020): 469–496.
27. Grasselli, Giacomo, Antonio Pesenti, and Maurizio Cecconi. "Critical care utilization for the COVID-19 outbreak in Lombardy, Italy: Early experience and forecast during an emergency response." *JAMA* 323, no. 16 (2020): 1545–1546.
28. ARIMA models for time series forecasting. Accessed from http://people.duke.edu/~rnau/411arim.htm.
29. Averaging and smoothing models. Accessed from http://people.duke.edu/~rnau/411avg.htm#HoltLES.
30. John Hopkins University dataset. Accessed from https://systems.jhu.edu/research/public-health/ncov/.
31. Harshil Jhaveri, Himanshu Ashar and Ramchandra Mangrulkar, "Leveraging Data Science in Cyber Physical Systems to Overcome Covid-19", *Journal of University of Shanghai for Science and Technology*, Vol. 22, no. 10, October 2020, pp. 1993–2015.
32. Heet Savla, Vruddhi Mehta and Ramchandra Mangrulkar, "Prediction and Diagnosis of COVID-19 using Machine Learning Algorithms", *International Journal of Recent Technology and Engineering (IJRTE)* Vol. 9, no. 3, September 2020, pp. 678–683.

23

Application of Deep Learning in Video Question Answering System

Mansi Pandya, Arnav Parekhji, Aniket Shahane, Palak V. Chavan, and Ramchandra S. Mangrulkar
Dwarkadas J. Sanghvi College of Engineering

CONTENTS

23.1 Introduction ... 354
23.2 Related Work ... 355
23.2.1 Visual Question Answering ... 355
23.2.1.1 Image Question Answering ... 355
23.2.1.2 VQA (Video Question Answering) ... 355
23.2.2 Emotion Detection ... 355
23.2.2.1 E.D. on Images ... 355
23.2.2.2 E.D. on Videos ... 356
23.2.3 Visual Captioning ... 356
23.2.3.1 Image Captioning ... 356
23.2.3.2 Video Captioning ... 356
23.2.4 Multitask Learning ... 356
23.3 AQAV ... 356
23.3.1 Overview ... 356
23.3.1.1 Emotion, Video and Question Embeddings ... 357
23.3.1.2 Vocabulary ... 358
23.3.1.3 Video Emotion Detector ... 358
23.3.2 Main VQA Route ... 358
23.3.2.1 Fusing Video, Emotion and Question Features ... 358
23.3.2.2 Feature Attention Techniques ... 359
23.3.3 Affective Route for Emotion Detection ... 361
23.3.3.1 Captioning-Module ... 362
23.3.3.2 Module – Text QA ... 363
23.3.4 Affective and Conventional Answers' Prediction ... 363
23.4 Experiments and Results ... 365
23.4.1 Video Dataset ... 365
23.4.2 Experiment Setup and Results ... 366
23.4.2.1 Optimization ... 366
23.4.2.2 Performance Analysis ... 366
23.4.3 Comparing with Baseline Models ... 366
23.4.4 Validating Attention Model ... 367
23.4.5 Accuracy Analysis of AQAV Conventional Answers ... 367
23.4.6 Accuracy Analysis of AQAV Affective Answers ... 368
23.4.7 Qualitative Analysis ... 369
23.5 Conclusion ... 370
References ... 370

23.1 Introduction

The vast domain of artificial intelligence has recently paid increased attention to Video Question Answering (VQA) which consists of Image[1–4] and VQA.[5–8] Tasks involving VQA involve concepts from computer vision and NLP (language processing). While there have been a myriad of papers, both experimental and review, on image question answering, the field of VQA is relatively new and untouched.[9] Unlike conventional VQA, the work presented here takes into consideration the emotion displayed by the video and this is done by combining features of VQA with emotion detection. The idea behind this is to make the generated answer more cogent by adding affective details to the video analysis. Adding this affective information to the VQA is beneficial in the fields of education, in health and in helping people with disabilities. The emotion adjectives in the final answer are added by feeding the attention model with emotion attributes, with the help of the AQAV model's emotion detector.

Actualizing AQAV however has some hurdles that need overcoming. These include some videos having too many emotions to process, answer analysis being a complicated task as answers could be completely or partially correct, or completely incorrect, and inaccurate detection of emotions potentially damaging performance. As displayed by the results in this paper, the AQAV model manages to take care of these challenges well enough to show how its benefits overshadow its challenges.

The standard VQA dataset contains descriptions which are in turn used to reform it by inserting labels indicating emotion, to the pairs of questions and answers. For different stages (training, testing and evaluating) of the model (AQAV), emotion labels are added to the existing dataset. The AFEW dataset consists of spontaneous videos and is used to pretrain the E.D. (Emotion Detector).

An important part of the VQA is attention, which is basically paying greater attention to certain factors when processing the data. A video comprises a large number of image frames. To solve all the challenges, only those frames need to be selected that contain information pertaining to the question while also taking care of redundancy between different frames. The question comprises tokens, each with varying weights. The tokens of the questions along with the emotion tokens got from the video are concatenated so as to facilitate the attention mechanism. Based on importance, different words in the video frames have different weights. For this reason, multiple attention techniques are required. These include frame- and token-based. These are then integrated together. For each token read, the token-based mechanism handles the relevant frames. On the other hand, each frame relevant tokens are referred to by the frame-based mechanism. The two techniques are then combined by following the methods mentioned in Xue et al.[9]

The major contributions made in the paper are as follows:

1. A new model, AQAV, has been proposed that adds the element of emotion to VQA tasks which leads to two answers being generated (conventional as well as affective).
2. Combined video and question attention along with attributes of emotion to build one common module of attention using LSTMs.
3. Parallelism is introduced resulting in the generation of conventional and affective answers along with the VQA and Affective paths, respectively.
4. Addition of emotion tags to the VQA dataset thus resulting in its customization for all the stages (train, test and evaluate).

VQA has multiple applications in the modern world including helping people with impaired vision, perceiving general crowd opinion and mobile-based communication. The range of these applications can be further expanded by the inclusion of emotion in the already existing VQA. They help in developing smart machines with certain human behavioural traits.

The following are the sections that will be encountered: Section 23.1 contains the related work, Section 23.2 includes the AQAV model features and architecture, Section 23.3 exhibits the experiments done on the dataset and the results obtained, and finally, Section 23.4 contains the conclusion.

23.2 Related Work

This section provides a brief outline of the different aspects of VQA systems while underscoring the pertinent research. It also provides a comprehensive comparison between systems with similar goals and gives a deeper understanding of the shortcomings of such methodologies.

23.2.1 Visual Question Answering

This task requires an in-depth study of the "Image Question Answering" and "Video Question Answering".

23.2.1.1 Image Question Answering

A CNN which is multi-labelled[10] with two disparate sections (Image-Analysis section and Caption Generation section) is another applied method used in VQA. Attention models like LSTM,[11] GRU,[12] BOW,[13,14] and Convolutional RNN[15] have been engaged. Contemporary techniques in the attention domain used for VQA comprise H.A.[16] (Hyperbolic Attention), J.A.L[17] (Joint Attention Learning), T.A.[18] (Triplet Attention) and M.M.A[19] (Multi-Modal Attention).

Supplementary exploration along with M.A. Networks[20] (Memory-Augmented Networks), cutback on the duplicacy in the dataset,[21] numerous datasets[22] and P.I.[23] (Paired Images) have been conceded in VQA. The N.C. Network[24] (Neural Central Network) for VQA spawns intramural questions' captions by means of a module, referred to as the "Caption Module", and thereafter, a Text Question Answering (TQA) Module produces an output answer in addition to a comprehensive description.

23.2.1.2 VQA (Video Question Answering)

"Image Question Answering (IQA)" handles stationary images, and "Video Question Answering (VQA)" deals with the manoeuvres from several video frames. VQA and IQA have different methods of feature extraction, with respect to the attributes (temporal) that are supposed to be scrutinized in VQA. Several VQA tasks[25,26] make use of the "Three-dimensional C3D" model.[27]

The "Reinforced Ranker-Reader"[28] is an accessible-domain question answering architecture that exercises reinforcement learning. Along with concurrently training the ranker with the answer-extraction reader, a ranked pipeline for accessible-domain question answering is utilized which allows the model to get better results.

A gated periodic unit with ranked spatiotemporal attention networks facilitates a VQA model to be trained on identifying the video and question features.[29] A well-informed scene graph in contingent VQA is used to compare video frames by the generation of the scene graph.[30]

Different Video QA datasets provide different functions and features:

a. TGIF-QA[31] consists of frame-based question answering. It also delineates three tasks: recognition of state conversion, estimating recurrence of a definitive action and disclosure of the frames which have actions which reappear.
b. Some of the well-known, most used VQA datasets include MSR-VIT,[32] VideoQA,[33] Youtube2Text[34] and Video-QA.[9]

23.2.2 Emotion Detection

Exploring ED on images and videos helps us establish a strong foundation for the task.

23.2.2.1 E.D. on Images

In order to train the models in the system, this research work uses a dataset with pictures that simulates natural mannerisms and facial expressions of humans. The "Static Facial Expressions in the Wild" dataset,[35,36] which encapsulates all these attributes, is used to train the models.

23.2.2.2 E.D. on Videos

Facial expressions that resemble legitimate observations from the real world are imperative for an accurate model. The "Acted Facial Expressions in the Wild" was the dataset that is chosen for training the model. The dataset comprises short videos from movies and furthermore utilizes a semi-automatic recommender framework which is based on captions. The fluctuations in the recognition of facial emotions are solved by using speech and lip movements.[37–38] The amount of information required to successfully train the CNN framework[39] can be significantly reduced by utilizing "Image Illumination", which simultaneously helps to simplify visual input.

In order to add on to the exquisite generalization potential of CNN,[40] "Advanced Transfer Learning" methods and pre-handling algorithms are used. To observe a favourable effect on the general execution of the "Video Question Answering" framework, we use a more gruelling dataset to train the emotion detector.

23.2.3 Visual Captioning

23.2.3.1 Image Captioning

Visual characteristics are encoded using CNNs by the "Neural-based" captioning[41–42] mechanism, and then, subsequently, using "Recurrent Neural Networks", the captions can be decoded. Another method might involve the usage of pre-characterized "Natural Language" formats in order to change recognized visual items, also termed as "Template-based" captioning.[43–45]

23.2.3.2 Video Captioning

By using an LSTM, a combination of RNN and CNN bodywork[46] can be created. The semantic qualities aggregated from the recordings and pictures would then be able to be taken care of into this joined bodywork. Right from one end to the other, "Video-to-Language" models can be prepared with a word locator. This procedure can be accomplished without the requirement for any outside information source.

Sentence generators, which use "Hierarchical RNNs", are used to generate an eloquent sentence for a short clip. Now, a paragraph generator, which incorporates the sentential embeddings produced by the sentence generator and the paragraph history, is used to create a more comprehensive form of the description by utilizing the guidelines specified in Yu et al.[12,47] Video captioning[48–49] can also employ various attention mechanisms to focus on the parts of the videos that encode the most crucial information.

23.2.4 Multitask Learning

When pertinent tasks run concurrently,[24] "Multi-task" learning techniques will facilitate the distribution of knowledge among disparate domains. Some applications of "Multi-task" learning are visual tracking,[50] speech synthesis,[51] machine language translation,[52] video concept detection[53] and image classification.[54–56] With the aid of video captioning, this paper primarily used it to combine emotion recognition and "Video Question Answering".

23.3 AQAV

This section takes a deep dive into the architecture of the affective question answering on video system. It offers an above board overview of different components of the system and elucidates the mathematical facets of different frameworks put in place. It successfully delineates the motivation for choosing specific procedures or techniques.

23.3.1 Overview

Producing an emotional answer from a video consisting of one or more people, based on a given question, is the primary purpose of this research paper. A preprocessed dataset (Video-QA) is used to perform the routine methods (training, evaluating and testing) using the architecture of the AQAV model given below.

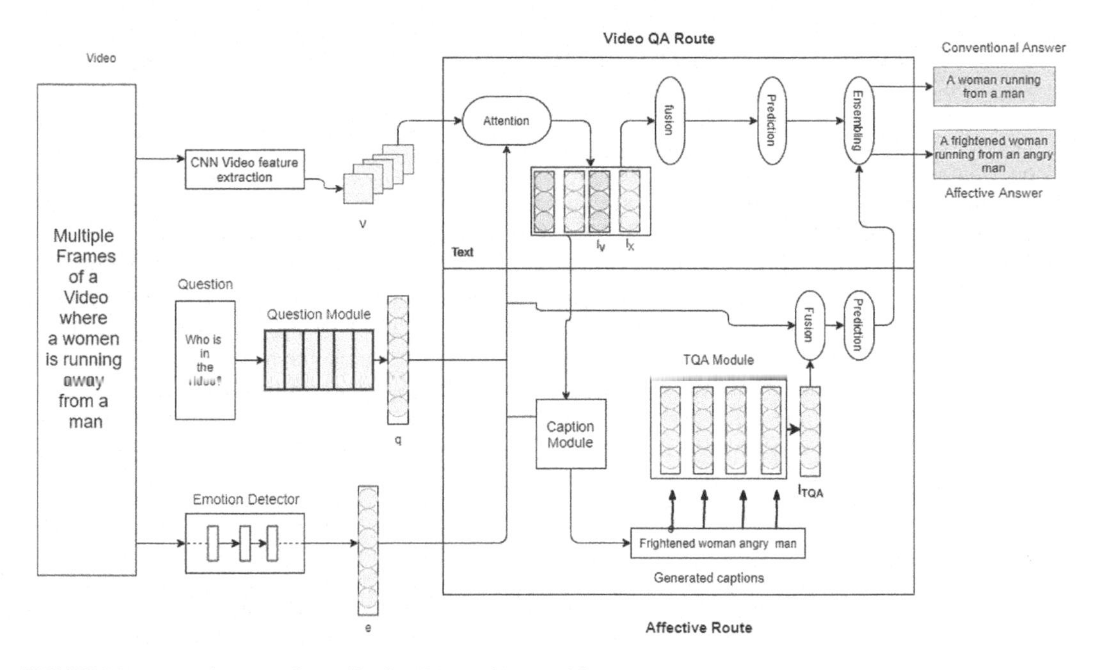

FIGURE 23.1 Architecture of the affective QA on video model.

As shown above (Figure 23.1), the model has two routes: the Video Question Answering route (VQA) and the Affective route.

In the VQA route, extracted features from the video (using concepts of CNN), the raw question and emotion features, are sent to the attention model. Its generated output is then sent to two areas of the model – First, to the fusion prediction area which makes predictions (i.e. generates the appropriate answer) based on the features of emotion, video and question and Second, to the Caption Module area. In the Affective route, the output of the attention model, question and emotion features are fed into the module referred to as the "caption module", which generates basic caption tokens. The output of the caption module is then fed into the TQA Module whose job is to generate meaningful sentences from the token generated in the caption module. The fusion prediction in Affective route is based primarily on the generated outputs of the caption module and the TQA module. In order to generate conventional as well as affective answers, the ensembling module requires prediction from both the routes, original question and emotion. This intersection between the VQA route and the Affective route gives us an improved answer. The conventional answers are not emotional answers but they include emotion tokens generated by the attention model in the VQA route, whereas the affective answers are emotional answers generated as a result of a multitasking process that enables the processes to occur simultaneously along both routes.

23.3.1.1 Emotion, Video and Question Embeddings

Before getting into the math and the different equations to be used, there are certain things that need to be taken care of to make the visual embedding task easier. They are as follows: ensuring that all the videos in the VQA dataset are exactly 40 frames and 244×244 in size. If frames are more than 40, then it is trimmed, and if frames <40, then the last frame is repeated until there are 40 frames in total. This specific necessity for 40 frames per video is done to help us get four different emotions per video clip.

Now, consider a video v with N frames as $v=(v_1, v_2, \ldots v_N)$, a question q with L tokens as $q=(q_1, q_2, \ldots, q_L)$, and an emotion string e with E tokens as $e=(e_1, e_2, \ldots e_E)$. The embedding of the v, q and e be learned as follows:

$$m_j = W_v\left[F\left(v_j\right)\right]+b_v,\ \ j=1,\ldots,N \tag{23.1}$$

$$m_{N+j} = U_q\left[\text{OH}(q_j)\right],\ j = 1,\ldots,L \tag{23.2}$$

$$m_{N+L+j} = U_e\left[\text{OH}(e_j)\right],\ j = 1,\ldots,E \tag{23.3}$$

In Equation 23.1, the function $F(.)$, which uses the layer referred to as the "FC7 Layer" of the "VGG-16 CNN" which makes use of the I.N. (ImageNet) dataset, is used to convert a frame v_j of a video v from a space of pixels to a "4096-Dimensional Feature Representation". After which it is converted to an "Embedding Space" (m_j) of d_v dimensions. In Equations 23.2 and 23.3, the function OH(.) is used to convert "Affective/Emotion Tokens" (e_j) and "Question Tokens" (q_j) into "One-Hot Encoded" features. The U_q matrix transfigures the vectors of q_j (which are one-hot encoded) into an "Embedding Space" (m_{N+j}) of d_q dimensions, and the U_e matrix applies transformations to the vectors of e_j (which are one-hot encoded) converting it into an "Embedding Space" (m_{N+L+j}) of d_e dimensions. All three of the dimensions $d_v d_q d_e$ have a linear measure of 512. The LSTM model is then fed with the embedding vectors $m_{1,\ldots,N,\,N+1,\ldots,N+L,\,N+L+1,\ldots,N+L+E}$. Here, it is clear that the first input token is that of the first frame of the video and the emotion token is the last one to be taken in.

23.3.1.2 Vocabulary

The vocabulary list has 3589 words from prominent answers that have appeared five times or more in the dataset and 14 affective descriptors (adjectives) in their suitable tenses which add up to form a total of 3603 words. There are some special tokens like Unknown which tell us that a certain word is absent from the list of words (vocabulary list) and EOS which demarcates the phrases of the output (answers). Chain of tokens $a = (a_1, a_2, \ldots, a_c)$ denote the generated answer which includes the output phrases and the video captions in written language from the list (vocabulary).

23.3.1.3 Video Emotion Detector

The Distributed Sensor Network (DSN) inspired the deeply supervised CNN which is used in the video emotion detector of the model (AQAV). The AQAV model has higher accuracy as compared to other standard models that have more parameters because they are tested to detect emotion from videos without audio. A multifaceted feature abstraction from distinct "Convolutional Layers" provides a forward rendition of E.R. (Emotion/Affective Recognition) which are collectively trained to upgrade the supervision of the whole framework. To produce an approximation of the sequence that would have been obtained by sampling the signal at a higher rate which is also known as upsampling, they use de-convolution techniques. To make the "Emotion Detector" more systematic, i.e. make it more cost-effective (efficient), they got rid of the V.E.R. (Voice Emotion Recognition) part. To scrutinize how the emotion detector reacts to different types of videos, the video emotion detector is run through the routine steps (training, evaluating and testing) on the architecture model mentioned above and VQA dataset.

23.3.2 Main VQA Route

The primary VQA assignment is done by the attention network shown in Figure 23.1. The attention mechanism consists of three parts: frame, token and frame based attention. Question features q, video feature maps v, and the emotion features e are fed as inputs to this attention mechanism and are processed by each of its parts. Finally, the integrated module is given in Figure 23.2.

The final answer is obtained when v is fused with q and e.

23.3.2.1 Fusing Video, Emotion and Question Features

The text string x of the AQAV model is formed by concatenating e (emotion features), q (question features) and v (visual features) together. The disparate features that are extrapolated into a collective semantic

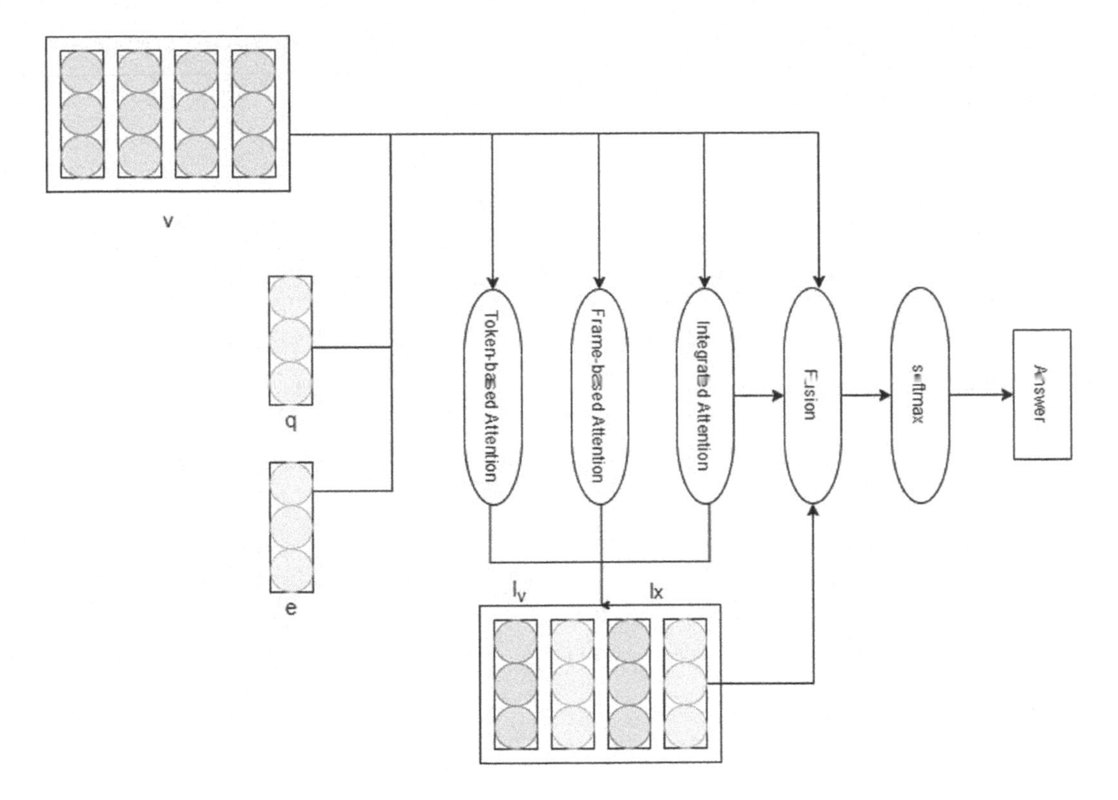

FIGURE 23.2 Video QA route.

space in the VQA route are combined together using the Hadamard product which is the component-wise multiplication of the same dimension and is represented by ⊙ symbol.

$$I = \sigma\left(W^T I_v \odot U^T I_x\right) \tag{23.4}$$

The above equation represents a fusion feature (I) that is obtained by applying the activation function $\sigma(\cdot)$ on two disparate features I_v and I_x.

23.3.2.2 Feature Attention Techniques

The integrated attention, frame-based and token-based activities are performed in the VQA route (Figure 23.3).

The attention network does the job of multitasking by retaining both the aggregate information and specific information with respect to applicable areas of the frames of videos. As the attention shifts between frames of a video and between tokens, the contents of the forget gate (f), usual input gate (i), input modulation gate (g), output gate (o) and the memory cell (c) are updated.

23.3.2.2.1 Token-Based Attention

The text string $x = q + e$, where q and e are question and emotion tokens, respectively, these question-emotion tokens are given, the video attention information, in sequential order of x. To confirm the relevance of question-emotion tokens to a particular frame feature, it is compared with all frames when picked. The term $r(j)$ represents the compiled attended video after it has been processed from 0 to j tokens and is produced by the weighted aggregate of the features of the frames of a video and prior gathered representation $r(j-1)$. The attention weights tell us the importance of distributing the tokens on all the frames of a video; by doing this, it can be understood which token the video is strongly inclined to which is done

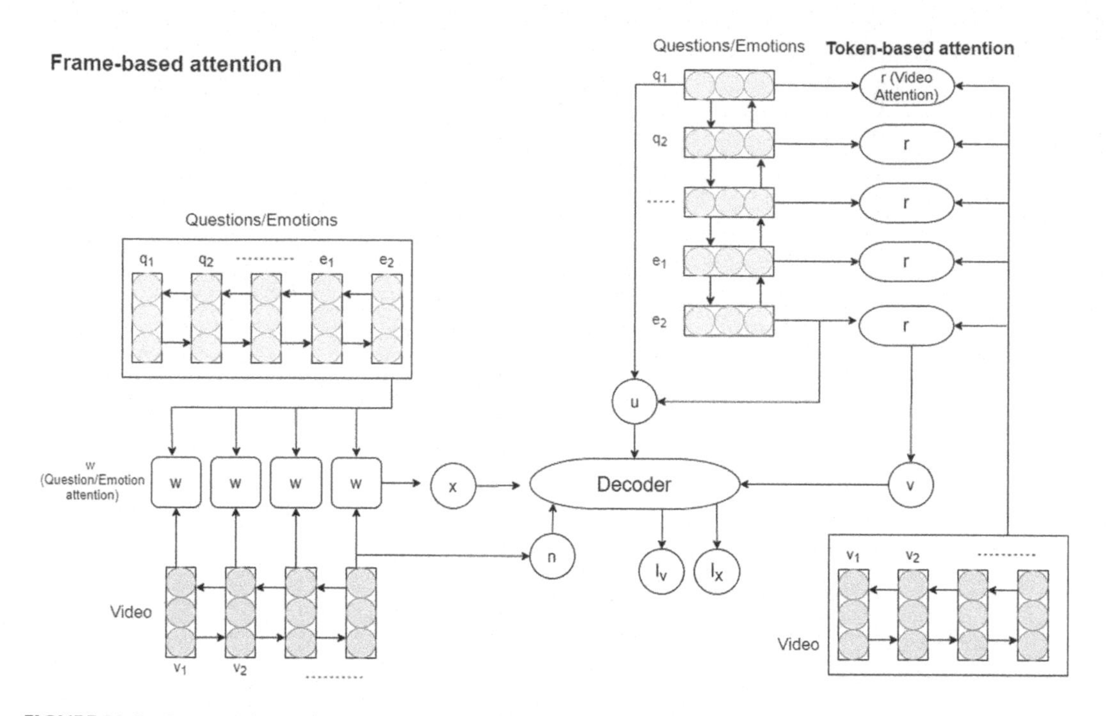

FIGURE 23.3 Integrated attention model.

by the attention mechanism. The following formulae tell us how the attention weight on the *k*th frame is produced by the *j*th token:

$$i(j,k)=\sigma\left(W_{iv}I_v(k)+U_{ix}I_x(j)+V_{ir}r(j-1)+b_i\right) \tag{23.5}$$

$$f(j,k)=\sigma\left(W_{fv}I_v(k)+U_{fx}I_x(j)+V_{fr}r(j-1)+b_f\right) \tag{23.6}$$

$$o(j,k)=\sigma\left(W_{ov}I_v(k)+U_{ox}I_x(j)+V_{or}r(j-1)+b_o\right) \tag{23.7}$$

$$g(j,k)=\tanh\left(W_{gv}I_v(k)+U_{gx}I_x(j)+V_{gr}r(j-1)+b_g\right) \tag{23.8}$$

$$c(j,k)=i(j,k)\otimes g(j,k)+f(j,k)\otimes c(j-1,k) \tag{23.9}$$

$$r(j,k)=o(j,k)\otimes\tanh\left(c(j,k)\right) \tag{23.10}$$

Conversion of the attended features gives us the real weights, shown by this equation:

$$s(j,\ k)_v\ \alpha\ \exp\ W_{cs}^T c(j,k) \tag{23.11}$$

$r(j)$ is the then updated using the compiled depiction $r(j-1)$ along with the above calculated weighted sum as shown below:

$$r(j)=I_v^T s_v(j)+\tanh\left(W_{rr}r(j-1),\ 1\le j\le\Delta\right) \tag{23.12}$$

The final encoded visual feature is the video attention accretion for each token $r(\Delta)$ and one that conserves the arrangement of the emotion and question string.

23.3.2.2.2 Frame-Based Attention

This system is consistent with the token-based module of attention and gathers question and emotion attention information for every frame of a video with respect to the temporal dimension. For any selected video frame, it is placed in juxtaposition with every available question and emotion token to perceive the significance of the features of the frame to the token. It is important that each video frame is distributed among all the text tokens so that with the help of weights it can figure out which attention mechanism accepts which particular frame of a video. Symbol $w(k)$ is the compiled attended representation of the question-emotion string after video frames from 0 to k have been observed. The following formulae represent how the weight on the jth token is produced by the kth frame when it is observed as $w(k)$:

$$i(k,j)=\sigma\left(W_{ix}I_x(j)+U_{iv}I_v(k)+V_{iw}w(k-1)+b_i\right) \tag{23.13}$$

$$f(k,j)=\sigma\left(W_{fx}I_x(j)+U_{fv}I_v(k)+V_{fw}w(k-1)+b_f\right) \tag{23.14}$$

$$o(k,j)=\sigma\left(W_{ox}I_x(j)+U_{ov}I_v(k)+V_{ow}w(k-1)+b_o\right) \tag{23.15}$$

$$g(k,j)=\tanh\left(W_{gx}I_x(j)+U_{gv}I_v(k)+V_{gw}w(k-1)+b_g\right) \tag{23.16}$$

$$c(k,j)=i(k,j)\otimes g(k,j)+f(k,j)\otimes c(k-1,j) \tag{23.17}$$

$$r(k,j)=o(k,j)\otimes \tanh\left(c(k,j)\right) \tag{23.18}$$

Conversion of the attended features gives us the real weights, shown by the following equation:

$$s(k,j)_x\,\alpha\ \exp\ U_{cs}^{T}c(k,j) \tag{23.19}$$

$w(k)$ is the then updated using the compiled depiction $w(k-1)$ and the above calculated weighted sum as shown below:

$$w(k)=I_x^{T}s_x(k)+\tanh\left(V_{ww}w(k-1)\right),\quad 1\le k\le N \tag{23.20}$$

The final encoded question and emotion feature is the accumulation of textual and visual data and one that conserves the temporal arrangement of the frames of a video.

23.3.2.2.3 Integrated Attention

In this attention mechanism, the question and emotion tokens and video frames are analysed in rotation and combined to produce sequential token-based and temporal frame-based attention output. This can be achieved by integrating the token- and frame-based attention methods.

23.3.3 Affective Route for Emotion Detection

Inputs given to VQA and the Affective routes are the same question, visual and emotion features, which are shown in Figure 23.1. Along with this, the inputs fed in the VQA course are likewise handed off to the Affective course. The emotion-aware captions generated by Affective route are redirected to answer generation process. The emotion-captioning component delivers an effective portrayal in layman terms, which transfers to the Text QA component for the planning of answer up-and-comer. After the conjecture, the appropriate response competitor will be sent to the VQA course for assembling. The Text QA emotion-captioning component and the emotion-captioning component are the major machineries of the Affective route.

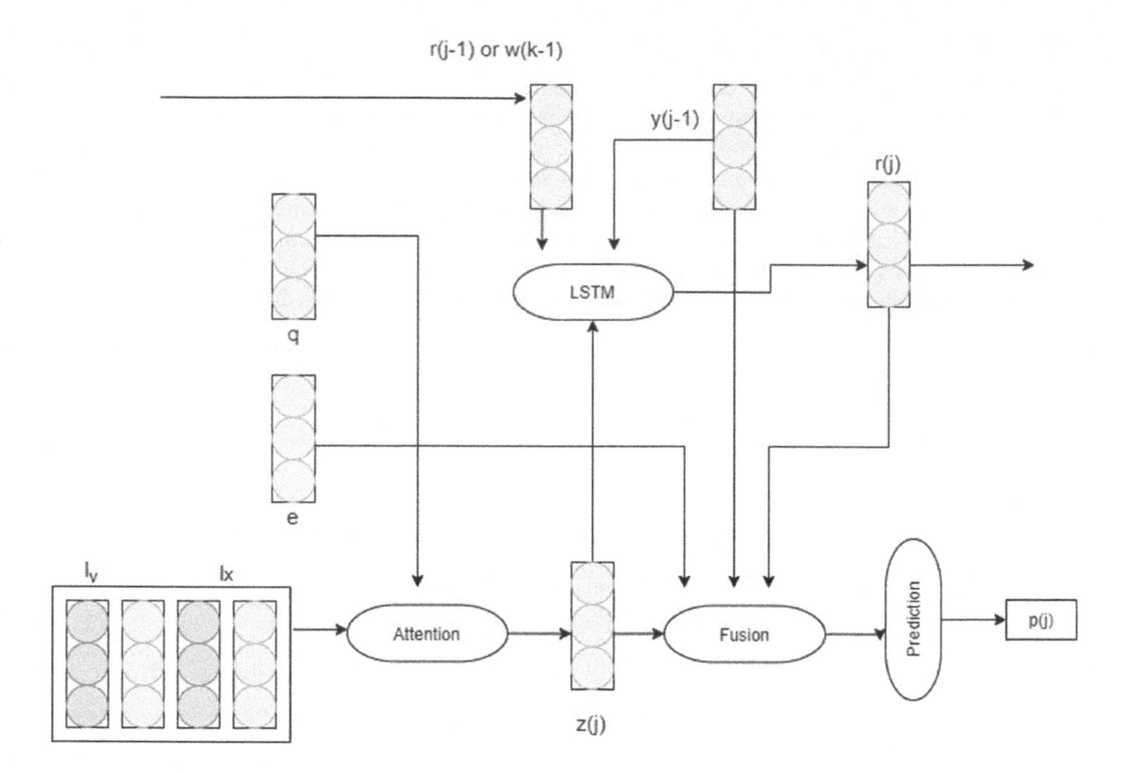

FIGURE 23.4 Emotion captioning module.

23.3.3.1 Captioning-Module

The tailored emotion-captioning component is demonstrated as shown in Figure 23.4.[51] They utilize Iv, which represents the attended visual feature, to instate the accumulated attended representation, r_0^d for token attention, W_0^d for frame attention, and z_0^d for the vector component of the LSTM decoder in the component for emotion-captioning as follows:

$$r_0^d = \sigma\left(W_r^T I_v\right) \tag{23.21}$$

$$z_0^d = \sigma\left(W_r^T I_v\right) \tag{23.22}$$

where W_s=predicted weight matrix.

The multi-task architecture, AQVA, is used so that a few procedures in the VQA course and in the Affective course occur simultaneously. Integrated attention includes the shift between audit of question/emotion features and of video features. The attended question/emotion feature I_x, the attended visual feature map I_v and the attended accumulated representation, for frame-based or token-based attention, are joined, while interpreting takes place in the Captioning module, so as to deliver the context vector of the LSTM as indicated by $z(j)$:

$$I_{rx} = \sigma\left(W_x I_x + W_r r_d^{j-1}\right) \tag{23.23}$$

$$\propto_i = \text{Softmax}\left(\sigma\left(W_r^T I_{rx} \odot U_z^T I_v^j\right)\right) \tag{23.24}$$

$$z_j = \sum_{j=1}^{k} \propto_j I_v^j \tag{23.25}$$

where W_x and W_r, that are the weight matrices, scheme the I_x, i.e. attended question/emotion feature and collected representation r_d to semantic space which is given as I_{rx}. Consolidating the attended question/emotion feature, the context feature, the implanting of last word, and the attended accumulated representation, the expressions of inscriptions are produced. The Softmax layer executes the forecasts. The *j*th word *p*(*j*) is deftly created as follows:

$$\beta_x = \text{Sigmoid}(W_x I_x) \tag{23.26}$$

$$\beta_z = \text{Sigmoid}\big(W_z z(j)\big) \tag{23.27}$$

$$\beta_y = \text{Sigmoid}\big(W_y y(j-1)\big) \tag{23.28}$$

$$I = \beta_x W_x I_x + W_r r_d^j + \beta_z W_z z(j) + \beta_y W_y y(j-1) \tag{23.29}$$

$$p(j) = \text{Softmax}(W_s I) \tag{23.30}$$

Here, β_s=different feature weights. The last word generated has an embedding vector given as $y(j{-}1)$. The data to use at each phase when creating words is decided by the model.

23.3.3.2 Module – Text QA

The captions generated, using the module for emotion-caption, are used for formulating the answer applicant in this module. Parameters are shared with both the routes and to all their parts. This is done by the LSTM module. The LSTM module is bifacial and forms the caption feature, i.e. I_z, combined with emotion feature, i.e. I_x. The answer this research work produces is given as follows:

$$I = \sigma\left(W_t I_x \odot U_t I_z\right) \tag{23.31}$$

$$a = \text{Softmax}(W_s I) \tag{23.32}$$

The appropriate response applicants anticipated from the module of Text QA along with the authentic question and its respective emotions are combined with the forecast from the VQA path to give the final answer.

23.3.4 Affective and Conventional Answers' Prediction

Section 23.2.1 mentioned the Hadamard Product Feature Fusion Scheme. This is used to combine I_v in the integrated attention mechanism and the emotion-based features that are indicated by I_x. I_v is used to denote the visual feature that was last attended. A layer with the Softmax function is used to predict the answer. This generates a conventional answer, as well as an answer with emotion, that is, an affective one. The VQA route is responsible for generating the conventional answer. On the other hand, the affective answer is the one that contains the special adjectives relating to emotion. That is generated along the AVQA route. The loss function is then described so as to train the model. It is essential to compute loss along both the routes, that is, the VQA path and the affective path. This is because the model architecture includes multitasking.

23.3.4.1 Loss along VQA Route

v and *q* represent the video and the question, respectively. For each of them, the sequence of tokens $\{h_1, h_2, \ldots, h_B\}$ make up the answer for the video in the dataset. The decoder produces the conformist solution $\{a_1, a_2, \ldots, a_D\}$. The loss function is represented as H_{con} $(h, Q\,(q, v))$. It is calculated here token-wise:

$$H_{\text{con}} = \sum_{j=1}^{B} 1\left[h_j \neq a_j\right] \tag{23.33}$$

where $1[h_j \neq a_j]$ is used to indicate a 0–1 function whose value is 0 in the case of $h_j = a_j$ and 1 in all other cases.

A conformist solution is measured right only when the authentic and generated answers are the same, i.e., $B=D$ and $h_j=a_j$ for j ranging from 1 to B. The kth sample of the dataset comprises an answer $\{h_k^1, h_k^2, \ldots, h_k^B\}$. The predicted answer for it is $\{a_k^1, a_k^2, \ldots, a_k^D\}$. Classification accuracy is denoted by Acc_{con}. The model performance is monitored by using this during the training process on the data's validation set:

$$\text{Acc}_{\text{con}} = \frac{1}{\phi} \sum_{k=1}^{\phi} \left(1 - \prod_{j=1}^{B_k} 1\left(h_j^k \neq a_j^k\right)\right) . \left(1 - 1\left(B_k \neq D_k\right)\right) \tag{23.34}$$

where ϕ indicates the quantity of samples (total number) and D_k and B_k indicate the length of the estimated answer and the authentic answer, respectively.

23.3.4.2 Loss along Affective Route

The two parts of affective answer, affective part (i.e. emotion string) and conventional part (no emotions), are divided by using the affective answer to fetch the emotion adjectives. The following emotion labels $\{\mu_1, \mu_2, \ldots, \mu_G\}$ are a string in alphabetical order that form the video dataset's ground truth emotions. The decoder generates $\{\lambda_1, \lambda_2, \ldots, \lambda_E\}$ that represent the emotions in alphabetical order. The loss function is calculated token-wise and represented by $H_{\text{aff}}(\mu, Q(e, v))$. The affective answer's affective part loss H_{emo} is calculated as:

$$H_{\text{emo}} = \sum_{j=1}^{G} 1\left[\mu_j \neq \lambda_j\right] \tag{23.35}$$

where $1[\mu_j \neq \lambda_j]$ is used to indicate a 0–1 function whose value is 0 in the case of $\mu_j = \lambda_j$ and 1 in all other cases. The answer's conformist part loss H_{ord} is formulated in a way similar to the loss of the conformist solution H_{con}. The total affective answer loss is computed by summing the 2 above-mentioned losses. It is represented as follows:

$$H_{\text{aff}} = H_{\text{ord}} + H_{\text{emo}} \tag{23.36}$$

If both the conventional part and the ground truth string of emotions are equal to the emotion string, then the affective answer is correct, i.e. $G=E$ and $\mu_j = \lambda_j$ for j ranging from 1 to G. kth sample in their dataset comprises emotion string $\{\mu_k^1, \mu_k^2, \ldots, \mu_k^G\}$. The expected answers' emotion string is $\{\lambda_k^1, \lambda_k^2, \ldots, \lambda_k^E\}$. Classification accuracy for the validation set is denoted by $\text{Acc}_{\text{aff},}$. It is used to find the maximum performance while training. Acc_{emo} denotes the affective part accuracy. It is given as follows:

$$\text{Acc}_{\text{emo}} = \frac{1}{\phi} \sum_{k=1}^{\phi} \left(1 - \prod_{j=1}^{G_k} 1\left(\mu_j^k \neq \lambda_j^k\right)\right) \left(1 - 1\left(G_k \neq E_k\right)\right) \tag{23.37}$$

Here, ϕ indicates the quantity of samples (total number), G_k represents the total number of emotion labels in the ground truth, and E_k represents the expected number of emotions in the answer. Acc_{con} and $\text{Acc}_{\text{ord,}}$ which are the accuracies of the conventional answer and the affective answer's conventional part correspondingly, are found out in a similar way. If the affective part and the conventional part are correct, the research work's estimated answer will be correct too.

$$\text{Acc}_{\text{aff}} = \frac{1}{\phi} \sum_{k=1}^{\phi} \left(\left(1 - \prod_{j=1}^{B_k} 1\left(h_j^k \neq a_j^k\right)\right) . \left(1 - 1\left(B_k \neq D_k\right)\right) \right) \cdot \left(\left(1 - \prod_{j=1}^{G_k} 1\left(\mu_j^k \neq \lambda_j^k\right)\right) \cdot \left(1 - 1\left(G_k \neq E_k\right)\right) \tag{23.38}$$

23.3.4.3 Loss Due to Multi-Task Learning

The loss incurred by the multitasking is calculated by summing the losses along the two routes, VQA and Affective (H_{aff}):

$$H = H_{con} + H_{aff} \tag{23.39}$$

23.4 Experiments and Results

This section aims to explicate the behaviour of the system under a myriad of conditions. It provides an elaborate juxtaposition of similar baseline systems and highlights the superior accuracy of this system (affective question answering on video). It also helps us in estimating the approximate behaviour of the model for different parameters (number of emotions and length of video). The section also highlights the qualitative analysis of the model for better assessment of its performance.

23.4.1 Video Dataset

For the desired holistic performance of this project, the initial dataset is modified so as to add labels to the pairs of questions and answers. The distribution of the questions used and their types are shown in the pie chart given in Figure 23.5.

The cleaned dataset does not contain any redundant images. This is done to make the dataset impartial. It is divided in the following percentages: training set – 78%, testing set – 12%, and validation set – 10%.

AFEW dataset is a video dataset that consists of random video clips based on real life that are relevant in detecting the emotion. It consists of seven emotions: angry, sad, scared, surprised, disgusted, happy and neutral. The videos are based on things like facial features and expressions and head movements. It consists of a large range of individuals with variations in gender, age, etc. The dataset is constructed by taking videos of length 0.3–5.4 seconds from various movies. This is an incredibly complex dataset thus giving great results while testing it on the VQA dataset. Thus, the AQAV model gets improved accuracy.

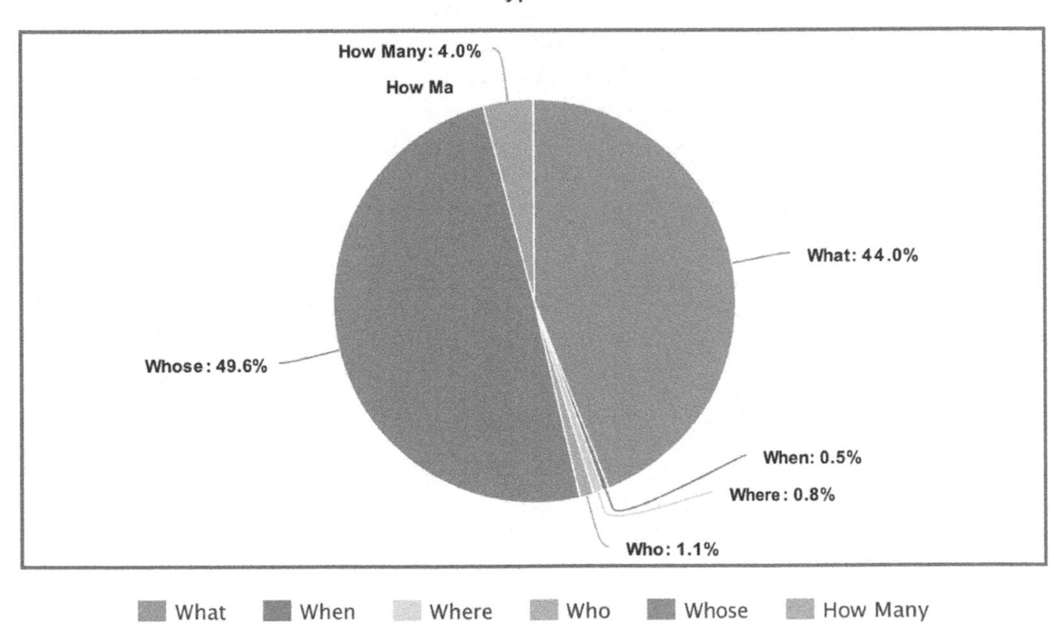

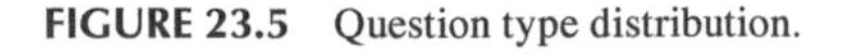
FIGURE 23.5 Question type distribution.

23.4.2 Experiment Setup and Results

AQAV architecture includes a lot of components. Thus, a lot of processing is required due to which the experiments are done over many days. Parameters relating to the video, the questions and their respective emotions all need to be shared between both the input layers and the output layers so as to ensure multitasking.

23.4.2.1 Optimization

All the models are trained using the Adam optimizer so as to make the loss as little as possible. The Adam optimizer is an extrapolation of the SGD algorithm and is well suited to problems with a sparse gradient such as this. The configuration parameters of the Adam optimizer are alpha (learning rate), betal (exponential decay 1), beta2 (exponential decay 2) and epsilon. The learning rate is set to 0.001. For the first and second moments, the exponential decay estimates are 0.9 and 0.999, respectively. Beta2 should be set close to 1.0 for problems with a sparse gradient and is thus set to 0.999. So as to make it as optimal as possible, epsilon is modified from le-8, which is the default, to 1.0. Epsilon is meant to avoid an error due to dividing by 0 during implementation.[1]

23.4.2.2 Performance Analysis

The performance analysis of a model is judged on the basis that the algorithm uses resources like memory, data and processor in addition to the training time. A Tesla K40 GPU is used to run the experiments. The training speed of the model is reduced due to the inclusion of the emotion labels inside the data as well as including all 3 features relating to the video, emotion and questions in one common LSTM module. In spite of this, its performance is still better as compared to the remaining baselines. In addition to this, other lesser effective reasons such as the LSTM being bidirectional could be responsible for reducing the performance of the model. The training duration for the different models is 7–9 hours for token-based, 8–10 hours for frame-based and 10–12 hours for the attenuation model.

23.4.3 Comparing with Baseline Models

The graph given below (Figure 23.6) compares the emotion detector built here with older models. The AFEW dataset is used to evaluate all these. OL_UC detects emotions from an audio-visual medium by training on small datasets. AIPL uses spatiotemporal techniques for feature learning.

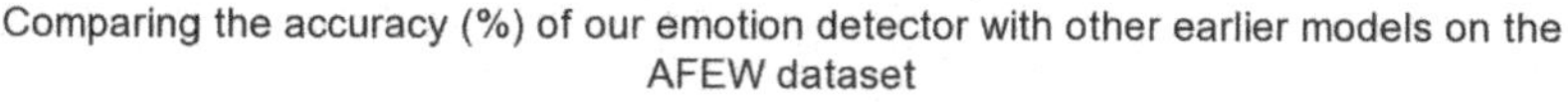

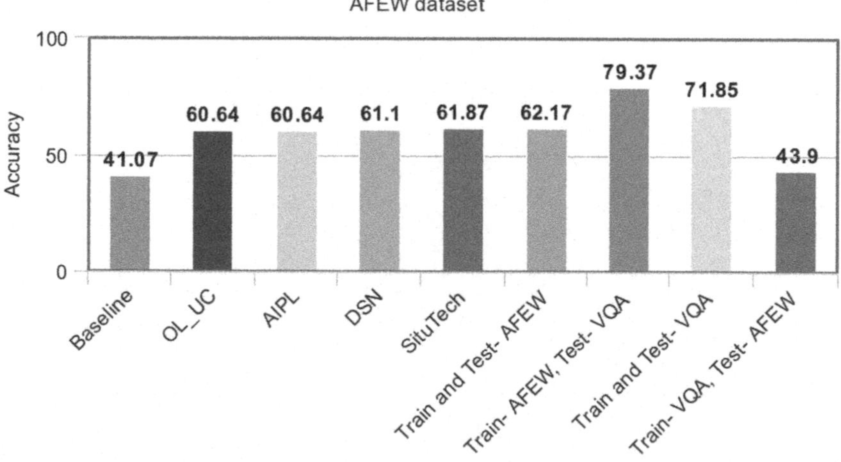

FIGURE 23.6 Accuracy percentage of the test results.

DSN uses a deep CNN for supervised learning on videos to extract emotions. SituTech uses two- and three-dimensional CNNs for extracting temporal information for detecting emotions in videos. The model created here is trained under four different situations with the best results being delivered when training was on the AFEW dataset and testing on the VQA dataset. This is due to the AFEW dataset having a lot of noise in the data whereas the VQA dataset has significantly lesser noise thus providing better results.

23.4.4 Validating Attention Model

The learning curves of each respective attention mechanism are plotted, which have been validated on the sampled batches. Following the training procedure, each model is independently validated beginning with the token-based, frame-based, and, in the end, the integrated model constituting both ones mentioned above. The token-based model performs the worst out of all followed by the frame-based model (3% increase). The integrated model is the best of the lot (8% increase). Thus, the integrated model is used while comparing with the baseline models.

23.4.5 Accuracy Analysis of AQAV Conventional Answers

The graph below (Figure 23.7) compares the accuracy of the model with other baseline models that have also been tested on the VQA dataset. It can be seen that the TB-Att, FB-Att and the Int-Att models all perform better than the other models. The Seq-V-Attr model collects attention from the video for each token of the questions while the Tmp-Q-Attr collects the attention for each frame of the video. The Unified model is an amalgamation of the previous two models mentioned. It is noticed that the Int-Att model averages an accuracy of 51.91% (14.26% mode than the Unified model). The individual TB-Att and FB-Att models also outperform their respective competitors by 4.18% and 2.34%. This improvement in the overall accuracy can be attributed to the addition of emotion. This helps in selecting the more relevant frames and regions in them. Further, the architecture is also simplified by using just one LSTM module. This module is responsible for attention and embedding of both the question and emotion-based features. While the answers do not include emotions as words, it is an important part of the attention process.

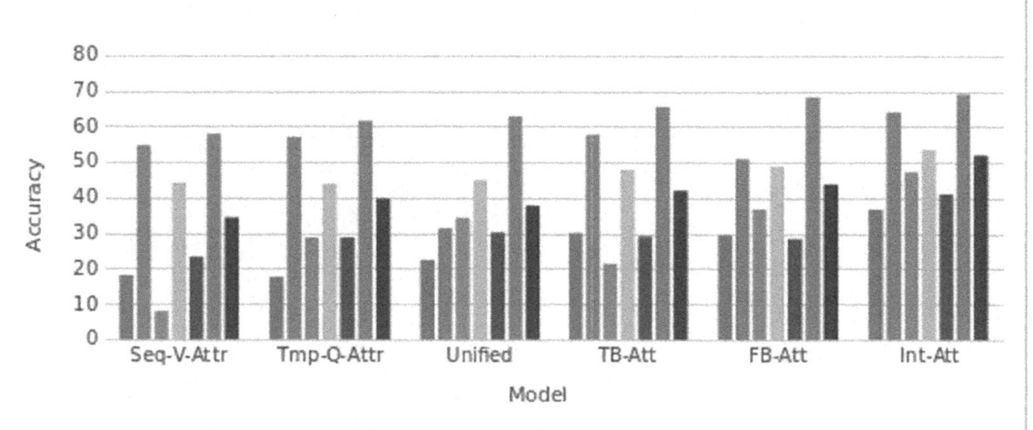

FIGURE 23.7 Graph comparing accuracy of the model.

23.4.6 Accuracy Analysis of AQAV Affective Answers

There are multiple modules in the AQAV affective answer generation system. So it is important to study the different modules and how they perform individually during the phases of testing. One important aspect of assessing the accuracy of the overall AQAV model is "error-propagation", which is essentially the idea that errors produced in one module will propagate and eventually contribute to the dip seen in the performance.

Since the answers generated have a component of effectiveness, the generated answers are scrutinized using that as a basis, during the testing. In order to keep things simple, they are divided into four broad categories. (i) First Category: Answer that is incorrect but the emotions detected are correct. (ii) Second Category: Answer that is correct but the emotions detected are incorrect. (iii) Third Category: Entirely incorrect answer. (iv) Fourth Category: Completely correct answer (Figure 23.8).

As it can be simply inferred from the definition of these categories, the answers present in the second category are incorrect entirely because the emotions that are detected by the "video emotion" detector are incorrect. Also, the answers yielded as defined by the third category are incorrect only partially due to the inaccuracy of the "video emotion detector". It might be tempting to infer, from the observations above, that these errors in the generation of answers should be solely assigned to the inconsistencies of the "video emotion detector". But, it is paramount to note that the errors produced in the overall model also originate from aspects such as the Convolutional Neural Network used for the "Softmax-Classifier" and the extraction of the features.

Some factors that can enhance the affective answer accuracy (in both the routes): "Triple Attention", "Flexible Captioning", "Multitasking Architecture" that allows multiple feature fusion and "Text QA".

The affective answers that are generated can have a wide range of the number of detected emotions, and generally, the number, that is observed, will range between 0 and 4 emotions per answer. The integrated attention mechanism's average accuracy is plotted against the number of detected emotions. Videos with somewhat ambiguous and neutral emotions yielded answers with zero emotions most of the time. The tests revealed that the inclusion of a single emotion in the answer produced the optimum results which proved that the emotions will improve the accuracy of the VQA model (Figure 23.9).

The detection of three emotions or more on a video that lasted for <2 seconds on average indicated that the video is either too complex to analyse or is extremely unclear. As a result, the plummet seen in accuracy should not be wrongly assigned to, only, the number of emotions. Lastly, the performance of the AVAQ model is compared to some other models. The models are as follows: (i) Hierarchical Relational

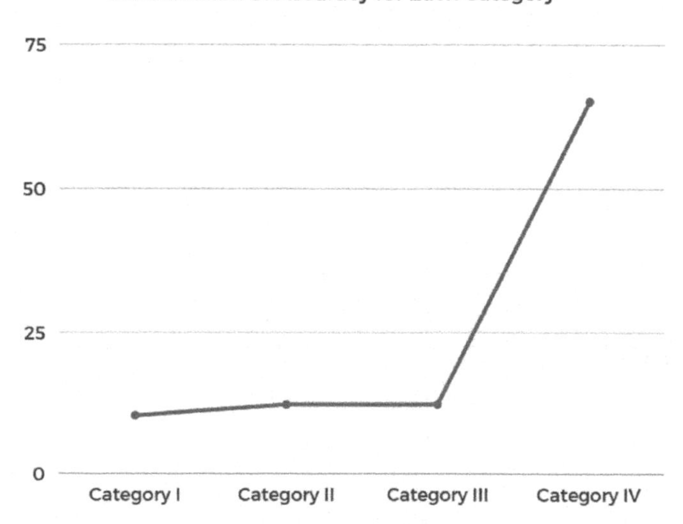

FIGURE 23.8 Classification of accuracy for each category.

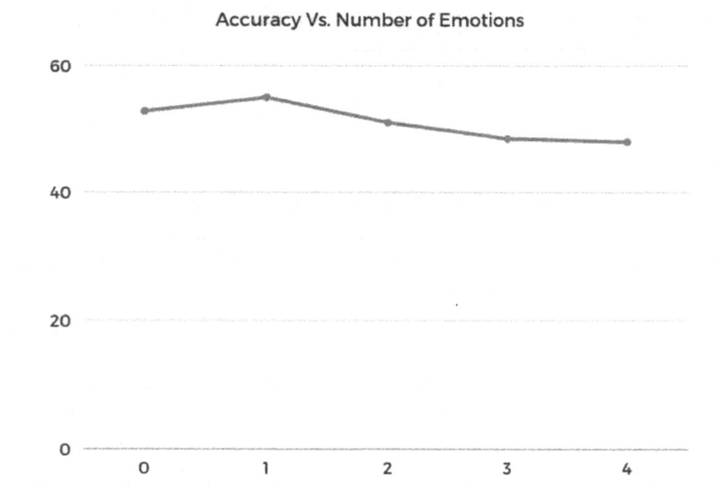

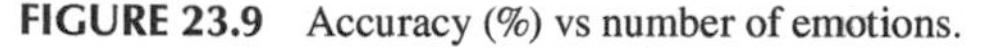

FIGURE 23.9 Accuracy (%) vs number of emotions.

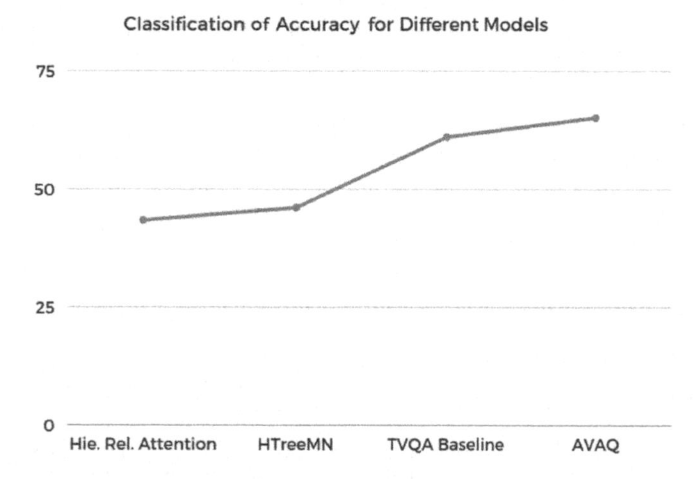

FIGURE 23.10 Classification of accuracy for different models.

Attention, (ii) HTreeMN, (iii) TVQA baseline, (iv) AQAV Affective Answers and (v) the results clearly display that the AVAQ model outdoes the best performing baseline by over 4% (Figure 23.10).

23.4.7 Qualitative Analysis

An example of an answer with emotions generated for a question is as follows: Question: What is happening? Answer: A happy woman and sad women looking at a disgusted woman. Question tokens included the following: What, happened. Emotion tokens included the following: Happy, sad, disgust.

Intensity curves are provided to show relevant frames based on emotion and question tokens. Most answers have only one emotion but in some cases, there can be multiple emotions involved as well, as in the case of the example mentioned at the start of the paragraph.

The algorithm generates curves for attention intensity which show the attention weights that have been computed. Relevant frames are identified, and localization is based on these weights. Both emotion and question tokens are important to the attention distribution. The curve goes up in cases where there is more relevance between the tokens and video frames. When the relevancy is less, the curve goes down. There is a high correlation between the questions and the emotions. The coefficient of correlation on the testing batch for the tokens comes out to be 87.38. This further corroborates the argument that AQAV performance is improved by the addition of emotions.

23.5 Conclusion

A model titled AQAV, a novel method of answering questions based on a video, was introduced. Two components that were previously thought to be distinct subdomains of machine learning (emotion detection and question answering based on visual data) were successfully merged in this technique, and a "multiple emotion detector" was created and trained on one of the most tricky datasets available, i.e. the "AFEW" dataset, which consists of short video clips which can be characterized as spontaneous or instinctive.

The dataset for question answering on videos is first tweaked by implanting emotion labels so that it can be trained, assessed and tested for the AQAV model as a whole. It is a characteristic multitasking architecture whose output is in the form of an answer that incorporates the emotions as well as the other elements identified from the video. The two routes that are employed are as follows: the "Video-QA" route which employs the conventional task of question answering based on the video and the "Affective" route that is important for the task of identifying the emotions. An ordinary answer and an answer with the identified emotions are generated by using the techniques of assembling and fusion which is made possible by concurrently running the two routes. The "Long Short-Term Memory" attention mechanism permits the model to concentrate on sections of the frames which correspond to specific aspects of the questions and emotions.

The proposed model has outperformed all the previous techniques used for "emotion recognition" and "video question answering". The common observation is that the enhancement of emotion recognition will result in a momentous advancement of the overall task of question answering based on a video. Consistent effort in the fields of CV, NLP and "Affective Computing" can result in a system that might surpass humans at the task of "affective question answering" on videos.

REFERENCES

1. P. Lu, L. Ji, W. Zhang, N. Duan, M. Zhou, J. Wang, R-VQA: Learning visual relation facts with semantic attention for visual question answering, 2, 2018. Accepted as an orl paper in SIDKDD. doi: 10.1145/3219819.3220036.
2. Y. Li, N. Duan, B. Zhou, X. Chu, W. Ouyang, X. Wang, M. Zhou, Visual question generation as dual task of visual question answering, IEEE, 2018. doi: 10.1109/CVPR.2018.00640.
3. Q. Li, J. Fu, D. Yu, T. Mei, J. Luo, Tell-and-answer: Towards explainable visual question answering using attributes and caption. *Proceedings of the 2018 Conference on Empirical Methods in Natural Language Processing*, Brussels, Belgium, 2018.
4. J. Liang, L. Jiang, L. Cao, L.-J. Li, A. Hauptmann, Focal visual-text attention for visual question answering, IEEE, 2018. doi: 10.1109/CVPR.2018.00642.
5. H. Xue, W. Chu, Z. Zhao, D. Cai, A better way to attend: Attention with trees for video question answering, IEEE, 27, 2018.doi: 10.1109/TIP.2018.2859820.
6. B. Xu, Z. ZHao, J. Xiao, F. Wu, H. Zhang, X. He, Y. Zhuang, Video question answering via gradually refined attention over appearance and motion, ACM, 2017. doi: 10.1145/3123266.3123427.
7. Y. Ye, Z. Zhao, Y. Li, L. Chen, J. Xiao, Y. Zhuang, Video question answering via attribute-augmented attention network learning, ACM, 2017. doi: 10.1145/3077136.3080655.
8. G. Jiao, Using flash platform to realize the video question answering system, 2014.
9. H. Xue, Z. Zhao, D. Cai, Unifying the video and question attentions for open-ended video question answering, IEEE, 26, 2017. doi: 10.1109/TIP.2017.2746267.
10. Q. Wu, C. Shen, P. Wang, A. Dick, A. van den Hangel, Image captioning and visual question answering based on attributes and external knowledge, IEEE, 40, 2017. doi: 10.1109/TPAMI.2017.2708709.
11. L. Kodra, E.K. Mece, Multimodal attention in recurrent neural networks for visual question answering. *Global Journal of Computer Science and Technology*, 17, Corpus ID: 53534699. 2018.

12. D. Yu, J. Fu, T. Mei, Y. Rui, Multi-level attention networks for visual question answering, CVPR, 2017.
13. M. Malinowski, M. Rohrbach, M. Fritz, Ask your neurons: A neural-based approach to answering questions about images, IEEE, pp. 1–9, 2015. doi: 10.1109/ICCV.2015.9.
14. K. Cho, B. Van Merriënboer, C. Gulcehre, D. Bahdanau, F. Bougares, H. Schwenk, Y. Bengio, Learning phrase representations using RNN encoder- decoder for statistical machine translation, EMNLP, 2014.
15. P. Harzig, C. Eggert, R. Lienhart, Visual question answering with a hybrid convolution recurrent model, ACM, pp. 318–325, 2018. doi: 10.1145/3206025.3206054.
16. C. Gulcehre, M. Denil, M. Malinowski, A. Razavi, R. Pascanu, K.M. Hermann, P. Battaglia, V. Bapst, D. Raposo, A. Santoro, et al., Hyperbolic attention networks, 2018.
17. D. Yu, J. Fu, T. Mei, Y. Rui, Multi-level attention networks for visual question answering, IEEE, p. 8, 2017. doi: 10.1109/CVPR.2017.446.
18. Z. Wang, X. Liu, L. Chen, L. Wang, Y. Qiao, X. Xie, C. Fowlkes, Structured triplet learning with pos-tag guided attention for visual question answering, 2018.
19. J. Xu, T. Yao, Y. Zhang, T. Mei, Learning multimodal attention LSTM networks for video captioning, *In Proceedings of the ACM on Multimedia Conference*, ACM, pp. 537–545, 2017. doi: 10.1007/s00170-017-0568-7.
20. C. Ma, C. Shen, A.R. Dick, A. van den Hengel, Visual question answering with memory-augmented networks, CoRR, 2017.
21. D. Teney, A.V.D. Hengel, Zero-shot visual question answering, 2016.
22. Z. Yang, X. He, J. Gao, L. Deng, A. Smola, Stacked attention networks for image question answering, IEEE, pp. 21–29, 2016. doi: 10.1109/CVPR.2016. 10.
23. Y. Goyal, T. Khot, D. Summers-Stay, D. Batra, D. Parikh, Making the V in VQA matter: Elevating the role of image understanding in visual question answering, 2016.
24. Y. Zhou, R. Ji, J. Su, Y. Wu, Y. Wu, More than an answer: Neural pivot network for visual question answering, ACM, pp. 681–689, 2017. doi: 10.1145/3123266.3123335.
25. K.-H. Zeng, T.-H. Chen, C.-Y. Chuang, Y.-H. Liao, J.C. Niebles, M. Sun, Leveraging video descriptions to learn video question answering, AAAI, pp. 4334–4340, 2017.
26. D. Xu, Z. Zhao, J. Xiao, F. Wu, H. Zhang, X. He, Y. Zhuang, Video question answering via gradually refined attention over appearance and motion, ACM, pp. 1645–1653, 2017. doi: 10.1145/3123266.3123427.
27. D. Tran, L. Bourdev, R. Fergus, L. Torresani, M. Paluri, Learning spatiotemporal features with 3d convolutional networks, IEEE, pp. 4489–4497, 2015. doi: 10.1109/ ICCV.2015.510.
28. S. Wang, M. Yu, X. Guo, Z. Wang, T. Klinger, W. Zhang, S. Chang, G. Tesauro, B. Zhou, J. Jiang, R3: Reinforced ranker-reader for open-domain question answering, AAAI, 2018.
29. Z. Zhao, Q. Yang, D. Cai, X. He, Y. Zhuang, Video question answering via hierarchical spatiotemporal attention networks, IJCAI-17, pp. 3518–3524, 2017. doi: 10.24963/ijcai.2017/492.
30. A. Ganesan, D. Pal, K. Muthuraman, S. Dash, Video based contextual question answering, 2018.
31. Y. Jang, Y. Song, Y. Yu, Y. Kim, G. Kim, TGIF-QA: Toward Spatio-temporal reasoning in visual question answering, CVPR, 2017.
32. J. Xu, T. Mei, T. Yao, Y. Rui, MSR-VTT: A large video description dataset for bridging video and language, IEEE, pp. 5288–5296, 2016.
33. M. Tapaswi, Y. Zhu, R. Stiefelhagen, A. Torralba, R. Urtasun, S. Fidler, MovieQA: Understanding stories in movies through question-answering, IEEE, 2016. doi: 10.1109/CVPR.2016.501.
34. S. Guadarrama, N. Krishnamoorthy, G. Malkarnenkar, S. Venugopalan, R. Mooney, T. Darrell, K. Saenko, YouTube2text: Recognizing and describing arbitrary activities using semantic hierarchies and zero-shot recognition, IEEE, pp. 2712–2719, 2013. doi: 10.1109/ICCV.2013.337.
35. S. Eleftheriadis, O. Rudovic, M. Pantic, Discriminative shared gaussian processes for multiview and view-invariant facial expression recognition, IEEE, 2015. doi:10.1109/TIP.2014.2375634.
36. A. Dhall, R. Goecke, S. Lucey, T. Gedeon, Static facial expression analysis in tough conditions: Data, evaluation protocol and benchmark, IEEE, pp. 2106–2112, 2011. doi: 10.1109/ICCVW.2011.6130508.
37. F. Ringeval, S. Amiriparian, F. Eyben, K. Scherer, B. Schuller, Emotion recognition in the wild: Incorporating voice and lip activity in multimodal decision-level fusion, ACM, pp. 473–480, 2014. doi: 10.1145/2663204.2666271.
38. J. Gideon, B. Zhang, Z. Aldeneh, Y. Kim, S. Khorram, D. Le, E.M. Provost, Wild wild emotion: a multimodal ensemble approach, ACM, pp. 501–505, 2016. doi: 10.1145/2993148.2997626.

39. G. Levi, T. Hassner, Emotion recognition in the wild via convolutional neural networks and mapped binary patterns, ACM, pp. 503–510, 2015. doi: 10.1145/2818346.2830587.
40. J. Schwan, E. Ghaleb, E. Hortal, S. Asteriadis, High-performance and lightweight real-time deep face emotion recognition, IEEE, pp. 76–79, 2017.
41. R. Kiros, R. Salakhutdinov, R. Zemel, Multimodal neural language models, *Proceedings of the International Conference on Machine Learning*, China, pp. 595–603, 2014.
42. J. Mao, W. Xu, Y. Yang, J. Wang, Z. Huang, A. Yuille, Deep captioning with multimodal recurrent neural networks (m-RNN), 2014.
43. A. Farhadi, M. Hejrati, M.A. Sadeghi, P. Young, C. Rashtchian, J. Hockenmaier, D. Forsyth, Every picture tells a story: generating sentences from images, Springer, pp. 15–29, 2010. doi: 10.1007/978-3-642-15561-1_2.
44. P. Kuznetsova, V. Ordonez, A. Berg, T. Berg, Y. Choi, Collective generation of natural image descriptions, *In Proceedings of the 50th Annual Meeting of the Association for Computational Linguistics*, Jeju Island, Korea, pp. 359–368, 2012.
45. G. Kulkarni, V. Premraj, S. Dhar, S. Li, Y. Choi, A. Berg, T. Berg, Baby talk: Understanding and generating simple image descriptions, *In Proceedings of the IEEE Conference on Computer Vision and Pattern Recognition, CVPR*, pp. 1601–1608, 2011. doi: 10.1109/CVPR.2011.5995466.
46. Y. Pan, T. Yao, H. Li, T. Mei, Video captioning with transferred semantic at- tributes, CVPR, p. 3, 2, 2017.
47. H. Yu, J. Wang, Z. Huang, Y. Yang, W. Xu, Video paragraph captioning using hierarchical recurrent neural networks, IEEE, pp. 4584–4593, 2016.
48. L. Gao, Z. Guo, H. Zhang, X. Xu, H.T. Shen, Video captioning with attention- based LSTM and semantic consistency, IEEE, 2017. doi: 10.1109/TMM.2017.2729019.
49. T.-H. Chen, K.-H. Zeng, W.-T. Hsu, M. Sun, Video captioning via sentence augmentation and spatio-temporal attention, *In Proceedings of the Workshops on Computer Vision – ACCV*, 2017. doi: 10.1007/978-3-319-54407-6_18.
50. T. Zhang, B. Ghanem, S. Liu, N. Ahuja, Robust visual tracking via structured multi-task sparse learning, Springer, 2013. doi: 10.1007/s11263-012-0582-z.
51. Z. Wu, C. Valentini-Botinhao, O. Watts, S. King, Deep neural networks employing multi-task learning and stacked bottleneck features for speech synthesis, IEEE, pp. 4460–4464, 2015. doi: 10.1109/ICASSP.2015.7178814.
52. D. Dong, H. Wu, W. He, D. Yu, H. Wang, Multi-task learning for multiple language translation, *Association for Computational Linguistics*, 1, pp. 1723–1732, 2015. doi: 10.3115/v1/P15-1166.
53. F. Markatopoulou, V. Mezaris, I. Patras, Deep multi-task learning with label correlation constraint for video concept detection, ACM, pp. 501–505, 2016. doi:10.1145/2964284.2967271.
52. X.-T. Yuan, X. Liu, S. Yan, Visual classification with multitask joint sparse representation, IEEE, 2012.
53. R. Ranjan, V.M. Patel, R. Chellappa, Hyperface: a deep multi-task learning framework for face detection landmark localization, pose estimation, and gender recognition, IEEE, 41, 2016. doi: 10.1109/TPAMI.2017.2781233.
54. Y. Yan, E. Ricci, R. Subramanian, O. Lanz, N. Sebe, No matter where you are: flexible graph-guided multi-task learning for multi-view head pose classification under target motion, IEEE, 2013. doi: 10.1109/ICCV.2013.150.

24

Implementation and Analysis of Machine Learning and Deep Learning Algorithms

Samip Kalyani, Neel Vasani, and Ramchandra S. Mangrulkar
Dwarkadas J Sanghvi College of Engineering

CONTENTS

24.1 Introduction 374
24.2 Variable Terminology 375
24.3 Experimentation with Regularization Algorithms 375
24.3.1 Ridge Regularization 375
24.3.2 Lasso Regularization 376
24.3.3 Elastic Net Regularization 377
24.4 Decision Trees 378
24.4.1 CART (Classification and Regression Trees) 379
24.4.1.1 Regression Trees 379
24.4.1.2 Classification Trees 380
24.4.2 Iterative Dichotomiser (ID3) 381
24.4.3 C4.5 and C5.0 382
24.4.4 Decision Stumps 383
24.4.5 M5 Model Tree 383
24.5 Ensemble Learning 384
24.5.1 Bagging 385
24.5.1.1 Random Forest 385
24.5.2 Boosting 386
24.5.2.1 AdaBoost 386
24.5.2.2 Gradient Boosting 387
24.6 Implementation and Analysis of Deep Learning Algorithms 388
24.6.1 Feedforward Neural Networks 388
24.6.2 Convolutional Neural Networks 390
24.6.2.1 Convolution 391
24.6.2.2 Pooling 392
24.6.2.3 Results and Applications 392
24.6.3 Recurrent Neural Networks 393
24.6.4 Long Short-Term Memory Units 395
24.6.4.1 Applications and Advantages 396
24.6.5 Generative Adversarial Networks 397
24.6.6 Deep Belief Networks 399
24.7 Conclusion 400
References 400

24.1 Introduction

The implementation and analysis of machine learning algorithms are crucial to understand the different approaches that solve real-life problems. The algorithms discussed in this chapter will help the reader analyze the concepts used in the machine learning and deep learning domains. This chapter includes many real-world machine and deep learning algorithms along with their fundamentals. Eventually, it provides algorithmic aspects to the machine learning fundamentals while covering their formulations, limitations, advantages, and use case scenarios. Section 24.2 describes linear methods for regression and classification that were developed in the pre-computer age of statistics which forms the basis of the machine learning domain. Since the decision boundaries of these algorithms are linear, they are used to tackle regression as well as classification problems. Section 24.2.4 covers support vector machines that provide an optimal separating hyperplane based on the samples near the margin. The algorithms discussed in this section form the basis in the field of pattern analysis and discovering trends prevalent in the market. When looking at data and visualizing the results, it is difficult to visualize beyond three dimensions of data and usually find two dimensions easier to interpret. Also, the dimensionality is an explicit factor for the computational cost of many algorithms. The real-world machine learning dataset contains a lot of features from which one needs to identify and extrapolate that are important and relevant to the machine learning model. The shrinkage methods described in Section 24.3 are used for extracting important features from a high-dimensional dataset using algorithms like lasso regularization, ridge regularization, and elastic net. Section 24.4 focuses on tree-based methods that include Classification and Regression Tree (CART), Iterative Dichotomiser (ID3), C4.5, C5.0, Decision stumps, and M5 model algorithm. These practical decision tree algorithms are applied to supervised as well as unsupervised machine learning problems. Section 24.5 emphasizes ensemble learning including bagging and boosting algorithms. Instead of using a single classifier or a regressor, we tend to use a set of them to make some predictions. Typically, the performance of classifiers is improved statistically rather than decreasing the variance of each of the classifiers. Ensemble learning is thus important to develop algorithms that have better empirical performance. The bagging algorithms include topics about bagged decision trees and random forest. The main idea behind bagging is to combine the results of multiple models to get a generalized outcome. The random forest algorithm helps to reduce the correlation between trees. Further, the concept of boosting is elucidated using AdaBoost and gradient boosting algorithms. If we take a collection of very poor learners, each performing only just better than chance, then by putting them together it is possible to make an ensemble learner that can perform arbitrarily well. This section would be important because many real-world machine learning solutions are based on ensemble learning. Section 24.6 starts with the idea of a simple perceptron which is then broadened to study various deep learning algorithms. This section includes topics feedforward neural networks, convolutional neural networks, recurrent neural networks, long short-term memory, generative adversarial networks, and deep belief networks. This section will thus cover the algorithms right from perceptrons up to deep learning. Convolutional neural networks (CNNs) are used extensively in computer vision. They are also used for modeling temporal data. The key idea in CNNs is to capture local patterns in the data through a convolution operation and then to downsample the resulting output with the pooling layer. CNN employs a series of convolution-pooling layers. Recurrent neural network (RNN) is an extension to traditional neural network frameworks where the output from the previous step is given as an input to the current step. In simple neural network frameworks, all the inputs and outputs are independent of each other. This is not helpful in situations wherein the classifier is expected to predict the next word based upon the previous word. Hence, RNNs are used to make the classifier remember the previous word, thus the context. Deep belief networks use a combination of restricted Boltzmann machines to outperform equivalent classifiers. A deep belief network has a stack of restricted Boltzmann machine layers and uses a different training approach than the classical machine learning perceptrons. This chapter helps to understand machine learning and deep learning algorithms, their implementation, and their fundamentals.

24.2 Variable Terminology

This chapter denotes the input variable by X. Continuous outputs are denoted by Y, and categorical outputs are denoted by G. Observed values of variables are written in lowercase letters. For example, i^{th} observed value of variable X is x_i (where x_i can be a scalar or vector). The output space is denoted by R. Matrices are represented as uppercase letters. The ith observation of X is x_i^T.

24.3 Experimentation with Regularization Algorithms

In the last section, recall the mean sum of squares error for estimating the $\beta_0, \beta_1, \ldots, \beta_p$ parameters. A real-world machine problem is not confined to a limited number of parameters. Practically, an actual problem contains a plethora of variables to be estimated. Now out of these variables, there is a need to determine which parameters are actually suitable to the problem and have an effect on the outcome of the prediction. The redundant parameters that have no effect on the outcome of the prediction model and hence are needed to be discarded. Since these parameters are abundant, parameter selecting methodologies are implemented for finding optimal parameters and discarding the rest. These methodologies are therefore called shrinkage methods or regularization methods. If the selection of parameters is done arbitrarily, then it may be required to drop out some important parameters and include trivial parameters that have minimal or no effect on the prediction model. Hence, there is a need to come up with a proper optimization formulation which allows us to shrink the unnecessary variables. The primary necessity of regularization for business executives is to analyze factors influencing their businesses and develop strategies accordingly.

Further, one has to keep in mind that while discarding parameters, although the interpretability of the model increases the prediction accuracy of the model should not decrease. But since the parameters are either retained or discarded, the prediction model has high variance. Hence, the prediction accuracy does not decrease.

24.3.1 Ridge Regularization

In ridge regularization, the sum of squares error formulation is optimized. In ridge regularization, the sum of squares error formulation is optimized and a penalty is imposed on the size of coefficients.

$$\text{Sum of squares error} = \sum_{i=1}^{n}\left(y_i - \sum_{j=0}^{p}\beta_j x_{ij}\right)^2 \tag{24.1}$$

The objective function is now formulated as,

$$\hat{\beta}_{\text{ridge}} = \underset{\beta}{\text{argmin}}\left\{\sum_{i=1}^{n}\left(y_i - \beta_0 - \sum_{j=1}^{p}\beta_j x_{ij}\right)^2\right\} \tag{24.2}$$

Subject to,

$$\sum_{j=1}^{p}\beta_j^2 \le t \tag{24.3}$$

This constraint is basically the l_2 norm constraint imposed on the vector space of β's. Making this equation unconstrained,

$$\hat{\beta}_{\text{ridge}} = \frac{\text{argmin}}{\beta}\left\{\sum_{i=1}^{n}\left(y_i - \beta_0 - \sum_{j=1}^{p}\beta_j x_{ij}\right)^2 + \lambda\sum_{j=1}^{p}\beta_j^2\right\} \tag{24.4}$$

where $\lambda \geq 0$ is called a tuning parameter. The term $\lambda\sum_{j=1}^{p}\beta_j^2$ is called a shrinkage penalty. This shrinkage penalty is used for penalizing the coefficients of the parameters. This penalizing of the coefficients of parameters is done mainly to reduce the range in which the parameters are confined. This overcomes the scenarios in which the models get overfitted. Consider an instance in which the model is overfitted, so the sum of squares error would be zero but due to the shrinkage penalty the overall optimization would not be equal to zero but some non-negative finite number this causes the model to further minimize and thus find a fit in which the model is not overfitted. Hence, the parameters are minimized.[1] Further, one must also take into account that β_0 is not actually included in the shrinkage penalty but is accounted separately. To understand this, consider a univariate regression problem. Now here, if β_0 is included in the shrinkage penalty it would make the intercept smaller but would increase the slope of the fit. This is exactly the converse of what ridge regularization is trying to achieve. And thus, a higher slope of the fit would increase overfitting. To overcome this, the parameters are centered around β_0. Note that when λ is 0 the penalty term is not applied and the optimization function reduces to the sum of squares of error optimization, and as the value of λ increases, the penalty value increases. The most optimal value for λ is determined computationally. Below is a code snippet for the above algorithm.

24.3.2 Lasso Regularization

Another similar implementation of shrinkage methods like ridge regularization is lasso regularization. Even in lasso, the sum of squares formulation is optimized along with a shrinkage penalty. The main difference between lasso and ridge regularization is the shrinkage penalty. The shrinkage penalty in lasso regularization is an l_1 norm constraint. The objective function is formulated as,

$$\hat{\beta}_{\text{lasso}} = \frac{\text{argmin}}{\beta}\left\{\sum_{i=1}^{n}\left(y_i - \beta_0 - \sum_{j=1}^{p}\beta_j x_{ij}\right)^2\right\} \tag{24.5}$$

Subject to,

$$\sum_{j=1}^{p}\left|\beta_j\right| \leq t \tag{24.6}$$

Making the above equation unconstrained,

$$\hat{\beta}_{\text{lasso}} = \frac{\text{argmin}}{\beta}\left\{\sum_{i=1}^{n}\left(y_i - \beta_0 - \sum_{j=1}^{p}\beta_j x_{ij}\right)^2 + \lambda\sum_{j=1}^{p}\left|\beta_j\right|\right\} \tag{24.7}$$

PSEUDOCODE:

1. Initialize the tuning parameter λ.
2. Set the shrinkage penalty to L2 constraint incorporating the tuning parameter.
3. Add the shrinkage penalty to the sum of squares optimization.
4. Train the dataset using the above formed optimization function.
5. Return the predicted values on the testing set.

PSEUDOCODE:

1. Initialize the tuning parameter λ.
2. Set the shrinkage penalty to L1 constraint incorporating the tuning parameter.
3. Add the shrinkage penalty to the sum of squares optimization.
4. Train the dataset using the above formed optimization function.
5. Return the predicted values.

From the above calculations, it is clear that ridge and lasso regularizations have the same formulations except for the shrinkage penalty. Lasso regularization penalizes the coefficient of the parameters toward zero. The lasso penalty forces some coefficients to exactly zero. Thus, lasso overcomes the drawback of ridge regression in which the final model can include every parameter.[2] This reduces the model's interpretability. But by shrinking coefficients to zero, lasso regularization selects important parameters for the estimate. Consider a machine learning problem of predicting house prices. Suppose the problem includes ten parameters for price estimation. Now what ridge regularization does, in this case, is that every coefficient of the parameter would be assigned a value how trivial it would be but lasso regularization on the other hand would consider only important parameters for prediction. Suppose there are four important parameters out of ten parameters, e.g., number of bedrooms, year of built, floor area, and type of locality, then only these would be considered by lasso regression and rest all parameters would be penalized to zero. Thus, lasso regularization acts as a method of feature selection. And these models produced are readily interpretable as compared to models generated by ridge regression. Refer to the below code sample for implementing lasso regression.

24.3.3 Elastic Net Regularization

As explained earlier the lasso and ridge regularization methods, these methods contain some powerful features that leverage themselves from their counterparts in specific problem scenarios. Lasso regularization offers highly interpretable models, whereas ridge regression produces models in which most of the features are equally important. But some real-world problems in specific domains like deep learning (which would be covered in the latter part of this chapter) often deal with millions of parameters wherein selecting a subset of parameters based on their importance for prediction is still a cumbersome task. Although lasso has been more successful than ridge regularization in specific problems, it is an inappropriate method of regularization in which the dimensionality of the problem is greater than the number of problems ($p > n$). In scenarios where there is a group of variables with high pairwise correlation, then lasso unreasonably selects anyone out of the two correlated parameters. Even in low dimensionality problems where $n > p$ having high correlation, the ridge outperforms the lasso. Due to the above reasons, there is a necessity of a hybrid model that performs like lasso regularization that can overcome the problems mentioned above.

Elastic net just like lasso and ridge regularization starts with optimizing the sum of squares error formulation and then combines the lasso penalty along with the ridge penalty. Altogether, elastic net combines the strengths of ridge and lasso regularization.[3]

$$\hat{\beta}_{\text{elastic}} = \frac{\arg\min}{\beta}\left\{\sum_{i=1}^{n}\left(y_i - \beta_0 - \sum_{j=1}^{p}\beta_j x_{ij}\right)^2\right\} \tag{24.8}$$

Subject to,

$$(1-\alpha)\sum_{j=1}^{p}|\beta_j| + \alpha\sum_{j=1}^{p}|\beta_j|^2 \le p \tag{24.9}$$

As before making the equation unconstrained,

$$\hat{\beta}_{\text{elastic}} = \frac{\text{argmin}}{\beta}\left\{\sum_{i=1}^{n}\left(y_i - \beta_0 - \sum_{j=1}^{p}\beta_j x_{ij}\right)^2\right\} + \lambda_2\sum_{j=1}^{p}|\beta_j|^2 + \lambda_1\sum_{j=1}^{p}|\beta_j| \quad (24.10)$$

The term $\lambda_2\sum_{j=1}^{p}|\beta_j|^2 + \lambda_1\sum_{j=1}^{p}|\beta_j|$ in Equation 24.10 is called an elastic net penalty. When $\alpha = 1$, the equation reduces to simple ridge regularization. And similarly when $\alpha = 0$, the equation reduces to lasso regularization. The lasso penalty in the equation generates sparse models while the ridge penalty stabilizes the path chosen by lasso regularization. One important thing to keep in mind while implementing the elastic net regularization is the different tuning parameters λ_2 for ridge and λ_1 for lasso regularization. The optimal values of the tuning parameters are determined by cross-validation. By combining lasso and ridge regularization, elastic net groups, and shrinks, the parameters associated with correlated variables and leave them in the equation or remove them all at once. See the below code sample for elastic net implementation.

The following Table 24.1 shows a comparison between the above regularization methods.

24.4 Decision Trees

Classification is the process of separating datasets into different labels or classes based on some conditions. Decision trees are one of the most famous and important techniques of classification. It is a graphical representation of all possible solutions to a decision taken after considering all conditions. Tree-based algorithms divide the feature space into a set of forms. Further, it fits a simple model in each set of forms. The algorithm is similar to the tree, just like a tree grows from roots, and the algorithm also starts from the root node which keeps on growing with an increasing number of decisions and conditions. Let us discuss one of the famous examples of the decision tree. Whenever one dials a toll-free number of a

PSEUDOCODE:

1. Initialize the tuning parameter λ.
2. Set the shrinkage penalty to L1 and L2 constraint incorporating the tuning parameter.
3. Add the shrinkage penalty to the sum of squares optimization.
4. Train the dataset using the above formed optimization function.
5. Return the predicted values.

TABLE 24.1

Comparison of Regularization Methods

Characteristics	Ridge Regularization	Lasso Regularization	Elastic Net Regularization
Type of weights produced	Distributed weights	Sparse weights	Sparse weights
Use case	Applicable in low as well as high dimensionality problems	Not applicable in low dimensionality problems	Applicable in low as well as high dimensionality problems
Penalty type	L1	L2	A mix of L1 and L2
Feature extraction	Cannot be used for feature extraction	Can be used for feature extraction	Can be used for feature extraction

certain company, it redirects us to intelligent computer assistance where it asks us questions like press 1 for English, and press 2 for Hindi. Now, once an option is selected it again redirects us to a certain set of similar questions. So it keeps on asking until the call is redirected to the right person. One might get caught in a voicemail trap, but the company was actually using the decision tree to get to the right person. Some of the terminologies of the decision tree are as follows:

- **Root Node**: It represents the entire population of the tree and can be separated into homogenous sets.
- **Leaf Node:** It is the last node which further cannot be segregated into further nodes.
- **Splitting**: Splitting is dividing the root node or sub-node into different parts based on certain criteria.
- **Branch tree**: A branch tree is formed by splitting the tree or node.
- **Pruning**: Pruning is the opposite of splitting. In pruning, unwanted branches from the tree are removed.
- **Parent and Child Node**: All the top node belongs to the parent node, and all the nodes that are derived from the top node are known as the child node. The root is the parent node, and all branches associated with the root node are known as a child node.

Decision trees are less stable compared to logistic regression, since deleting one node in a tree will lead to different decision trees altogether. Lot of times this problem of instability can be solved by combining all possible decision trees to generate one tree for prediction.[4]

24.4.1 CART (Classification and Regression Trees)

24.4.1.1 Regression Trees

Consider a regression problem having continuous output y and p inputs represented as $x_1, x_2, \ldots, x_p$. The feature space can be partitioned by the line parallel to the coordinate axes. The partitioning can be made simple using a recursive binary partitioning method. The main advantage of a recursive binary partition is that it is easy to interpret. The leaf node of the regression tree is always numeric. There being p inputs and suppose there are n observations, then in (x_i, y_i), $x_i = (x_{i1}, x_{i2}, \ldots, x_{ip})$, and suppose the region is partitioned into m regions then the response constant as c_m for each region would be:

$$f(x) = \sum_{m=1}^{M} c_m I(x \in R_m) \tag{24.11}$$

Here in each region, the average of the output of all observations in the given region is as follows:

$$\hat{c}_m = \text{ave}\left(y_i \mid x_i \int R_m\right) \tag{24.12}$$

Since each leaf node corresponds to the average value in a different region of observations, the tree reflects more accurate data than the straight line. Regression trees can easily handle complicated data, i.e., having more inputs. Now the main task is to find the correct split points. The sum of squares of residuals is calculated considering each data, i.e., using the greedy approach until the minimum sum of squares of residuals for the data point is achieved. That data point is considered as the split point. Consider j as a splitting variable and s as a split point:

$$R_1(j,s) = \{X_j \le s\} \text{ and } R_1(j,s) = \{X_j \le s\} \tag{24.13}$$

$$\min_{j,s}\left[\min_{c1} \sum_{x_i \in R_1(j,s)} (y_i - c_2)^2 + \min_{c2} \sum_{x_i \in R_2(j,s)} (y_i - c_2)^2\right] \tag{24.14}$$

A very large tree leads to overfitting the data since the model has no bias, but very high variance. There are a bunch of techniques to prevent overfitting of the data. The simplest way is to split observations when there are a certain minimum number of observations. If the number of observations is more than the minimum number, then allow for a split. The cost complexity pruning is used to overcome overfitting:

$$Q_m(T) = \frac{1}{N_m} \sum_{x_i \int R_m} \left(y_i - \hat{c}_m\right)^2 \tag{24.15}$$

The first step in cost complexity pruning is to calculate the sum of squared residuals of each tree. Cost complexity criterion:

$$C_\alpha(T) = \sum_{m=1}^{|T|} N_m Q_m(T) + \alpha|T| \tag{24.16}$$

Now to get the value of '$\alpha|T|$', the weakest link pruning technique is used. The main task is to achieve the minimum value of cost complexity, but with every 'α', the different sequences of subtrees are generated giving the minimum value of cost complexity. So fix the value of 'α' by using a five- or tenfold cross-validation sum of squares. Now, the tree that corresponds to the value of 'α' is selected. The selected subtree will be the final pruned tree. Refer to the below code for regression tree implementation.

24.4.1.2 Classification Trees

In contrast to regression trees, classification trees are used when the output is not continuous, i.e., it is a class label for each region. The sum of squares of residuals is not suitable for classification.[5]

The probability of region m belongs to class k with N_m being the total number of observations in region R_m is given by $\hat{p}_{mk}$. First, check whether the training data point in R_m belongs to class k, if so then it will assign value 1, else 0. Now take the sum and divide it by the total number of observations to get the probability of the region.

$$\hat{p}_{mk} = \frac{1}{N_m} \sum_{x_i \int R_m} I(y_i = k) \tag{24.17}$$

Class labels are assigned to that region which gives maximum probability value. Here in the classification tree, the Gini index is used as an error measure to classify observations. Gini index equation is calculated in the following manner:

$$\sum_{k \neq k'} \hat{p}_{mk} \hat{p}_{mk'} = \sum_{k=1}^{K} \hat{p}_{mk} \left(1 - \hat{p}_{mk}\right) \tag{24.18}$$

PSEUDOCODE – REGRESSION TREE

1. X is the input matrix that gets the value of the predictor column from the dataset.
2. Y is the target column that consists the value of the expected class.
3. Initialising tree Regressor.
4. Find the split set on X that minimizes the mean squared error and it is then used to split the node into two child nodes.
5. Continue step 3 until the max depth is reached i.e. the stopping criteria.
6. Prune the tree and return the prediction.

PSEUDOCODE – CLASSIFICATION TREE:

1. *X* is the input matrix that gets the value of the predictor column from the dataset.
2. *Y* is the target column that consists the value of the expected class
   Initialising tree Classifier
3. Find the split set on *X* that minimizes the Gini index and it is then used to split the node into two child nodes.
4. Continue step 3 until the max depth is reached i.e. the stopping criteria.
5. Prune the tree and return the prediction.

Further, the Gini index is given as,

$$\left(\hat{p}_{m1} + \hat{p}_{m2} + \cdots + \hat{p}_{mk}\right) - \hat{p}_{m1}^2 - \hat{p}_{m2}^2 - \cdots - \hat{p}_{mk}^2 \tag{24.19}$$

$$1 - \sum_{k=1}^{K} \hat{p}_{mk}^2 \tag{24.20}$$

The region with the minimum Gini index is selected as the root node. Further, the Gini index is also used in cost complexity pruning. See the below code for classification tree implementation.

24.4.2 Iterative Dichotomiser (ID3)

ID3 is one of the renowned algorithms that is used to generate decision trees, and the pseudocode is discussed below.

So, from the pseudocode, there is a need to stress on selecting the best attribute. The selection criteria of the best attribute are based on the entropy value.

$$H(D) = -\sum_{k=1}^{K} \hat{p}_{mk} \log \hat{p}_{mk} \tag{24.21}$$

The attribute with the lowest entropy value is selected as the root node. Let *D* be the current dataset for which the entropy is being calculated. Further, this attribute is used to split the dataset *D* by imposing certain conditions on the dataset. A constant quantity has zero entropy value; i.e., the class labels are the same or the distribution is completely known. Entropy zero specifies that the probability is one. As probability decreases, the entropy increases, and as probability increases, the entropy decreases.

There is another concept called information gain which is the measure of the difference in the entropy of the attribute before splitting and after splitting of dataset *D*.

$$IG(D,A) = H(D) - \sum_{t \in T} \hat{p}_{mt} H(t) = H(D) - H(D \mid A) \tag{24.22}$$

Where,

- $H(D)$ – the entropy of dataset *D*
- *T* – the subsets generated after splitting dataset *D* by attribute *A*
- $\hat{p}_{mt}$ – the probability of the number of elements in '*t*'
- $H(t)$ – the entropy of subset *t*

Information gain can be used instead of entropy after getting the first attribute. The attribute with the largest information gain is used for splitting the dataset *D*.

PSEUDOCODE

Define a root node.
If all data points are '+ve',
Then return the root node having a 'positive' label.
Elif all data points are '–ve',
Then return the root node having a 'negative' label.
Elif the number of predicting attributes == null,
Then return the root node having label=target attribute having the most common value among the data points.
Else
A ← selection of the best attribute that categorizes the data points.
The root node of the decision tree=A
For every value x_i of A,
A new branch is added to the root node, such as the test node $A = x_i$.
From examples select a subset Examples (x_i) having value x_i for A
If the subset== null
Then add a leaf node to this new branch with label=target attribute having the most common value among the examples.
Else
Add the subtree ID3 (Examples (x_i), Target_Attribute, Attributes – {A}) to this new branch
end
return Root

The ID3 algorithm sometimes leads to overfitting. So, when to stop? The following are stopping criteria.

- When all attributes are used to build a decision tree and there are no more attributes, then stop splitting the attribute.
- When the entropy of the attribute is zero, i.e., the class label of all data points is the same, then stop splitting even if attributes are present.
- If the attribute contains very few data points in the tuple, then stop splitting the dataset, and splitting of the dataset might result in overfitting.

24.4.3 C4.5 and C5.0

C4.5 is one of the most popular algorithms in machine learning that is used to generate decision trees. This algorithm further improves the ID3 algorithm. C4.5 builds a decision tree from a training dataset just like the ID3 algorithm using the concept of information gain.[6]

The pseudocode seems similar to the ID3 algorithm. Some of the improvements from ID3 are as follows:

- C4.5 handles both discrete and continuous attributes. For continuous attributes, the algorithm sets a threshold and then splits the dataset into the data points those whose attribute value is above the threshold value and those that are less than or equal to it.
- The algorithm can easily handle missing values, and missing values are not used to calculate entropy and information gain.

The algorithm checks the whole tree when it is created and tries to delete the branch that overfits the data by substituting it with the leaf node. C4.5 prunes the tree after it has been created.[7]

PSEUDOCODE:

- For every attribute in the dataset, find the information gain ratio from splitting on the attribute in the dataset. The attribute having the largest information gain value is selected.
- Let *A* be the attribute selected with the normalized information gain.
- Generate a decision tree that splits on attribute *A*.

Recurse on the dataset obtained by splitting on the attribute *A*, and add those nodes as children of the attribute.

Further C5.0 algorithm was created which proved to improve C4.5 in terms of

- **Speed:** C4.5 is slower than C5.0.
- **Memory usage**: C5.0 uses memory efficiently.
- **Small Decision Trees:** C5.0 gives equivalent output with smaller decision trees as compared to C4.5.
- **Supports Boosting**: The overall accuracy of the trees is improved by boosting. The concept of boosting will be covered in the later sections.
- **Winnowing**: C5.0 automatically removes the attributes that may be unhelpful.

24.4.4 Decision Stumps

Decision stump is an algorithm that generates a one-level decision tree; i.e., its root node is immediately connected to the terminal nodes. That means the decision tree makes predictions based on one input feature. Many variations can be done depending on the type of input feature. Features are as follows:

- **Nominal Feature**: Builds a stump which contains a leaf for each input feature or can have two leaves, one corresponds to one category and another leaf to all other categories.
- **Binary Feature**: Two leaf nodes are used to distribute binary features, and missing values can be treated as another category.
- **Continuous Feature**: The threshold value is selected, and a stump of two leaves is generated, one for a value less than equal to the threshold and the other for a value greater than the threshold value. Sometimes, multiple threshold values can be selected but in this case the number of leaf nodes increases. Decision stumps mostly used as bagging and boosting in ensemble techniques. Bagging and boosting will be discussed in the next section. Refer the code below for implementing classification trees.

24.4.5 M5 Model Tree

M5 is a decision tree algorithm used for regression problems to predict the output numeric value. It is a binary decision tree having a linear regression function at its leaf node that predicts continuous outputs. The M5 algorithm performs the same approach as CART using squared error as an impurity function.

The generation of the M5 model tree requires two steps. First, the splitting criterion needs to be selected to split the dataset. The splitting criterion reduces the standard deviation error of each class's values. The aim is to calculate the reduction in the error for each attribute, and the attributes having the highest reduction error are selected as the current node. By splitting, the data in the child node get a lower standard deviation value as compared to the parent. After checking all possible splits, the algorithm finalizes the split which maximizes the error reduction. Sometimes, this method results in large trees which cause overfitting.

PSEUDOCODE:

1. *X* is the input matrix that gets the value of the predictor column from the dataset.
2. *Y* is the target column that consists the value of the expected class
   Initialising tree Classifier
3. Here the input feature is the only parameter used for the prediction. Hence the max-depth of the tree is one.
4. Return the prediction.

TABLE 24.2

Comparison of Decision Tree Algorithms

Characteristics	CART	ID3	C4.5 or C5.0	M5 Model Tree	Decision Stumps
Splitting criteria	MSE, Gini index	Information Gain	Gain ratio	Standard deviation error	Gini index
Missing values	Handles missing value	Does not handle missing values	Handles missing value	Handles missing value	Handles missing value
Pruning strategy	Cost complexity pruning	No pruning is used	Error-based pruning	Standard regression error-based pruning	No pruning is used
Type of attributes	Handles both numeric and categorical values	Handles only categorical value, i.e., works only for binary attributes	Handles both numeric and categorical values	Handles both discrete and continuous attributes	Handles both discrete and continuous attributes

To avoid overfitting, the second step is very important that prunes the tree and replaces it with a linear regression function. This is the M5 method of forming the model tree that splits the dataset and builds a linear regression model in each of them.

There are many advantages to using the M5 model algorithm.

- **High dimensionality**: The algorithm can handle very high dimensionality like hundreds of attributes.
- **Cost-Efficient**: Since it can handle high dimensionality apart from other regression tree algorithms whose cost increases as trees dimension increases.

The Table 24.2 below differentiates the above-discussed algorithms based on common characteristics.

24.5 Ensemble Learning

In machine learning, the ensemble method uses multiple algorithms to create a model that provides better prediction than by using a single algorithm model. The main goal of using ensemble methods is to reduce the variance and that improves the empirical result. Although the ensemble method is better in predictive performance, the computing speed is slightly low since multiple algorithms are used to make a model. This algorithm proves to give much more accuracy, precision, and less error rate compared to other algorithms.

The number of learners in an ensemble is known as base learners. The output of an ensemble learning method is generally more accurate than that of the output given by a single base learner. Ensemble learning is very powerful because it can promote weak learners that perform just better than random guesses to strong learners that predict very accurate results. Therefore, 'base learners' are known as 'weak learners'. Base learners are generated from training data using a base learning algorithm which could be a neural network, decision tree, regression, or some other type of algorithm. Homogenous base learners could be produced by ensemble methods using a single base learning algorithm.

Ensemble Techniques:

- Bagging
- Boosting

24.5.1 Bagging

Bagging is also known as bootstrap aggregation. The idea of bagging is simple, just fit several independent models and average their prediction to create a model that lowers the variance. However, it is not just practically possible to fit all independent models since it would require a great amount of data. So bootstrapping will help in the procedure.[8]

Re-sampling dataset from the original database with replacements which is also known as bootstrap samples. These samples have pretty good statistical properties, and they can be seen as being drawn both directly from the true underlying data distribution and independently from each other. Bootstrap samples are also used to evaluate variance or confidence intervals of statistical models. In most cases, gathering true independent samples would require too much data as compared to the amount of data available. Thus, the bootstrap samples can be used to collect data independently from each other and representative.

The classifiers are trained on these samples such that *N* classifiers will be trained on the *N* datasets. All the classifiers are combined to form an ensemble classifier. The generated ensemble classifier is strong compared to all *N* classifiers. The voting classifier is used by the ensemble model to predict the class of the test case.

In the bagging method, the different learners are fitted, independent from each other so that they can be trained concurrently. Now, consider a regression problem and fit a model to the training data given as $Z = \{(x_1, y_1), (x_2, y_2), \ldots, (x_n, y_n)\}$ which obtains the prediction denoted as $f\ (x)$ on the input x. Bagging averages the prediction values over bootstrap samples, which further reduces the variance. For every sample $Z * b$, where $b = 1, 2, \ldots, N$, fit the model and predict $f\ {*b}(x)$. The bagging estimate is given as,

$$f_{\text{bag}}(x) = 1N_b = 1N_f * b(x) \tag{24.23}$$

In this type of regression example, simple average works well while in classification problems the bagged classifier selects the most votes from the bagged classified tree as discussed earlier.

Suppose the tree generates a classifier *g*(*x*), where $g(x) = \arg\max_k f(x)$ for *K* class response and underlying vector function *f*(*x*). The class having the maximum number of votes is selected by the bagged classifier,

$$g_{\text{bag}}(x) = \arg\max_k \left(f_{\text{bag}}(x)\right) \tag{24.24}$$

24.5.1.1 Random Forest

Random forest combines the simplicity of decision trees with flexibility resulting in a vast improvement in the accuracy. The main idea in bagging is to average many unbiased models and to reduce the variance, and hence, trees are ideal candidates for bagging. Now, let us make a random forest.[9]

1. **Create a Bootstrapped Dataset**: To create a bootstrapped dataset that is of the same size as the original, just randomly select samples from the original dataset. The important detail is that it is allowed to pick the same sample more than once.
2. Create a decision tree using this dataset, but only use a random subset of variables at each step.
3. Repeat steps 1 and 2 multiple times.

This method of using a bootstrapped sample and considering only a subset of the variables at each step makes the random forest more effective than individual decision trees. Take the data and run it down in the first tree that was created earlier. Note the output, and pass the data to the second tree and repeat for all trees, and keep track of output. After running the data down all of the trees in the random forest, the option received the most number of votes is interpreted.

24.5.2 Boosting

Boosting is a sequential ensemble. As discussed in ensemble learning earlier, a weak classifier has an error rate just better than random sampling. The boosting algorithm sequentially implements the weak classification algorithm to continuously modify the data which results in the generation of a sequence of weak learners $M_n(x)$, where $n=1, 2,\ldots, N$. The predictions of each of these learners are combined by a weighted majority vote to give the final predictions.[10]

$$M(x) = \text{sign}\left[\sum_{n=1}^{N} \alpha_n M_n(x)\right] \tag{24.25}$$

24.5.2.1 AdaBoost

Boosting is a sequential process where multiple weak classifiers are combined to get a big ensemble strong classifier. The first step of AdaBoost is to initialize weights denoted as $w_1, w_2,\ldots, w_W$ to every training observations and set $w_i = 1/W$, $i=1, 2,\ldots, W$, where W is the total training points.

For the first iteration $n=1$, the classifier is trained on the data using the weights. For each successive iteration $n=2, 3,\ldots, N$, the observation weights are modified individually and the classification algorithm is again applied to the weighted observations. For a step n, the weights of those observations that were misclassified by the previous classifier $M_{n-1}(x)$ are increased, whereas the weights are decreased for those that were classified correctly. Thus as the iterations proceed, the observations that are difficult to classify correctly get greater importance. Hence, each successive classifier is forced to concentrate on those training observations that are missed by the previous classifier.

Now, let us understand the AdaBoost algorithm,

1. Initialize the weights $w_i = 1/W$, $i=1, 2,\ldots, W$.
2. For each $n=1$ to N:

   a. $M_n(x)$ classifier is fitted to the training data using weights w_i.
   b. Calculate the error rate

   $$e_n = \frac{\sum_{i=1}^{W} w_i I(yi \neq Mn(x))}{\sum_{i=1}^{W} w_i}$$

   c. Calculate $\alpha_n = \log\left(\frac{1-e_n}{e_n}\right)$
   d. $w_i \leftarrow w_i e^{\alpha_n I(yi \neq Mn(x))}$, $i=1, 2,\ldots, W$

   return Output

$$(x) = \text{sign}\left[\sum_{n=1}^{N} \alpha_n M_n(x)\right] \tag{24.26}$$

AdaBoost can be extensively used to solve a variety of real-world problems like prediction and classification.

24.5.2.2 Gradient Boosting

When a gradient boost is used to predict continuous value, like weights, it is called as 'Gradient boost for regression'. After understanding AdaBoost, a lot of gradient boost will seem very similar. So let us briefly compare and contrast AdaBoost and gradient boost.

AdaBoost starts by building a very short tree called a stump from the training data, and the amount of saying that the new stump has on the final output is based on how well it compensated for those previous errors. Then, the AdaBoost builds the next stump on the errors that the previous stump made. It advances to generate stumps in this manner until it reaches the required stumps, or it has a perfect fit.

In contrast, gradient boost is built using a single leaf, rather than a tree or a stump. This leaf denotes guesses for the weights of all the samples. Here, the aim is to predict continuous values like weights, and therefore, the first estimate is always the average value. The gradient boost generates a tree by avoiding the errors caused by the previous trees, and it is generally larger than a stump.[11] The limit of the maximum number of leaves of the tree is often between 8 and 32. Unlike AdaBoost, gradient boost scales all trees by an equal amount. Gradients continue to build trees in this fashion until it has made the number of trees that are required, or the additional trees fail to improve the fit.

ALGORITHM

1. Initialize model with a constant value $f_0(x) = \text{argmin}_\gamma \sum_{i=1}^{N} L(y_i, \gamma)$, here '$\gamma$' is predicted value and 'y_i' is observed value and L is the loss function and the 'argmin' over 'γ' means there is a need to find a predicted value that minimizes the sum. Using all this data initial predicted value '$f_0(x)$', is generated that means the predicted value is just the leaf.
2. For $m=1$ to M, which means M trees are generated
   a. Compute

$$r_{im} = -\left[\frac{\partial L(y_i, f(x_i))}{\partial f(x_i)}\right]_{f=f_{m-1}}$$

This part is just the derivative of the loss function with respect to the predicted value, where r is the residual, i is the sample number, and m is the tree that needs to be built.

   b. Fit a regression tree to the rim values and create terminal regions R_{jm}, for $j = 1 \ldots J_m$. This means a regression tree is built to predict the residuals instead of the weights. 'j' in R_{jm} is the index for each leaf in the tree.
   c. For each $j = 1 \ldots J_m$, calculate

$$\gamma_{jm} = \text{argmin}_\gamma \sum_{xi \in R_{jm}} L(yi, f_{m-1}(xi) + \gamma)$$

   Here, the output values are calculated for each leaf.
   d. Update

$$f_m(x) = f_{m-1}(x) + \sum_{j=1}^{J_m} \gamma_{jm} I(xi \in R_{jm})$$

Here, a new prediction is made for each sample. Final output $\hat{f}(x) = f_M(x)$.

From the main similarities and differences between gradient boost and AdaBoost, let us understand the algorithm of gradient boost. Input: Dataset $\{(x_i, y_i)\}, i = 1$ to n and a differentiable Loss Function $L(y_i, f(x))$. Here, the loss function that most commonly used when doing regression in gradient boost is $12 * (\text{observed} - \text{predicted})^2$.

The Table 24.3 below shows the features of bagging and boosting discussed above.

24.6 Implementation and Analysis of Deep Learning Algorithms

24.6.1 Feedforward Neural Networks

Before looking into feedforward neural networks, let us look at the basic fundamental unit of the artificial neural networks, i.e., a perceptron. A perceptron is a basic unit that combines inputs given to it and provides an output if the combined output exceeds a certain threshold (Figure 24.1).[12]

The activation function used above could be any one of the known activation functions, e.g., sigmoid, ReLU, and tanh. The problem with a single perceptron is that it is only able to solve problems with a single decision boundary. But a combination of perceptrons can provide an effective decision boundary to data which are non-linear. This single perceptron might seem only able to perform multiplication of inputs and parameters but these combined layers of perceptrons called multilayer perceptrons can solve many complex deep learning problems. These multilayer perceptron layers are called artificial neural networks (Figures 24.2 and 24.3).[13]

Consider a two-variable problem of predicting whether a car belongs to a high mileage class or a low mileage class with input variables the number of cylinders and horsepower of the engine. The first neuron in the first layer is connected with the inputs x_1 and x_2 having weights w_{11} and w_{12} along with

TABLE 24.3

Comparison between Bagging and Boosting

Characteristics	Bagging	Boosting
Weights	Equal weights are assigned to the model	Weights are assigned to the model based on its performance
Training input	Training data are randomly drawn from the training dataset with replacement	Training input contains the data that were misclassified by the previous model
Building models	Every model is built independently	The model is influenced by the performance of the previous model
Application	It is used to solve the problem of high variance, i.e., overfitting	It is used to solve the problem of high bias
Famous algorithms	Random forest	AdaBoost, gradient boost

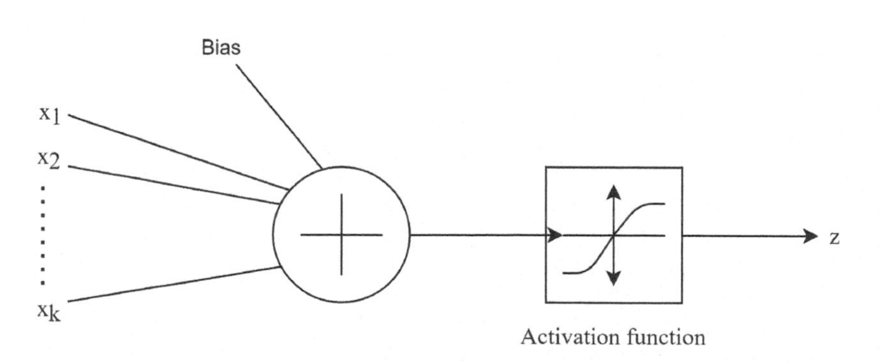

FIGURE 24.1 Single perceptron.

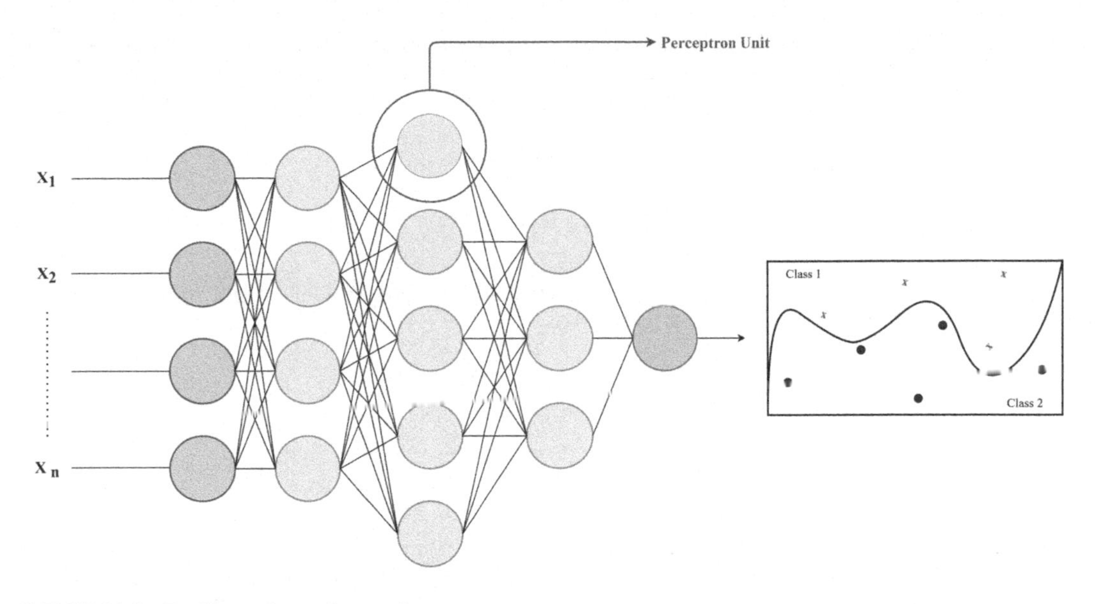

FIGURE 24.2 Feedforward neural network.

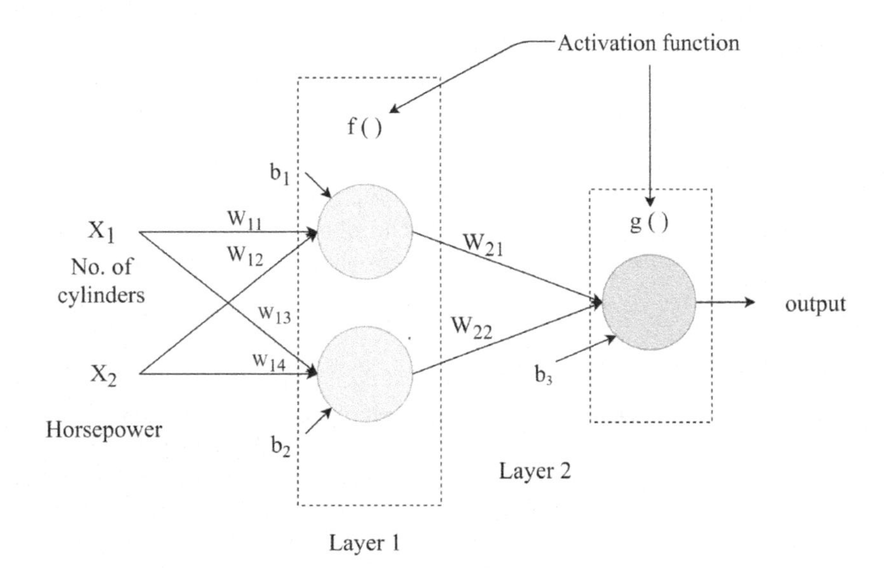

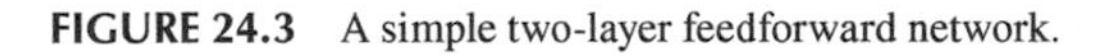

FIGURE 24.3 A simple two-layer feedforward network.

bias b_1. Similarly, the second neuron in the first layer is connected with the same two inputs with weights w_{13} and w_{14} along with bias b_2. The output of the first neuron is represented as

$$O_1 = f(x_1, x_2) \tag{24.27}$$

Here, $f(.)$ is the activation function. Suppose the activation function is a sigmoidal function, then the output of that neuron would be represented as,

$$O_1 = \frac{1}{1 + e^{-(w_{11}x_1 + w_{12}x_2 + b_1)}} \tag{24.28}$$

Since the second neuron is in the same layer and has the same activation function, its output is given as,

$$O_2 = f(x_1, x_2) \tag{24.29}$$

$$O_2 = \frac{1}{1+e^{-(w_{13}x_1+w_{14}x_2+b_2)}} \tag{24.30}$$

The second layer takes the output of the previous layer and weights as inputs along with the bias. The output of the second layer of the network is given as,

$$O_3 = g(\ O_1, O_2) \tag{24.31}$$

$$O_3 = \frac{1}{1+e^{-\left(w_{21}\left(\frac{1}{1+\ e^{-(w_{11}x_1+w_{12}x_2+b_1)}}\right)+w_{22}\left(\frac{1}{1+\ e^{-(w_{13}x_1+w_{14}x_2+b_2)}}\right)\right)}} \tag{24.32}$$

This artificial neural network introduces many more parameters than in a single perceptron which makes it capable of achieving complex decision boundaries. For the multi-class classification problems, the number of neurons in the final layer will be equal to the number of classes. But since the above problem was a binary classification, only one neuron is included in the final layer. Such simple neural networks are extended to more layers, neurons, and parameters in practical deep learning models that solve real-world problems. In the previous example, the cross entropy loss is computed for the predicted output. This loss is used for determining the best possible value of parameters using the gradient descent algorithm. The fundamental goal in this gradient descent algorithm is to minimize cross entropy loss for the predicted output. The parameters are updated in multiple iterations. Initially, the weights are assigned randomly. Based on the value of the loss, the parameters of the models are updated such that the current loss would be less than the previous loss. And this is repeated over multiple iterations, and when the value of the current loss does not change from the previous value, the model stops doing iterations since it has found the best value of parameters.[14] The following code shows an implementation of a simple neural network. The pseudocode provided below shows the general method to implement a feedforward neural network.

24.6.2 Convolutional Neural Networks

CNN was introduced in 1998 and used in many applications in computer vision from image classification to audio synthesis and WaveNet. CNN is used for processing data with grid-like topology. The key idea of CNN is to detect the local patterns in the data using the convolutional operations. It then downsamples the data into the pooling layer. CNN employs a series of convolution and pooling layers. Let us understand convolution and pooling concepts through the image classification example.[15]

PSEUDOCODE:

1. Add the input layer to the neural network having the same number of nodes as the number of features.
2. Add the hidden layers according to the problem.
3. Add the output layer having a suitable number of nodes as per the problem. A k class classification problem would have $k-1$ nodes.
4. Train the model by providing the loss function which the neural network has to optimize.
5. Return the values predicted by the trained neural network.

24.6.2.1 Convolution

Here, CNN takes an image as input and classifies them. Suppose, the input is a gray-scaled image hence the depth is 1, while height and width both are 28 and thus input shape is (1, 28, 28). The convolutional operation involves a filter that captures local patterns and applies to the image. The convolutional operation defines a filter of specific height and width, and let us say our filter is of size 3 * 3.

The filter is taken and positioned it at each different place and slide the filter across the length and breadth of the image until the last patch. The filter has nine entries, and each entry has weight $w_1, w_2, \ldots, w_9$. So initially, the filter is taken and positioned it to the first patch of the image and the image has the following inputs: $x_1, x_2, \ldots, x_9$. The linear combination of the weight in the filter with the feature value in the patch,

$$Z = \text{relu}(w_1 x_1 + w_2 x_2 + \cdots + w_9 x_9 + \text{bias}) \tag{24.33}$$

It is similar to a feedforward neural network. Z is the scalar quantity for a given positioning of a patch on the image.

Similarly, for the colored image, there are three channels and 3D filter gives

$$Z = \text{relu}\left(b + \sum_{i=1}^{27} w_i x_i\right) \tag{24.34}$$

The strides are set while sliding to the next patch, and in the given example, the value of stride=1 so that filter shifts by 1 column. Once it reaches the last column, the filter is shifted by 1 row. In this way, the filter slides to the image and matches the pattern. In this example, input consists of 28 * 28 * 1 image size and filter of size 3 * 3 * 1, so there is 26 * 26 possibility to position the filter on the image. Therefore, 26 * 26 * 1 output of the image is obtained for 1 filter, and for n such filters, the output of size 26 * 26 * 1 is generated.

Convolutional layer has a number of key features. Convolutional layer learns translational invariant patterns. Using this property, CNNs are able to recognize a certain pattern considerably. They can learn special orders of patterns, and the first convolutional layer will learn small patterns like edges. The second layer will learn larger patterns that are formed from features of the first layer, and the rest of the layers are trained in a similar manner. This will help to learn the complex structures and concepts accurately (Figure 24.4).[16]

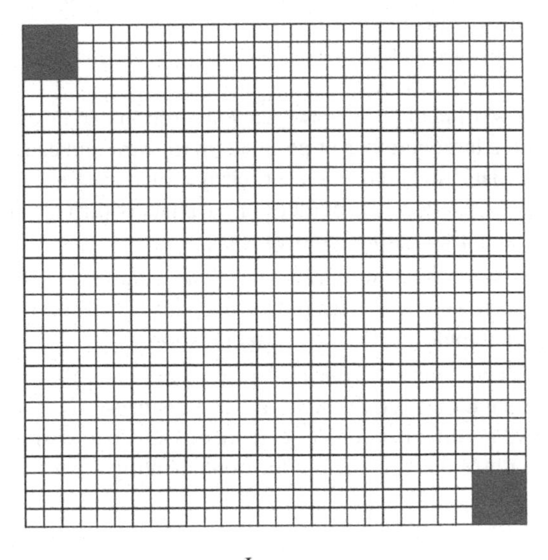

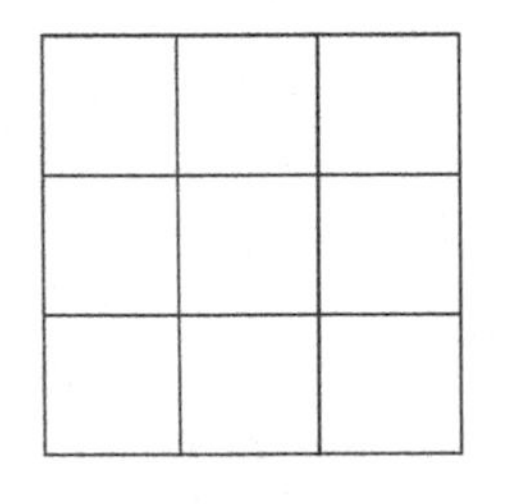

Filter

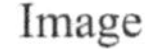

FIGURE 24.4 Convolution.

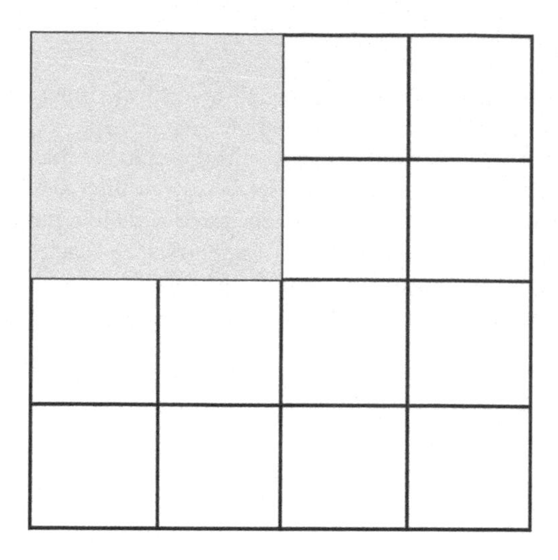

FIGURE 24.5 Pooling.

24.6.2.2 Pooling

The next important concept is pooling. Pooling tends to downsample the output of convolution adversely. It is conceptually similar to the stride convolution. It consists of extracting the specific window from the input feature and compute the output based on the pooling policy. Pooling is usually done with a window of size 2 * 2 with stride of 2 (Figure 24.5).

In the above example, pooling defines the window of size 2 * 2, so this is the position of the pooling window on convolution output. Then, apply the pooling policy on the window and select the number based on that policy. The pooling operation is applied on each channel separately; for example, the output of a convolution is in the form of multiple channels, while in pooling it is applied on each channel separately. Two pooling policies are max pooling and average pooling. Max pooling will return maximum from the pooling window. Average pooling returns the average of the pooling window.

So, 2 * 2 output is output for 4 * 4 convolution input and thus results in the downsampling of the output of convolution. Max pooling does not have any parameters.

The three main features reason for using CNN are as follows:

- **Sparse Interaction**: In typical feedforward neural networks, each neuron in one layer is connected to the next layer. This results in a large number of parameters to learn and cause main issues like the greater need for training data, and convergence time also increases and may end up with an overfit model. Hence, CNN can reduce the number of parameters through indirect interactions. One more advantage is computing output requires less time.
- **Parameter Sharing**: It controls the number of weights as well as parameters in CNN. Previously in other neural networks, each weight was used only once but CNNs on the other hand use one feature multiple times to compute a different spatial position.
- **Equivariant Representation**: Object detection is invariant to the changes in illumination, change of position, but internal representation is equivariant to these changes. The modifications in illumination and position have no effect on object detection. However, it affects the internal representation of the system. Refer the code below for implementing a convolutional neural network.

24.6.2.3 Results and Applications

Following are the advanced research and applications of CNN in some of the fields:

PSEUDOCODE:

1. Creating Sequential model.
2. Adding first convolutional layer
   Conv2D layer with filters=32, kernel size=(3,3), activation=relu
3. Add Pooling layer
   MaxPooling2D with pooling size=(2,2).
4. Repeat steps 2 and 3.
5. Add Dense layer where activation function=softmax.
6. Compile the model using adam as the optimiser and sparse categorical cross entropy for evaluating loss.
7. Train the convolution network on training sets and known target values.
8. Predict the test values on a testing dataset.

- **Face Recognition:** The face recognition system is now embedded in various security and surveillance systems. Facebook started with the DeepFace program in 2014 that can classify if photographs belong to the same person with an accuracy of more than 97.25%. Similarly, in 2015, Google came up with FaceNet with an accuracy of over 99.63%.
- **Image Classification**: As there is a daily increase in data, the different image classification models, ZFNet (2013), GoogLeNet (2014), VGGNet (2014), ResNet (2015), DenseNet (2016), etc., yield accurate results on ImageNet (14,197,122 images and 21,841 classes) dataset.
- **Object Detection**: The object detection method not only accurately classifies but also provides the exact location of the object in the image. Recently, YOLO and OpenCV methods are widely used and have increased the efficiency of detection systems.
- **Video analysis and Semantic Segmentation**: CNN is used to analyze the role of each pixel in semantic segmentation, and it has been used in major computer vision projects like self-driving cars.

24.6.3 Recurrent Neural Networks

In the previous sections, various algorithms were discussed for solving classification and regression problems. A huge dataset of animals can be provided to the feedforward neural network and can predict an animal in a particular image. In training the model with the dataset, the classification of a given image has no effect on the classification of the next image. For instance, during training, if the model classifies an image as an elephant and the next image as a cat, the prediction in the later image as a cat is not dependent on the previous image as an elephant. This is because outputs are actually independent. But this assumption is not valid for every problem.

Deep learning also deals with problems in which each output is actually dependent on each other. Consider a scenario in which the machine learning model has to provide a summary of a document. To do this task, the machine learning model has to essentially first understand the meaning of the document itself and so its underlying semantics and context. If the model were to solve the problem as before, then it would understand the meaning of a word and so on move to the next word accordingly but would forget the previous word. Such a model would not be able to remember the previous words of the sentence.

To solve this problem of considering previous elements in a sequence, recurrent neural networks, i.e., RNNs, are used. They are called recurrent since they perform the same task on every element of the sequence. In recurrent neural networks, every unit is a simple artificial neural network. To overcome the above problems, RNNs share the parameters across the whole network. These shared parameters help the later block to take into consideration the previous output and generate its output accordingly.[17] For understanding the working principle of RNN, consider a text processing problem. The input given to the RNN would be text as words denoted as x_1, x_2, x_3,…,x_n. The output for the RNN would be denoted as o_1, o_2, o_3,...,o_n. The working of the RNN described below is not specific to a problem but instead is a general implementation. Hence, the output words o_1, o_2, o_3,...,o_n are specific to the type of problem (Figure 24.6).

As seen in Figure 24.6, the RNN has layers of artificial neural network that generate two outputs, one is the output o_t for an input word x_n and the weight W that is given as input to the next artificial neural network. Based on the above transition of weights, a state of any particular artificial neural network could be summarized as follows (Figure 24.7):

The state S of the block is determined using the following equation:

$$S_t = f(Ux_t + WS_{t-1}) \tag{24.35}$$

This state of the block is used to determine the output at the current timestamp as shown.

$$o_t = f(VS_t + b) \tag{24.36}$$

Here, $f(.)$ is the activation function. U and V are the weights for the simple artificial neural network.

From the previous section, it is evident that the backpropagation algorithm is used for estimating the gradients. Since the parameters are shared in recurrent neural networks, the gradient not only depends upon the current output but also on the outputs of previous timestamps. The loss can be computed for

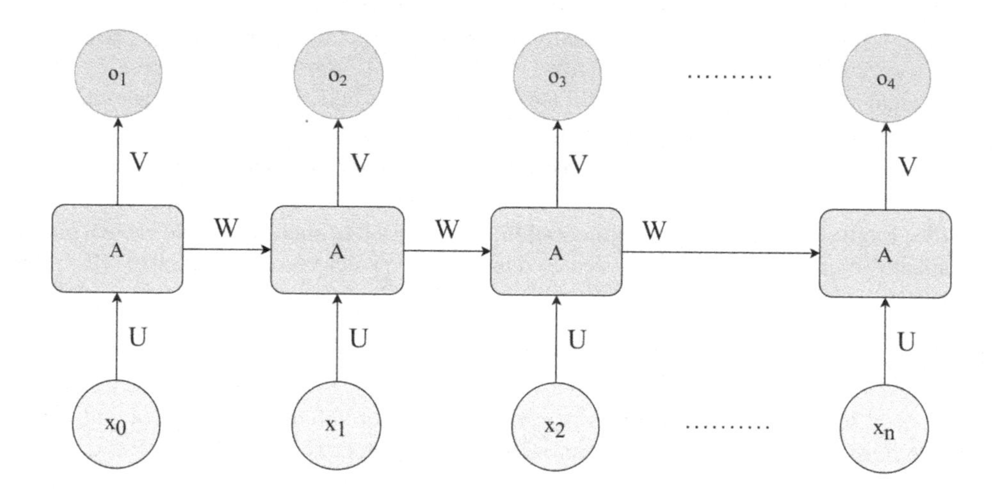

FIGURE 24.6 Recurrent neural network.

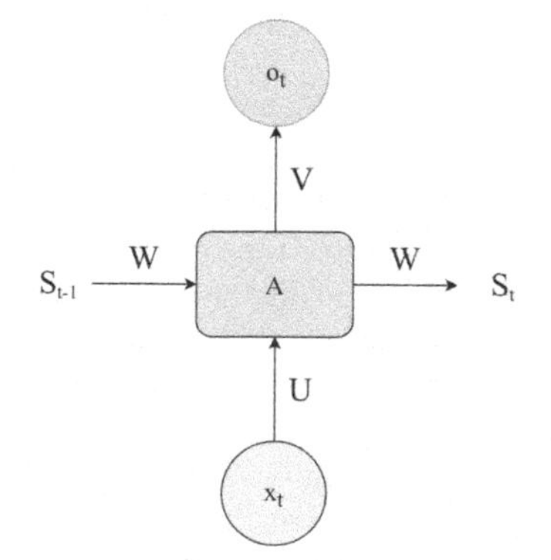

FIGURE 24.7 A single recurrent unit of network.

PSEUDOCODE

1. Initialize zero vector for initial state.
2. Randomly assign weights for *W*, *U* and *b*.
3. Iterate over each timestamp and do:
   i. Combines input with the current state to obtain the current output.
   ii. Update the state for next timestamp.
3. Return output.

each of the output word as l_1, l_2, l_3,...,l_n. The individual loss is computed by comparing the output generated word with the desired target word y_t.These individual errors are then added together to get a combined error.

$$L(x,y,W,V) = \sum_{t=1}^{T} l(y_t,o_t) \quad (24.37)$$

This combined error is backpropagated to every unit of the recurrent neural network for each timestamp. For each unit in the network, the gradient is calculated with respect to the weight parameter. Since the weight in recurrent neural networks is shared and has the same weights, the gradients can be combined for each timestamp. RNN has varied applications in text processing domain such as providing summary to a user document, providing user-based recommendations on search engine, and assessing consumer-based reviews. In case of e-commerce applications, RNN is fundamental in developing conversational chatbots communicating with consumer. See the code below for implementing a recurrent neural network.

24.6.4 Long Short-Term Memory Units

RNN has a problem of vanishing gradient, and to solve the problem, modified RNN is implemented known as LSTM (long short-term memory). As explained in RNN, gradients are calculated using backpropagation algorithm and are used to update the weight of further layers of the network. As the number of layers in network increases, the gradient value decreases. The gradient value being very low will become harder for the network to update weights and further take longer time to get the final result.

Example

The dog, which absolutely ...., was full.

The dogs, which absolutely ...., were full.

In the given example, the model needs to remember whether the sentence is in singular or plural form. In the sentence between 'dog' and 'was', there can be many words and RNN cannot remember which results in the output. RNN is not able to calculate very long dependency, and it also suffers from exploding gradient problems that make it unusable. The main principle of LSTM is to avoid the long-term dependency problem by remembering the information for a long period of time.

To solve this problem, LSTM was introduced with the carry unit 'C_t' which carries the information from the previous unit. Let us better understand LSTM through the diagram (Figure 24.8).

The network takes three inputs: X_t is the input of the current unit, 'H_{t-1}' is the output of the previous LSTM unit, and 'C_{t-1}' is the memory or carry from the previous units. Here, the memory 'C_{t-1}' plays an important role, and as for output, 'H_t' is the output and 'C_t' memory of the current unit.[18] Therefore, LSTM makes decisions based on current input, previous output, and previous memory; generates new output; and alters the memory.

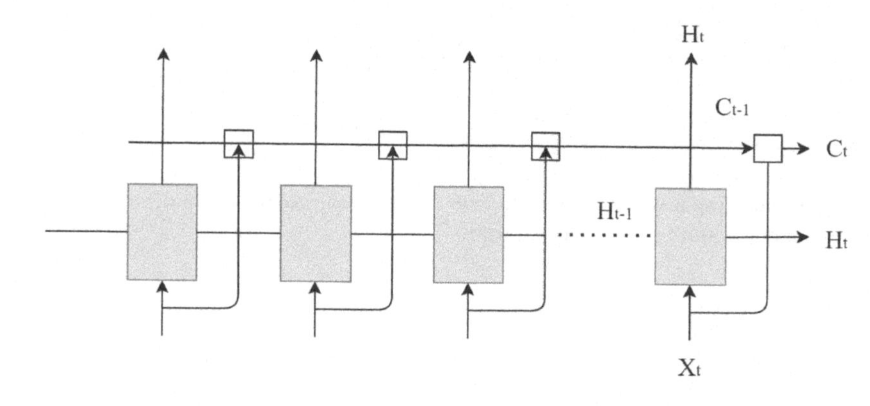

FIGURE 24.8 LSTM network.

LSTM is efficient for any sequential processing task, and hence, it is assumed that a hierarchical decomposition may be available but does not know the decomposition in advance. This algorithm is used to achieve state-of-the-art results in many problems like handwriting recognition and its generation, language modeling and translation, and synthesizing audio and its analysis. Another important concept is bidirectional LSTM, a combination of bidirectional RNNs with LSTM. Unlike LSTM, bidirectional LSTM uses the previous output from both directions to get the output. The sequence-to-sequence LSTM is also known as encoder-decoder LSTM. It is used in several sentiment analysis systems, where the encoder maps a sequence having different length to a vector of fixed size, and the decoder then maps this vector to a varying output sequence. The code below shows an implementation of an LSTM model.

```
import keras
def lstm(dataset, predictors, epochs, testing_set):
lstm = keras.layers.Sequential():
lstm.add(keras.layers.LSTM(32))
lstm.add(keras.layers.LSTM(64))
lstm.add(keras.layers.Dense(64,activation="relu"))
lstm.add(keras.layers.Dense(1))
lstm.fit(dataset[predictors], dataset['y'], epoch=epochs)
y_prediction = lstm.predict(testing_set)
return y_prediction
```

24.6.4.1 Applications and Advantages

LSTM is massively used in the applications where the data is in sequential format like time, text, and sentences. The different types of LSTM models depend on the input and output format are as follows:

- **One to One**: Image labeling is used for giving labels to the image. Here, image is the input and text label is the output.
- **One to Many**: Image captioning is used to give description to the image. Here, the sequence of words or complete sentence acts as the output.
- **Many to One**: In sentiment analysis, input is text sentence and output is rating. Predicting a word in an incomplete sentence consists of an incomplete sentence as input and prediction word as output.
- **Many to Many**: In music generation, input consists of music notes and output is audio generated. Another is stock market prediction which is based on past data extracted.

24.6.5 Generative Adversarial Networks

Generative adversarial networks are deep learning methods used in generative modeling; i.e., they are used to 'generate' data based on the training data. From its advent, generative adversarial networks have been widely used for generating images, videos, portraits, music, and in wide other applications. The main purpose of GANs (generative adversarial networks) is to generate data. This could be any type of data as previously mentioned. But its application is quite broad not only in the field of deep learning but also in reinforcement learning. A GAN consists of two models which are trained simultaneously. First is a generator *G* that uses the given data distribution to generate data. Second is a discriminator *D* that estimates the probability that a particular sample is sampled from the original data distribution rather than some generator *G*.[19]

The generator and discriminators are multilayer perceptrons, which are trained by the backpropagation algorithm. The generator *G* which generates data is of distribution over data, the prior which is a normal or uniform distribution on the input noise variables. Thus, the network with parameters θ_g defines a mapping as $G(z;\theta_g)$.

Actually, *z* represents the features of the output data. For example, consider generating images using GANs, and then, *z* would typically be the features of that image. It could represent any feature of that image, color, or shape. In the training of the generator *G*, the effect of *z* on the distribution is learned. The value of *z* is not manually controlled so as to have a difference in a feature of the image; i.e., it is not estimated that how *z* determines the color in the picture or the width of the hair in the picture. Instead, the generator *G* is used to upsample *z* to produce an image. If the generator just uses the input distribution to produce images, then the output would be random. Hence, GANs have a discriminator *D*. The discriminator *D* is a second multilayer perceptron which outputs the probability that *x* comes from the original data or some other model distribution. For estimating this probability, the discriminator uses the training data to determine which features in an image make it real. Then by using these features, it determines whether an image given by the generator is real or fake. Then, discriminator provides this feedback to the generator which makes *G* to produce an improved image that can fool *D*. Thus, the two models play a minimax game improving themselves along with the training and eventually reaching a solution where the probability of whether an image is real or not estimated by *D* is 0.5. Thus, the network is in sense an 'adversarial' network wherein the generator and discriminator play a minimax game with respect to each other. Hence, this network is called as 'generative adversarial network' (Figure 24.9).

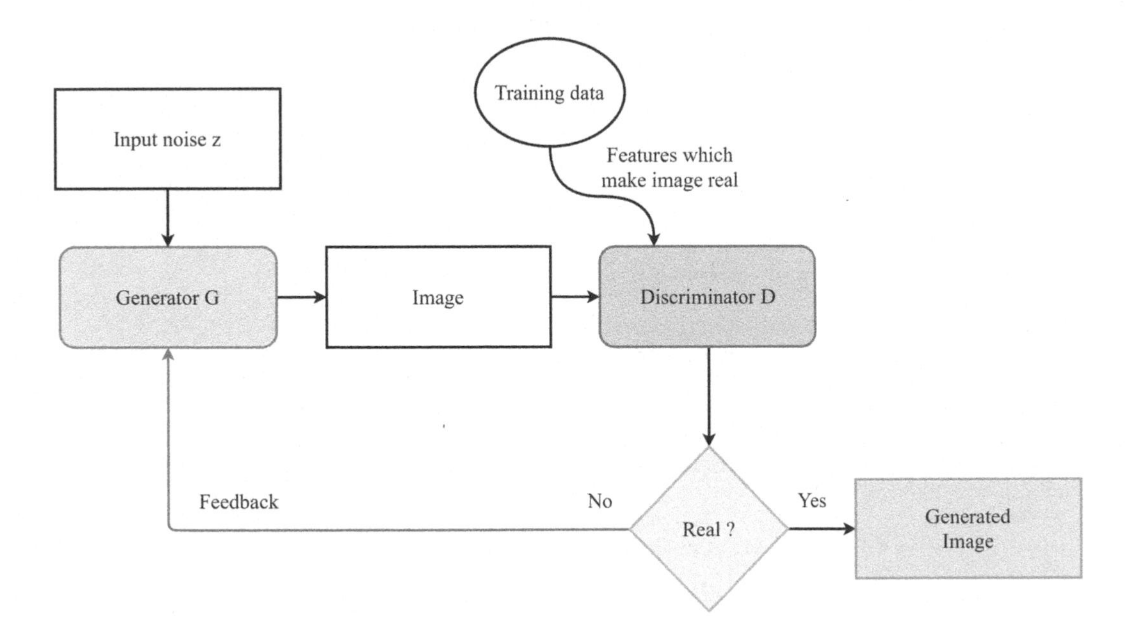

FIGURE 24.9 Flowchart of generative adversarial network.

For a discriminator, its objective is to maximize the probability to recognize the image as real or fake. Cross entropy $p\log(q)$ is used as an evaluation measure.

$$\frac{\max}{D} V(D) = E_{x\sim p_{\text{data}(x)}}\left[\log D(x)\right] + E_{z\sim p_z(z)}\left[\log\log\left(1 - D(G(z)\right)\right] \tag{24.38}$$

For generator, its objective function is,

$$\frac{min}{G} V(G) = E_{z\sim p_z(z)}\left[loglog\left(1 - D(G(z))\right)\right] \tag{24.39}$$

This increases the probability to fool the discriminator. The combined objective which is a minimax game is given as,

$$\frac{\min}{G}\frac{\max}{D} V(D,G) = E_{x\sim p_{\text{data}(x)}}\left[\log D(x)\right] + E_{z\sim p_z(z)}\left[\log\log\left(1 - D(G(z))\right)\right] \tag{24.40}$$

The generator is kept fixed and performs an iteration on the discriminator. Then, the discriminator is fixed and the generator is trained for the next iteration. Thus, the models are trained in an alternate

```
# Input noise for generator model
Z = tf.placeholder(tf.float32, shape=[None, 100])
#input image for discriminator model
X = tf.placeholder(img)
def generator(z):
  with tf.variable_scope("generator", reuse=tf.AUTO_REUSE):
    x = tf.layers.dense(z, 128, activation=tf.nn.relu)
    x = tf.layers.dense(z, 784)
    x = tf.nn.sigmoid(x)
  return x
def discriminator(x):
  with tf.variable_scope("discriminator", reuse=tf.AUTO_REUSE):
    x = tf.layers.dense(x, 128, activation=tf.nn.relu)
    x = tf.layers.dense(x, 1)
    x = tf.nn.sigmoid(x)
  return x
# Generator model
Gen-sample = generator(Z)
# Discriminator models
Disc-real = discriminator(X)
Disc-fake = discriminator(Gen-sample)
# Loss function
Disc-loss = -tf.reduce_mean(tf.log(Disc-real) + tf.log(1. - Disc-fake))
Gen-loss = -tf.reduce_mean(tf.log(Disc-fake))
# Selecting parameters
Disc-vars = [var for var in tf.trainable_variables() if var.name.
startswith("disc")]
Gen-vars = [var for var in tf.trainable_variables() if var.name.
startswith("gen")]
# Optimize
Disc-solve = tf.train.AdamOptimizer().minimize(Disc-loss, var_list =
Disc-vars)
Gen-solve = tf.train.AdamOptimizer().minimize(Gen-loss, var_list =
Gen-vars)
```

manner. GANs have proven to be applicable in a wide variety of domains and have proven one of the significant achievements in deep learning for the last decade. See the code below for training, optimizing, and generating data by a general adversarial network.

24.6.6 Deep Belief Networks

Before going to the deep belief network, let us first understand its important element: The restricted Boltzmann machine. RBM is an undirected graphical model that achieves state of the art in collaborative filtering. Many hidden layers can be learned efficiently by composing RBM using the feature activation of one and training data for the next. Unlike Autoencoder, which is a three-layer neural network, restricted Boltzmann machine is a two-layer neural network but instead of deterministic it uses a stochastic approach with a particular distribution (Figure 24.10).

From the diagram, it is clear that each node of the visible layer is connected to each node in the hidden layer. However, no nodes of the same layer are connected to each other. RBM has two biases: hidden bias and visible bias. Hidden bias helps to provide the activation to forward pass, while the visible bias is useful to learn the reconstruction on the backward pass.

In the first step, the weights of the hidden node will be multiplied with the input. Then, a corresponding bias is added to the vector multiplication which is given as,

$$\text{Activation } f\left(\left(\text{weight } w * \text{ input } x\right) + \text{bias } b\right) = \text{Output } a \tag{24.41}$$

Similarly, for multiple inputs, multiple outputs are generated using multiple activation functions.

Now in the reconstruction phase, activation of the hidden layer becomes the input and they are multiplied by the same weights. The sum of these products is added to visible bias, to each visible node, and the output obtained is the reconstruction. A deep belief network is visualized as a bundle of restricted Boltzmann machines wherein a hidden layer each RBM is a visible layer of the previous RBM above it.[20] Training of a deep belief network is given below:

1. The first RBM of the deep belief network is trained for accurate input reconstruction.
2. The hidden layer for the first RBM is the visible layer for the second RBM, and the output of the first RBM acts as an input for the second RBM.
3. The same process is repeated for each layer in the network.
4. In the deep belief network, each RBM layer learns the entire input. In other kinds of models, like the convolutional network, an early layer detects simple patterns and later layers recombine them.
5. Unlike other models like CNN where the first layers of the model detect simple patterns in the input which is then recombined by further layers, every layer in the deep belief network learns the entire input.

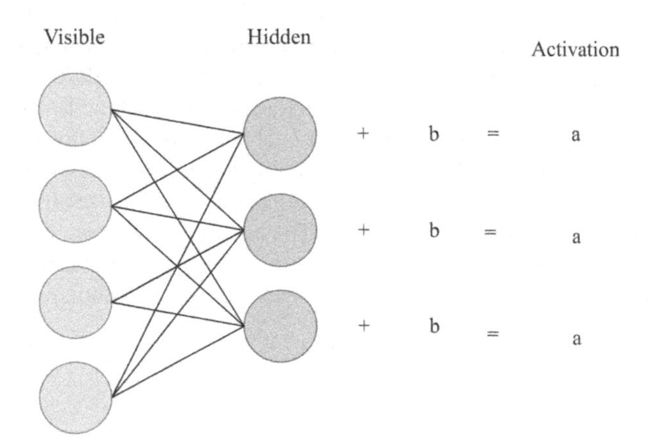

FIGURE 24.10 Restricted Boltzmann machine.

```
def Deep_belief_network(full-dataset):
    # constructing training and test data from the given full-dataset
    (train-x, test-x, train-y, test-y) = train_test_split(full-dataset.
data / 255.0, full-dataset.target.astype("int0"), test_size = 0.33)
    # Here input is of 784 units, 500 hidden layers and 10 output layers
    dbn-layer = DBN([train-x.shape[1], 500, 10], learn_rates = 0.3,
learn_rate_decays = 0.9,epochs = 10,verbose = 1)
    dbn-layer.fit(train-x, train-y)
    # Predicting the output using dbn-layer model
    y_preicton = dbn-layer.predict(test-x)
    # Comparing predicted output with test data
    compare-classification = (test-y, y_prediction)
```

TABLE 24.4

Comparison of Different Neural Networks

Characteristics	FeedForward Neural Network	Convolutional Neural Network	Recurrent Neural Network	LSTM	GAN	Deep Belief Networks
Primary usage	Storing information	Image recognition	Text-based learning	Text-based learning	Computer vision	Data classification
Parameter sharing	No	Yes	Yes	Yes	No	No
Vanishing gradient	Yes	Yes	Yes	No	Yes	Yes
Advantages	The network can generalize	Very powerful for extracting features on images	Useful on sequenced data	Remembers information for a long time	Powerful in the accurate distribution of input and then generate data	Powerful in image classification based on different images

Hence, it is clear that the stack of RBM will outperform a single unit, just like the multiple layer perceptron that outperforms the single perceptron.[21] The pseudocode below shows the implementation of a deep belief network.

The Table 24.4 below compares the different neural networks studied above along with their use cases and advantages.

24.7 Conclusion

In this chapter, the authors have explained the machine learning and deep learning algorithms and their advances that are currently being used by industries to achieve business goals, obtain significant insights, and promote innovation by developing new products and services. Machine learning and deep learning algorithms are fundamental units in technology that enable training machines to perform tasks that humans can do. Specifically, the algorithms in the domain of regression/classification, data regularization, decision trees, ensemble learning, and deep learning are studied. These algorithms are represented along with pseudocodes and real-world applications. The chapter is structured in a way that how algorithms are developed based on the predecessor. There is a discussion at the end of each section regarding the analysis and caveats of the algorithms. This encourages the reader to anticipate the different fields in which a particular algorithm could be applicable.

REFERENCES

1. Hoerl, Arthur E., and Robert W. Kennard. "Ridge regression: Biased estimation for nonorthogonal problems." *Technometrics* 42, no. 1 (2000): 80–86. Accessed August 27, 2020. doi:10.2307/1271436.

2. Tibshirani, Robert. "Regression shrinkage and selection via the lasso." *Journal of the Royal Statistical Society: Series B (Methodological)* 58, no. 1 (1996): 267–288.
3. Zou, Hui, and Trevor Hastie. "Regularization and variable selection via the elastic net." *Journal of the Royal Statistical Society. Series B (Statistical Methodology)* 67, no. 2 (2005): 301–320. Accessed August 27, 2020. http://www.jstor.org/stable/3647580.
4. Quinlan, Ross. "Induction of decision trees." *Machine Learning* 1 (2004): 81–106.
5. Breiman, Leo, Jerome Friedman, Charles J. Stone, and Richard A. Olshen. *Classification and Regression Trees*. CRC Press, Boca Raton (1984).
6. Salzberg, Steven L. "C4. 5: Programs for machine learning by J. Ross Quinlan. Morgan Kaufmann Publishers, Inc., 1993." (1994): 235–240.
7. Pandya, Rutvija, and Jayati Pandya. "C5. 0 algorithm to improved decision tree with feature selection and reduced error pruning." *International Journal of Computer Applications* 117, no. 16 (2015): 18–21.
8. Breiman, Leo. "Bagging predictors." *Machine Learning* 24, no. 2 (1996): 123–140.
9. Breiman, Leo, "Random forests." *Machine Learning* 45, no. 1 (2001): 5–32.
10. Freund, Yoav, and Robert E. Schapire. "A decision-theoretic generalization of on-line learning and an application to boosting." *Journal of Computer and System Sciences* 55, no. 1 (1997): 119–139.
11. Friedman, Jerome H. "Stochastic gradient boosting." *Computational Statistics and Data Analysis* 38, no. 4 (2002): 367–378.
12. Rosenblatt, Frank. "The perceptron: A probabilistic model for information storage and organization in the brain." *Psychological Review* 65, no. 6 (1958): 386.
13. Popescu, Marius-Constantin, Valentina E. Balas, Liliana Perescu-Popescu, and Nikos Mastorakis. "Multilayer perceptron and neural networks." *WSEAS Transactions on Circuits and Systems* 8, no. 7 (2009): 579–588.
14. Sazlı, Murat. H. A brief review of feed-forward neural networks (2006).
15. Gu, Jiuxiang, Zhenhua Wang, Jason Kuen, Lianyang Ma, Amir Shahroudy, Bing Shuai, Ting Liu et al. "Recent advances in convolutional neural networks." *Pattern Recognition* 77 (2018): 354–377.
16. O'Shea, Keiron, and Ryan Nash. "An introduction to convolutional neural networks." *arXiv preprint arXiv*:1511.08458 (2015).
17. Sherstinsky, Alex. "Fundamentals of recurrent neural network (RNN) and long short-term memory (LSTM) network." *Physica D: Nonlinear Phenomena* 404 (2020): 132306.
18. Hochreiter, Sepp, and Jürgen Schmidhuber. "Long short-term memory." *Neural Computation* 9, no. 8 (1997): 1735–1780.
19. Goodfellow, Ian, Jean Pouget-Abadie, Mehdi Mirza, Bing Xu, David Warde-Farley, Sherjil Ozair, Aaron Courville, and Yoshua Bengio. "Generative adversarial nets." *In Advances in Neural Information Processing Systems*, Montreal Convention Center, Canada, pp. 2672–2680 (2014).
20. Hua, Yuming, Junhai Guo, and Hua Zhao. "Deep belief networks and deep learning." *In Proceedings of 2015 International Conference on Intelligent Computing and Internet of Things,* Harbin, China, pp. 1–4, IEEE (2015).
21. Doshi, F., Doshi, P., Gandhi, J., Dwivedi, K., & Mangrulkar, Dr. R. (2020). "Image Modification using Text with GANs". *International Journal of Computer Applications Technology and Research*, 9, no. 11 (2020): 287–294.

25

Comprehensive Study of Failed Machine Learning Applications Using a Novel 3C Approach

Neel Patel, Prem Bhajaj, Pratik Panchal, Tanmai Prabhune, Pankaj Sonawane, and Ramchandra S. Mangrulkar
Dwarkadas J Sanghvi College of Engineering

CONTENTS

25.1 Introduction 403
25.2 Literature Review 405
25.3 3C Approach 407
25.4 Consolidation 408
25.5 Classification 411
25.6 Case Study and Failure Analysis 412
25.7 Results and Discussion 418
25.8 Conclusion 418
25.9 Future Scope 418
References 418

25.1 Introduction

The machines have had quite an aggressive evolving curve in the past few decades in many terms. They have evolved towards higher performance, smaller sizes and ability of computing even in distributed environments. Along with this very evolvement, the world has seen emergence and growth of technologies to which this evolvement was just arguable, a "right thing at right time" type of coexistence. One such technology is machine learning.

One of the initial definitions of machine learning was given as, "Field of study that gives computers the capability to learn without being explicitly programmed" by computer scientist Arthur Samuel in the field of AI. In simpler terms, it can be explained as automating and improving the learning process of a computer based on past experiences; i.e., there is no such need of programming or human assistance. A computer is said to be learning from experiences (E) with respect to some class of tasks (T) and performance measure (P), if its performance at T, as measured by P, improves with experience E. To generalize, any machine learning problem can be assigned to one of two broad classifications, supervised learning and unsupervised learning. If one understands these two types with some examples: clustering of the data which can be done by machine without any need of supervision that is unsupervised learning, it can also be used to identify any pattern(s) in the data without any label(s). In contrast, if we feed a machine with data and some expected output such an approach is called supervised learning. This type of learning is restricted to a limited set of output values; for example, the mail received is spam mail or not [1]. The first computer learning program, "Game of checkers" was written by Arthur Samuel in 1952 where the IBM computer improved at the game the more it played, studying the moves that made up winning strategies and incorporating those moves into its program. Further, psychologist Frank Rosenblatt designed the first neural network for computers which simulated the thought process of the human brain [2].

Since its inception, it has evidently excited various industries by giving them hope of achieving automation to such scales and complexities along with exhibiting some notion of humanness which was only possible in movies back then. ML has been observed to have penetrated diverse scientific areas such as bioinformatics, biochemistry, medicines, meteorology, economic sciences, robotics, food security and climatology. Interestingly, agriculture sector's environmental functions and its various related data processes can be understood and analysed using ML, since it is indeed growing on the lines of creating new opportunities there as well [3].

Intriguingly in 2017, the total number of machine learning deals reached 91 globally, with the entire business value of 16.9 billion dollars [4]. According to Deloitte's somewhat shocking prediction, 100,000 legal positions will be automated by 2036. As a result, automation and machine learning will make a huge difference in many human lives in multiple ways. Also, it will change the way we work in unprecedented ways. Following is the survey of companies at least $500M in sales who are benefited with ML brought such statistics [4]:

- 76% of leaders share their experience that they got higher sales growth by using machine learning. New technology helped them to predict better user preferences and behaviour, optimize processes, and lead upsell and cross-sell.
- More than 50% of enterprises are applying machine learning to refine marketing issues.
- 38% believe ML allowed them to gain better sales performance metrics.
- A few European banks have already improved new product sales by 10% while decreasing churn 20%. A recent McKinsey study figured out that a lot of European banks shifted from statistical modelling approaches to ML. It helped them to refine customer satisfaction.

All aforementioned advantages indicate that a wide spectrum of applications are available where ML can potentially solve critical problems. Although very useful & powerful, just like any other, ML has another side of it too. The ML systems that do not perform as per the expectations are called failed ML systems. Following is a list of somewhat classic factors that are critical to be solved to a considerable level so that ML ends up scaling up on the beneficial aspects of its applications and not the harmful ones:

- **Data-Related Issues**: Even at the rudimentary stages like data collection and preprocessing of any ML application development process, getting clean and useful data is itself a major challenge. Many of the times the data which is collected from different sources contains noise and is also not structured. There are chances that the data can be biased causing the result to be biased. For example, feedback data collected from the mass population might have a large number of random values which can impact the result drastically.
- **Increased Data Privacy Concerns**: As ML is all about data, a lot of data is exposed during the data collection phase(s) which solidifies the data privacy concerns. It is a popular belief among the civilians that their personal data is not quite safe on the Internet, the best example for which would be using a map application where we feed in current location and destination, which ultimately uncovers our daily movements.
- **ML Algorithm-Related Issues**: After data collection, the next step is to choose appropriate algorithms. To do so, initial tasks might involve analysing what the input is and what output is expected. For example, the task is to simply predict which team is going to win? Classification algorithms are a way to go whereas if the prediction has to be done on the score then Regression algorithms would be a best choice [5].
- **Algorithm Fooling**: As ML is used in various fields, the question arises how safe it is? Can these autonomous systems be hacked or manipulated in some way? If yes, then to what degrees?

Remaining chapter commences with a literature review section, followed by description of the approach chosen by the authors (3C approach), to explain the flaws present in various machine learning applications. The authors justify the use of such an approach by specifying its benefits and demonstrating how it can help identify the precise source of the error. Next, the authors move on to the classification scheme of the flaws by describing how exactly the flaws can be categorized, i.e. three levels of hierarchy: Services, Domains and Applications. Finally, actual case studies are consolidated and analysed using the aforesaid approach, to find out what exactly went wrong in each application.

25.2 Literature Review

North Carolina State University's Tam N. Nguyen in his paper focuses on challenges that can occur while developing ML applications w.r.t Software-Defined Networks (SDNs). The author has also provided few theoretical methods for attacking ML Models. He recommends ML network security designers to integrate some form of threat modelling into their SDN implementations so as to make them more secure [6]. Some other specific suggestions given in the paper are that the ML network security engineers should:

- Invest time on attack modelling or define threat model
- Work towards designing *audit-able* ML model(s)
- Plan and follow such a development processes that is secure
- Design operational cost modelling [6]

As per the author, peculiar issues with "ML-Based Security" in SDN include the following:

- Finding organic training data is a challenging task
- Between prototype and practical real-world deployment, there exists a semantic gap
- Varying normality problem (hard to find valid and invalid traffic flows since there exists enormous variability in input traffic)
- Cost of errors has to be measured and minimized
- Other evaluation problems [6]

Author Roman V. Yampolskiy, in his paper published in the year 2019, suggests that both the frequency of future failures and seriousness towards failures will steadily increase. He went on to classify AI safety into two parts based on the criticality of the damage they cause if failed. First classification is *narrow AI;* i.e., these AI applications do not result in catastrophic events if failed, and the second classification is *general AI*, which does cause fatal results in case of failure(s). The author rightly suggests that no system can be 100% secure. He also claims that his work was the first attempt to assemble a public data set of AI failures and should render extremely valuable to AI safety researchers. The paper studied the first few ML failures that were recorded or documented, to eventually find the cause(s) behind those. According to the paper, AI failure means those situations when the AI did not perform up to its expected level. Additionally, he has presented some interesting case studies where AI failed and given partial solutions that can be used to make AI safer. Definition of narrow AI w.r.t discussed paper is any program with accomplishing specific tasks and having at least one "if" statement, i.e. some decision-making aspect is present [7].

Researchers Elizamary de Souza Nascimento et.al in their paper shed light on the development phase problems that ML application developers come across. It studies 3 small companies' ML application development cycle and their challenges and suggest a checklist-based solution to partially solve the challenges if not eliminate them completely. Forexample, the provided checklists called CheckBM (checklist for business modelling) and CheckDP (checklist for data processing) are used to avoid missing out important considerations, tasks, etc in the "Identifying Business Model or Metric" & "Data Handling Phase" respectively. The authors emphasize a few specific considerations multiple times in their paper. According to them, the process that companies they studied used more or less the following procedure for ML application development [8]:

1. Problem Understanding
2. Data Handling

3. Model Building
4. Model Monitoring

The primary development phase problems that employees mentioned to the researchers are as follows[8]:

1. Business Metric Identification
2. No process standardizations
3. Database design is difficult

Moayad Alshangiti et al. in their paper presented the challenges faced by the ML developers. They have also discussed the increasing requirement of software developers and the cost required to acquire the required skill set. The research was conducted by studying and investigating a set of posts published on Stack Overflow web platform that were related to the ML application development process. In their analysis, they attempted to answer five main research questions (RQs). The RQs are as follows:

- RQl is regarding the difficulty of any particular ML-related question compared to other traditional development-related questions present on Stack Overflow.
- RQ2 tried answering whether there are an adequate number of ML experts who can answer ML-related queries on Stack Overflow?
- RQ3 tries to understand which specific ML application development phases often seem challenging to software developers.
- RQ4 attempted finding those ML subtopics which are more popular and challenging compared to other ML subtopics.
- RQ5 results in a discussion about the knowledge areas one needs to be well aware of in order to properly answer the issues developers face while using ML.

As a consensus of the results of overall investigation conducted on Stack Overflow posts and answers of RQs, the authors put forth the following conclusions: (i) ML-related queries on Stack Overflow are mostly left unanswered. (ii) No expert guidance on Stack Overflow is available for ML implementation-related queries. (iii) Mainly the challenges fall between data preprocessing and model deployment phases. (iv) ML implementation knowledge is what is required to address such kinds of issues rather than just conceptual understanding. Additionally, they provide an analysis table that shows all common types of questions that arise at various phases in any ML application development process. They have also stated that data preprocessing and manipulation, and model deployment are the two phases which get overlooked. They are found to be the main reasons for a majority of challenges faced by any developer or present in any application [9].

Author Georgios Mastorakis, in his paper, has attempted to find diverse ways of training models in the constantly evolving field of machine learning and at the same time attempted to connect the currently used methods to the ones available out there in nature, e.g. the way humans learn and animals learn. The author attempts to compare and contrast various training methods. Particularly, he also mentions some crucial limitations about the "data-driven training approach". One such limitation is that it works appropriately only when complete or adequate data is given whereas in reality data related to all possible scenarios is not always available. The paper focuses on two types of issues, i.e. related to data and algorithm. When it comes to data issues, the data availability issue is discussed by identifying that some of the other researchers suggest having more data for better performance whereas some say less is more. Also, issues like occlusions, sensor location and data quality have been addressed under the same category. In terms of algorithmic issues, specific and popular issues like

- selecting of right set of features
- classifier – one-fits-all approach
- learning is as good as the data is

have been delineated. Overall, the discussion about various approaches using which machines can learn proves to be a good source to identify the pros and cons of various training techniques. Additionally, it gives some ideas on how many different learning approaches can be found from observing nature as well [10].

Ram Shankar Siva Kumar et al. in their research discuss the failures on ML specifically due to adversarial attacks which can be done on either the algorithm or data used for training. The authors of this paper have further classified the failures into two different categories, (i) Intentional failures: essentially the attacks done by any means to deliberately produce the wrong output and (ii) Unintentional failures: these failures occur when the model is producing output which seems correct but is completely unsafe to proceed with, i.e. act of deceiving. Throughout the paper, the authors have discussed various types of attacks, their definition and examples. They have also structured the attacks, in a way they represent what among Confidentiality, Integrity, and Availability (CIA) is the attack targeting, the knowledge which is required to perform the attack, i.e. either a blackbox or whitebox [11].

25.3 3C Approach

The below Figure 25.1, i.e. an inverted pyramid structure, describes the way 3C approach is designed so as to scrutinize the reports that have articulated the ML failures and have even context of the applications come into the discussion since they play a pivotal role in any particular failure existing and in the way it is. The top block is the consolidation block, followed by classification block, and finally case study by case study, various ML failures have been studied and discussed.

- **Planning and Refining Goals**

  The authors of the chapter decided to lay a case study selection criteria or methodology after the classification step mentioned in Figure 25.1. The primary criteria should be such that the case studies getting selected are those that have had considerable impact on the lives of civilians either directly or almost directly. According to the methodology defined, the first step was to gather information from various sources that were relevant to the topic of Machine Learning Failures. The types of sources considered were as follows: news articles, blogs and forums, and technical papers and chapters.

  Major industries where machine learning has found its application were found out. Based on the industries selected, information from the above identified sources gathered initially was filtered out, so that the authors only have data about the selected industries. The industries chosen were unique since that would widen the scope of study to some extent.

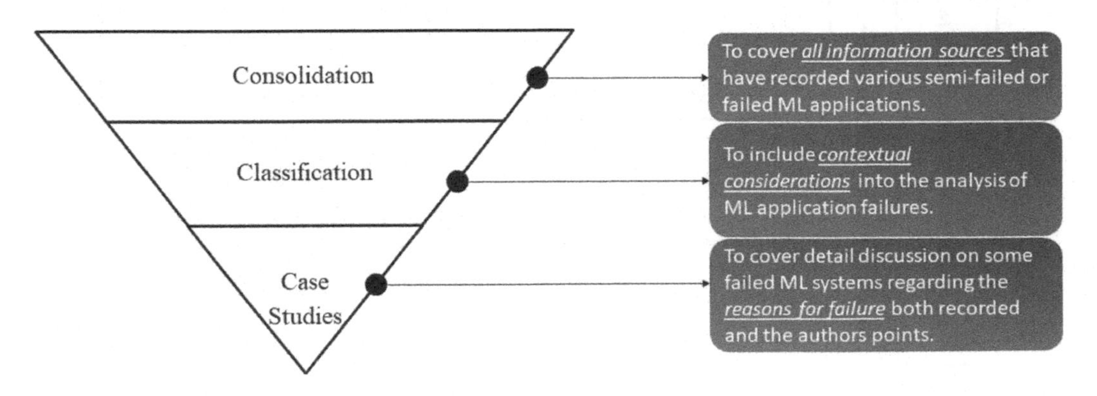

FIGURE 25.1 3C approach of studying failed ML applications.

25.4 Consolidation

Category of sources is mentioned in Figure 25.2. The authors worked towards studying various sources of data related to occurred or predicated ML failures and their speculated as well as justified reasons. The actual criteria of the selection are mentioned and discussed in detail in the below section.

- **Criteria of Selection**
  A set of parameters have been identified to be considered for selection. If all the parameter values were satisfied, then the case study was selected for detailed analysis in the consequent "Case Study" section, whereas other case studies that also more or less fulfilled the parameters are categorically analysed for any pattern or association being present or not w.r.t failing of that ML application below in this section (Figure 25.3).

The selection criteria covered the following metrics and corresponding accepted value sets:

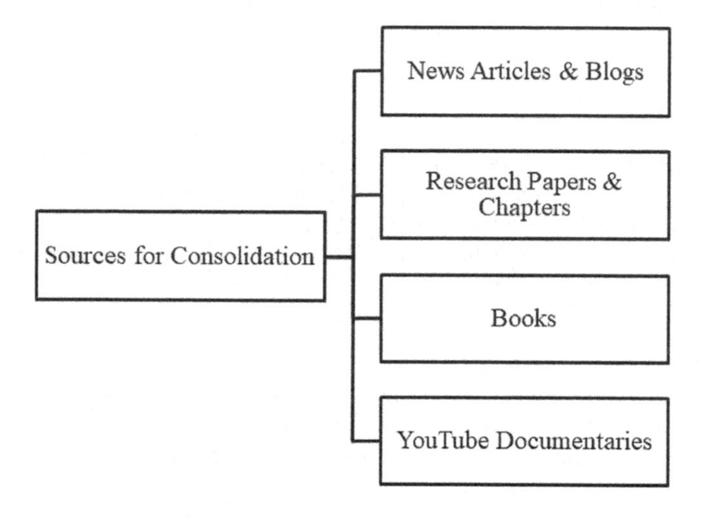

FIGURE 25.2 Sources of information for consolidation of ML failed application.

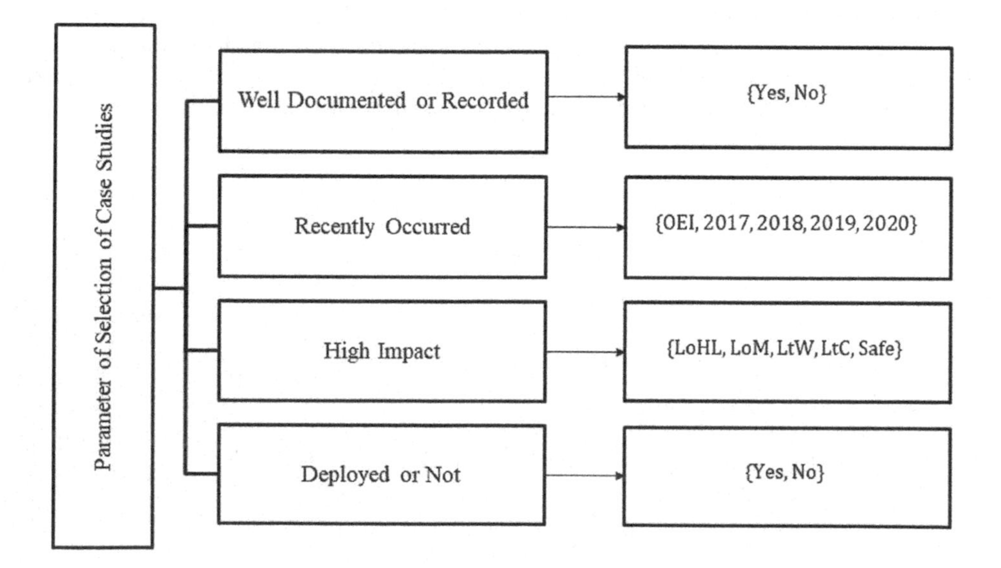

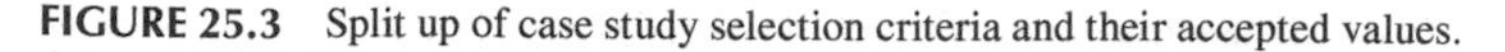

FIGURE 25.3 Split up of case study selection criteria and their accepted values.

a. **Parameter 1**: Well Reported or Documented
   - *Corresponding Value Set*= {Yes, No}
   - **Detail Criteria***:* The case study which has at least 2 news articles from well-known sources and lesser-known one written about the same. Also, it was examined whether the case study being looked upon has already been analysed in the past or not and if yes, then to what degree. The methodology of analysis was not restricted to any specific one, i.e. news article and blog-based analysis. As a result, the aforementioned considerations were done and each case study was checked for whether it is well documented or not and the ones with a "Yes" for "well documented" parameter were checked whether they satisfy other criteria or not.

b. **Parameter 2**: Recently Occurred
   - *Corresponding Value Set*= {OGEI,2017,2018,2019,2020}
     - OGEI: Old but Good Enough to Include
   - **Detail Criteria**: While studying ML failures, it is crucial to also have currency of data-related failed applications brought into consideration. Since the field is evolving very rapidly, the problems that were present in the initial phases of ML's emergence have now been partially solved if not completely. Hence, it is important to take into account the recent ML application failure-related case studies. In other words, the notion being followed is essentially that, the older the case study is, the higher the chances are that the error or failing point that was found to exist is now either obsolete or analysed and solved to a considerable extent. Hence, the lower bound was kept to 2017. Even then, the case studies dated in before 2017 were considered only if they were high on other criteria and hence were ending up being impactful.

c. **Parameter 3**: Level of impact to human lives directly and indirectly
   - *Corresponding Value Set*= {LoHL,LoM,LtW,LtC,Safe}
     - **LoHL**: Loss of Human Life
     - **LoM**: Loss of Money
     - **LtW**: Leading to War
     - **LtC**: Leading to Cultural/Community Related Bias or Problems
   - **Detail Criteria**: Motivation behind such criteria is the understanding that no matter which engineering method or technology is trying to attempt to help individuals, group of people, organization, or nation, in various aspects of their existence, it should act as influence for things that would catastrophes like war or loss of human life. Some might argue that technologies that are fed to do specific things do that only, and it is illogical to correlate them to other phenomena and say that they influence any other thing like war or pandemic or any other highly catastrophic event. But, when it comes to a technology like machine learning, two points come into considerations; firstly, it evidently exhibits a dynamic nature in terms of its output, and secondly, majority of ML applications, at least in the current times, do not have any ethical filter using which they are learning keeping ethical and moral rules into consideration. As a result, the authors identified few such factors that ML should be made to handle since there are concepts like equality and safety related to human lives that, ideally, no technology should hamper.

d. **Parameter 4**: Deployed Project or Not
   - *Corresponding Value Set*= {Yes, No}
   - **Detail Criteria**: This binary criterion checks whether the case study is related to a product that was deployed and accessed by end users or not. For such a case study that is related to a deployed application, the value for this parameter was made "Yes", otherwise "No". This parameter is very crucial, since deployed ML product(s) with a failure would be even more disastrous than the ones that were in their experimental environments only (Tables 25.1 and 25.2).

TABLE 25.1

ML Services to Be Referred to in the Table 25.3 Have Been Assigned Symbols for Ease of Understanding

No.	ML Service Name	Symbol
1	Classification	C
2	Image recognition	IR
3	Data evaluation	DE
4	Diagnosis	D
5	Prediction	P
6	Learning associations	LA
7	Face recognition	FR
8	Extraction	E
9	Natural language processing and generation	NLPG

TABLE 25.2

ML Application Sector to Be Referred to in the Table 25.3 Has Been Assigned Symbols for Ease of Understanding

No.	Sector Name	Symbol
1	Crime and security	C&S
2	Automobile	AMB
3	Healthcare	HC
4	Recruiting	RC
5	Personal assistant	PA
6	Media	ME
7	Chatbot	CB
8	Robotics	RB
9	Smart home	SH
10	Advertising	AD
11	Sports	SP
12	Instant messaging	IM

The analyses are as follows:

1. **Personal Assistants**

   As the data populated in the Table 25.3, clearly indicates that the failure rate in Personal Assistants (PAs) is high compared to other domains (or sectors). But, if we considered "severity of failure" as a consolidation factor, the failures reported in PA devices are comparatively less harmful to human lives. In other words, Personal Assistants are less likely to directly cause life threats. One inference can be drawn in this case is, since PAs are widely spread among almost all end users (or masses) is the reason that they reported more failures and that they are too quiet and diverse in nature.

   Some failures also cost financial loss to the user, which later got refunded the company in charge [22], whereas one case study showed that the Google AI messed up a user's photo that was clicked on a snowy mountain [16], and such things could be avoided to some extent and were evidently not life-threatening. They might have been mood spoiling for that person though. But undoubtedly, such a consequence is less harmful compared to getting hit by a self-driving car. Evidently, some failures affected the user privacy [21]; e.g., in one scenario

TABLE 25.3

ML Case Studies

No	Company/ Institution/ Organization	Case Study	Domain	ML Services
1	Chinese Gov	Chinese billionaire's face identified as jaywalker [12]	C&S	C
2	Uber	Self-driving car kills a pedestrian [13]	AMB	IR, C, DE
3	IBM	IBM Watson comes up short in healthcare [14]	HC	D, P
4	Amazon	AI recruiting tool is gender biased [15]	RC	P, LA
5	Google	Google photos confuses skier and mountain [16]	PA	LA, C
6	Microsoft	AI Chatbot corrupted by Twitter trolls [17]	ME	LA, E
7	Apple	Face ID defeated by a 3D mask [18]	C&S	FR
8	Facebook	AI smart bot unintentionally spreads hate [19]	CB	NLPG
9	Amazon	Alexa brings the party with her in Germany [20]	PA	LA, E
10	Google	Google Home Minis spied on their owners [21]	PA	LA, E
11	Amazon	Alexa orders all the dolls houses [22]	PA	LA, E
12	Hanson Robotics	Sophia says she wants to destroy humans [23]	RB	LA, E
13	Amazon	Face recognition falsely matched 28 members of congress with mugshots [24]	C&S	LA, C
14	Google	Google Home 100% failure rate reported [25]	PA	LA, E
15	Keolis	A self-driving shuttle got into an accident [26]	AMB	IR, C, DE
16	LG	Robot Cloi gets stage fright at its unveiling [27]	SH	LA, E
17	Facebook	Enabled advertisers to reach "Jew Haters"[28]	AD	C, DE
18	Unanimous A.I.	AI misses the mark with Kentucky Derby predictions [29]	SP	P, LA
19	Google	Offensive chat app responses highlight AI fails [30]	IM	P, LA

Google Home Mini continuously recorded all the conversations which were happening in the room.

2. **Crime and Security**

   "Crime and Security" related three failures that were reported and studied also raised privacy concerns. The first one is where the Apple iPhone face ID-based phone lock completely bypassed the facial recognition system [18]. However, this failure was comparatively less severe because of physical access control still being organically present to some extent. However, the other two cases were such that it could damage an individual's or organization's reputation. It was found that in the other case, a Chinese government deployed traffic law-related ML application falsely identified a rule breaker as one of the public figures in China [12] and matching anyone with a mugshot data set [24].

3. **Automobile**

   If we look at the automobile domain, then there are two failed cases the authors have analysed, out of which one was not deployed. Having said that, as per analysis w.r.t severity of the damage they cause, each application in this domain is directly linked to life-threatening consequences. Hence, here the ML failure might cost the loss of human lives, recovery of which is impossible [13].

25.5 Classification

The authors have included this step to also include in the discussion, contexts of various ML-based failed and semi-failed applications. It is crucial to understand whether any failing point had its root cause as any feature or attribute that was exclusively present in that particular application domain (e.g. medical,

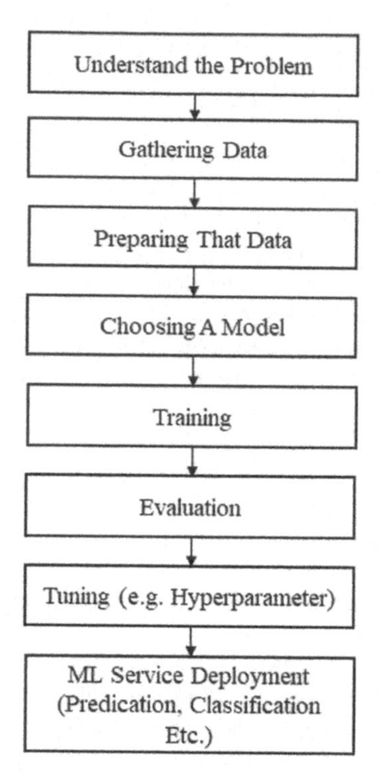

FIGURE 25.4 ML application development phases.

crime and security, automobile, etc.) and absent when the same technique is applied in another domain(s). [31]. Essentially, the goal is to find whether the failing point was domain independent or dependent. A hypothetical example ofa specific failing point in medical-related ML applications can be: not involving human expertise i.e. doctors in the learning process of a machine, which will filter out any adversarial or biased training data involved.

This step is the simplest among the 3C's methodology, but it is not quite uncommon to see developers ignoring the context-level dependencies that the "domain of application" brings in. While in the subsequent sections, the authors have presented the case study breakdowns keeping in consideration the domain in which they have classified, e.g. health care or crime or any other.

The results of involving contextual aspects in the study have been evidently effective. This seems to be true because data science rooted applications (e.g. machine learning and big data analytics) tend to have considerable influence on their domains. Even the data science industry as a whole is gradually seen to become more specific towards the fields they are being used for. In other words, people who have experience in applying machine learning in the healthcare domain need not get an instant job in the crime domain. This is because the application is not just about the algorithms being used but largely about various aspects of data involved.

25.6 Case Study and Failure Analysis

Figure 25.4 shows the generalized ML application development process that is assumed to be used by the case studies elaborated below. The assumption is that, if not exactly same, largely the same procedure would be followed by all ML application developers. Reason for the assumption is the fact that this process is very close to the process described by the ML engineers in seminars, blogs and papers [9,32,33]. Below mentioned are four selected case studies that are further explored for finding and discussing the failing point and related aspects.

Case Study 25.1: Facebook Targeted Advertisement

Facebook allows the world's largest social network enabled advertisers to direct their pitches to the news feeds of a category of people. Whatever users describe in their profile appears to advertisers as probable groups of potential customers to which ads could be targeted so as to somewhat guarantee increase in sales. In 2017, Facebook allowed advertisers to target a particular group of people who showed common interests such as "History of 'why Jews ruin the world,' " "Jew hater", or "How to burn Jews". It did not only let such anti-Semitic categories be created but when tested with promoted posts if such ad categories were real it approved the ads within 15 minutes. According to Facebook, these categories are autogenerated by algorithms and have no human involvement. The company assured that it will work to fix this issue and also limit the number of categories available to advertisers or scrutinize the ads before they are displayed to buyers [19] (Figure 25.5).

FAILURE ANALYSIS

According to the authors of the chapter, the system should have been trained to exclude such anti-Semitic categories, which was not the case here. If the training phase had been designed to exclude such categories, then there are high changes it would have been possible to avoid spreading hatred against people belonging to certain communities, races, showing certain interests, etc. Hence, it was a critical flaw in the training phase of the ML system development process. Alternatively, another issue that can

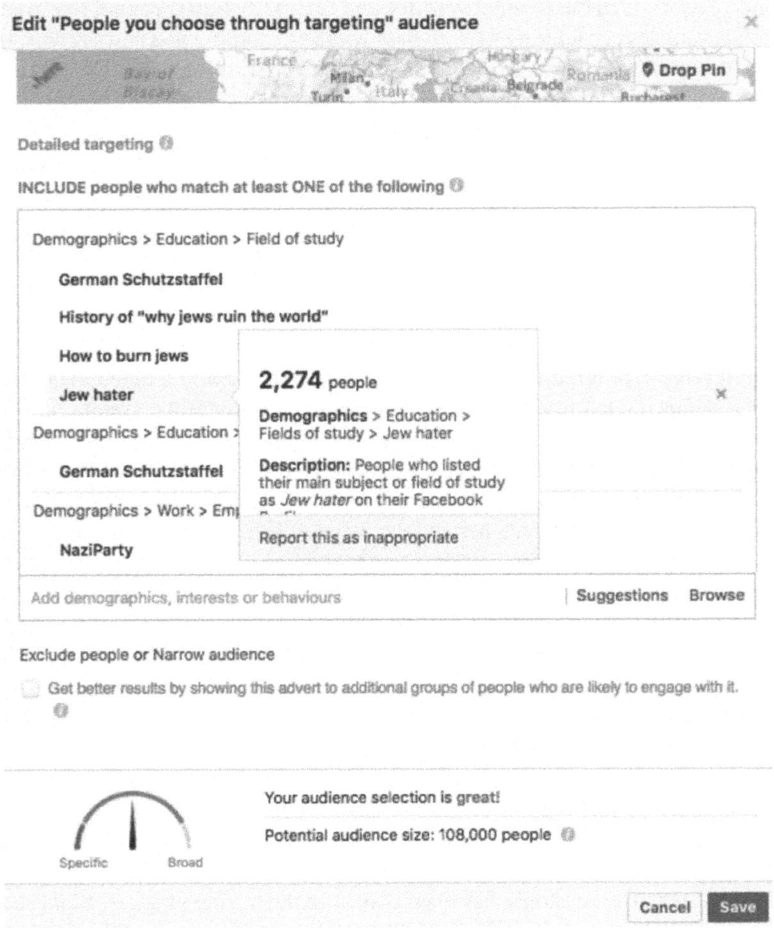

FIGURE 25.5 Screenshot of the screen showing above-mentioned anti-Semitic group creation [19].

be highlighted is that the model was not tested properly against inputs that can potentially be unethical causing certain communities or races to be subjected to hatred. This points out the fact that the testing phase was not done adequately considering all possible scenarios of the input. Therefore, this shows negligence in the evaluation phase as well. If the evaluation was done in an ideal manner, improvement, such as taking bias removal measures to leave out certain categories, would have avoided the problem to some extent or rather temporarily. A possible corrective measure could be applying a scrutinizing process to remove categories like anti-Semitic ones, before displaying it to the users. This is the same approach that Microsoft's popular chatbot "Zo" was acquainted with. The chatbot used a filter to ignore posting the messages that revolved around religion and politics [17]. The authors assert that such a solution is actually like the following analogy, "If one's kidney doesn't work, removing it can be a solution rather than treating it". It is because still the ML application is not in a position that it would not generate such messages, posts or advertisements altogether rather just suppress it from coming in front of audiences.

Case Study 25.2: Uber Autonomous Car

In March 2018, an autonomous Uber car, Volvo XC90, hit and killed Elaine Herzberg, a 49-year-old woman on a street in Tempe, Ariz, despite having an emergency backup driver. On March 18 at around 10 pm, Elaine was crossing the street with her bicycle when the car speeding at over 40 mph struck her leading to a catastrophe. The preliminary investigation showed that the car did not slow down before smashing into her, even though the weather was dry and clear [13]. Furthermore, the street was ideal for testing autonomous vehicles considering its sufficient width. According to a news report, the self-driving car was not programmed to respond to pedestrians found jaywalking [34]. The autonomous vehicle detected Elaine 5.6 seconds before the dash but repeatedly kept misclassifying her as other objects such as a bicycle and vehicle and estimated the trajectory/path of the recognized object each time it reclassified her. As a consequence, it was rendered incapable to precisely predict the actual path of the detected object. The SUV finally classified Herzberg as a woman a mere 1.3 seconds before running into her and concluded it must brake [35,36]. However, since the present automatic emergency braking system of the car was disabled by Uber before testing started, manual intervention was needed to stop the SUV. The backup driver failed to take control over the car in a second's interval, leading to the catastrophe.

FAILURE ANALYSIS

Since the system was not programmed to detect jaywalkers, it majorly seemed to be a training-related issue, as the car was not trained to deal with such a situation. Though the system had detected Elaine Herzberg around 5.6 seconds before the crash, it repeatedly kept misclassifying her as other objects and then accordingly kept interpolating the object's trajectories. There can be various reasons why the ML system misclassified Elaine, one of which could be that she was crossing the street with her bicycle, and the vehicle could not distinguish between her and the bicycle clearly, as both appeared together. Another potential reason could be the orientation of Elaine and the bicycle. Machine learning models often find it difficult to classify objects when the orientation of the objects changes, sometimes even causing the model to not recognize the objects at all or with very low confidence. This issue of orientation is called the Object Orientation (pose) problem [35]. Tracing to the roots, the authors are of the opinion that a suitable model was not chosen. Additionally, since the system misclassified Elaine Herzberg multiple times, the authors believe that the model was not trained with an adequate amount of possible scenarios and data for each scenario. Hence, it is highly likely to be a flaw present in the "training phase". Despite deciding to apply brakes 1.3 seconds prior to the fatal crash, it could not do so due to the disabled SUV's default automatic emergency braking system, which means the brakes had to be applied manually by the operator. However, due to lack of attentiveness, the operator could not apply the brakes [34]. This was an issue in the evaluation phase as the testing environment set-up for testing was improper from a human safety point of view. Enough care should have been taken in the testing phase to avoid such an accident, since it is not new for ML to go wrong in ways the developer would never imagined it to.

Case Study 25.3: Amazon Automatic Resume Filtering

In 2014, Amazon.com, Inc. built an AI-powered system to look at a miscellany of resumes and identify the best candidates. The intention was to develop a system that could quickly crawl through the web and recognize the candidate's worth recruiting. The tool allotted scores to each candidate, from a range of one to five stars. To train the system, the company fed the system its database containing resumes of applicants they had received in the past 10 years. The tech team made over 500 computer models based on occupational roles and locations. The team taught each model to recognize over 50,000 words that were commonly found on past candidates' resumes. However, in 2015, Amazon came to know that its new AI-powered system was rating candidates for certain technical jobs in a gender-biased way. When investigated, the root cause for the same was that the computer models were trained to choose candidates by observing patterns in resumes submitted to the company over a decade, containing many resumes belonging to male candidates [15]. The phenomenon of the data set containing more men than women's CVs is described as Class Imbalance in data science terms [37,38]. Having too little data from a specific demographic can seriously be a hindrance to a machine learning model [37,38]. Consequently, the AI system taught itself that male candidates were desirable. For instance, it undermined resumes that contained the word "women's", as in "women's chess club captain". [15] Similarly, it also penalized graduates of two women's colleges [15]. Also, the technology preferred applicants who described themselves using verbs more commonly found on male candidates' resumes.

FAILURE ANALYSIS

Since most of the resumes provided to the system were of male candidates, this led the system to find an implicit and unwanted pattern that suggested it was preferable to select male candidates over female candidates, leading to a biased outcome. This was a blunder made in the first stage, i.e. gathering the data for training the model. If the data gathered had a proper ratio of male: female, or even if the resumes belonged to one demographic, i.e. either male or female but not both, then the model would not have learned such a pattern.

Possibly, the data being fed into the system could be a result of the biased mindset of the CV reviewer or HR manager, which again points out a flaw in gathering the data, as the source selected was not unbiased [37]. Also, while testing the model to check its accuracy, such flaws should not have been missed by the eye. If the prior mentioned mistake renders to be true, it would not be wrong to conclude that the mistake was made in the evaluation phase.

Corrective measures such as hyperparameter tuning and cross-validation could have been implemented if the issue was figured out in the evaluation phase. A possible solution that some authors propose could be to boost the score of the discriminated gender to correctly train the model on the biased data [37]. Even if the authors zoom out of this specific application to others like this where the data set is rendering the output as biased towards one class of people, they come to know that this specific problem has many aspects to it.

Few possible facets of the problem can be as follows:

- Data volume-related problem
- Lack f data available for the class that is facing issue(s) due to bias (e.g. male dominance in the current case study)
- Problem of expecting ideal output, which is practically impossible to achieve

The nature of that situation or application is such that bias is real and already present in the system (e.g. among males and females, males are given preference for say military personnel jobs). As a result, the data that it is generating is anyways replicating the real-world scenario, and hence, neutralizing it might render the ML application irrelevant, unrealistic or even worse, harmful. The point is, the humanity was not always free from problems like discrimination, but it evolved over time. Hence, to feed the data of the problematic past to ML applications for training organically carry forward those problems. In such cases, data currency should be considered at the same time the above-discussed point should be kept in mind.

Case Study 25.4: Automatic Jaywalker Recognition Failure

A facial recognition (FR) camera falsely displayed a famous businesswoman, Dong Mingzhu, president of an air-conditioning firm, for jaywalking after its AI system detected her face crossing an intersection [12]. The incident took place in 2018, in Zhejiang, south of Shanghai. The face displayed was captioned with the false allegation that she had broken the law. It also listed her name and a part of her government ID number, though misidentified her last name as "Ju" [12]. The AI-powered system operates with a camera designed to trigger when anybody passes through the crossing during a red signal. The camera captures a photo of the rule breaker's face, then sends it to the big screen and matches it with the records in the police database [39]. The system counts how many times an offender has violated traffic rules in the past. However, in reality, the FR camera picks an advertisement showing the woman's face on the side of a bus. The police soon acknowledged that identifying the woman as an offender was an error made by the FR system and also stated that the issue had now been resolved through an upgrade. This incident proved that the AI-powered system can get tricked in multiple known and unknown ways rendering its operations towards producing false outputs (Figure 25.6).

FAILURE ANALYSIS

The authors are of the impression that the system was not designed to check whether the detected face was 3-dimensional or 2-dimensional. For example, checking whether coordinates of the facial features are in the same 2-dimensional plane or not. In other words, is there a notion of depth in the face image or not. If so, then the face is just an image and not a real face. This seemed to be an issue in the model selection phase. The model was not designed to capture parameters indicating the dimensionality of the detected face. If the system was designed to detect just 3D faces, there is high probability that this issue would not have arisen. Also, there could be a possibility that the testing team did not evaluate the system by providing 2D images as input, suggesting that there was a potential flaw in the evaluation phase as well. If the evaluation had been carried out considering all the possible inputs, the issue would have been found out before deploying the system, and corrective measures could have been taken to resolve the issue.

The authors also think that the system should have been trained to identify other characteristics of a human body such as hands, legs, walking speed, and movement, as well, and thereby not fully relying only on the face. The reason for this is that the picture on the vehicle contained just a face, and no other body part. Also, the image was still and showed no signs of movement. Furthermore, the speed of the bus could have been faster than the usual walking speed of humans. All these considerations would have aided the system to not mistaken images for actual human beings. Even techniques like liveness detection could render efficient solutions to check whether the face recognized is an image or real person [40].

FIGURE 25.6 Jaywalker detection-related ML failure's CCTV footage [12].

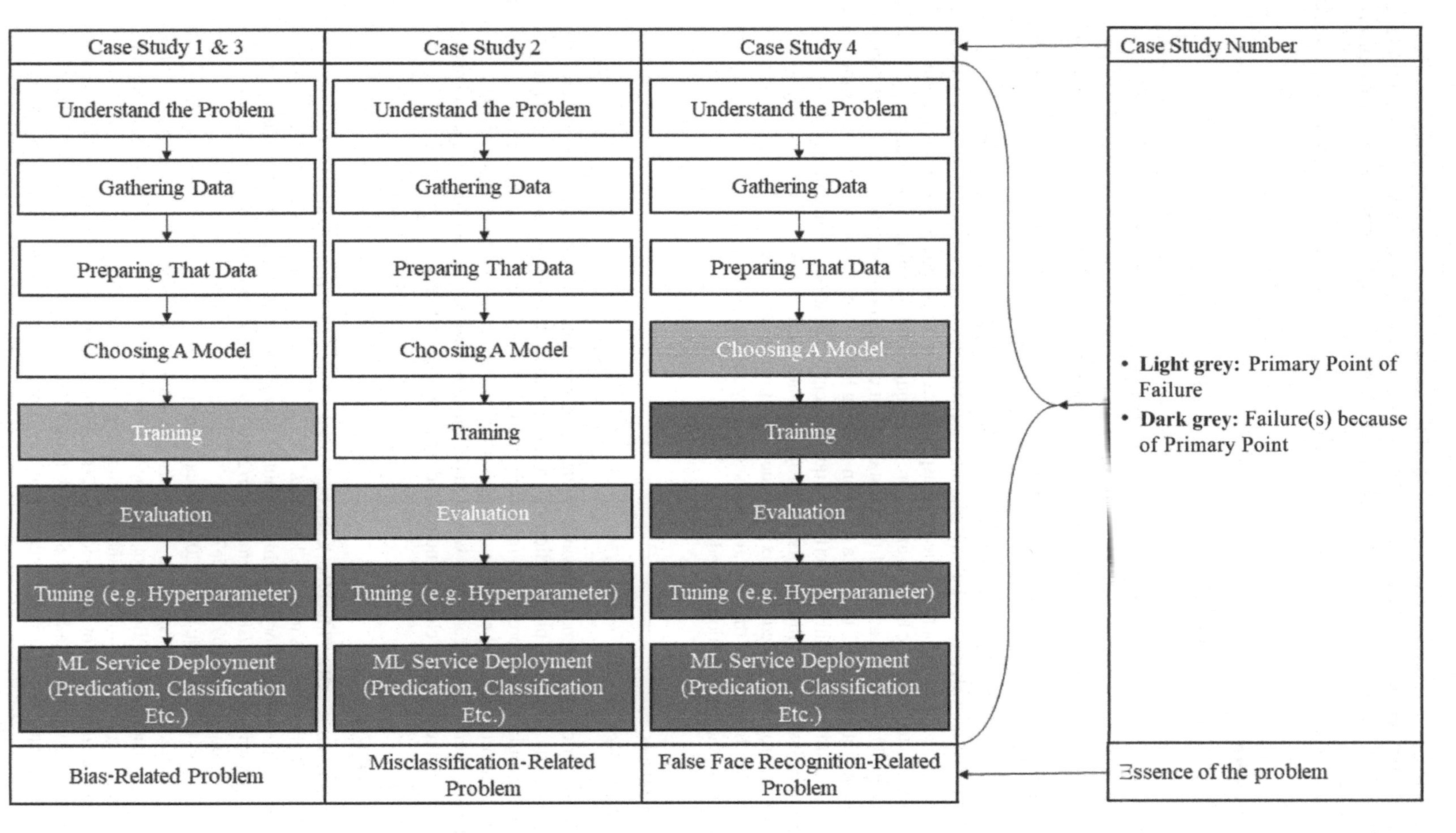

FIGURE 25.7 A collective analysis as to where the failing point could have most potentially existed.

Moreover, the issue that the detected face's identity was not mapped correctly implies that there was an issue in the training phase as well. The model was not trained sufficiently to identify the jaywalkers' faces with high accuracy. Specifically, this case study showed multiple failing points, some primarily ML related but even others.

25.7 Results and Discussion

Figure 25.7 shows the failing points of the case studies as per analysis. The light grey blocks are identified to be highly likely the primary failing point, and then, dark grey blocks will as a ripple effect render problematic when say the trigger or case that it is not trained for occurs. In the figure, the steps of ML application development process are viewed to find the points that contributed towards the failure of ML application.

25.8 Conclusion

Machine learning is evolving day by day. Machine learning just like every other emerging technology has two sides to it. One of course is the good aspect; i.e., it is helping to solve problems that involved computers having to think like humans and where it is required for machines to learn things on itself and act accordingly. This renders as one of the most efficient and accurate way when otherwise one would have chosen an "intuitive way" to code a decision-making logic. Hence, we put forth the failing points of various well-documented, highly impactful and quite recent applications of machine learning. We used a 3C approach to classify the failing points to appropriate buckets. The main reason for doing so was to be able to apply appropriate remedies to fix the issue(s). Following categories of case studies were selected and analysed for finding out the primary and secondary failing points: bias related, misclassification related and false face recognition related.

25.9 Future Scope

The 3C approach can be extended towards having a more framework like structure for analysing and finding out exactly where the problem lies with a set of steps to be followed by the ML engineers. Such a framework would help improve the resilience in ML applications being developed. In case such a framework is appropriate and effective enough for being a de facto failure analysis framework among ML engineers, then in those cases this would make the communication related to failures between ML professionals and researchers clear, standardized, easy and extensible.

REFERENCES

1. Royalsociety.org. 2017. "Machine learning | royal society". [Online] Available: https://royalsociety.org/topics-policy/projects/machine-learning/.
2. B.Marr,2020."Ashorthistoryofmachinelearning:Everymanagershouldread."Forbes.[Online]Availableat: https://www.forbes.com/sites/bernardmarr/2016/02/19/a-short-history-of-machine-learning-every-manager-should-read/#2093f95a15e7.
3. Technostacks Infotech Pvt. Ltd. 2020. "How can machine learning helps in agriculture industry: ML apps in agriculture?" [Online] Available at: https://technostacks.com/blog/machine-learning-in-agriculture/.
4. Hackernoon.com. 2020. "Benefits of machine learning for your business." [Online] Available at: https://hackernoon.com/benefits-of-machine-learning-for-your-business624c7297a3af.
5. K. Nimavat, and Society of Data Scientists, 2020. "What algorithms can be used to predict the outcome of a cricket match?" Data Science Stack Exchange. [Online] Available at: https://datascience.stackexchange.com/questions/11159/what-algorithms-can-beused-to-predict-the-outcome-of-a-cricket-match.

6. T. N. Nguyen, 2019. "The challenges in ML-based security for SDN," *2018 2nd Cyber Security in Networking Conference (CSNet)*, pp. 1–9. doi:10.1109/CSNET.2018.8602680.
7. R. V. Yampolskiy, 2019. "Predicting future AI failures from historic examples," *Foresight*, vol. 21, no. 1, pp. 138–152. doi: 10.1108/FS-04-2018-0034.
8. E. De Souza Nascimento, I. Ahmed, E. Oliveira, M. P. Palheta, I. Steinmacher, and T. Conte, 2019. "Understanding development process of machine learning systems: Challenges and solutions," *International Symposium on Empirical Software Engineering and Measurement*, vol. 2019, pp. 1–6, doi: 10.1109/ESEM.2019.8870157.
9. M. Alshangiti, H. Sapkota, P. K. Murukannaiah, X. Liu, and Q. Yu, 2019. "Why is developing machine learning applications challenging? A study on stack overflow posts," *International Symposium on Empirical Software Engineering and Measurement*, vol. 2019. doi: http://10.1109/ESEM.2019.8870187.
10. G. Mastorakis, 2018. "Human-like machine learning: Limitations and suggestions," pp. 1–24, [Online] Available: http://arxiv.org/abs/1811.06052.
11. G. Li, P. Zhu, J. Li, Z. Yang, N. Cao, and Z. Chen, 2018. "Security matters: A survey on adversarial machine learning," vol. 07339. [Online] Available: http://arxiv.org/abs/1810.07339.
12. L. Dodds, 2020. "Chinese businesswoman accused of jaywalking after AI camera spots her face on an advert." The Telegraph. [Online] Available: https://www.telegraph.co.uk/technology/2018/11/25/chinese-businesswoman-accused-jaywalking-ai-camera-spots-face/.
13. D. Wakabayashi, 2018. "Self-driving Uber car kills pedestrians in Arizona, where robots roam", Nytimes.com.[Online] Available: https://www.nytimes.com/2018/03/19/technology/uber-driverless-fatality.html.
14. C. Farr, 2017. "Don't let IBM's blunders in health care blind you to the huge benefits of technology in medicine", CNBC. [Online] Available: https://www.cnbc.com/2017/09/07/ibms-struggle-with-watson-in-health-care-doesnt-mean-tech-cant-help.html.
15. D. Meyer, 2018. "Amazon killed an AI recruitment system because it couldn't stop the tool from discriminating against women" Fortune. [Online] Available: https://fortune.com/2018/10/10/amazon-ai-recruitment-bias-women-sexist/.
16. M. Zhang, 2018. "Google photos' AI panorama failed in the best way", PetaPixel. [Online] Available: https://petapixel.com/2018/01/23/google-photos-ai-panorama-failed-best-way/.
17. K. Leetaru, 2016. "How Twitter corrupted Microsoft's tay: A crash course in the dangers of AI in the real world", Forbes. [Online] Available: https://www.forbes.com/sites/kalevleetaru/2016/03/24/how-twitter-corrupted-microsofts-tay-a-crash-course-in-the-dangers-of-ai-in-the-real-world/#673e84426d28.
18. T. Brewster, 2017. "Apple Face ID 'Fooled' By $150 Mask -- But Big Questions Remain", Forbes. [Online] Available: https://www.forbes.com/sites/thomasbrewster/2017/11/13/apple-face-id-hacked-by-vietnamese-researchers-mask/#6746aaa04987.
19. J. Novet, 2017. "Facebook AI researcher slams 'irresponsible' reports about smart bot experiment", CNBC. [Online] Available: https://www.cnbc.com/2017/08/01/facebook-ai-experiment-did-not-end-because-bots-invented-own-language.html.
20. J. Huggler, 2017. "Alexa, Nein! Police break into German man's house after music device 'held party on its own'", The Telegraph. [Online] Available: https://www.telegraph.co.uk/news/2017/11/08/alexa-nein-police-break-german-mans-house-music-device-held/.
21. S. Burke, 2017. "Google admits its new smart speaker was eavesdropping on users", CNNMoney, [Online] Available: https://money.cnn.com/2017/10/11/technology/google-home-mini-security-flaw/index.html.
22. A. Liptak, 2017. "Amazon's Alexa started ordering people dollhouses after hearing its name on TV", The Verge. [Online] Available: https://www.theverge.com/2017/1/7/14200210/amazon-alexa-tech-news-anchor-order-dollhouse.
23. CNBC, 2016. "This hot robot says she wants to destroy humans". [Online] Available: https://www.cnbc.com/video/2016/03/16/this-hot-robot-says-she-wants-to-destroy-humans.html.
24. J. Snow, 2018. "Amazon's face recognition falsely matched 28 members of congress with Mugshots", American Civil Liberties Union. [Online] Available: https://www.aclu.org/blog/privacy-technology/surveillance-technologies/amazons-face-recognition-falsely-matched-28.
25. A. Jain, 2017. "Google Home Outage: 100 Percent failure rate reported", ValueWalk. [Online] Available: https://www.valuewalk.com/2017/06/google-home-outage-failure-rate/.

26. N. Statt, 2017. "A self-driving shuttle in Las Vegas got into an accident on its first day of service", The Verge. [Online] Available: https://www.theverge.com/2017/11/8/16626224/las-vegas-self-driving-shuttle-crash-accident-first-day.
27. L. Kelion, 2018. "CES 2018: LG robot Cloi repeatedly fails on stage at its unveil", BBC News. [Online] Available: https://www.bbc.com/news/technology-42614281.
28. P. Martinez, 2017. "Facebook allowed advertisements to target "Jew haters," report says", Cbsnews.com. [Online] Available: https://www.cbsnews.com/news/facebook-ads-targeting-jew-haters-propublica-report/.
29. D. Morris, 2017. "Artificial intelligence fails on kentucky derby predictions", Fortune. [Online] Available: https://fortune.com/2017/05/07/artificial-intelligence-kentucky-derby-predictions/.
30. S. Larson, 2017. "Offensive chat app responses highlight AI fails", CNNMoney. [Online] Available: https://money.cnn.com/2017/10/25/technology/business/google-allo-facebook-m-offensive-responses/index.html.
31. J. Brownlee, 2013. "Practical machine learning problems", Machine Learning Mastery. [Online], Available: https://machinelearningmastery.com/practical-machine-learning-problems/.
32. M. Mayo, 2018. "Frameworks for approaching the machine learning process - KDnuggets", KDnuggets [Online] Available: https://www.kdnuggets.com/2018/05/general-approaches-machine-learning-process.html.
33. B. Qian, J. Su, Z. Wen, D. N. Jha, Y. Li, Y. Guan, D. Puthal, P. James, R. Yang, A. Zomaya, O. Rana, L. Wang, and R. Ranjan, 2019. "Orchestrating development lifecycle of machine learning based IoT applications: A survey." [Online] Available: https://www.researchgate.net/publication/336550812_Orchestrating_Development_Lifecycle_of_Machine_Learning_Based_IoT_Applications_A_Survey.
34. A. Hawkins, 2019. "Serious safety lapses led to Uber's fatal self-driving crash, new documents suggest", The Verge. [Online] Available: https://www.theverge.com/2019/11/6/20951385/uber-self-driving-crash-death-reason-ntsb-dcouments.
35. L. Eliot, 2019. "Machine learning ultra-brittleness and object orientation poses: The case of AI self-driving cars - AI trends", AI Trends. [Online] Available: https://www.aitrends.com/ai-insider/machine-learning-ultra-brittleness-and-object-orientation-poses-the-case-of-ai-self-driving-cars/.
36. D. Lee, 2019. "Uber self-driving crash 'mostly caused by human error'", BBC News. [Online] Available: https://www.bbc.com/news/technology-50484172.
37. J.Dastin,2018."Insight:AmazonscrapssecretAIrecruitingtoolthatshowedbiasagainstwomen",Reuters.[Online] Available: https://in.reuters.com/article/amazon-com-jobs-automation/insight-amazon-scraps-secret-ai-recruiting-tool-that-showed-bias-against-women-idINKCN1MK0AH.
38. J. Lauret, 2019. "Amazon's sexist AI recruiting tool: How did it go so wrong?" Medium. [Online] Available: https://becominghuman.ai/amazons-sexist-ai-recruiting-tool-how-did-it-go-so-wrong-e3d14816d98e.
39. L. Vaas, 2018. "Facial recognition traffic camera mistakes bus for famous woman", Naked Security. [Online] Available: https://nakedsecurity.sophos.com/2018/11/29/facial-recognition-traffic-camera-mistakes-bus-for-famous-woman/.
40. W. Bao, H. Li, N. Li and W. Jiang, 2009. "A liveness detection method for face recognition based on optical flow field," *2009 International Conference on Image Analysis and Signal Processing*, Taizhou, pp. 233–236. doi: 10.1109/IASP.2009.5054589.

Index

Note: **Bold** page numbers refer to tables and *italic* page numbers refer to figures.

- activation function
  - hyperbolic tangent function 197
  - ReLU 198
  - sigmoid function 197
  - softmax function 198
- AdaBoost 386, 387
- adaptive gamma correction 150
- adaptive network-based fuzzy inference system (ANFIS) 322
- adoptive moment estimation (Adam) 152
- affective question answering on video (AQAV)
  - accuracy analysis
    - affective answers 368, *368, 369*
    - conventional answers 367, *367*
  - affective and conventional answers' prediction
    - loss along affective route 364
    - loss along VQA route 363–364
    - loss due to multi-task learning 365
  - architecture *357*
  - emotion, video and question embeddings 357–358
  - emotion detection (*see* emotion detection)
  - feature attention techniques
    - frame-based attention 361
    - integrated attention model *360,* 361
    - token-based attention 359–360
  - fusing video, emotion and question features 358–359
  - overview 356–358
  - video emotion detector 358
  - vocabulary 358
- Afroz, S. **171**
- Agrawal, R. 26
- agriculture 53–54
- Ahmed, S. 187
- AI *see* artificial intelligence (AI)
- Akaike information criteria (AIC) 332
- Akhare, R. 187
- Alshangiti, M. 406
- Amazon automatic resume filtering 415
- Amazon wildfires 209–210
- Ambarkar, S. 187
- ANFIS *see* adaptive network-based fuzzy inference system (ANFIS)
- ANN *see* artificial neural network (ANN)
- anticancer drug prediction 54–55
- anti-social activity 186
- antisocial elements 185
- Antoniou, C. 302
- applications
  - cartography 270
  - CNN 392–393
  - data acquisition layer 5–6
  - deep learning 14
  - machine learning 14
  - sentiment analysis 296
  - 3C approach 404
- Apriori algorithm 26–27
- AQAV *see* affective question answering on video (AQAV)
- ArcFace 201
- ARIMA *see* autoregressive integrated moving average (ARIMA)
- artificial intelligence (AI)
  - computer gaming 257
  - intervention 100
  - ML 14
  - resurgence 13
- artificial neural network (ANN) 22–24, *23,* 41
  - architecture *110*
  - backpropagation 190
  - confusion matrix **111**
  - layers 190, 197, *197*
  - stress recognition 109, *110*
  - validation testing *110*
- Australian bush fires 210
- autoencoder 28–29, *29*
- automated hematology counter 78
- automated sentiment analysis 298–299
- automatic Jaywalker recognition failure 416, 418
- autoregressive integrated moving average (ARIMA)
  - actual and predicted values of deaths *334*
  - after first difference *329*
  - after second difference *330*
  - autocorrelation plot *331, 338, 344*
  - confirmed cases during lockdown *327*
  - daily active count forecasting 333, **334,** 340
  - daily confirmed cases count forecasting 340, 342, **347**
  - daily death count forecasting 327, 332–333
  - data analysis
    - five-phased lockdown, India 323–324
    - highest fatality rates 322
    - population to confirmed cases ratio 323
    - state-wise death plot *323*
    - state-wise fatality rate *324*
    - ten states with fatalities 322
  - exponential smoothing methods 322
  - lockdown testing *326*
  - mathematical explanation 324–327
  - number of deaths *vs.* days *328*
  - objective 321
  - parameters 324
  - partial autocorrelation plot *332, 339, 345*
  - plot after second differencing *337*
  - predicted and actual confirmed cases *347*
  - summary *333, 340, 346*
  - time

autoregressive integrated moving average (ARIMA) (*cont.*)
vs. log, active cases count *335*
vs. natural values, active cases *336*

backpropagation algorithm 24
backward feature elimination 68–70
Badia, A. P. 312
bagging 385–386
algorithms 374
vs. boosting **388**
Baidu 201–202
Bayesian information criteria (BIC) 332
Bayesian network 300
BCE *see* binary cross-entropy (BCE)
BCELogItsLoss 263
Beck's depression inventory (BDI) 100–102
Benevenuto, F. **192**
benign/malignant tumors 245, 252, *254*
Bharat, A. 246
Bhat, W. A. 235
BIC *see* Bayesian information criteria (BIC)
big data 3
algorithm computational efficiency **44**
characteristics *44*
massive datasets 43
variety
data heterogeneity 47
data locality 46–47
unclean and noisy 47
velocity
concept drift 48
data availability 48
independent and identically 48
streaming/real-time processing 48
veracity
data source 49
data uncertainty 49
unclean and noisy 49
volume
Bonferroni's principle 46
class imbalance 45
dimensionality curse 45
feature engineering 45–46
modularity curse 45
nonlinearity 46
processing efficiency 43–45
variance and bias 46
binary classifier 19
binary cross-entropy (BCE) 228, 263
bioinformatics 39, 52–53
BIRADS *see* breast imaging-reporting and data system (BIRADS)
bird species 223
Bledsoe, W. W. 195
blockchain technology
implementation 51
infrastructure 51
memory 51
privacy 51
quantum resilience 51
research challenges *52*
security 51
suitability 50–51
Bonferroni's principle 46
boosting
AdaBoost 386–387
gradient 387–388
bootstrap aggregation 141, 385
Bovis, K. 140, **141**
Braud, C. **171**
breast border detection algorithms
gray-level thresholding 129
iterative optimal thresholding 129–130
merits and demerits **130**
minimum cross-entropy thresholding 130
Otsu's optimal thresholding 130
thresholding techniques 128, *129*
breast cancer 126
attributes 253, *254*
benign/malignant tumors 252, *254*
dataset 246
health and medicine 255
heat map 252, *253*
implementation
classification algorithms 252
confusion matrix 251, *251*
performance metrics 251–252
literature survey 246–247
onset 245
performance 253, *255*
proposed methodology
classification model *247*
dataset 247–248
machine learning algorithms 248–250
breast imaging-reporting and data system (BIRADS) 126
Breiman, L. 301
Brennan, M. **171**
Brockman, G. 312
Brown pelican *230*
Brox, T. 258, 260

Caetano, J. A. **192**
caption module 355
cardiac arrhythmia
data preprocessing *117,* 117–118
detection module *119,* 119–120
ECG 115
echo state network approach 117
electrical activity 115
evaluation parameters and measures 120–121
existing systems comparison 121, **123**
experimentation and results 120
fast Fourier transform 116
medical and technological domains 116
model development, detection 118–119
neural network 116
prediction 116
proposed machine learning implementation 117
regular and irregular heartbeat *116*
spectrogram image *118, 122*

training and experimentation 119
Xception model 117
cardiovascular diseases (CVDs) 115
cartography
applications 270
computer vision 257
data preprocessing 259–260
dataset 258–259, *259*
experimentation 264–269
generative models
GANs *262*, 262–263
residual neural networks 263–264
UNet 260–262, *261*
limitations 258
reasons 258
results 269
cascaded fake news 172
catchment basins 134–135
CBC test
bioengineering techniques 78
blood count 78
clinical pathology 77
information 78
C4.5 algorithm 277
Chang, R. F. 135
Chaniotakis, E. 302
Chaplot, D. 313
checklist for business modelling (CheckBM) 405
checklist for data processing (CheckDP) 405
Chhabra, S. **192**
CIA *see* confidentiality, integrity and availability (CIA)
Ciritsis, A. 139, **141**
classical learning–based approach 105, *105*
classification and regression trees (CART) 379–381
classification trees 380–381
classifiers
challenges 106
data mining techniques 108
logistic model tree decision tree algorithm *108*, 108–109, **109**
parameters 107
questionnaire format 108
stress recognition, artificial neural network 109, *110*
class inequality 45
clustering 52–53, *63*
clusters 148
CNN *see* convolutional neural network (CNN)
Codella, N. C. F. 46
computer vision 56–57
conditional exponential classifier 300
confidentiality, integrity and availability (CIA) 407
convolutional neural network (CNN) 31–33, *32*, 53
activation function 79
architecture *80*, *199*, *225*
array irregularities and distortion 232
classifier 231
convolutional function *79*, 79–80, 391, *391*
dataset 225–226
deep (*see* deep CNNs)
flattening operation *81*
fully connected layer 81
grid-like topology 390
image processing 198
K-fold validation 226–227
layers 198
max-pooling operation *80*
MBD
activation function 138
baseline (density map), breast density prediction 139
classification layer 138
convolutional layer 138
dropout layer 138
input layer 138
kernel selection 139
max pooling layers 138
measures 228
min-pooling operation *81*
multilayer perceptrons 224
pooling function 80–81, 392, *392*
receptive field 224
results and applications 392–393
SIANN 79
speech signals
architecture *103*
behavioral-based mental health detection system 102
experiment result and conclusion 103
performance metrics **103**
working 102–103
training and validation
accuracy *227*
F_1 score *229*
loss *227*
precision *228*
recall *229*
UNet 260–261
visual imagery 224
covariance matrix *65*
Crop Water Stress Index (CWSI) 235, 237, **237**
Cutler, A. 301
CVDs *see* cardiovascular diseases (CVDs)
cybercriminals 186
cyborg users 186

Dahl, G. B. 47
daily infection rate (DIR) 322
DARPA *see* Defense Advanced Research Projects Agency (DARPA)
data acquisition layer
applications 5–6
databases 5
data ingestion services 6
data integration services 6
devices 6
diagram 4, *5*
enterprise data gateway 6
field gateway 6
file systems 5
data analysis
five-phased lockdown, India 323–324
highest fatality rates 322

data analysis (*cont.*)
  population to confirmed cases ratio 323
  state-wise death plot *323*
  state-wise fatality rate 324
  ten states with fatalities 322
  cartography 270
data collection approaches
  CNN 102–103
  standard questionnaire–based evaluation 101–102
  TBERS 103–106
data-driven training approach 406
data ingestion layer
  data processing layer 8–9
  data storage layer 7–8
  sources 6, *7*
data lake 3
data mining techniques 39, 108
data processing layer
  data processing engine 8–9
  data processing programs 9
  scheduling engines 9
  scheduling scripts 9
data provenance 49
data quality/cleansing layer
  data quality check programs 10
  data refining 10
  diagram *9*
  MDM referencing programs 10
  MDM system 10
  rejected/quarantined layer 11
data storage layer
  cleansed layer 8
  landing layer 8
  processed layer 8
data visualization 62
data warehouses 4
DBSCAN *see* density based spatial clustering of applications with noise (DBSCAN)
decision stumps 383
decision tree (DT) 21–22, *22,* 74, 156, 158–159, *159,* 190, 250, *250*
  CART 379–381
  C4.5 and C5.0 382–383
  comparison **384**
  decision stumps 383
  ID3 381–382
  M5 model tree 383–384
  pseudocode 378
  terminologies 379
deep belief networks 374, 399–400
deep CNNs
  base model 226
  data augmentation 226
  implementation 226
DeepFace 199–200, *200*
deep learning (DL)
  algorithms 14
  ANN 27
  autoencoder 28–29, *29*
  CNN 31–33, *32*
  criticisms 57
  description 14
  domains and applications 14
  facial recognition (*see* facial recognition)
  feature extraction 28
  MBD (*see* mammographic breast density (MBD))
  medical fields 57
  ML *vs. 28*
  RNN 29–31
  user similarity test 57
deep learning–based approaches *105,* 105–106
deep neural networks (DNNs) 53
deep Q-network architecture *311*
Defense Advanced Research Projects Agency (DARPA) 196
deforestation control 209–211
  dataset required 216
  density 2004 *218*
  density 2007 *219*
  density 2010 *219*
  density 2013 *220*
  images processing and analysis 216
  model creation, logistic regression 216–220
  original image, 2001 *218*
DenseNet algorithm 161, *161*
density based spatial clustering of applications with noise (DBSCAN) 156
Deokar, A. V. **171**
depression
  algorithms used 148
  architecture 146, *147*
  light sensor data 146
  preprocessing 147
  psychological illnesses 153
  questions and features **150**
  score generation 148, 150
  smartphones 146
  user physical activities 146
Devi, S. S. 129, 132, 133
diabetic retinopathy (DR)
  algorithms used
    testing phase 152–153
    training phase 152
  block diagram 150, *151*
  color fundus photographs 150, *151*
  lifestyle diseases 146
  ocular disorder 150
  preprocessing 150–152
digital media 166
dimensionality reduction techniques 62–63, 75
Dinis, G. 302
DIR *see* daily infection rate (DIR)
discriminator module 267–268
dissemination 166, 172
DL *see* deep learning (DL)
DNNs *see* deep neural networks (DNNs)
Dooshima, M. P. 107
Dov, D. 157
DR *see* diabetic retinopathy (DR)
DT *see* decision tree (DT)

ECG *see* electrocardiogram (ECG)
educational technology 58

EHRs *see* electronic health records (EHRs)
Eigenfaces method 196
elastic net regularization 377–378
electrocardiogram (ECG) 115
electrodermal responses 109
electronic health records (EHRs) 57
Elf Owl *231*
Elmoufidi, A. 135
ELT *see* extract load transform (ELT)
emerging architectures 53
emotion detection system *54*
    affective route
        captioning module *362*, 362–363
        text QA 363
    images 355
    videos 356
endocrine system 155
Ensari, T. 246
ensemble learning 374
    bagging 385–386
    base learners 384
    boosting 386–388
    goal 384
ensemble tree 74
entropy 250
Euclidean distance
    contrast loss 203–204
    triplet loss 204
exponential smoothing methods 322
extract load transform (ELT) 8
extract transform load (ETL) 8

FAA *see* Federal Aviation Administration (FAA)
fabricated news 168
Facebook targeted advertisement 413–414
FaceNet 200–201
facial recognition
    DL
        ArcFace 201
        Baidu 201–202
        comparison 203
        datasets 202–203
        DeepFace 199–200, *200*
        FaceNet 200–201
        introduction 197–199
        loss functions 203–204
    Eigenfaces method 196
    scope and challenges 204–205
    skin texture 196
    task 205
    3D analysis 196
    two-dimensional recognition problem 196
    unique identification systems 195
fact-checking
    automatic 169–170
    physical 169
fake news detection
    architecture *189*
    comparative study 191–193, **192**
    COVID-19 virus 166
    data pre-processing 187–188
    data retrieval 187
    data visualization 188
    definition 167
    dissemination 166, 172
    exponential growth 181
    feature extraction 188
    integrity 173
    life cycle *167*
    literature survey 186–187
    machine learning algorithms
        ANN 190
        classification *190*
        decision tree 190
        Naive Bayes classifier 189
        SVM 189
    open research challenges
        cross-domain 181
        DL 181
        identification of trustworthy contents 181
        redetection 176. 180
        social media 181
    overview 168–169
    performance measure parameter **174**
    political and psychological factors 166
    propagation-based 168
    proposed work
        accuracy *180*
        design and workflow diagram 175, *176*
        experimental results and analysis 176
        methodology 174–175, *175*
        Naive Bayes *vs.* *179*
        objectives 174
        performance measures **179**
        random forest *vs.* *180*
        support vector machine *vs.* *179*
    public attention 166
    tokenization 188
    traditional approaches
        automatic fact-checking 169–170
        physical fact-checking 169
    training and testing model 190–191
    types 168
    user-based 168
    writing mode 167–168. 170–172
fake news propagation 172
falsehood-oriented lexicon-based procedures 167
faster RCNN
    architecture 84, *85*
    confusion matrix **96**
    directory structure *87*
    global_norm and global_steps *92*
    Google Colab *86*
    ground truth *95*
    implementation 85–96
    IoU operation *91*
    loss *vs.* steps *93*
    precision *vs.* steps *91*
    prediction result *95*
    problems, solutions and suggestions 96–97
    recall *vs.* steps *92*
    RPN 84, *84*

faster RCNN *(cont.)*
sample CSV files *88*
subnetworks 83
TensorBoard *89, 93*
training cell *90, 91*
viability solution 97–98
WBCs, RBCs and blood platelets **96**
fast RCNN 82–83, *83*
feature importance score *74*
Federal Aviation Administration (FAA) 223
feedforward neural networks 23, 388–390, *389*
FFDM *see* full-field digital mammography (FFDM)
fibroglandular tissue segmentation methods **136**
Filip, D. 107
fine needle aspiration (FNA) 159
Fischer, P. 258, 260
flower dataset **67**
forest ecosystems 210
forest fire calamities
control deforestation density 221
deforestation control 216–220
detection methods 212–215, *213, 214*
human activities 210
LiDAR technology 210
parameters 210
prediction methods 215
reasons and causes, wildfires 211–212
society/environmental need 210–211
solution 211
Forest fire Weather Index (FWI) 215
forward feature construction 70
Francois-Lavet, V. 311
Fukushima, K. 257
full-field digital mammography (FFDM) 133
fuzzy C-means algorithm 134
FWI *see* Forest fire Weather Index (FWI)

GA *see* genetic algorithm (GA)
galvanic skin response (GSR) 109
gated recurrent unit (GRU) 30
Gaussian thresholding in image processing 150
generative adversarial networks (GANs) *397,* 397–399
adaptive switcher 262
BCELogItsLoss 263
learning rate *vs.* loss *269*
modules 262
super-resolution 262, *262*
training
combined 268–269
discriminator module 267–268
generator module 266–267
generator module
loss function 267
normalization 266
ranged sigmoid function 267
self-attention *266,* 266–267
genetic algorithm (GA) 116
geographic information system (GIS) 216
global warming 233
Goap, A. 54
Goldstein, A. J. 53, 195
gradient boosting 387–388
graph-cut algorithm 134
graphical user interface (GUI)
category dropdown 176, *178*
search page 176, *177*
source dropdown 176, *178*
user login 176, *177*
Graves' disease 156, 162
gray-level thresholding 129
Greedy snake algorithm 224–225
Greenstadt, R. **171**
GRU *see* gated recurrent unit (GRU)
GSR *see* galvanic skin response (GSR)
GUI *see* graphical user interface (GUI)
Gupta, M. **192**
Gurav, S. 186
Gurton, K. P. 196

Ha, R. 140, **141**
hadoop distributed file system (HDFS) 5
hair-like histogram of oriented gradient-support vector machine (HoG-SVM) 55
Hamilton depression rating scale 101
Haque, A. 116
Hashimoto's disease 156
Hauch, V. **171**
HDFS *see* hadoop distributed file system (HDFS)
health care 39, 41
health data linkage 57
heart rate 109
hemocytometer 78
Hermanto, D. 300
Herzberg, E. 414
heterogeneous backhaul/fronthaul management 49
HIDS *see* host-based intrusion detection systems (HIDS)
high correlation filter 64–66
Ho, T. K. 301
HoG-SVM *see* hair-like histogram of oriented gradient-support vector machine (HoG-SVM)
host-based intrusion detection systems (HIDS) 274
hyperthyroidism 156–159, 161, 162
hypothyroidism 156–159, 161, 162

IBk classifier 281–282
IBk–K-nearest neighbor classifier **284**
ICA *see* independent component analysis (ICA)
ILSVRC *see* ImageNet Large Scale Visual Recognition Challenge (ILSVRC)
ImageNet Large Scale Visual Recognition Challenge (ILSVRC) 57
image question answering 355
IMs *see* induction motors (IMs)
inception-v3 model 153
independent component analysis (ICA) 67–68, *69*
induction motors (IMs) 68
integrity-based fake news detection
headline news 173
news comments 173
news outlets 173
spreader news 173
intelligent system games 109, 111

iterative dichotomiser (ID3) 381–382
International Agency for Research on Cancer, The 126
Internet of Things (IoT) 51–52, 55–57
intrusion detection system
    HIDS 274
    NIDS 274
    proposed system
        C4.5 algorithm 277
        evaluation 281–286
        features 276–277
        J48 classifier 277–278
        multilayer perceptron classifier 278–281, **280**
        requisition 275
        server log record **276**
        stages 275
        study stage architecture *275*
        Weka 277
    web spider 273, 274
IoT *see* Internet of Things (IoT)
iterative optimal thresholding 129–130
Ivakhnenko, A. 257

Jain, D. 302, 305
J48 classifier 277–278, **283**
Jesus, C. 107

Kaiser, L. 312
Kallenberg, M. 140, **141**
Kau, L. J. 45
Kaur, S. **192**
Keller, B. M. 134
Kesorn, K. 302
keyword spotting approach *104,* 104–105
Khuriwal, N. 246
kids rating for games **71**
Kim, E. 157
K-means algorithm 135, 146
K-means clustering graphs 25–26, *27*
    poor appetite/overeating 148, *149*
    trouble concentrating on things 148, *149*
K-nearest neighbor (KNN) 158, *158*
Koch, J.-A. **171**
Krizhevsky, A. 199
Kumar, I. 41
Kurnianggoro, L. 41

labeled faces in the wild (LFW) 202–203, **203**
Lample, G. 313
Lan, K. 53
Lanctot, M. 313
Lasso regularization 376–377
lazy learner algorithm 158
Lee, D. 313
Levenberg–Marquardt function 109
Li, R. 50
lifestyle disease 146
Lin, X. 246
linear discriminant analysis algorithm 156
line integral 132
Liu, L. 132
Liu, Q. 132, 133
Lizzi, F. 140, **141**
logistic model tree decision tree algorithm *108,*
        108–109, **109**
logistic regression 248–249, 300
    binary classifier 19
    implementation 19
    sigmoid curve 19, *20*
long short-term memory (LSTM) 105, 321, 342–343, 348,
        348, 349, 395–396, *396*
Lopez, M. C. R. 302
low variance filter 64

Ma, L. 157
machine learning (ML)
    algorithms 14
    approach 38–40
    breast cancer
        decision trees 250, *250*
        logistic regression 248–249
        random forest *249,* 249–250
        reinforcement learning 248
        supervised learning 248
        types *248*
        unsupervised learning 248
    challenges (*see* ML challenges)
    classification
        diagram *15*
        reinforcement learning *16,* 16–17
        semi-supervised learning 16
        supervised learning 15
        unsupervised learning 15–16
    components 14
    description 14, 38
    vs. DL *28*
    domains and applications 14
    DR (*see* diabetic retinopathy (DR))
    issues 40–42
    MBD (*see* mammographic breast density (MBD))
    methodology *42*
    psychology (*see* psychology)
    thyroid prediction (*see* thyroid disease prediction)
    training data 38
Maglogiannis, I. 246
Maitra, I. K. 131
Makarov, I. 312
malicious web spider 274
mammographic breast density (MBD)
    bootstrap aggregation 141
    classes 126
    concept *126*
    DL
        bootstrap aggregation 140
        CNN design 138–139
        comparative analysis **141**
        estimation 139
        generalized approach *137*
        hyper parameter optimization 140
        phases **140**
        preprocessing 138
        validation and testing 139, *139*
    fibroglandular tissues 126

mammographic breast density (MBD) (*cont.*)
ML
breast border detection algorithms 128–130
classification 137, **137**
feature reduction 137, **137**
image processing pipeline *127*
MLO view *128*
pectoral muscle removal *128*
pectoral muscle removal algorithms 130–132
preprocessing 127–128
segmentation 133–135
statistical feature extraction 135, *136*
semiautomatic and automatic methods 127
survey 140
manipulated news 168
Markov decision making algorithm 318
Martin, B. A. 303
Martynenko, A. 235
master data management (MDM)
referencing programs 10
system 10
Mastorakis, G. 406
maximum entropy classifier 300
maximum margin separators 20
MBD *see* mammographic breast density (MBD)
McKinsey 404
McNair, J. 313
MDM *see* master data management (MDM)
mean squared error (MSE) 66
medical disease diagnosis 55
Medio lateral oblique (MLO) 126
mental disorder 99
mental health diagnosis
awareness 100
biological impairment 100
challenges 100
cognitions 111
data collection approaches 101–106
intelligent system games 109, 111
ML algorithms 101
proposed model
classifiers 106–109
model architecture 106, *107*
standard questionnaire format 106
psychological and emotional state 100
metric learning *202*
Meyyappan, T. 302, 304
M5 model tree 383–384
micro-batching 3
Mingzhu, D. 416
mini-Mammographic Image Analysis Society (MIAS) 132
minimum cross-entropy thresholding 130
Mishu, S. Z. **192**
missing value ratio **63,** 63–64
ML *see* machine learning (ML)
ML challenges
accuracy 53
BD 43–49
bioinformatics 52–53
blockchain 50–51
complexity 53
IoT 51–52
practitioner 42–43
wireless sensor networking 49–50
MLO *see* Medio lateral oblique (MLO)
MLP *see* multilayer perceptron (MLP)
MSE *see* mean squared error (MSE)
Mughal, B. 132
Mukhdoomi, A. 135
multilayer neural networks 224
multilayer perceptron (MLP) 156, 278–281, **280, 284,** 322
multiple emotion detector 370
multivariate/multiple linear regression 17
Mustra, M. 132
multitask learning 356

Naive Bayes classifier 189, 299–300, 301
Narasinga Rao, M. R. 156
natural language processing (NLP) 56, 169
Naveen, R. K. S. 246
neocognitron 257
network-based fake news detection 172
network intrusion detection systems (NIDS) 274
network loss functions
Euclidean distance 203–204
softmax loss and variants 204
neural networks (NN) 77, **400**
neuron 197, *198*
neuroscience 41
Ng, A. Y. 116
Nguyen, T. N. 405
NIDS *see* network intrusion detection systems (NIDS)
NLP *see* natural language processing (NLP)
NN *see* neural networks (NN)
nonproliferative diabetic retinopathy (NPDR) 146

O'Brien, N. 187
Olsen, C. 135
online shopping sites 61
OpenAI Gym 312, 313, 318
OpenSpiel 313
osteoporosis 156
Otsu's optimal thresholding 130, 135
overactive thyroid 156
Ozdalilib, S. 116

Pak, J. **171**
Papoudakis, G. 312
parody news 168
Patient Health Questionnaire-9 (PHQ-9) 102
Pawar, S. 140
PCA *see* principal component analysis (PCA)
PCs *see* principal components (PCs)
PDR *see* proliferative diabetic retinopathy (PDR)
pectoral muscle removal algorithms
comparative analysis **133**
concept 130, *131*
image-based approach 131–132
model-based approach 132
personalized tourism information service (PTIS) 302
PHQ-9 *see* Patient Health Questionnaire-9 (PHQ-9)
PhysioNet 117

Pikkemaat, B. 303
polynomial curve fitting 132
Potthast, M. **171**
principal component analysis (PCA) 66–67, *68,* 150, 160, 210
principal components (PC) 66, 160
probabilistic classifier 299
proliferative diabetic retinopathy (PDR) 146
psychology
  depression detection 145–150
  lifestyle disease 146
PTIS *see* personalized tourism information service (PTIS)

Qazvinian, V. **192**

radial basis function (RBF) 156
Rafiuddin, S. M. **192**
Rahman, M. A. 132
Ramachandran Nair, A. 247
Ramanathan, V. 304
Rampun, A. 132
Rana, M. 246
random forest (RF) 22, *23,* 74, 190, 215, *249,* 249–250, 301, 385–386
random tree classifier 282–283, **284,** 285–286
Raoof, I. 157
Razia, S. 156
RBM *see* restricted Boltzmann machine (RBM)
RCNN *see* region-based convolutional neural network (RCNN)
recurrent neural networks (RNN) 342–343, 348, 393–395, *394*
RDDs *see* resilient distributed datasets (RDDs)
rectified linear unit (ReLU) 152, 161, 198
recurrent neural network (RNN)
  bioinformatics 53
  feedback mechanism 29, *30*
  implementation 30–31
  internal structure *31*
  issues 29–30
  LSTM 30
  structure 29, *30*
Reddy, U. S. 107
reference architecture
  AI/ML data processing *2*
  analytical data store 4
  analytics and reports 4
  batch processing 3
  data sources 3
  data storage 3
  machine learning 4
  orchestration 4
  real-time message ingestion 3
  stream processing 3–4
region-based convolutional neural network (RCNN) 33
  architecture 82, *82*
  classify data and detect object 82
  fast 82–83
  faster 83–85
region-growing algorithm 133, 135
region proposal network (RPN) 84, *84*
regression
  classification
    ANN 22–24, *23*
    DT 21–22, *22*
    RF 22, *23*
  code snippet 20, 24–25
  distinctive property 17
  implementation 17–19
  issues 17
  multivariate/multiple 17
  training data points 17
  trees 379–380
  visualisation 17, *18*
regularization algorithms
  comparison **378**
  elastic net 377–378
  Lasso 376–377
  parameters 375
  ridge 375–376
reinforcement learning 248
  description 16, 309–310
  flowchart *310*
  frame generated after preprocessing *311*
  hypothesis function 17
  literature 311–314
  problem formulation 310–311
  proposed method/design
    architecture *314*
    documentation *316, 317*
    epsilon trend *317*
    experimental setup 315–316
    max-pooling method 314
    results 316–318
    rewards trend *317*
    streamline process flow, DQN architecture *316*
  scope 318
  working method *16*
reinforced ranker-reader 355
rejected/quarantined layer 11
ReLU *see* rectified linear unit (ReLU)
remote sensing 216
residual neural networks 263–264
resilient distributed datasets (RDDs) 44
ResNet 261, *264*
ResNet-34 convolutional layers *240*
respiration rate (RR) 109
restricted Boltzmann machine (RBM) 399, *399*
RF *see* random forest (RF)
Ribeiro, M. T. 57
ridge regularization 375–376
RNN *see* recurrent neural network (RNN)
Ronneberger, O. 258, 260
Roopaei, M. 53–54
root mean square propagation (RMSprop) 152
Rosenblatt, F. 403
RPN *see* region proposal network (RPN)
RR *see* respiration rate (RR)
rule-based sentiment analysis 298

Samuel, A. L. 257, 403
Sapate, S. G. 131

satellite image processing technique 212
satire news 168
Schmude, J. 302
SDNs *see* software-defined networks (SDNs)
segmentation
  concept *133*
  fuzzy C-means 134
  graph-cut algorithm 134
  K-means and region-growing algorithms 135
  Otsu's optimal thresholding 135
  region-growing algorithm 133
  watershed algorithm 134–135
Selwal, A. 157
semi-supervised learning, description 16
sensory ratio function 170
sentiment analysis
  applications 296
  block diagram 296–297, *297*
  classification techniques *299,* 299–300
  datasets 297–298
  influential factor 296
  methods
    automated 298–299
    rule-based 298
  models
    logistic regression 300
    Naive Bayes classifier 301
    random forest 301
    SVM 301
  procedure 296
  research
    air quality 302
    categories 301
    customer feedback 302
    environment 303
    language 304
    locations and hotels *303,* 303–304
    objective 301
    parameters **304**
    PTIS 302
    SEWS 302
    transportation 302
    trip length 305
  scopes 296
  types
    aspect-based 297
    emotion detection 297
    fine-grained 297
    intent analysis 297
SEWS *see* socioeconomic well-being score (SEWS)
SFC *see* social face classification (SFC)
SGD *see* stochastic gradient descent (SGD)
Shao, K. 314
Sharma, S. 246
Sherly, E. 132
shift-invariant artificial neural network (SIANN) 79
ship data **66**
Shojaee, S. **171**
shrinkage methods 374
Siering, M. **171**
sigmoid curve 19, *20*
Simonyan, K. 153
simulated policy learning (SimPLe) 312
single feature regression **70**
single perceptron *388*
single shot detector (SSD) 85
singular value decomposition (SVD) 71–72, *73*
Siva Kumar, R. S. 407
social bot 186
social credibility 186
social face classification (SFC) 200, 203
social networks 166
society/environmental need
  animal, insect attack 211
  borders national security and illegal smuggling 211
  deforestation 210–211
  forest fires 211
socioeconomic well-being score (SEWS) 302
Søgaard, A. **171**
software-defined networks (SDNs) 405
de Souza Nascimento, E. 405
Sreedevi, S. 132
Srikant, R. 26
Srividya, M. 107
SSD *see* single shot detector (SSD)
Stein, B. **192**
stochastic gradient descent (SGD) 152
stock market analysis 55
strengths, weaknesses, opportunities and threats (SWOT) 302
Subashini, T. S. 132
supervised learning 248
  classification
    logistic regression 19
    regression 20–25
    SVMs 20, *21*
  decision tree 158–159, *159*
  description 15
  KNN 158, *158*
  linear regression 17–19
  SVMs 159, *160*
support vector machine (SVM) 20, *21,* 69, 156, 159, *160,* 189, 215, 300, 301
SVD *see* singular value decomposition (SVD)
Szczypiński, P. M. (symbol) *136*

Taifi, K. 132
Taigman, Y. 199
Tao, Y. 302, 303
Teigar, H. 313
term frequency-inverse document frequency (TF-IDF) 188
text-based emotion recognition system (TBERS)
  computational linguistics 104
  definition 104
  implementation
    classical learning–based approach 105, *105*
    deep learning–based approaches *105,* 105–106
    keyword spotting approach *104,* 104–105
text question answering (TQA) 355
TF-IDF *see* term frequency-inverse document frequency (TF-IDF)
Thelwall, M. 302

Thool, V. R. 131
Thota, A. 186
3C approach
    applications 404
    case study and failure analysis 412–418
    classification 411–412
    consolidation *408,* 408–411
    failed ML applications *407*
    literature review 405–407
    machine learning 403
    ML application development phases *412*
    planning and refining goals 407
    primary and secondary failing points 418
    scientific areas 404
    scope 418
    statistics 404
thresholding
    gray-level 129
    iterative optimal 129–130
    minimum cross-entropy 130
    Otsu's optimal 130
thyroid disease prediction
    advantages 162
    cytopathology 157
    DBSCAN 156
    diagnosis 163
    impact/case study 162
    implementation 161–162
    menstrual cycles 155
    ML
        algorithms 157
        sectors 157
        supervised learning 157–159
        unsupervised learning 159–160
    MLP 157
    parathyroid glands 156
    scope 163
    SPECT images 157
    stages 156–157
    types 156
    WSI 157
thyroid gland 155
thyroid-stimulating hormone (TSH) 156
Tiwari, G. 302, 305
Torrado, R. 313
Tortajada, M. 130
tourism
    defined 290
    industry (*see* tourism industries)
    sentiment analysis (*see* sentiment analysis)
    types 290–291
tourism industries
    challenges
        competition 294
        culture 293
        economy 293
        environment 293–294
        globalization 292
        infrastructure 293
        marketing 292–293
        pandemic 294
        price 294
        security 293
        social media 294
        taxation 292
        terrorism 293
    cultural industries 292
    DL 295
    hotels 291
    ML 294–295
    restaurants 291
    retail and shopping 292
    tour operator 292
    transportation 292
    travel agencies 292
TQA *see* text question answering (TQA)
training data 38
transpiration 233
trollers 186
TSH *see* thyroid-stimulating hormone (TSH)
Tsou, P. 157
Tureac, C. E. 290
Turtureanu, A. 290
Tyagi, A. 156
Tzikopoulos, S. D. 130

Uber autonomous car 414
UNet
    architecture *261*
    CNN 260–261
    data flow *261*
    learning rate *vs.* loss *269*
    ResNet 261
    training 264–265
    upsampling path 261
uniform resource locators (URLs) 273
unsupervised learning 159–160, 248
    Apriori algorithm 26–27
    description 15–16
    k-means clustering 25–26, *27*
URLs *see* uniform resource locators (URLs)
user segmentation 39

variable terminology 375
variational autoencoder (VAE) 312
variational mode decomposition (VMD) 68
video question answering (VQA); *see also* affective question answering on video (AQAV)
    baseline models 366–367
    concepts 354, 355
    contributions 354
    dataset 354, 365
    emotion detection 355–356
    experiment setup and results
        optimization 366
        performance analysis 366
    multitask learning 356
    qualitative analysis 369–370
    question type distribution *365*
    validating attention model 367
    visual captioning 356
Vidivelli, S. 129, 132, 133

Vikhe, P. S. 131
visual captioning
  image 356
  video 356
VMD *see* variational mode decomposition (VMD)
Volkova, S. **171**
VQA *see* video question answering (VQA)

Waikato environment for knowledge analysis (Weka) 277
Wang, K. 132
watershed algorithm 134–135
water-stressed crops
  application user interface *243*
  comparative analysis **241**
  CWSI 235
  electromagnetic spectrum *234*
  infrared spectrum 234
  invasive strategies 233
  proposed methodology
    architecture *240*
    augmented images *239*
    data augmentation 237
    data set 237
    environmental conditions 235, *236*
    feature extraction 238–240
    flowchart *236*
    pre-processing, thermal images 237–238, *240*
    steps 237
    thermal image dataset **237,** *238*
  results 241, *242*
  stressed tomato leaf *234*
  thermal imaging 234, 235
WBC count 77–78
WDBC *see* Wisconsin Diagnostic Breast Cancer (WDBC)
Webb, S. 53
web spider 273, 274, 287
weight *vs.* height *62*
Weka *see* Waikato environment for knowledge analysis (Weka)
well-behaved web spider 274
WHO *see* World Health Organization (WHO)
whole slide images (WSI) 157
wireless sensor networking
  infrastructure update 49
  ML-based 49
  network slicing 50
  standard datasets and research environments 50
  theoretical guidance, algorithm implementation 50
  transfer learning 50
Wisconsin Diagnostic Breast Cancer (WDBC) 247
Wolf, L. 203
Wolfe, J. N. 126
Wong, Y. D. 302
workflows 6
World Health Organization (WHO) 99, 245
writing mode–based approach
  characteristics **171**
  deception 167–168, 171–172
  falsehood detection and analysis 170–171
  hyperbiased style 168
  objectivity oriented 168
WSI *see* whole slide images (WSI)

Xception model
  accuracy curve *121*
  architecture 117
  inception architecture 119
  loss curve *120*
  training dataset 120
Xenou, K. 313
Xiang, H. 48
Xue, H. 354

Yampolskiy, R. V. 405
Yang, Y. **171**
Yasar, H. 139, **141**
Yosinski, J. 45
YouTube faces (YTF) 203
Yu, D. 356

Zafarani, R. 167, 186
Zhang, J. 187
Zheng, Y. 313
Zhou, L. **171**
Zhou, X. 167, 186
Zhou, Y. 55
Zisserman, A. 153
ZuEissen, S. M. **192**